GREAT BRITAIN

GEOGRAPHICAL ESSAYS

GREAT BRITAIN

GEOGRAPHICAL ESSAYS

EDITED BY

J. B. MITCHELL

formerly Fellow of Newnham College and
Lecturer in Geography in the
University of Cambridge

MAP EDITOR

M. A. MORGAN

Lecturer in Geography in the
University of Bristol

CAMBRIDGE
AT THE UNIVERSITY PRESS
1962

Published by the Syndics of the Cambridge University Press
Bentley House, 200 Euston Road, London NW1 2DB
American Branch: 32 East 57th Street, New York, N.Y.10022

ISBN: 0 521 05739 6

First published 1962
Reprinted 1967, 1972

First printed in Great Britain by Latimer, Trend & Co., Ltd.,
Plymouth
Reprinted in Great Britain by photolithography by
Unwin Brothers Ltd., Woking and London

CONTENTS

I. GREAT BRITAIN: INTRODUCTION

II. ENGLAND AND WALES

CONTENTS

PREFACE

Great Britain displays an astonishing variety of scene: mountains and moorlands above 3,000 ft, fenlands below sea-level; isolated crofts and hamlets still inaccessible to the motor-car, great cities choked with traffic; farms and factories; all are to be found within one small island. It is therefore a reflexion of reality that the essays collected here are very different in their themes; to impose uniformity would be to distort. Each author has selected a theme or themes to bring out the essential character of his area as he sees it; a formal analysis of regional frontiers and sub-divisions has been deliberately avoided as, on the ground, they rarely exist. The areas treated in each essay have been chosen in part for convenience; some areas overlap, some have gaps between. Many have a clear geographical entity, but no claim is made that each is a geographical region, still less that the areas selected for study here are in any sense *the* geographical regions of Great Britain. To stress this fact no map of Britain showing the areas is printed.

This collection of essays is not a new edition of *Great Britain: Essays in Regional Geography*, ed. A. G. Ogilvie, Cambridge, 1928. That collection stands enduring, a testimony to the Britain and to the geographers of its period. Readers will find much of value in it.

The Land of Britain: The Report of the Land Utilization Survey of Britain, ed. L. Dudley Stamp, London, 1936–46, is also a most useful work of reference. It is cited here to avoid repetition in chapter bibliographies of many county reports.

It is assumed that readers will use the general maps of Great Britain in one of the larger atlases, and where necessary the local sheets of the ¼-in. Geological and Ordnance Survey maps. Bartholomew's *Road Atlas of Great Britain* with layer coloured maps on a scale of 5 miles to 1 inch is a useful collection, and the series of 10 miles to 1 inch published by the Ordnance Survey gives much valuable information. *The National Atlas of Great Britain* containing a splendid series of maps on a scale of one in two million is to be published shortly.

The figures for height of land are, unless otherwise stated, height above Ordnance Datum; the population figures are from the Preliminary Report of the Census, 1961, except when a year is given. Place-names that are Welsh are in the form determined by the Board of Celtic Studies of the University of Wales, and as they will appear on the 8th edition of the 1 inch maps of the Ordnance Survey.

The maps and diagrams in this book have been designed by their authors to give information not found on easily accessible maps. All contributors, and especially the general editor, owe a debt of gratitude to M. A. Morgan who as map editor has given untiringly his generous help and advice in their preparation, and has drawn almost all the maps for publication.

As editor I thank most warmly all the contributors to this book for their unfailing co-operation and patience, and I am most grateful to the officials of the Cambridge University Press, and to the referee, for all the aid that they have given to me.

JEAN MITCHELL

NEWNHAM COLLEGE
1 March 1961

PART ONE

GREAT BRITAIN: INTRODUCTION

CHAPTER I

THE RELIEF OF GREAT BRITAIN[1]

B. W. Sparks

Great Britain, although hardly exceeding 750 miles in latitudinal extent and 375 miles in longitudinal extent, possesses a wide diversity of landforms. In general terms they may be assigned to the operation of the factors of structure, process and time, but it is probably true to say that structure is the dominant one of the three because of the great variety of structure and rock type displayed in Great Britain. Added to this is the fact that Britain as a whole is in a stage of late youthful dissection, if one may use the terms of the Davisian cycle for an area with so complex a recent geological history. At such a stage in the cycle of erosion the effects of structural and lithological variety should be most apparent in the relief. Yet process, too, has played its part: most of the landforms have probably been developed under the so-called normal subaerial cycle of erosion and modified by glaciation, but the nature of the glaciation varies from severe erosion in the highlands through deposition in the lowlands to periglacial activity beyond the margins of the ice-sheets. In some ways glaciation has accentuated differences due to structure and lithology and so increased the basic importance of these two in the relief of Great Britain.

The dominant structural trend in Great Britain is north-east to south-west, a reflexion partly of the fact that the oldest rocks are in the north-west and the youngest in the south-east and partly of the importance of the Caledonian fold-trend in the structure. The Highlands of Scotland, the Southern Uplands, the Lake District and most of Wales show this Caledonian trend, partly in their folding, but often more obviously in their fracturing. Renewal of Caledonian trends during the later Hercynian folding is responsible for structures such as the down-faulted Central Lowlands of Scotland and the Rossendale and Bowland Forests, west of the Pennines. A comparable

[1] It is suggested that the 1:625,000 geological map and the ¼-in. topographical map are used when reading this essay.

north-east to south-west graining of the country is shown in the south-eastern scarplands, although this is not due to Caledonian folding but to greater uplift of the north-west than of the south-east. Yet there are considerable areas where the trend is different, especially in the whole of southern England, which is dominated by east–west structures and relief, and in the Pennines with their north–south alinement.

Just as Europe may be regarded as a series of structural zones with Alpine structures south of Hercynian structures and both south of the Caledonian regions, so may Britain be considered as a series of structural belts flanking an ancient crystalline nucleus in the far north-west. In the Outer Hebrides and along parts of the north-west coast of Scotland between Skye and Cape Wrath are exposed the oldest rocks in Britain: these are the Lewisian, a series of heavily glaciated metamorphic rocks, mostly coarse gneisses, resembling in structure and relief the Laurentian shield of Canada, to which the area has been likened.

Flanking the Lewisian and generally occupying the next most north-westerly position in Great Britain are the Caledonian mountains: the Highlands and Southern Uplands of Scotland, the Lake District, and Wales north of the Brecon Beacons. The continuity of the zone is broken by the down-faulted Central Lowlands of Scotland, by the Eden valley, and by the Triassic lowland of Cheshire and Lancashire. The mountains resemble each other not only in structure, but also to a great extent in lithology. Within each area lithological diversity is so great as almost to constitute geomorphological homogeneity. The Highlands of Scotland differ from the other areas in being composed almost entirely of crystalline metamorphic rocks, a great series of gneisses, schists, granulites and quartzites included in the Moine and Dalradian series. The rocks are coarse, micaceous and flaggy, shattered by fractures and scored by the Pleistocene glaciations. In the other Caledonian areas the rocks are less altered Lower Palaeozoic sediments and include predominantly shales, slates, mudstones, greywackes, and flags with variety supplied, at least in the Lake District and North Wales, by important accumulations of ancient volcanic rocks. Fracture and glaciation are again prominent, though probably to a lesser degree than in the Highlands of Scotland.

South and east of the Caledonian zone lies the Hercynian zone, involving rocks of Devonian and Carboniferous ages. North of Pembrokeshire, Britain was outside the zone of strong Hercynian folding, which runs east–west across the centre of Europe, and in this region the Hercynian rocks form plateaux, cuestas and lowlands of relatively simple structure and relief. These include the low plateaux of Caithness and the Orkneys, the Pennines, the plain of Hereford and the Black Mountains. But even in this more northern region exception must be made of the down-faulted Central Lowlands of Scotland, where the rocks are shattered by faults and where igneous activity, especially volcanic activity, has been extremely important. In Pembrokeshire and in Devon and Cornwall the rocks have been much more intensely folded and these areas form exceptions to the generalization that the Hercynian areas, unlike the Caledonian areas, are regions where structure and relief seem to be fairly simply related. They are much more like the highly folded Ardennes and Rhine Block to which they are connected beneath the Mesozoic and Tertiary rocks of south-eastern England.

The south and east of Britain, farthest away from the Lewisian shield, is formed of a series of less resistant rocks, ranging in age from Permian to Pleistocene, and consisting mainly of clays, sandstones and limestones. North of London the prevailing general dip towards the south-east is reflected in the dominant scarpland pattern. South of London, the trend is mainly east–west, a reflexion in part of the effects of the buried Hercynian structures on the Alpine folding, which produced here a wealth of minor east–west asymmetric folds and fractures. The folds are usually simple except in the far south, where the rocks are vertical in the northern limb of the Isle of Wight monocline, while local thrusting has occurred on the northern side of the Weymouth anticline. But structures on the whole are much simpler than those of Caledonian and Hercynian Britain, a fact reflected in the less complicated pattern of relief, while the lesser resistance of the rocks is the cause of the highest elevations rarely exceeding 1,000 ft.

The geomorphological development of these various zones of Britain has been very similar except in so far as they themselves have caused differences in glaciation, through their position and elevation, and been affected by differences in glaciation, owing to their position.

Whatever the age of the rocks and of the folding, the whole of Britain appears to have been worn down and flooded by the sea by the end of the Cretaceous, according to the widely held hypothesis that a cover of Cretaceous rocks formerly extended over most of Britain. The Tertiary era, on the other hand, was dominantly a period of uplift, not in one single movement, but in a series of movements which gave Britain the main outlines of its present form.

Only in the south-eastern parts of Britain may the phases of uplift and their extents be judged with any degree of confidence, for it is in this region that Tertiary sediments have been laid down and later affected by folding and uplift. At the end of the Cretaceous a series of gentle earth-movements defined for the first time the major units of an uplifted Weald and down-warped Hampshire and London basins, though not in their present extent and form. But this does represent, at least, the beginning of modern relief for, at earlier dates, the London Basin had stood up, often above the sea, while the Weald had been down-warped. Other evidence for the emergence of Britain at the beginning of the Tertiary comes from the other end of Britain, from the north-west of Scotland, which was part of a great province of Eocene igneous activity including Antrim, the Faeroes, Iceland and Jan Mayen Island. The associated lavas, the products of subaerial eruptions, imply that, at the time of their formation, this whole area was dry land. Yet, today, the originally more extensive areas of igneous rocks have been separated by fracturing and foundering into a series of isolated remnants, notably in Skye, Rhum, Mull, Arran and the Ardnamurchan peninsula. The date of this second phase of movement is most likely to have been the middle of the Tertiary era, either the late Oligocene or the early Miocene.

The evidence for this again comes principally from southern England, where the Eocene and Oligocene sediments of the Hampshire Basin were involved in the Alpine folding to produce the series of structures already referred to. Outside these parts of Britain, where the earth-movements may be dated by the sediments affected, the blocks of older rocks are often held to have been uplifted as plateaux. Such uplift cannot be proved, because of the lack of young sediments to show it, but an argument by analogy may be derived from the uplifted Hercynian blocks of Europe. The fracturing and uplift of such areas as the Central Plateau of France, where there are Lower Tertiary sediments to demonstrate it, were completed at the time of the Alpine folding. Some difficulty arises from the distribution of the intensity of uplift, which, judging from the Central Plateau and other European features, was greatest nearest to the Alps. Yet in Britain the hypothesis of mid-Tertiary uplift requires gentle folding in the south and strong uplift in the north farthest from the Alps.

Whatever the difficulties, however, it is likely that the emergence and uplift of Great Britain was accomplished in two main phases in the Tertiary, though it must not be presumed that all was quiet between, for geological evidence can be advanced to show the probable development of a tilt towards the east in southern Britain in the intervening period. The period between the two phases of uplift may well have been one of predominant erosion and the formation of surfaces of faint relief, which are perhaps now just detectable as the approximately accordant summits of the highlands of western and northern Britain.

The available evidence seems to point not to one catastrophic phase of uplift, but to a slow and intermittent emergence, involving not only the old blocks, but also the younger folded areas of southern and eastern England. Many of those interested in the landforms of Britain have brought forward evidence for an intermittently falling base-level in the shape of successions of terraces and terrace-like forms extending to elevations as high as 2,000 ft or more, though often confined to lower elevations. Opinion has not been unanimous that these features are, in fact, terraces formed by erosion, neither is it finally agreed, assuming that these are terraces, by what particular processes of erosion they have been formed. The arguments as to the responsible processes are inconclusive, in view of the lack of fossiliferous deposits and the dangers of arguing from form alone, but for present purposes they are not of great significance. Whether a terrace at 1,500 ft is interpreted as subaerial or marine in origin does not affect the general conclusion that base-level must have been some 1,500 ft higher than at present during the formation of that terrace. Thus the idea of a sea-level falling since the middle of the Tertiary era from a level high up our present mountains, or of the intermittent and almost uniform uplift of the whole of Britain in the same period, cannot be escaped. As the terraces and terrace-like forms appear to be virtually horizontal, the former interpretation has been the favoured one in Britain.

The intermittently falling base-level has had two important effects on the relief of Britain. It has encouraged down-cutting at the expense of lateral erosion and back-wearing, thus ensuring the preservation of a youthful appearance in much of the relief. In addition, the probable emergence of more and more of Britain from beneath the sea is a useful hypothesis to account for the superimposed nature of the drainage of many parts of Britain, for example the Lake District, South Wales, and the Hampshire Basin.

The evolution of the relief of Britain was disturbed violently by the advent of the Ice Age. Although opinion on the number and relative importance of glacial and interglacial periods is open at present and awaits evidence provided by deep borings, it seems to be beyond question that Britain was glaciated more than once, although there is considerable doubt whether the classic pattern of four glaciations can be read into the available evidence. But this is a stratigraphical rather than a geographical problem.

The presence on the western side of the land of the uplifted plateaux of old rocks directly in the path of winds from the Atlantic led to the accumulation of the main ice caps on the Highlands of Scotland, the Lake District and Wales. Thus Britain shows a repetition on a smaller scale of the pattern of ice caps in Europe, with the largest one in Scandinavia and smaller ones to the east on the various mountains of the U.S.S.R. where the precipitation was insufficient for their best development. The location of the ice caps is presumptive evidence for the continuation in the Ice Age of an atmospheric circulation not vastly different from the present one and against the assumption of a vast, permanent glacial anticyclone with outblowing easterly winds.

The ice flowed, with velocity depending on its surface gradient, from these centres of accumulation into the eastern and southern lowlands where it spread out, slowed down and finally halted, when wastage equalled supply. The structural and relief divisions have thus been accentuated by an agent of erosion, the distribution of which they helped to control. The western highlands not only possessed ice caps for the longest periods, but were also subjected to the greatest glacial erosion for ice was moving rapidly over them. Ice erosion was dominant here and has left a characteristic imprint which subsequent subaerial erosion has had little effect in modifying.

The lowlands of Britain are areas of glacial deposition rather than glacial erosion with the form of the deposits varying to some extent with the age of the glaciation and their proximity to the highlands. The areas nearest to the highlands are characterized by those forms depending on the moulding of drift by rapidly moving ice, notably drumlins, while areas near the ice margins are commonly buried beneath featureless sheets of boulder clay. Well-preserved depositional forms are found mainly in the area covered by the last glaciation, the area of the Newer Drift, and so are not found south of York except in coastal Lincolnshire and north Norfolk.

Beyond the ice margins, that is in the area of southern Britain characterized by an east–west trend in structure and relief, periglacial activity was the only Pleistocene effect. But periglacial processes were not confined to this region, for they spread back and forth across the country as the ice-sheets waxed and waned.

Glacial diversions of drainage may be found in any of the glaciated areas. In the highlands they may have been due to ice becoming so congested in narrow valleys that it spilled over the divides and ground out an escape route which was later followed by the post-glacial rivers. Alternatively, in both highlands and lowlands, proglacial lakes dammed between the ice front and escarpments or ranges of hills, may have risen in level and overflowed at the lowest available cols, the violent escape of meltwater ripping these out into typical overflow channels. Glacial diversion or superimposition provides the explanation of most of the anomalies between Britain's structure and drainage.

The Ice Age was a period of oscillating sea-level, as water was alternately abstracted from the oceans to form the ice-sheets and returned to the oceans as the ice-sheets melted. Interglacial sea-levels were high and glacial sea-levels low, but they were not equally high nor equally low. The lack of uniformity between the high interglacial sea-levels may have been due to the continued tendency for intermittent negative movements of base-level to occur, or to differences in the extent to which the ice caps thawed out in different interglacials, or to a combination of both factors. Whatever the exact truth, the general effects on the coastal regions of Great Britain may be seen in the frequent occurrence of raised beaches, buried channels and submerged forests, and the possibility of sea-level having reoccupied the same level more than once makes it especially difficult to assign particular features to particular periods.

Neither may warping be excluded from the recent physical history of Great Britain. The Highlands of Scotland appear to have been domed gently upwards, as they were relieved of the weight of the ice cap and restored to isostatic equilibrium: the evidence is provided by the up-warped '100-foot' and '25-foot' raised beaches, both of them no older than the last glaciation. This is a small-scale repetition of the 1,000 ft or more of up-warping which has occurred since the melting of the ice cap in Scandinavia. Warping of a different type has affected the extreme east of England, principally the counties of Norfolk, Suffolk and Essex, for these are sufficiently near to the down-sinking southern North Sea to have shared in its subsidence, though to a much smaller degree than the Netherlands.

The themes, then, of the recent physical development of Britain have been emergence and glaciation and these have accentuated the differences between the different structural and relief zones of Britain.

Two major regions are commonly recognized in Britain, highland and lowland, with the dividing line formed by the base of the Coal Measures. Like all great generalizations this creates difficulties, especially in the diversity of landforms placed in the highlands. Probably the only really satisfactory solution is to avoid classification and to treat all areas individually, but on a broad scale a case may be made for a threefold division into what may be termed highlands, uplands, and lowlands.

Highlands are formed essentially of Pre-Cambrian and Lower Palaeozoic rocks folded mainly in the Caledonian earth-movements. The structures are so complex and the lithology so variable yet so similar that structure is less important in the relief here than in the uplands and the lowlands. Over large areas of central Wales, the Southern Uplands and even parts of the Highlands of Scotland the relief is monotonous, uplifted, dissected plateau. The general impermeability of the rocks, allied with the heavy rainfall deriving from their westerly position, together with the general lack of limestone, results in infertile, leached, acidic soils. Having been the centres of the main ice caps of Britain the highlands bear signs of more severe glacial erosion than the other regions of Britain. Yet there are differences within and between the areas.

The most extensive of them, the Highlands and islands of Scotland, is by far the largest area of crystalline metamorphic rocks in Britain. Over the whole area occupied by the Moine and Dalradian rocks the relief is due mainly to fracturing and glaciation and not to lithology. This is the highest region in Britain, the summits usually exceeding 3,000 ft and often 4,000 ft. It is deeply dissected, often along fracture lines, by broad glacial troughs, many of them now occupied by lochs some of which, for example Ness and Morar, occupy greatly overdeepened basins. All the signs of severe highland glaciation are here: valley steps, roches moutonnées, hanging valleys, corries, recessional moraines, and streams rising either in corries or on imperceptible divides in through valleys. In this extensive area of foliated crystalline rocks few rocks have individual effects on the relief with the possible exception of the quartzites, which are held responsible for the more pointed form of some of the peaks, for example Schiehallion.

Lithological diversity increases in importance near the west coast, where, beyond the thrust planes which terminate the outcrop of the Moine series, there are tracts of Lewisian gneiss, Torridonian sandstone and very much younger Tertiary igneous rocks. On the Lewisian is developed monotonous, lowlying, bare, hummocky glaciated relief, showing in many places numerous lakes and indeterminate drainage, especially on the mainland north of Ullapool and in the Outer Hebrides, notably in Uist and Benbecula. Along the north-west coast of the mainland the Torridonian mountains rise above the Lewisian in the greatest contrast, slabs of near-horizontal, ancient, resistant sediments often weathered, as on Stac Polly and Suilven, into steep and ragged ridges. It is curious how the Lewisian and also the Moine rocks on the coast farther south around Arisaig give the appearance of having been almost ground out of existence by the strength of the glaciation, while the Torridonian rocks, both in the mountains and in lowlying outliers on the Lewisian, do not.

The Tertiary igneous rocks, although so different in age, have many features in common with the Pre-Cambrian rocks: like them they are crystalline, while the relief on some of the major intrusions, especially the gabbro of the Cuillin Hills of Skye and the granite of northern Arran, provides superb examples of mountain glaciation. Sheets of plateau basalt, especially in northern Skye and the dyke swarms of Skye, Mull and Arran are distinctive elements in the relief, though both lava-sheets, for example the Devonian andesites south of Oban, and dykes of earlier ages are not unknown in the Highlands.

None of the other highland regions is quite like the Highlands of Scotland. In general the relief is smoother, although they have the characteristic of high average elevation in common. The greatest sharpness of relief seems to be provided by a combination of igneous rocks and severe glacial erosion. Snowdonia provides a good example: a series of lavas and ashes, predominantly rhyolitic in character, and a number of intrusions, mark this region off geologically from much of the rest of Wales. Heavy glaciation has produced on these rocks the corries of the northern side of the Glyders, glacial troughs such as the Nant Ffrancon, and a number of lakes, features which make this region, in spite of its much smaller size, comparable with the Highlands of Scotland. Cader Idris shows similar corried escarpments preserved on doleritic sills, while, in the Lake District, on the andesitic Borrowdale volcanic rocks ice has produced similar glaciated highland. Outside these districts, that is in much of the Southern Uplands, the northern part of the Lake District, and central Wales, the hills are big but rounded. Glacial features occur, but they have not the angularity typical of those of the volcanic outcrops, while the more common occurrence of sheets of drift seems to have softened the landscape.

Although the highlands as described in the preceding section seem to possess certain features in common, notably the absence of a simple structural pattern and its reflexion in a neat relief pattern as well as a general lack of limestones, there is one area which is completely atypical. It is true that the Cambrian Durness Limestone of the far north-west of Scotland introduces karst features on a small scale, but it does not form an area so different from the rest as do the alternating limestones and shales of the upper part of the Silurian in the Welsh borderland. Between and around Wellington and Ludlow the comparatively gentle dips of the Wenlock and Aymestry limestones and interbedded shales have resulted in a pattern of high cuestas, ridges and vales much more characteristic of what will be described below as upland Britain than of true highland Britain. The presence of abundant limestone results in a brightness of landscape not normally found in the highlands and allies this area to the uplands, in spite of the sombre effects introduced by the extensive planting of conifers on the limestones.

The upland areas of Britain are the areas of Devonian and Carboniferous rocks folded in the Hercynian earth-movements. Sandstones and limestones, disposed in simple structures, are the principal rock types. The relief, therefore, typically has a pattern, whether it be high-level plateau as in the Pennines, low coastal plateau as in Caithness and the Orkneys, or high escarpment as in the Brecon

Beacons, the Black Mountains and the hills inland from Brora in Sutherland. Compared with the highlands the signs of glaciation are far less pronounced. Mantles of drift are common enough and seem to intensify the smoothness of many of the hillsides, but there are, for example, few indisputable corries: the soft rounded forms that occur could be spring-heads, or corries badly preserved in not particularly resistant rocks, or spring-heads modified by snow-patches. Although many parts of the Old Red Sandstone produce acidic, impermeable and infertile soils, the more marly sections form fertile lowland, while the Carboniferous Limestone adds a type of relief, soil and vegetation almost absent from highland Britain. Regional differences occur within the uplands, but the essence of the relief is comparative simplicity of structure allied to a relief of bold cuestas and plateaux.

The Pennines are typical of the high plateaux of upland Britain. Slight dips and great lithological differences both between the major divisions of the Carboniferous rocks and within the Yoredale facies of the Carboniferous Limestone ensure the prevalence of structural surfaces. Whether these surfaces are truly structural or structures approximately stripped by erosion is not fundamental in a general review of this type, for the net result is the same, the dominance of the landscape by tabular relief. In addition, the Carboniferous Limestone is unique in possessing the clearest and most complete karst features in Britain and these are nowhere better developed than in parts of the Pennines, notably around Ingleborough.

Plateau at a much lower elevation is to be found in the coastal areas of Caithness and in the Orkneys, where the rocks responsible for it are the closely bedded Devonian Caithness Flags. A more monotonous scenery than that of parts of inland Caithness would be difficult to imagine, but the coast, where marine erosion has exerted its full effect on jointed and faulted horizontal rocks to produce vertical cliffs and spectacular stacks and inlets, is magnificent.

Where a simple regional dip affects the rocks, striking escarpments and dip slopes, such as those of the Brecon Beacons and Black Mountains, occur. These are impressive features, the crests of the escarpments reaching nearly to 3,000 ft in the former and to 2,500 ft in the latter upland. In form they are magnified versions of the cuestas of lowland Britain, but, apart from the question of scale, they differ from these in the number of steps and scars caused by hard beds, and in the dull colours of their acid moorland vegetation. To the south the Pennant series of grits and sandstones is a very important relief-forming unit in the synclinal upland of the South Wales coalfield: it is responsible for the moorland plateau between the industrial valleys and for a high escarpment overlooking the dissected vale developed on Lower Coal Measures in the northern part of the region.

Like the highlands, the uplands of Britain contain areas which are not typical. North-east of the Black Mountains lies the Red Marl plain of Herefordshire, recalling much more the Triassic plains of lowland Britain than the uplands to which it belongs, except in the discontinuous ranges of bold hills which cross it in a south-west to north-east direction. Devon and Cornwall are also exceptional: superficially they are plateau and thus resemble the Pennines, but this plateau surface is an erosion form cut across a variety of rocks and complex structures, for this region was near enough to the main belt of Hercynian folding to be strongly affected and the rocks largely altered to slates. In lithology and structure it resembles highland Britain, but in elevation and surface form it is more like upland Britain. It differs from both in the absence of glaciation and the smaller extent of periglacial activity.

But the most exceptional Hercynian region by far is the Central Lowlands of Scotland, an area shattered by faults into a pavement of blocks of Devonian and Carboniferous rocks. Further, the lithology of the Carboniferous Limestone is abnormal, that bed containing almost more coal than limestone, while the area is characterized by frequent masses of igneous rocks. They cause a great variety of relief from the jointed plug on which Stirling Castle is built to the great fault-line scarp truncating the andesitic Ochils on their southern side. These and other forms, such as exhumed laccoliths, dykes, volcanic necks and plateau basalts around the lower Clyde, ensure a complex pattern of relief different from that found elsewhere in Britain.

In lowland Britain there is simplicity, a predominant dip to the south-east affecting rocks from the Permo-Trias upwards, a patterned relief of cuestas and vales reflecting the alternating beds of sandstone, limestone and clay, a relief generally below 1,000 ft in maximum elevation, and an area where features of glacial deposition are far more marked than features of glacial erosion.

The oldest beds concerned, the sandstones and marls of the Trias, wrap around the sides of the Pennines and form the great triangular lowland of the Midlands, where diversification of relief is provided principally by the upthrust horsts of Pre-Cambrian and Lower Palaeozoic rocks, such as the Malverns, Charnwood Forest and the Nuneaton ridge. Some of the sharp relief formed on these rocks is in striking contrast to the gentle character of Triassic relief.

The areas of Jurassic and Cretaceous rocks north of London provide considerable contrasts in scenery, because of lithological variations, especially within the Jurassic. In Yorkshire, a great development of sandstone is responsible for the North Yorkshire Moors, a region more akin in lithology, vegetation and elevation to upland Britain than to the rest of the lowlands. Over much of Lincolnshire and the East Midlands the relief is not great as massive limestone beds are lacking in the Jurassic. In Lincolnshire the lowlands are narrow, but, to the south, a general decrease in regional dip is responsible for the great spread of low plateaux through the East Midlands and East Anglia. In the Jurassic rocks of this area limestones are not well developed unlike clays the predominance of which in the Upper Jurassic has facilitated the excavation of the wide plain in which the Fen deposits have been laid down. Both here and in East Anglia a general spread of glacial deposits, mostly boulder clay, has added to the smooth monotony of the relief. Farther south, the main Cotswold and Chiltern section of the lowlands, with Oxford at its centre, forms much more characteristic scarpland, largely because of the strong development of limestone in the Middle Jurassic and the higher relief of the Chalk in the Chilterns, for which there is no obvious lithological cause. This area differs from the lowlands to the north not only in its relief, but also in the far smaller importance of glacial drift in the landscape.

The area south of London differs from the rest of lowland Britain in its generally east–west structural and relief trends. It has less extensive outcrops of Jurassic rocks, but much wider ones of Lower Cretaceous and Tertiary sands and clays. In addition, the finest development of chalk scenery is to be found here, for, not only does the Chalk outcrop over wide areas such as Salisbury Plain, but it is also less obscured by superficial deposits so that it exerts an immediate effect on landforms, soils and vegetation. The prevalence of minor folds causes the pattern of escarpments to be more complicated than it is to the north of London: in areas where these folds are pronounced, for example near Weymouth and in the western Weald, the escarpments are more intricate in plan than in areas such as the North Downs, where folds are far less prominent. Finally, unlike the rest of the lowlands, this region was not glaciated, although periglacial activity was responsible for some locally extensive developments of solifluxion deposits and to a certain, though arguable, degree for modification of landforms.

Although the threefold division of Great Britain used here has been made primarily for physical description, it may prove useful in the human geography of the land. The highland regions are areas of low population density, infertile soils and mainly pastoral farming; the upland regions include all the significant coalfields, apart from their concealed sections, and hence most of the major old industrial regions; the lowlands are the most fertile regions and the areas of the greatest uniformity of rural population spread.

Selected Bibliography

Department of Scientific and Industrial Research: Geological Survey and Museum. *British Regional Geology.* 18 vols. London and Edinburgh (H.M.S.O.), 1935–54.

[These are the regional accounts of British geology and are more geological than geomorphological in general approach.]

T. N. George. 'British Tertiary landscape evolution.' *Sci. Prog.*, **43**, 1955, pp. 291–307. [A comprehensive paper with an excellent list of references for further reading.]

H. J. Mackinder. *Britain and the British Seas.* London (2nd ed.), 1907.

T. G. Miller. *Geology and Scenery in Britain.* London, 1953.

L. D. Stamp. *Britain's Structure and Scenery.* London, 1946.

J. A. Steers: *The Coastline of England and Wales*, Cambridge, 1948; *The Sea Coast*, London, 1953; *The Coastline of England and Wales in Pictures*, Cambridge, 1960. [These three works offer a comprehensive survey of a part of the physical geography of Britain, which has necessarily been neglected in this chapter.]

A. E. Trueman. *The Scenery of England and Wales.* London, 1938.

The following works are considerably more detailed, but cover major parts of the country:

E. H. Brown. *The Relief and Drainage of Wales.* Cardiff, 1960.

S. E. Hollingworth. 'The recognition and correlation of high-level erosion surfaces in Britain.' *Quart. Journ. Geol. Soc. Lond.*, **94**, 1938, pp. 55–74.

S. W. Wooldridge and D. L. Linton. *Structure, Surface and Drainage in South-east England.* London (2nd ed.), 1955.

CHAPTER 2

CLIMATE, VEGETATION AND SOILS

A. AUSTIN MILLER

For a country ranging from 50° to 60° North the dominant feature of Britain's climate is its mildness, and the quality that characterizes its weather is its variability. For the wind brings the weather and Britain is visited, at times, by air-masses varying from polar to tropical, blowing sometimes over thousands of miles of ocean, sometimes, though much more rarely, from the heart of the European landmass. But even the continental air, hot and dry in summer, cold and dry in winter, must cross the narrow seas and its extremes are more or less tempered in its passage. Britain is a battleground invaded from time to time by one of these distinctive air-masses, soon to be reconquered by another. Each brings its own type of weather and the battleground is not infrequently a 'front' with a sequence of wet and sometimes stormy weather changes that follow one of a number of patterns. The four chief masses are Tropical maritime, Tropical continental, Polar maritime and Polar continental.

The essential properties of each of these air-masses can be deduced from their source regions, in anticyclones, either permanent or temporary, and the regions over which they pass on their way. Broadly speaking Tropical maritime air comes from the west, south-west or south over the warm waters of the Atlantic Ocean. It is warm, damp and stable, for the lower layers are being slowly cooled. It brings stratus cloud and, especially in summer, sea fog. Rising up windward slopes it makes hill fog and if forced to rise suddenly over mountains or by strong convection, and particularly over a cold air-mass along a warm front, it may cause heavy rain. Polar maritime air comes mainly from the north-west or north, though occasionally from the south-west (Polar maritime air returning). Initially cold, its lowest layers become warmed over the warm waters of the North Atlantic Drift and it becomes unstable, causing cumulus cloud, sometimes growing to cumulo-nimbus which brings sudden heavy showers, but between the showers are bright intervals of clear blue sky. In the spring the showers are often of sudden hail.

Polar continental air affects Britain only in the winter months, coming from the continental high pressure to the east or north-east. It is a cold, dry, biting wind and brings black frosts, sometimes severe and prolonged. Being very cold it keeps to the ground giving rise to an inversion of temperature, beneath which fog in the country and smog in industrial towns are trapped. But if the cold air is deep the sky is clear, blue and hard.

Tropical continental air comes from the south or south-east in summer only. It is hot, dry and stable, giving a low pearly haze. In summer anticyclones the upper (superior) air is descending and, being further heated by compression, gives some of Britain's rather rare heat-waves.

At no season of the year is all Britain without risk of invasion by one of these air-masses, but clearly the climate depends on their frequency. Thus, in winter, when the Icelandic low is at its deepest and the continental high is at its strongest, the winds blow mostly from the south-west bringing rain, especially to the hilly west, and mild, cloudy, damp weather. Frontal depressions are never far away and, especially in early winter, they may cross the islands in procession bringing an alternation of rain followed by bright fine weather for a day or so, or there may be a strong surge of Arctic air behind the cold front. But towards the end of winter the continental high often sends a wave of cold dry polar continental air. Between the polar air, affecting the eastern counties mostly, and the damp oceanic air to the west, snow may fall, sometimes thick and soft, occasionally as a fierce blizzard building drifts. February and March are the months with most snow, though it can fall as late as June in the Scottish Highlands.

Spring is the driest season; the weather is often anticyclonic, with cold, dry easterly winds or with calm and clear skies and strong radiation at night, giving damaging frosts into May. But it may be otherwise, with depressions crossing the islands, bringing south-west winds, mild temperatures and rain.

By early summer the wind is more westerly and sometimes the Azores High sends fine sunny weather with blue skies in which cumulus clouds build up by day and die away at night. Or they may, especially if encouraged by a shallow depression coming from the south over France, build up into great thunder-heads. Thunderstorms are most frequent now and the eastern counties especially get their highest rainfall in this manner at this season. But still the rain is heavier in the hills and it is generally here that occasional cloudbursts cause great floods, the run-off concentrated, of course, in narrow valleys where it does the worst damage.

Calm anticyclonic conditions often occur into early autumn, especially after a fine summer, but the air is damp and as the sun sinks lower, the temperature falls and mist and valley fog form in the evening after a fine day. At first they melt in the warmth of the early morning sun but as the days get shorter they persist later and form earlier until they last throughout the day. At other times strong winds, associated with depressions, bring stormy weather with heavy rain and gales out of the south-west, usually backing to west and north-west as the colder air succeeds the warmer air when the cold front passes through.

Thus the fluctuations of the weather, related to the air-masses that bring it, or to the passage of the fronts that separate them, blur the passage of the seasons. They bring back winter cold when spring seems already to be here, or halcyon summery days return in mid-October (St Luke's summer), gilding the autumn leaves already sealed for their winter rest.

However, there are other permanent factors that make for contrast in British weather, the effects of which are much more predictable—proximity to the sea, the height of land, and the exposure to the wind. The highlands and the mountains lie mainly on the west and bear the brunt of the westerly winds off the oceans, the plains lie mainly on the east, nearest and most vulnerable to continental influences of frosts and drought. The contrasts in many places are spectacularly abrupt; Snowdon has 200 in. of rain in the year, Rhyl, only 25 miles away to leeward and on the coast, has only 26 in.

On the whole, such is the persistence of warm moist winds off the warm waters of the North Atlantic Drift that mildness characterizes the whole of the western coastal areas so that the January mean isotherm of 40° F. runs up the west coast as far as Cape Wrath and the Shetland Isles. Devon and Cornwall and the south-west of Ireland have average temperatures of 43° to 45° F. in the coldest month, cattle can graze the grass that continues to grow in most years, early vegetables are grown and the Scilly Isles produce

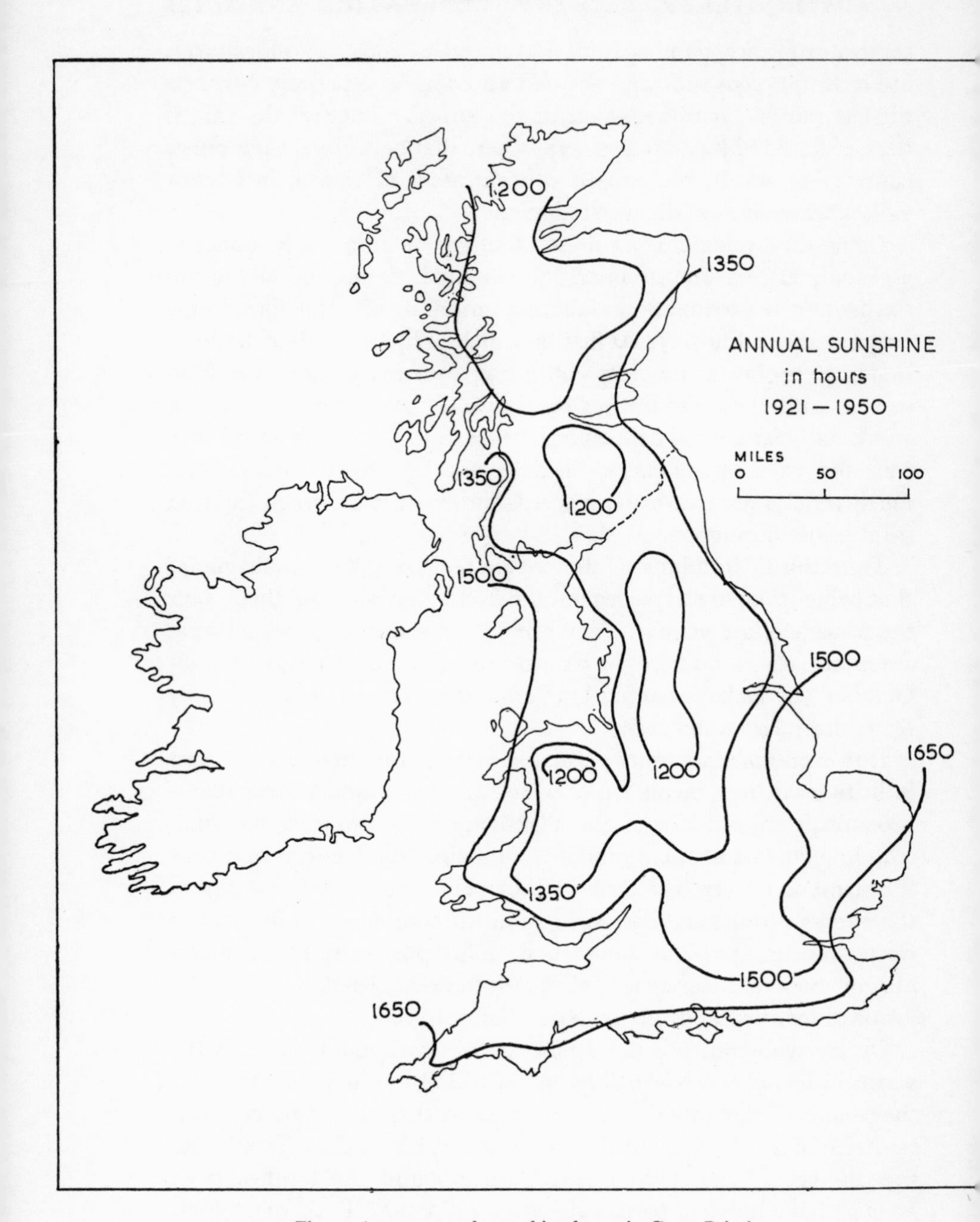

Fig. 1. Average yearly sunshine hours in Great Britain.

flowers at Christmas and daffodils for St David's Day (March 1st).

The most noticeable feature in the map of sunshine hours is the general parallelism of the lines to the coast (fig. 1). The south coast has an average of 1,650 hours a year (an average of 4½ hours per day) but the Welsh mountains, the Peak of Derbyshire and much of the Scottish Highlands average less than 1,200 hours (3¼ hours a day). Where the mountains come close to the sea in North Wales or Scotland, one may sometimes sunbathe on the coast all day and watch the mountains covered by orographic cloud.

For such a number of reasons, then, the weather is about as variable as it could be in such a relatively small region but the extremes are hardly ever severe. The temperature rarely exceeds 90° F. or falls below zero, the heavy rain that falls in the mountains runs off quickly down steeply graded valleys where it can be stored in reservoirs which provide water for the lowland towns and cities. Only rarely does the floodwater reach higher than the alluvial flood-plain and the river's rise is thereafter gradual, giving adequate warning to those who have unwisely built on land labelled 'liable to flood'. Only when an unexpected 'cloudburst' fills a narrow valley, as at Lynmouth in 1952 or at Louth in 1920, are lives lost, and property wrecked. Droughts occur, but crops are never a complete loss, nor do animals perish. The occasional little whirlwind (a 'twister') can unroof houses, heavy snowfalls can immobilize traffic locally and cause great inconvenience, as can the rare glazed frost and the much commoner icy roads, but fog is probably the greatest potential disaster that can happen, causing collisions and death on roads and railways. These minor crises fill the newspapers and provide a conversational gambit, but in contrast with the heat-waves in New York, or blizzards on the prairies, or floods in China or droughts in Australia, or hurricanes in Florida or tornadoes in Kansas with the roll of deaths and the cost of damage done, British weather seems indeed moderate. This degree of moderation is borne out by the statistics given in Table I.

Station	Altitude	Mean Temperature of Warmest Month	Mean Temperature of Coldest Month	Mean Daily Max. Temp. of Warmest Month	Mean Daily Min. Temp. of Coldest Month	Extreme Maximum Temperature	Extreme Minimum Temperature	Mean Duration of Growing-Season (Mean Temp., 42° F.)	Average Number of days with Frost	Average Date of last Spring Frost	Average Date of first Autumn Frost	Average Duration of Frost-free Period	Daily Mean Sunshine	Mean Annual Rainfall	Mean Annual Potential Evapo-transpiration	Average number of days with Snow	Average number with Snow lying	Average number with Hail	Average number with Thunder	Average number with Fog at 09 G.M.T.	Annual Percentage Frequency of Wind Direction at 09 G.M.T.								
																					Calm	N	NE	E	SE	S	SW	W	NW
	feet	°*F.*	°*F.*	°*F.*	°*F.*	°*F*	°*F.*	*days*	*days*			*days*	*hrs.*	*in.*	*in.*	*days*	*days*	*days*	*days*	*days*	%	%	%	%	%	%	%	%	%
WICK	119	55	39	59	35	80	8	237	52	May 9	Oct. 24	168	3·7	30	14	38	15	30	3	10	2	7	5	5	13	19	18	16	15
DALWHINNIE ..	1176	55	33	62	29	86	—5	203	117	Jun. 8	Aug. 12	64	2·9	47	14	51	58	3	2	3	9	9	11	2	11	17	27	8	6
DUNDEE	147	59	37	67	33	86	8	234	66	Apr. 25	Oct. 22	179	3·7	27	15	19	15	6	1	12	13	5	7	9	6	5	25	23	8
BRAEMAR ..	1111	56	34	64	29	85	17	186	112	Jun. 2	Aug. 25	83	3·1	37	13	46	62	1	3	3	22	7	7	7	3	4	28	11	10
TIREE	29	57	f41	61	38	79	20	310	20	Mar. 31	Dec. 1	244	4·0	45	15	12	3	28	5	3	4	12	7	6	15	18	14	13	11
RENFREW ..	29	59	38	66	34	86	0	243	57	May 17	Sep. 30	135	3·3	41	14	22	11	12	9	33	21	2	8	14	4	6	19	19	6
ESKDALEMUIR ..	794	56	35	64	31	85	—1	199	106	Apr. 27	Sep. 29	154	3·3	62	14	49	25	16	13	17	21	11	13	4	5	13	18	9	6
DURHAM	336	59	38	68	33	87	3	241	67	May 25	Sep. 30	127	3·6	26	14	20	16	5	9	26	13	9	8	4	3	18	17	16	12
SCARBOROUGH ..	118	61	40	67	35	90	16	279	31	Mar. 23	Nov. 28	249	3·8	26	17	13	9	4	7	23	0	9	5	3	12	10	18	21	22
GORLESTON ..	5	62	40	68	36	89	10	268	31	Apr. 7	Nov. 10	216	4·4	24	17	20	9	13	16	16	3	9	9	7	8	11	19	19	15
CAMBRIDGE ..	41	63	39	72	33	96	1	268	67	May 6	Oct. 17	163	4·1	22	17	14	11	5	13	19	8	7	14	4	7	9	24	13	15
SUTTON BONINGTON ..	157	62	38	71	33	91	0	254	71	May 5	Oct. 13	160	3·7	22	17	17	13	2	10	33	1	11	13	8	3	8	14	37	6
BIRMINGHAM (EDGBASTON) ..	535	61	39	69	35	94	11	260	42	Apr. 9	Nov. 11	215	3·6	29	19	26	14	11	15	37	3	9	10	7	8	15	20	15	13
ROSS-ON-WYE ..	223	62	40	70	36	91	—1	277	47	Apr. 24	Sep. 23	151	4·0	28	20	13	8	5	13	31	5	9	9	9	4	11	27	19	7
BUXTON	1007	68	36	64	32	88	—1	264	78	May 8	Oct. 12	156	3·1	49	15	28	30	5	11	43	1	3	11	11	11	9	23	9	22
KEW..	18	64	40	72	35	94	13	276	37	Apr. 7	Nov. 5	211	4·1	24	20	16	7	5	16	35	9	10	12	8	3	10	26	15	7
READING	148	63	40	72	34	95	9	280	44	Apr. 12	Nov. 2	204	4·2	26	19	13	3	3	11	30	2	7	15	10	7	8	23	15	13
HASTINGS ..	149	62	41F	69	36	90	12	286	32	Mar. 28	Nov. 26	242	4·9	29	19	9	7	4	9	7	2	9	19	4	10	7	25	8	16
KESWICK	254	59	39F	66	34	91	0	262	60	May 1	Oct. 15	166	3·3	58	16	19	12	9	8	3	NOT AVAILABLE								
SOUTHPORT ..	35	61	40	67	35	91	7	265	50	Apr. 15	Nov. 1	199	4·1	34	17	15	7	19	12	24	6	7	6	11	16	13	10	19	12
ABERYSTWYTH ..	452	59	40	65	36	91	13	314	35	Apr. 14	Nov. 19	118	4·1	37	19	9	4	8	5	11	5	8	7	6	8	23	18	16	9
LLANDUDNO ..	13	61	42	66	38	90	10	365	25	May 22	Nov. 20	181	4·1	29	18	11	3	15	7	2	1	8	7	13	5	16	25	22	4
CARDIFF	202	61	41	69	36	91	2	287	39	Apr. 10	Nov. 12	215	4·3	42	18	6	3	2	5	18	0	5	20	11	5	7	23	22	7
BATH	67	63	41	70	35	93	5	282	57	Apr. 27	Oct. 23	178	4·1	31	19	12	6	8	12	21	9	5	11	15	5	6	15	24	9
FALMOUTH	167	62	44	68	40	85	18	365	14	Mar. 3	Dec. 15	286	4·6	42	23	5	2	14	6	10	3	13	10	7	7	10	19	16	15

In the classification of climates Britain falls generally into the cool temperate, humid type. This suggests that it has a surplus of rain, and over the whole year this is true. In fact the rain is so abundant that if it could all be caught and saved there would be sufficient to supply every man, woman and child with 2,000 gallons per head per day.

The present average domestic consumption is about 60 gallons per day by all those who have piped supplies, though some rural districts are still without mains. Water engineers base their calculation of necessary storage on the dryest three consecutive years in the records, and do their best to match the supply with the demand. Occasionally there is a shortage; reservoirs run low and water has to be rationed. Such, in fact, is the increase in demand for water for industrial processes, for example, for washing and cooling, especially of atomic power-stations, that alarm is felt for the future. It is becoming more and more necessary to provide additional storage, or to consider, for example, ways in which the wetter west can be called in to redress the balance of the drier east, where drought is a growing danger, not to survival, of course, but to the maintenance of adequate supplies for industrial and domestic purposes and for agriculture. Water is needed, too, for flushing the streams to remove pollution which is an increasing danger to life in the rivers through lack of oxygen. The remedy here seems to be a much more vigorous control of effluents, for not all the available water is enough to deal with the foul state of many rivers, especially in industrialized areas.

Generally speaking, about half the annual rain falling in Britain finds its way to the sea. In lowland areas the percentage is less, but in the highlands the water runs off faster, and quickly reaches the sea, unless there is industry farther downstream to consume and foul it; but it is, of course, available for water power. In days gone by it worked the mills by the river-banks, now it drives the dynamos of hydro-electric power stations and, thanks to the grid system, contributes more to the nation's power. The combination of steep slopes and heavy rainfall, needed together, limit the power stations to the highland zone and especially to the glaciated highlands (fig. 2).

Some of the rainfall percolates into the ground and some of this re-emerges as springs, but some goes to replenish the underground supplies in water-bearing strata such as the Chalk, the Greensand, the Triassic 'waterstones', and a number of other geological forma-

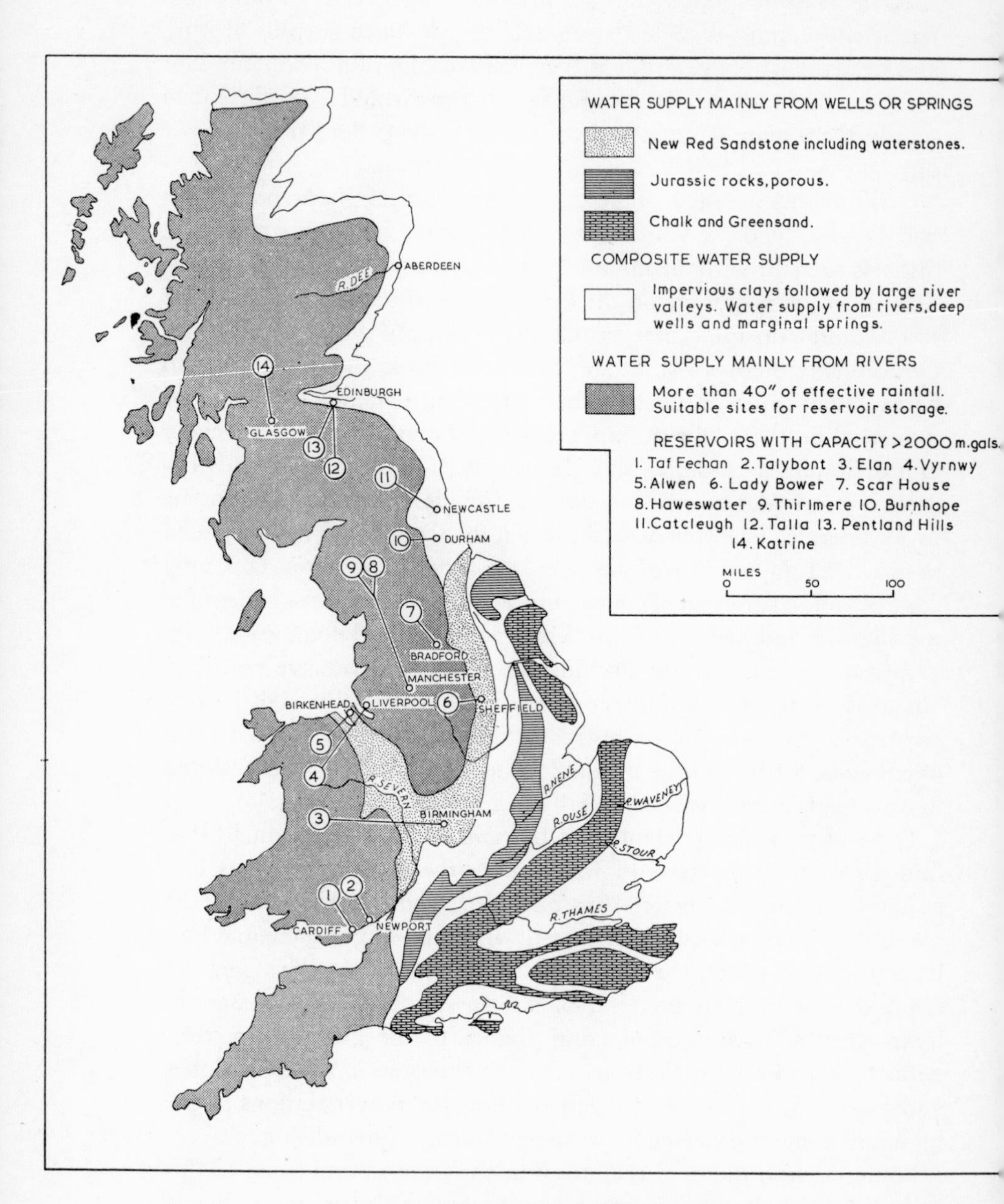

Fig. 2. Sources of water supply in Great Britain.

tions from which water can be extracted from wells. Though this is regarded as geological water, stored through the ages, it is being pumped out at an alarming rate and the replenishment is a meteorological matter. The great underground reservoir that lies beneath London was once 'artesian' and spouted under hydrostatic pressure through the fountains in Trafalgar Square. It has long ceased to be so, and so much pumping of 'free' water has gone on that the level is falling, in some places as much as 15 ft a year. No further wells may be sunk without special permission and ways are even being sought to replenish the reservoir by pumping river water back into the Chalk.

But the main reason why rivers fail to carry all the rainfall of their catchment areas to the sea is, of course, evaporation. Not only is there great loss from the surface of rivers and reservoirs, especially in the summer months, but the vegetation transpires enormous quantities through its vascular system and stomata, which by itself often exceeds the summer rainfall income. It is for this reason that rivers are at their lowest in summer and autumn and for the opposite reason that the season of extensive floods is in the end of winter (flash floods due to 'cloudbursts' are quite another thing), and since many crops and grasses are shallow rooted they exhaust the moisture of the surface layers. The soil falls below 'field capacity', replenishment by capillary rise from below is very slow indeed, and a state of partial, though often concealed, drought exists. This is now well known and it is becoming increasingly appreciated that some supplementary irrigation will increase the yield of crops and the stock-carrying capacity of grass in all but the wettest years over at least the south lowland areas of England (fig. 3).

It is generally agreed that the natural vegetation, without interference by man, comes in the long run to reflect the climate, or at least to reach a biotic climax, but Britain has been interfered with since the earliest historical times. In all probability the natural vegetation, recovering from an Ice Age that came to a gradual end in Britain only about 15,000 years ago, was forest, deciduous for the most part. Some small areas, for example, the Chalk Downs, were perhaps edaphic grasslands, the ill-drained swampy regions were fen and marsh, and the highlands, above a tree-line at about 1,000 ft, were moorlands deteriorating to arctic Alpine formations on the very tops. The snow-line, it is said, is only just above the summit of Ben Nevis and the Cairngorms. Patches of snow sometimes linger here (on

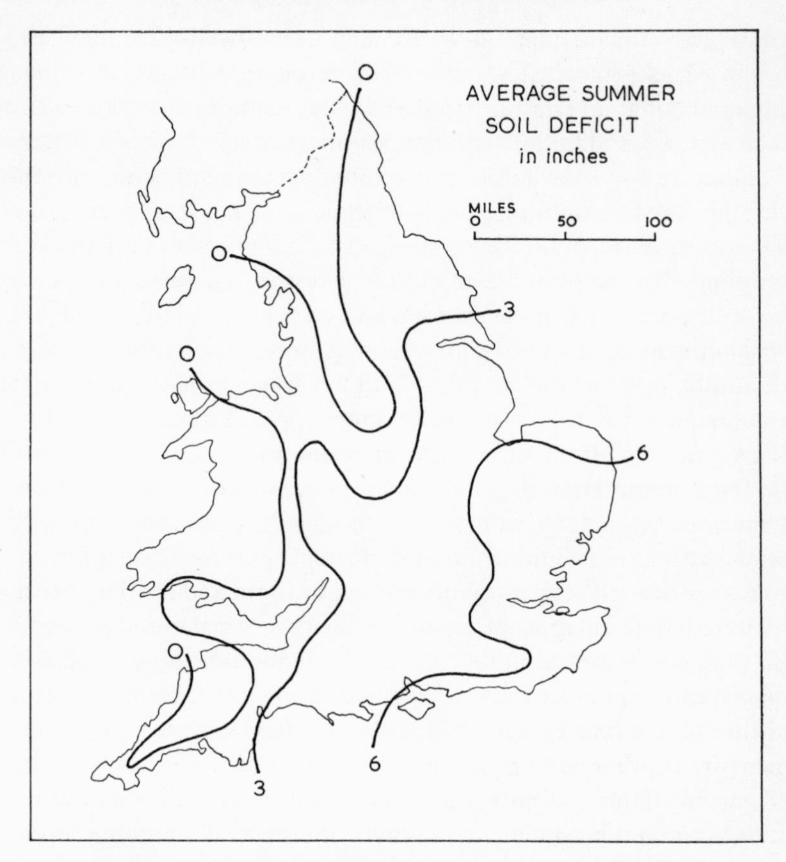

Fig. 3. Average summer soil deficit (in inches) in Great Britain.

northern slopes) throughout the year, and in most years ski-ing is to be had in the Scottish hills, even into April.

Little is left, however, of the natural vegetation, the forests have been cleared for agriculture and pasture is now to be found, either permanent or in short leys, enjoying a climate that is luxuriantly wet for natural grass (grassland is a semi-arid formation in the natural state). Cereal crops, too, are cultivated grasses and, under the cultivation conditions of today, produce yields that would startle their ancestors. Nowhere is it too dry for them and in few places too wet, but oats grow better in the wet western and barley does best in the drier eastern counties, though harvest weather can be critical. But temperature limits crops more severely, and the effective northern limit of successful wheat cultivation lies along the line of the northern boundary fault of the Midland Valley from Dumbarton to Stonehaven. As would be expected each crop has need of a certain quantity of heat during its growth and although temperature is not a unit of heat a helpful guide to the crop-growing potentiality of the land is given by the map of day degrees or accumulated temperatures above a threshold figure (42° F.) at which most cool-temperate crops begin actively to grow. There are many ways of calculating this, but the simplest is to sum the temperature (above 42° F.) for each day of the year. Clearly in the far north the early sunrise of summer takes the temperature above the threshold figure earlier in the morning and holds it there later in the evening. Thus there would seem to be a good reason for counting the 'hour-degrees'. But in fact this is allowed for in using the mean temperature of the day. Many crops and especially the cereal crops, require a minimum number of day-degrees between sprouting and ripening, but as growth clearly depends partly on the amount of solar energy received by the plant the approach can only give an approximation to the truth. The matter is complicated, but in English practice it is usual to regard the average duration of the period between the passing of the threshold (42° F.) in spring and the repassing of this value in autumn, not without good reason, as defining the growing season. This is shown in fig. 4 and clearly depends mainly on latitude and altitude.

It works fairly well for crops that are insensitive to frost, but the Americans prefer to define the growing season as the interval between the last killing frost of spring and the first of autumn. Since Durham, for example, has only 127 days between the last frost on 25 May and the first on 20 September, but has 241 days above 42°, there is an obvious difference between these two concepts, though of course the frosts are not necessarily 'killing'.

In Britain the dates of killing frosts vary greatly from year to year. A late spring frost when buds are sensitive and even young fruit may have formed spells disaster for the growers of sensitive fruit and largely accounts for the great fluctuation in the annual yield of fruit and the price at harvest. Thus 1935 had bad frosts as late as 16–17 May and the apple crop fell below 100,000 tons, but 1936 was a bumper crop of 385,000 tons. A disastrous night with spring frosts (9–10 May 1938) gave very low yields, less than 100,000 tons, but 1939 was a frostless year that sent the production up to 450,000 tons, a record. The incidence of frost varies greatly with the site of the orchard, for late

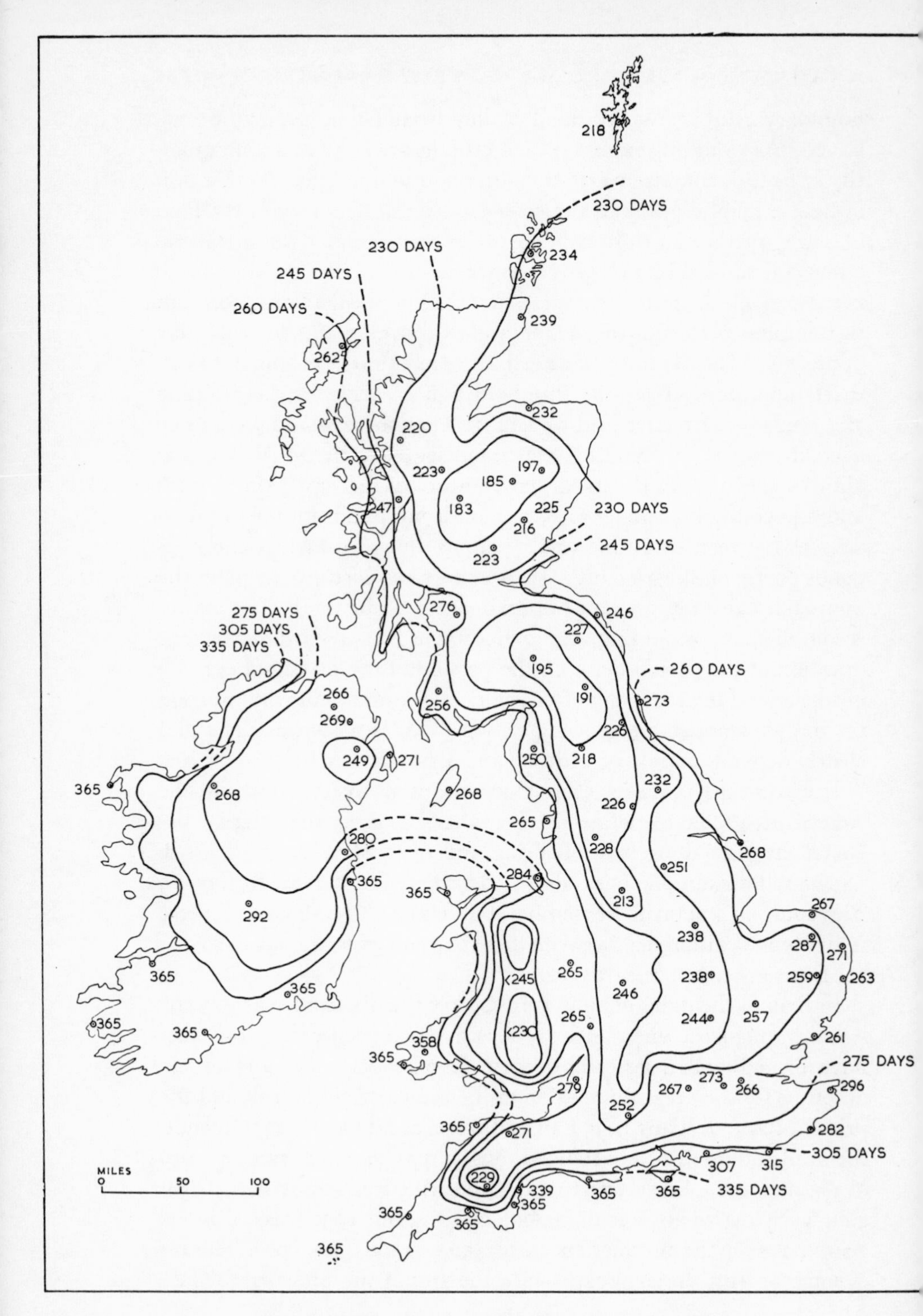

Fig. 4. Duration (in days) of growing season in Great Britain.

frosts are always caused, or intensified by nocturnal radiation; the cold heavy air sinks downhill into 'frost pockets' in the calm anti-cyclonic night.

Slightly exotic crops face an annual risk. Maize, for example, requires more warmth than is generally available in Britain, but now quick-ripening varieties are bred and some is grown as a vegetable for cutting while still soft and sweet. Tomatoes are a chancy crop in the open, even when the young plants are grown under glass; they fail to ripen in years of cool summer. Most of the soft fruits do quite well but commercial fruit-growing shows a preference for sunny climate and long autumns and the 'gardens of England' mostly lie within the area of more than 1,400 hrs of sunshine in the year. Some quality of the climate in fact sets a limit to most crops at some point in Britain, in latitude or altitude, or in local conditions of exposure. By long experience the professional knows the limits of safety; only the amateur can afford to take a chance and pay for it in a bad year.

It is not suggested that climate is the only cause of crop distribution. Relief limits commercial production to flattish lands or to valley floors, and soil too has its effects. One of the most striking features noticeable in the agricultural atlas is the concentration of potatoes on the fen and warp soils of the Wash and Humber, while sugar-beet likes light soils. Yet in the dry summer of 1959 the yield of sugar-beet was doubled in certain areas where supplementary irrigation was available to keep the soil at field capacity.

But soil itself is partially determined by climate. Though there are many local soils, Britain has only two zonal types, podsols (humid) and brown forest soils (sub-humid). The former dominate in the cold north and the wet west where a heavy excess of winter rain with small evaporation loss leaches out the soluble salts leaving an impoverished grey soil beneath a black layer of raw humus and, under coniferous forest, of resinous pine-needles. In the lowland zone is a less leached brownish soil of generally higher base-status, in a climate which is naturally fitted to deciduous forest. Both soils lose fertility under cultivation; the bases and their soluble salts, lime and potash, which constitute important crop-foods, need constant replenishment, either with artificial fertilizers or farmyard manure or by a sound system of crop rotation, including treading by stock, the ploughing-in of humus, and the introduction of leguminous crops to fix atmospheric nitrogen. It is only because of the limited range of the climate, and therefore of soils that British soil scientists often refer to soils in terms of the parent material, but, given time, the end-product would be a reflexion of the climate. But nearly all British soils have had centuries of management; natural soils are only to be found in uncultivated regions, uncultivated mainly for reasons of geology, for example, sandy heathlands, swampy marshes and blanket bogs. Even in such places the Forestry Commission has found an economic use for barren soils and makes trees grow. Its activities have in the forty years of its existence revolutionized what was formerly waste land. Generally speaking some trees, many of them exotic, will give a commercial yield and the regimented rows of conifers cover about 1¼ m. acres of land in Scotland, about ½ m. in Wales and 1¾ m. in England. For reasons of quick financial return conifers are generally preferred and the native oak forests on the claylands and to a less extent beech on the drylands are being replaced. Coppices decline slowly, but still survive to provide local timber for gates, fences and poles and to give cover for the pheasants and foxes that still provide picturesque but uneconomic sport for the countryman. The hedgerow trees, elm, ash, horse chestnut and sycamore, keep their place and give variety and charm to the 'bocage' lands of the English counties and the Welsh marches.

In such a brief account as this it is not possible to deal with the local peculiarities of climate, for the almost infinite variety of 'micro'-climates almost defies description. They result from differences of aspect, of exposure, of the quality of the soil, its wetness or dryness, its colour, its conductivity, of plant cover, or of buildings. Woodland, grassland, ploughland, pavement and water-bodies have their different coefficients of absorption (of insolation), of radiation, of latent heat, which affects the degree and the time of maximum and minimum temperatures, but their consequences can be of great importance. The contrasts are most obvious in calm, anticyclonic weather; when the wind blows and the air is turbulent the layers of air get mixed and the contrasts disappear.

Cities especially modify their own climate. Narrow confined streets obstruct and deflect the wind, houses and pavements absorb radiation and the issue of hot air from houses, furnaces and restaurants generally cause towns to have temperatures a few degrees higher than the surrounding country. Snow goes to slush under the wheels of the traffic and is swept away or melted down the gutter. But perhaps the most important feature of the urban climate is its degree of atmospheric pollution from domestic chimneys, from factories, from chemical industries, from gasworks. The smog spell of 4–10 December 1952 put up the death-rate in London from 300 per day to nearly 1,000 and altogether 4,000 died, mostly of bronchial and respiratory diseases. By a sad coincidence the Dairy Show was on and several prize cattle died. All this atmospheric pollution (200 tons of soot fall on each square mile of central London in a year) is one of the great sins of civilization against which a powerful campaign for 'smokeless zones' is being urgently pressed. Meanwhile it reduces the sunshine, absorbs the sun's rays, checks the escape of heat like a glasshouse, smothers the grass, blights the miserable trees and does immeasurable harm to the health of the town-dweller. When we consider how many of our climatological stations are urban it would seem likely that they give a slightly distorted statement of our climate, flattering the town perhaps at least as far as temperature is concerned but depressing the sunshine figures. But then most of our population is urban and spends its holidays by the sea.

Man's reaction to the climate is as variable as his physique or his purse allows. Refrigerators are in increasing use but there are few who find it worth while to cool their houses by air-conditioning, preferring to open the windows to bring in the air-temperature of outdoors, which is cool enough. Some heating is needed for at least six months of the year and now techniques of central heating by oil or solid fuel are making good progress, aiming at temperatures in the sixties rather than, as in America, in the seventies. More could be done in this direction, not only against winter chill but also against damp which is damaging; there is no doubt that Britain's climate, especially in winter, is everywhere too humid for good health and rheumatic diseases are too common.

It is worth while to remember that Britain's climatic network, its professional observers supplemented by a host of volunteers, is one of the closest in the world and reaches the highest degree of accuracy. The Meteorological Office, a department of the Air Ministry, supplies the press and the B.B.C. with weather forecasts, gale, frost, fog and snow warnings and with regular shipping forecasts. Its organization includes a climatological branch with a huge reservoir of statistical data for the whole world. Information is available for farmers, gardeners, town- and country-planners and geographers of all kinds and interests on those matters that concern the atmosphere and its ways.

Selected Bibliography

E. G. Bilham. *The Climate of the British Isles*. London, 1938.
R. Geiger. *Climate near the Ground*. London, 1950.
A. Austin Miller and M. Parry. *Everyday Meteorology*. London, 1958.
L. P. Smith. *Farming Weather*. London, 1958.
A. G. Tansley. *The British Isles and Their Vegetation*. Cambridge, 1939.

CHAPTER 3

THE POPULATION OF GREAT BRITAIN

J. B. MITCHELL

A brief examination of the essential features of the population of Great Britain as a whole must be added to that of its relief and climate to complete the setting for the more detailed study of its parts. No area can be considered as a separate entity, each is an integral part of the island whole.

Diversity within unity is one characteristic of the population pattern. The variety of relief that is contained within the unity of the island form, the variability of the weather within the equability of the oceanic climate, these are paralleled by the diversity of the economy and society within the uniformity of this densely peopled and highly industrialized land. The census of 1961 recorded 51,298,245 people living in Great Britain and the Isle of Man which gives an average density of 573 persons per square mile, a figure exceeded only in Japan, Belgium and the Netherlands. Some 80% of Britain's inhabitants live in towns, and nearly half of these in Greater London and the six provincial conurbations. The British are shopkeepers and office workers, miners and factory workers; only one man in every two dozen farms the land. But such figures give a wholly false impression of uniformity. There can be few greater contrasts within so small an area than between some highland parishes of Scotland, northern England or Wales inhabited by two or three hundred folk, and the streets of Glasgow, Birmingham or London with their thronging thousands.

Change within stability is a second key note. Within the span of man's occupation of the land, the patterns of relief and climate remain in broad essentials stable; what alters is human knowledge and evaluation of them. Efforts to use the opportunities of the physical environment well and to overcome the limitations effectively bring about change now slow, now rapid, and often albeit painful, in the patterns of population. There is change in the total numbers: a great increase, though at varying rates and not an uninterrupted one, has taken place since the time of the Domesday Survey. The information collected by the Domesday Commissioners in 1086 suggests that about 2 m. people lived then in Britain to the south of the Ribble and the Tees. A slow increase, with perhaps a marked fall as a result of famine, pestilence and war in the later fourteenth and early fifteenth centuries, brought the total to some 5½ m. in England and Wales and 1 m. in Scotland when Gregory King made his estimate at the end of the seventeenth century. The later eighteenth century and the nineteenth century was a period of rapid increase; the birth-rate remained high at about 34‰ until the 1870's, the death-rate fell steadily and population rose to 8·8 m. in 1801, 27·3 m. in 1851 and 43 m. in 1911, a fourfold increase within the nineteenth century.

There is change in the population pattern between country and town: in England 80% of the population in 1770 was rural; in 1850 the balance between town and country was even; by 1931 only 20% of the population is recorded in rural districts, and although in 1961 the figure has not fallen further, this probably represents an increase of townsfolk living in the country, rather than an increase of true country folk. In the proportion of town to country dwellers, as in the total size of the population, stability may be approaching. Migration has taken place not only from country to town, but also into and out of Britain. It is estimated that during the nineteenth century some 25 m. people left Britain to settle in Europe, America and the Colonies. This number includes foreigners who sailed from British ports and also British who returned later, and is offset in part by immigrants into Britain, but since 1871 the net loss by migration has been about 3½ m. The 1951 census showed for the first time a net gain in Britain by migration of about ½ m., the net gain of about ¾ m. in England being offset by a net loss in Scotland. Emigration and immigration in 1957 were approximately balanced, about 20,000 persons moving each way; and the estimates for 1960 suggest a slight gain by immigration. Stability is here again approached, but, if migration from the West Indies, India and Africa continues to increase, the long-established outward flow may be reversed.

Within the island the pattern of distribution of population densities shows both stability and change: throughout, the highland areas have been sparsely and unevenly peopled, with large areas empty; throughout, the fat agricultural lowlands have been well populated with many villages evenly spaced. But since the industrial revolution, dense urban populations, in many areas over 1,000 per square mile, have grown upon the coalfields; today, there are trends, sufficiently well established, to suggest that the old industrial and urban concentrations are weakening, that modern industry is becoming more widely distributed, and thus a new pattern, or rather the revival of an older one, with the centre of gravity of the population to be found in the Midlands, somewhere approximately in the middle of the belt of country stretching from the Wash to the Severn, is becoming re-established.

The population of Britain in its physical types, its numbers, and its distribution is greatly affected by the position of Britain. Britain is an island but is not 'a world by itself' or 'a land estranged', it is a part of Europe, a part of 'one world'. In differing degrees this has always been so.

Position as an off-shore island of Europe gave to Britain in the first place its varied stocks, and isolation on the outer fringe of the then known world gave time to absorb the many elements. There is considerable variety of physical types within Britain today but the ancestral stocks from which the British spring were, with a few exceptions, island colonizers before the Norman Conquest, and many, perhaps most, before the Roman Conquest.

Mediterranean groups bringing agriculture arrived in the early part of the third millennium B.C., colonizing along the southern and western coasts of Britain, probably mingling with more northerly groups already established who were as yet mere collectors of the fruits of the field, hunters and fishers. The descendants of these first agriculturalists with small bones, dark hair and eyes still form a noticeable element in the population of Britain, and particularly so in Cornwall and Wales. Soon to follow came from Spain through France the Beaker folk with their herds, and then from western Alpine lands the Celtic groups with their light ploughs. By the first century A.D. cultivated fields and grazing-grounds around villages and farms had replaced the wild on great stretches of the lighter soils in the lowland east, and on the coastal lands and in the lower valleys of the highland west.

The Romans must have found if not a polyglot people, groups of varied cultures. During the centuries of Roman rule if the highlands remained not unwitting but largely unheeding of Roman fashions, the lowlands with Romano–British towns, villas and farms became, for the first time perhaps, culturally and administratively one, sharing if somewhat remotely in the civilization of the Mediterranean world.

The Anglo–Saxons called in to help the Romans to maintain order in this distant frontier province, staying to conquer, then to colonize were certainly not beginning life anew in a virgin land. How great the break with the past is debatable. The survival of Celtic speech—of Gaelic and Welsh until the present day, of Cornish until the nineteenth century—is ample proof of the continuity of culture from the pre-Roman period in the west of the island. In this highland zone, some of the villages and farms and the fields and pastures around them may have been in continuous use for near four thousand years. On the other hand in this hard environment colonization proceeded slowly, reclamation continued late, and thus some of the most recently established farms, villages and towns are here too. It is in southern and eastern lowland Britain that the Anglo–Saxon imprint is dominant, but upon how many villages and towns it is but an overstamp cannot now be determined. New men with new ideas and new tools, equipped as they were with a more effective plough, may well have preferred new sites, but in many places they may have settled down in older villages gradually changing the look of them and even their names.

The Vikings, the Danes and the Norsemen, arrived in the eighth and ninth centuries. Scandinavian elements in place-names and Scandinavian physique in the population are still distinct in eastern districts devastated by them and re-settled, and on the then frontiers of settlement in areas isolated by fen or marsh, in upland valleys and on western coasts. If Scandinavian colonists were few in relation to men of other stocks already established, the Norman–French settlers who came with King William were fewer still, and in physical type were closely related to groups long established.

Position close to Europe fostered from early times strong trading links with the Continent. Until the fourteenth century Britain's economy was that of a young society: a valuable raw material, wool, was traded for the luxury manufactures of the Low Countries, Italy and Byzantium, and for the fruits, the wines and spices, and the furs, the wax and the timber that Mediterranean and Hanseatic merchants brought. Britain lay on the narrow strait where sea-routes from the Mediterranean and from the Baltic met, and were joined across the water by the land-routes from the valleys of the Rhône and the Seine, the Alpine passes, the Danube and the Rhine, and the north European plains. By the end of the fourteenth century Britain's economy had changed: cloth of high quality to be finished abroad, and cheaper kinds dyed and finished at home, had replaced raw wool as a staple export. At the beginning of modern times British merchants were seeking even wider markets, and it was in part the need for new markets for their cloth that attracted British interest and capital to compete for a share in the new lands and ocean-routes explored by Spain and Portugal.

Position on the Atlantic shore gave freedom of movement on ocean-routes throughout the world, and from the late sixteenth century onwards, without losing the advantages of position where European sea-routes and land-routes met, the advantages of the new situation were pursued. Overseas trade grew, a great overseas empire was built, by the middle of the eighteenth century Britain lay at the nexus of the ocean routes, mistress of the seas. The commercial and financial capital of the world had shifted north-westward across Europe: from Byzantium to the Italian cities in medieval times, to Antwerp then to Amsterdam in the early modern world, to London in the eighteenth century.

The new position encouraged both immigration and emigration. Into Britain came the Flemings, the Dutch and the Huguenots in the sixteenth and seventeenth centuries, Jewish groups in many periods, Italians, Germans, Russians and Poles in the nineteenth and early twentieth centuries to share an expanding economy. They were, however, as were the Norman–Frenchmen before them, too few in numbers to affect significantly the British stock, although locally some for long formed groups culturally distinct. From Britain went forth explorers and many more settlers. From the seventeenth century when it is guessed that some 80,000 left Britain to settle the western seaboard of the Atlantic, a great stream of emigrants has moved overseas to find new homes: in the eighteenth century in the United States and Canada, in the nineteenth and twentieth centuries in South Africa and tropical Africa, in Australia and New Zealand too. This movement of population is of great importance to the geography of the island in that it affects the number and distribution of people at home, and, through the links they make, fosters trade abroad. Expanding trade allowed not merely the maintenance of increasing numbers in Britain, but also an increasingly higher standard of living.

During this period the volume and value of overseas trade steadily and rapidly increased though the items most important in it changed. It was not until the second half of the nineteenth century that food became a major item: in 1840 Britain still produced about 90% of the wheat consumed. From 1850 onwards increasing quantities of wheat were imported, until in 1939 Britain produced only 13% of the wheat needed at home. The export trade at the beginning of the eighteenth century was dominated still by the export of woollen cloth which provided about half the total value of it; by 1800 though the export of woollen cloth had increased greatly it was outstripped by the export of cotton textiles: until 1850 cotton provided more than 40% of the export total. In 1850 Britain was already producing half the world's pig iron, and in the last half of the nineteenth century Britain dominated the world market in iron and steel, as railways largely built of British metal were flung across Europe and America, and machines, tools and implements of British iron and steel were demanded throughout the world. Coal was mined in increasing quantities: production rose from 4·7 m. tons in 1750 to 56 m. tons in 1850 to 287 m. tons in 1913, and coal exports rose from 3·8 m. tons in 1850 to 94 m. tons in 1913. The decades 1870–90 were those in which Britain's world position was most powerful, the value of its foreign commerce was greater than that of any other country; by the turn of the century the industries of Germany and the United States were beginning to compete, but not until the First World War was British supremacy seriously threatened. In 1921 the foreign commerce of the United States surpassed that of Britain. In the nineteen-twenties and thirties the world demand for British iron and steel decreased, the export value of textiles fell seriously as India and Japan not only satisfied their own demand but also became competitors in other eastern and in African markets. Coal exports declined too as mines in newer countries came into production, and the advantages of oil over coal became more and more obvious.

Now, in the mid-twentieth century, yet another aspect of Britain's position should perhaps be emphasized—the situation of the island between two great continents. Its position is now less favourable, and equivocal too, for Britain retains some of the links and something of the roles that belong to earlier periods. It is still an island off the shore of Europe; 24% of Britain's exports go to European countries. But it is a part of Europe without being in Europe, and the delicate negotiations to establish free trade in industrial products between the countries of western Europe and Britain are, in mid-1961, still not complete. Britain is also of the Atlantic world, a member of the North Atlantic Treaty Organization thus strategically as well as economically closely linked to America. The United States takes the largest single share of Britain's exports and provides the greatest single share of its imports. But Britain is not only of the Atlantic world. As the senior member of the Commonwealth its links stretch out across the Pacific and the Indian oceans; the Commonwealth takes 40% of Britain's export trade, and much British capital is invested in Commonwealth countries. Responsibilities to, and interests in, the Commonwealth cross its interests in Europe and emphasize the position at the centre of world ocean-routes. Britain has the largest merchant marine in service; about 17% of the world's shipping tonnage of 100 gross tons and over is registered in the United Kingdom, and Britain carries about one-fifth of the world's international sea-borne passengers and goods traffic. But Mackinder's assessment of the relatively greater strength in a railway age of land powers in comparison with sea powers is perhaps now being realized when air-routes too annihilate the effects of distance. The advantages of continentality, of huge areas organized as one for politcal and economic strength is evident, and, in an air age, the two great 'heartlands' of North America and the Soviet Union face each other across the Arctic rather than across the Atlantic Ocean. Britain's position may once more be seen as an off-shore island on an outer edge, and nearer the heartland of the Communist world than the heartland of the democratic world.

Britain is, however, still the world's second largest trading nation, although its share has fallen from 35% at the beginning of the century to 17% in 1959. It is surpassed only by the United States. Low-grade iron-ore and coal are the only major natural resources of the island; the ore, since the growth of modern iron and steel industries, has never been sufficient for home consumption, and there is now little demand for coal in overseas markets. The export of coal by 1959 had fallen to just under 6 m. tons. Britain therefore imports nearly all the raw materials needed for her industries including large quantities of metals, wool, cotton and petroleum, and about half the food its people eat. It must pay for these by an export of manufactures, and about 30% of its manufactures is exported. It is still one of the four largest producers of steel, but raw steel and textiles no longer dominate the export trade, engineering products, including vehicles and electrical engineering parts, account for over 40%. Without the advantages of the huge resources of the United States and the Soviet Union, Britain must rely to keep ahead in the new industrial revolution, on making goods in world demand, that need its knowledge and skill rather than raw materials peculiar to it, to produce them. But competition is severe; in invention, in design, in production per man-hour the race is to the far-seeing, the bold, and the persistently industrious. It is not easy to keep abreast let alone ahead.

All Britain is affected by the external situation, its agricultural, mining and manufacturing activities, and thus the density and distribution of its population is to a large extent a reaction to it. The opportunity to draw upon the world for supplies of food and raw materials in return for satisfying the needs of areas overseas for manufactured goods and technical and financial aid, has allowed continuous growth of population and the steady increase of the proportion that is urban. But the reaction is no longer wholly individual or local. Government policy and planning, through agricultural subsidies and price-fixing, grants to aid and powers to limit industrial expansion in certain areas, and plans to control building and thus movement of population, play an important role.

The pattern of distribution of the population before the days of a world economy when Britain fed its people from its own acres reflected essentially the agricultural value of the land. The pattern before 1750 so far as it can be reconstructed from taxation returns and parish registers, checked by calculations from later census figures, shows a broad belt of country stretching from the Wash to the Severn estuary as most densely peopled. London and its immediate environs already housing some three-quarters to a million people, or nearly a sixth of the total, lay almost in the middle of the southern edge of this belt. To the south and east of it, large tracts of forest, Epping and Hainault to the north of the Thames, the Weald to the south, lowered the density of population, and to north and west, even in the lowlands, population thinned rapidly. The uplands and the highlands were but poorly peopled.

Although the greater part of the population was throughout engaged in agriculture, as time passed industry occupied an increasing number. But even in 1750 industry was widely dispersed, and, broadly speaking, industrial distribution too was related to the fertility of the soil. This was true not only of such industries as milling and malting, leather working, and the building of carts and waggons, carriages and coaches, and even ships, but also of the two major industries, iron working and textile manufacture. Industry played the biggest part in the economy first, where people were most numerous, therefore specialization easiest and local markets largest, that is where soil was most fertile; and secondly where soil was infertile and thus where there existed a stimulus to eke out the resources of the land by making things to sell to richer neighbours. Within the rich lowlands, Norwich was the centre of the worsted industry, and the Stroud, the Windrush and the Wiltshire Avon valleys held pride of place in the broadcloth industry. On its fringes Exeter was renowned for serges, Lancashire for its cottons, and West Yorkshire for its kersies and shalloons. In the remote valleys of Central Wales and around Kendal in the Lake District, hill farmers made fatter their living by making and marketing cloth. In the Forest of Dean, in the Stour and the Severn valleys and in Derbyshire and Yorkshire dales furnaces and forges were particularly numerous and ironmasters busy. However, nowhere as yet did industry dominate the population pattern though locally it increased the density.

By 1801, the date of the first census, change was apparent, the modern design was foreshadowed, and by the middle decades of the nineteenth century the regional distribution of population had wholly altered and so had its relation to industry. High density of population reflected no longer agricultural prosperity and fertility of soil, but industrial activity and the accessibility of coal. The great succession of inventions of the late eighteenth and of the nineteenth centuries from the flying-shuttle, the spinning-jenny and the water-frame to the spinning-mule and the power-loom, from Darby's successful smelting of iron with coke and Cort's puddling process to Bessemer's converter and Thomas and Gilchrist's basic process, these, together with Watt's steam-engine, revolutionized industry, bringing into being the factories and the plants, and giving great advantages to those situated on or near the coalfields. The improvement of roads and of river navigations, the cutting of canals and above all the building of railways and of steamships, made it possible to bring together in one place large quantities of food and raw materials, and to distribute far and wide the finished products of industry, and thus allowed the growth of great cities. West Yorkshire soon outpaced the West Country in the production of broadcloths and East Anglia in worsteds, and dominated both the home and export market; Lancashire outdistanced all other areas in cotton manufactures, importing raw cotton and exporting finished goods through Liverpool. The iron and steel industries of the Midlands, South Wales, Lowland Scotland and North-East England expanded rapidly. The last three grew at an especially rapid rate since a coastal situation made easy the import of the foreign ores, free of phosphorus, needed to make steel by the Bessemer process, and gave great advantage in marketing products in nearby shipyards, and in exporting to countries overseas voracious for the rails and railway-engines and machinery of all sorts that they had not yet learnt to make for themselves. Coal production soared, more was mined than could be used at home and much was exported, most of it from South Wales and North-East England.

During the first half of the nineteenth century, the upsurge of population affected the country as well as the town. In fact death-rates in the country fell faster than in the towns, in many of which the state of housing and sanitation kept death-rates high. The towns grew by migration to them of country people rather than by natural increase, and because of this movement the rural population grew less rapidly than the urban in spite of a flourishing agriculture encouraged by the growing demand for food of a rapidly increasing population. Arable farming reached its greatest extent in the 1880's when over 14 m. acres were under the plough. But even in the middle of the nineteenth century, after the repeal of the Corn Laws in 1845, the import of foreign wheat, which averaged about 4·5 m. quarters in 1852–59 when home production was about 14·3 m. quarters, was causing many farmers, especially on the heavier soils of the wetter west, to lay down arable acres to grass. By 1876 the import of foreign wheat had reached 13·7 m. quarters and home production had fallen to 11 m. quarters. By the last two decades of the nineteenth century some rural districts were losing population absolutely, and many villages and small country towns reached their maximum size in the 1870's or 1880's. In 1911 for the first time the agriculturalists were no longer the single largest group in the occupational census; they were surpassed in number by those engaged in mining and metal-working. By the end of the nineteenth century some 75% of the population was living within the boundaries of urban administrative regions, and the conurbation—the modern unit, wider than a town or city—had come into being.

The essential features of the patterns of population seen in the late nineteenth century have persisted throughout the early twentieth century, in spite of, and in certain places perhaps because of, the needs of two world wars (fig. 5). Though some elements of the present pattern are stable and deeply rooted, there are evident signs of change in others now.

In agricultural areas the age-old pattern of population remains; for the most part it still reflects, in general throughout Britain and in detail within each district, essentially the fertility of the soil. Superimposed upon this, however, in particular places there is another pattern that reflects accessibility to the farmers' market, and both where distance is a very great disadvantage and where proximity is a very great stimulus, fertility of soil may be overshadowed.

Since agriculture occupies only about 1 m. out of over 51 m. people but some 48 m. out of 60 m. acres, the density of population in agricultural districts is low. In few areas is it more than one person per acre, and in great stretches of highland and upland Britain it is less than one per ten acres.

The division of Britain, not into highland and lowland, but, as Sparks suggests in discussing relief, into highland, upland and lowland, is a significant one in considering rural population. The true highland areas, the North-West Highlands and Islands, the South-West Highlands and the Grampians, much of the Southern Uplands of Scotland, most of the mountains of the Lake District and of central Wales have large tracts virtually uninhabited. The greater part of the area is rough pasture, the forest-planted land is only a very small proportion of the whole and still less is under crops and grass. People live in groups of crofts or on small scattered farms rather than in villages and sheep farming dominates the economy. These areas have

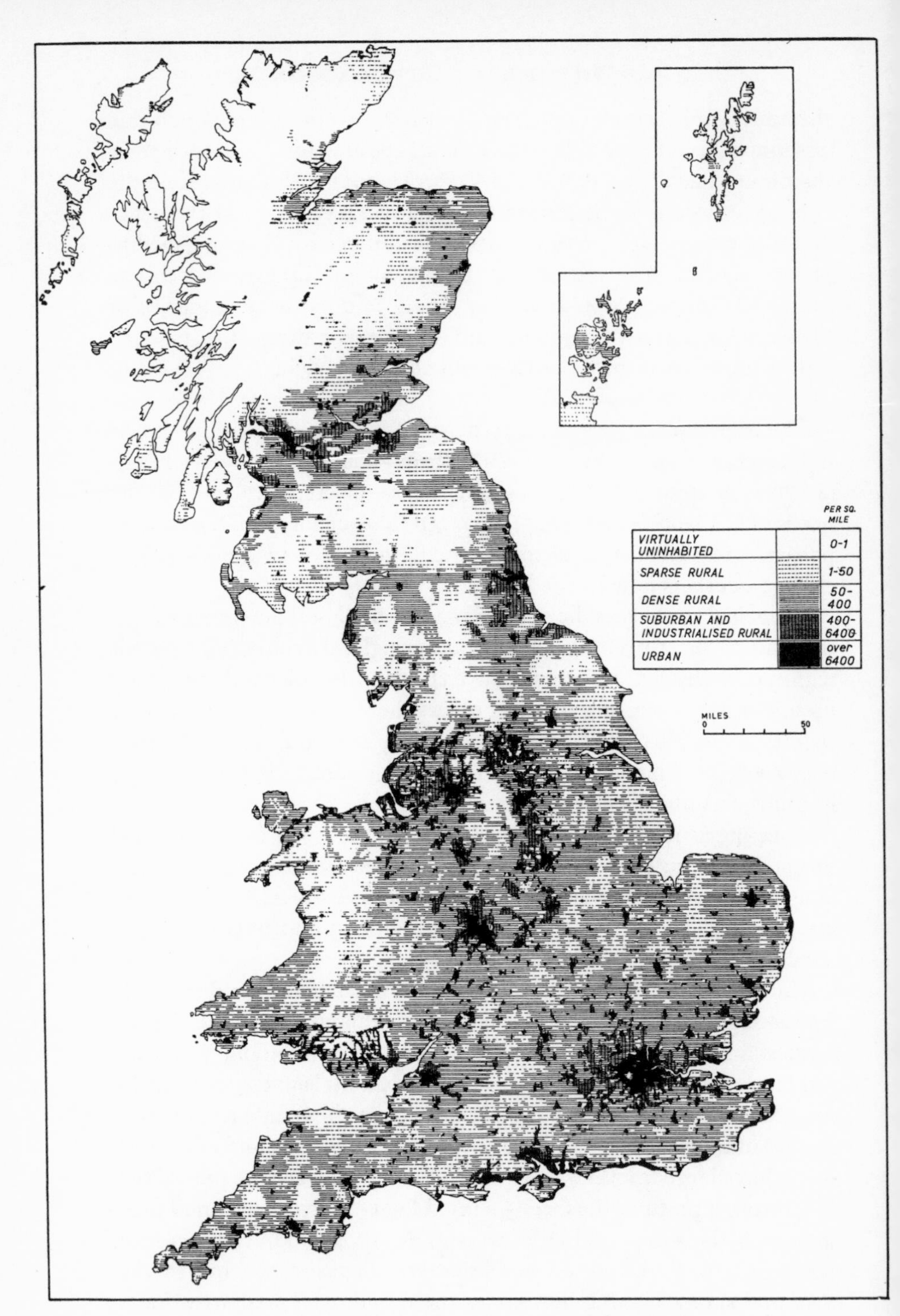

Fig. 5. Distribution of Population in Great Britain, 1951.

not only a thin population, very unevenly distributed but also a falling population (fig. 6). All the counties of highland Scotland, with two exceptions, show a loss of population rapid at the beginning of the century, slowed or halted during the two world wars and the years of industrial depression in the thirties, but accelerating again in the last decade. Birth-rates are relatively high and death-rates in the younger age-groups low, but migration to the industrial areas of Scotland, and still more to England and overseas, more than offsets the natural increase. Long-continued emigration, since it is the younger men and especially women who go, brings in time a decrease in the rate of natural increase. The decline of population is not of course universal or everywhere the same. The building of the dams or power-houses of the hydro-electric schemes, the draining and planting of Forestry Commission land, the making or improvement of roads, brings in workers, many temporarily, a few permanently, to certain areas. Caithness and Dumfriesshire show a total rise in population since the 1951 census, a rise accounted for by the attraction of workers to the atomic energy stations of Douneray and Chapelcross. Central Wales is also losing population in a very similar way.

The worth of these highland areas should not be measured in terms of the density of population alone, nor the rate of decrease be taken as necessarily a sign of declining usefulness. They may be more valuable for purposes other than agricultural, and their economic health may well improve with a decrease in population. Areas poorly farmed may be better used as great forests, or solely as gathering grounds for water for hydro-electric enterprises or city supplies, or even as playgrounds for city populations. None the less, declining population in these remote areas poses problems. People may emigrate because the isolation is too great and the social and economic amenities too few, roads too poor, markets, shops and schools too distant, water supplies inadequate, electric power and light wanting. Crofts and farms are then abandoned. But the fewer the people in an area the more difficult does it become to keep schools going, to induce postmen and delivery vans to make their rounds, the more uneconomic to improve roads, bring piped water and electric cables. There are a few who seek solitude in which to live, and a few more who enjoy it as a respite on holiday, but they are, and will probably increasingly be, the minority.

The upland agricultural areas have a higher density of population

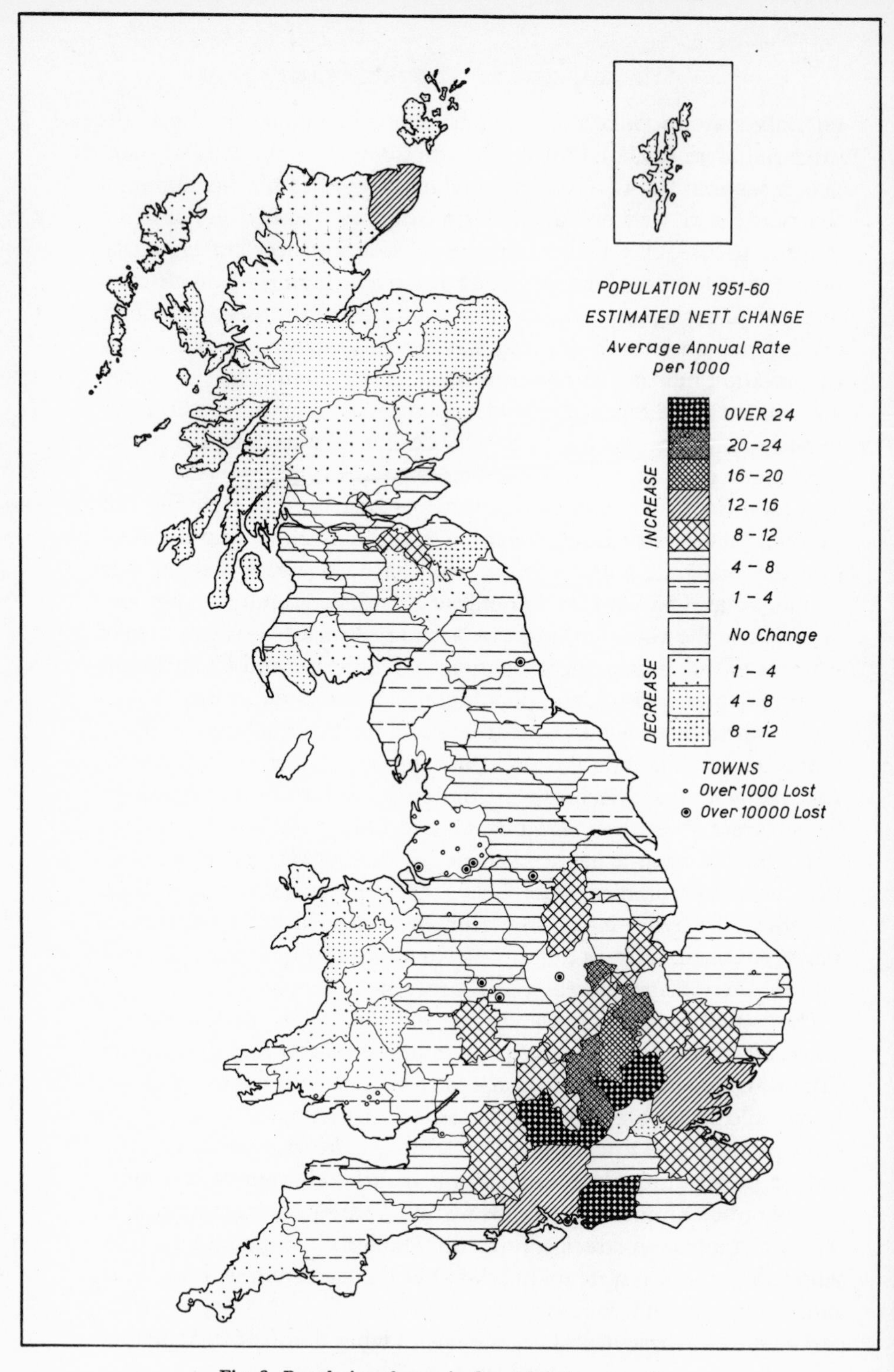

Fig. 6. Population change in Great Britain, 1951–60.

partly because many of them lie adjacent to the nineteenth-century industrial areas gathered to the uplands by the easily accessible coal. In fact these areas may be divided both in density of population and in rate of growth into those that directly serve the nearby industrial areas and those that do not. Those that do not share many of the characteristics of the highlands. The coastal counties of north-east Scotland between the firths of Moray and Tay, with a considerable arable acreage growing oats and barley, and a large acreage under fodder crops, temporary and permanent grass supporting many beef and some dairy cattle, support a denser population than the highlands to the west, and in the better parts employ as much labour, between three and four workers per 100 acres, as in the grazing counties of the English lowlands. However, the remote situation and the distance from markets is such that all these counties are losing population by emigration as rapidly as the Highlands (fig. 6). From the northern Pennines, and the uplands of Devon and Cornwall, the loss by emigration also more than balanced the natural increase, in fact in Cornwall emigration from many areas has been sufficient to make the population in 1961 more than 3,600 less than in 1951.

In contrast the uplands bordering the Pennines, the Lake District and Wales because of their position are increasing in population many of them faster than the neighbouring industrial areas (fig. 6). Dairy farming, poultry farming, and market gardening are intensively carried on and maintain a rural population of considerable density and even spread. However, most of the increase is the result of outward movement of towns-people from the industrial areas; industry dominates the scene and determines the nature of the agriculture.

The lowlands of southern and eastern England, the scarplands and the vales between them, are the lands of the traditional agricultural village. They fall into three parts: a western area on the margin of grassland and arable farming, an area to the north and east of the metropolitan region which retains much of its traditional character, and an area to the south and east which has almost entirely lost it.

In the west both north and south of the Thames mixed farming is the rule: crops and stock support the farmer. It was this area that showed the greatest increase in arable during the Second World War, and the increase has been largely maintained though a considerable part of the ploughed acres grow fodder crops. The method of farming rather than the products have changed. Agricultural labour is intensively used though less so than farther east; two to three workers per acre is a normal figure.

In the East Riding, Lincolnshire and East Anglia, the growing of wheat and especially barley, with sugar-beet, and locally potatoes, vegetables and special crops, dominate the farming. The area remained essentially arable even during the long agricultural depression of the later nineteenth and early twentieth centuries, but even here there was a marked increase in the 1940's in the proportion of arable to grassland under the stimulus of war conditions, and since the war the increase has been maintained. Increased mechanization on arable farms, though it has brought about some saving of labour, has not on the whole led to a decline in the number of farm workers, since the addition of dairy herds or of more vegetable, fruit and flower growing, has increased the intensity of farming. In the fens of Lincolnshire and Cambridgeshire the truly rural population has increased steadily since 1921 and is still increasing. In much of Lincolnshire and throughout East Anglia more than two workers are employed per 100 acres, and in Fenland, Cambridgeshire, and the loam district of north-east Norfolk, the figure rises to more than three.

Around London and to the south and east of it, the density of agricultural workers per 100 acres farmed is very high. In north-east Kent, in much of Surrey, Buckinghamshire, Bedfordshire, Hertfordshire and southern Essex the figure is more than three and in much of it over four. Here it is not cereal farming but the growing of fruit, vegetables and flowers for the London market that occupies most of the rural population: market gardens, hop gardens, and glasshouses are much in evidence. Among country dwellers there are many, however, whose income comes not from the surrounding acres directly or indirectly, but from capital invested in industrial or commercial enterprises at home or overseas. This is also a rural area with a growing population, but only a part is growth in agricultural population and that part is probably highly specialized.

The rural population of areas readily accessible to urban and industrial districts is threatened as greatly as that of the most remote areas, but in another way. Here competition for land becomes acute between groups in need of building land and groups cultivating land. The best land is coveted by both, and, even if the less valuable land is used for building, the built-up area brings about changes in the environment both physical and social that are often deleterious to the farmer. Smoke pollutes the air and blackens his crops, industrial waste fouls the rivers and poisons his livestock, his labourers are lured to industry and town, his market becomes highly specialized and perhaps capricious. The interests of the townsman, not the countryman, dominate the neighbourhood. The rural population too declines fast, if not in absolute numbers, certainly in relative importance.

In contrast to the rural population the urban and industrial population is as yet strongly localized in limited areas in which very high densities are reached (fig. 5). It is these areas superimposed upon the rural pattern that dominate the population map, and the largest ones, apart from Greater London distinguished by its metropolitan position, are still on the upland margins of the highland zone where the Coal Measures outcrop. But the link between industries and power based on steam coal, that Mackinder noted sixty years ago was weakening, may well be broken altogether before the present century ends. Change is the order of the day: the different urban areas are growing at very different rates, and in each of them there is a strong movement out from the centre to the edge. If the industrial revolution of the nineteenth century led to concentration of industry, that of the twentieth century seems to be encouraging dispersion once again.

Some industrial areas are now growing only slowly, and are growing by natural increase rather than by migration to them from rural areas. Some of these are now exporting population, thus the rate of gain by natural increase will lessen too. The areas dependent on the heavy industries and the textile industries have felt most the changes that have encouraged one-time buyers in the overseas markets to become first self-supporting, and then competitors. The towns of the Rhondda Valley lost population more rapidly than any other group in the decades 1931–51, but the rate of decrease has declined, suggesting that adjustment to new conditions has here already been made. In Lancashire, fourteen out of seventeen county boroughs, and forty-five out of the ninety-five municipal boroughs and urban districts show a loss of population if the 1961 are compared with the 1951 census figures. In 1959 about 100,000 people were employed in spinning and doubling and 93,000 in weaving, that is only about

50% of the number employed in 1937. The South-East Lancashire and the Merseyside conurbations have as a whole been almost stationary in population for thirty years and more. The woollen manufacturing towns of West Yorkshire show decrease in the last decade, although the numbers employed in the woollen industry have not fallen markedly since 1939. Tyneside was declining during the depression of the 1930's but showed as a result of the stimulus of war-time an increase of 1·2% by 1951 and the 1961 census shows continuing growth of 2·0% over the 1951 figure. The belt of counties in Lowland Scotland from Ayrshire and Dunbarton to the Lothians and Fife all show a rise in population, a slow rise in 1951 and a continuing rise in 1961, but the area as a whole is losing by emigration to England, America and the Commonwealth.

The rate of decline in these areas is slowed by government intervention; schemes are designed to help them in three ways. First, compensation is given where mines, factories or plants are closed down because on a long-term view their rehabilitation or renewal is judged uneconomic. Secondly, grants are made where it seems profitable to reorganize, re-tool, and modernize existing units carrying on the traditional industry: for example to install new machinery and equipment in those coalfields with better seams and prospects of a longer life; to rebuild and expand, or build anew, steelworks of modern design as at Ravenscraig, near Motherwell, at Middlesbrough, or at Margam in South Wales, or to put new machinery into old cotton mills to convert them to mills to spin and weave new fibres. Thirdly, help and encouragement is given to set up new and varied industries to use the labour, and if possible the buildings, left idle by the decay of the old.

Other areas in contrast are growing fast. The conurbation of the West Midlands shows marked increase in population: the 1951 census showed an increase of 7·6% over the mid-1939 estimates, and the 1961 census shows an increase again of 4·8% over the 1951 figure. The East Midlands is increasing fast; Nottingham is among the eighteen British cities of over a quarter of a million inhabitants and is one of five of them increasing steadily. Above all, the metropolitan region continues to expand in spite of all efforts of planners to control it, and the prognostications of those who consider that urban areas if they grow too big choke themselves (fig. 5). In these areas new resources, new sources of power, available labour and modern transport encourage new industries, and thus foster growth of population.

The new industries now expanding most rapidly are the manufacture of cars and other vehicles, electrical engineering, the making of radio, television and electronic apparatus, scientific instruments and nuclear plant. These depend for success on the new inventions and new skills of the scientific mid-twentieth century, and the position chosen by firms engaged in them depends on available space to build, on an adequate supply of labour skilled or capable of profiting swiftly by training, on the availability of suitable power, and on good communications to collect the varied raw materials and distribute the finished products. The new industries may find these advantages where a declining industry is releasing building and labour, but they depend on electrical power rather than steam power and thus find no special advantages on or near coalfields. The building of thermo-electrical stations in the Trent valley is one example of the loosening out of the pattern of distribution of power-supplies that in turn attracts industry to new areas. Though the new industries use a great variety of raw materials many of them do not need these in great bulk, and the finished products are valuable in relation to their weight, so that many firms rely on road transport with the great advantages of flexibility in routing and timing. The building of the motorways is a reflexion of this, and the gathering of the motorways into the Midlands is emphasizing once again the nodal position of this area: Coventry was a route centre second in importance only to London in the road system of the seventeenth century, and a great network of routes is now again centring on the Midlands. Above all, the new industries are attracted by the market, and thus the great pull of Birmingham, the second city of Britain, and above all of London, the metropolis.

Changes are taking place not only in the relative importance of the industrial areas one to another, but also in the distribution of population within them. The centres of great cities show a persistent and in some cases marked decline: this is true of Glasgow and Edinburgh, of Newcastle and Durham, of Manchester and Liverpool, Leeds and Sheffield, of Birmingham and of course of London. It is a movement begun before the war, and thus though evacuation and bombing may have accelerated the trend temporarily it did not create it, nor in fact have the long-term effects of war increased it. As the bombed areas have been cleared and rebuilt, though in some offices and factories have risen on them, in many, in accordance with modern ideas of town-planning, houses, often in terraces, and flats, sometimes in large and high blocks, have been built alongside the places where their dwellers may be supposed to find work. But this high-density housing has not reversed the decline of residents at the heart of great cities. The suburban growth so characteristic of the period between the wars has continued even farther until it has reached, on the outskirts of many cities, the green belt where building is restricted. The competition for land is here acute; the city workers take houses in existing villages in the green belt, or build houses in the country beyond it. Industry, too, is moving out of the centre of cities. New enterprises find room on trading estates built in hitherto rural districts to take overspill populations. These are scattering industry, albeit as yet in small units compared with the size of those that grew uncontrolled in the nineteenth century.

The planning and development of 'new towns', and the planned expansion of selected old ones, to take the increasing populations of the cities, is designed to distribute people more widely and more evenly. Glasgow plans to settle her citizens with industries to occupy them in towns throughout the length and breadth of Scotland from Wick in Caithness to Stranraer in Galloway. But these widely scattered new units are but plans; the new towns in being in Scotland are all within the industrial belt. East Kilbride, Lanarkshire, and Cumbernauld, Dunbartonshire, are close to Glasgow, and Glen Rothes designed also to take surplus population from Glasgow is primarily for the miners of the new coalfield in Fife. In England the new towns by virtue of the space available are nearer other and older centres; they collectively make outer rings beyond the green belts of their parent cities. Corby, like Glenrothes, houses a new heavy industry; Peterlee and Newton-Aycliffe in the North-East and Cwmbrân in South Wales have been established to relieve congestion and give diversity to industry in the old areas depending on heavy industry. Manchester and Merseyside have many schemes in embryo to build new communities in rural Lancashire and Cheshire, but here as in West Yorkshire new towns are needed to make slum-clearance easier rather than to meet the needs of rapidly growing populations. It is around Birmingham, and above all around London, that new towns are most needed and most numerous and successful. The movement of population out from Birmingham and the Black Country is largely to new centres, planned and unplanned, to south and east of the conurbation, and with the exception of Crawley, London's group swings in a great arc from Basildon on the east through Harlow, Stevenage, Hemel Hempstead on the north to Bracknell on the west. Haverhill and Thetford are being given a new life farther out still, and plans are afoot to colonize in many other East Anglian towns.

The deliberate planting of new towns to house the overspill population is the modern attempt to control and direct the outward flow. The plans needed are vast, the foresight demanded of the planners great. It is clear that the increase of population, at least in lowland Britain, is greater than present plans for redevelopment in existing cities and for overspill into new centres can accommodate. Planning must outstrip growth if all the plans are not to be set at nought. With every year that passes the problem of achieving the good of a balance between man and his environment in this crowded island grows. The population is increasing but it is increasing at an uneven rate; between highlands and lowlands the disparity in population density is increasing, within the lowlands the differences are evening out.

Migration from remote highland glens and islands, and upland farms and isolated lowland villages, brings continuing decrease. Coastal towns, too, are losing population. The fishing industry is less flourishing than it was, and now the holiday-makers, who fostered the growth of many fishing-villages into fashionable or popular resorts in the nineteenth century, are deserting them for places abroad. From the one-time-favoured industrial areas, from Clydeside, Tyneside and South Wales with their great heavy industries focused on the export market, from Lancashire and West Yorkshire once producing the greater part of the world's demand for textiles, people, especially young people, are moving away to towns with new industries where skill possessed or acquired will earn a good living. The area that lies to the south and east of the old coalfields and to the north and west of London on that indeterminant frontier between Mackinder's Industrial Britain and Metropolitan Britain offers all the advantages. Into this area migration is strong, and since the migrants are young the birth-rate is high; the rate of population-growth is rapid (fig. 6).

If some elements in the geographical pattern of distribution of population are old—the emptiness of the true highland regions, the even density of the rural population on the fertile soils of Gloucestershire, Worcestershire or East Anglia, and the dominant size of London for example—the striking areas of high density in upland Britain date but from the nineteenth century and are the legacy of the cumulative forces of the industrial revolution. The forces that brought this pattern into being are mostly spent, the conditions that generated them have passed, but the pattern that they created is deeply imprinted. But not indelibly; change comes slowly but inexorably to meet a new age, and it is coming full circle: the new pattern of population density emerging bids fair to re-establish the old. Before the industrial revolution the centre of gravity of Britain's population lay in central England: the pattern emphasizing the flanks of the Pennines is, historically, a new one and may well prove historically to be an ephemeral one.

Selected Bibliography

H. C. Chew. 'Changes in Land Use and Stock over England and Wales.' *Geographical Journal*, **122**, 1956, pp. 466–70.

A. Demangeon. *The British Isles*, translated E. D. Laborde. 3rd edition, London, 1952.

H. J. Mackinder. *Britain and the British Seas*, London, 1902; *Democratic Ideals and Reality*, London, 1919.

Wilfred Smith. *An Economic Geography of Britain*. London, 1949.

E. C. Willatts and M. G. C. Newson. 'The Geographical Pattern of Population Changes in England and Wales, 1921–51.' *Geographical Journal*, **119**, 1953, pp. 431–54.

F. D. V. Spaven, Esq., of the Department of Health for Scotland most kindly provided much material on Scotland.

PART TWO

ENGLAND AND WALES

CHAPTER 4

THE LONDON REGION

M. J. Wise

The majestic story of the growth of London on the banks of the Thames has been told many times by geographers and historians. Though each writer has his own tale to tell, one mighty theme sounds above all others, that of the site and situation of the noble city, first in the land, 'so commodiously pitched' at the meeting-place of land and sea upon this small, but great, river. London on the Thames stands and has stood principally by the advantages of its site at the lowest crossings of the river. It was from the medieval gatherings of people under the protection and leadership of tower, church and court that the modern expansion of London began along and, in time, away from the stream.

Medieval London had two distinct and clearly separated parts, Westminster and the City. Little by little, the Strand way, joining the two, was lined by shops and houses, and from it stairways led down to the mud flats and the great thoroughfare of the Thames. In its modern guise London has still its separate parts. Though the ancient centres of power and trade still dominate, within the continuous built-up area that extends for 10 miles and more from Charing Cross to north and south of the river, areas of special character and function have arisen to provide for the life and work of the great city. Even beyond the 10-mile limit the towns and villages of the Home Counties are linked socially and economically with the ancient centres of London, and with the shops, offices and places of entertainment that have grown around them.

The Thames retains a living role in the life of London and its region; even the dwellers in the far suburbs, in Barnet or in Purley, and in the towns beyond, are in large measure dependent upon the work accomplished along its banks. The Houses of Parliament and the government offices in Westminster and Whitehall stand upon the river's brim, while facing them rise the massive offices of County Hall. Along the former marshlands of Thames-side stand great industrial plants, and through the heart of London ply the tugs with their tows of lighters bringing coal, oil and timber. The Port of London remains the major port of the country. The strength of the forces that create the region and determine its changing character derives momentum from the site on the river-bank at the crossing-place of routes of national, indeed of international, importance. And not only the region but also the whole nation, through its capital, enjoys the geographical advantages of this favoured site for its own internal government and trade and for contact and commerce with Europe and the world beyond.

The theme of this essay is the changing human geography of the London region in the mid-twentieth century. The growth of employment, population, and the outward expansion of the built-up area are continuing characteristics. The increasing hold that London has exerted on the nation's life is another. At the beginning of the twentieth century, Sir Halford Mackinder showed the extent of London's influence on Metropolitan England, that part of the country which lay to the south and east of a line from the head of the Severn estuary to the Wash; an influence that he ascribed primarily to the long-established convergence of the main routes at London. In 1928 Hilda Ormsby emphasized 'the ever-tightening grip of London . . . gradually welding, not only the London Basin, but all south-eastern England into a geographic and economic whole'. The size of London, and the strength of its grip, have become so great as to arouse anxiety that London has become of disproportionate size and importance in the life of Great Britain. The effects, direct and indirect, of the deliberate attempts to check and to control growth and to re-shape the pattern of urban life that is the end-product of centuries, are new characteristics, but increasingly important ones.

In maintaining this theme no attention is given to the physical conditions that influenced the site and early growth of London. These have been discussed at length by Ormsby and others. Less excuse exists, perhaps, for the failure to record the results of recent studies of the geomorphology of the London Basin, studies that have brought a new understanding of the origins of landforms and soils. It is easy in discussing the growth of a great urban region to dismiss as irrelevant the physical conditions of rock, gravel and soil on which its buildings stand. Such a conclusion is erroneous for 'the soil on which London rests is still a factor in its life and future'. Varied conditions of relief, slope, drainage and soil reveal themselves closely in the pattern of land use of the urban region and its green setting.

The expansion of London in the inter-war years was rapid. Between 1921 and 1937 the total population of Great Britain increased by 7½%, from 42·75 m. to 46 m. people. But the increase was unevenly distributed. In London and the Home Counties the increase was about 18%, in marked contrast to the low rates of increase in such areas as the West Riding and mid-Scotland and the actual declines in parts of South Wales, and Northumberland and Durham. During the same period at least 1 m. people migrated to London and the Home Counties. This growth of population in and around Greater London reflected the stability of employment in the region during the years of the Great Depression and the subsequent economic recovery. In the same period Greater London absorbed no less than one-third of the total annual increase of employment in Great Britain. Between 1923 and 1937 the number of persons insured against unemployment rose from 1,950,000 to 2,650,000, or by 36·1%, compared with a general increase for Great Britain of only 22·3%. Not only was London increasing rapidly as an area of employment, but it was becoming of greater importance in relation to other industrial areas. While some areas suffered unemployment and social distress, Greater London prospered.

The demand for housing resulting from this substantial increase of population changed the face of London as the suburban housing-estates, some built by local authorities, others by speculators, spread around the terrace houses and villas of Edwardian London. Outward expansion of London into the adjoining counties of Middlesex, Essex, Kent and Surrey was prompted also by the movement of people from areas of congested housing. In part this was the result of a search for more attractive living conditions on the London fringe: partly, also, a planned movement associated with slum-clearance in inner London. It was encouraged by the improvement in motor-bus services, the electrification of suburban railway lines, and the extension of some 'underground' railways. The peak population of the area of the Administrative County of London had, in fact, been reached as early as the beginning of the century. Thereafter a slow fall in population had set in, though in 1931 the numbers were only 140,000 less than at the peak of a little over 4½ m. After 1931 the process quickened and in the next seven years the total fall in population exceeded 334,000.

Dramatic changes occurred during this period in the landscape of London's edge where, it was said, 'the countryside is being overwhelmed in an alarming way'. New arterial roads and by-passes appeared; around the road junctions, as at Park Royal and Wembley, sprang up new, quickly built factories. Many of these were on speculatively built industrial estates and provided splendid conditions for the birth of new industrial concerns and the rapid growth of young ones. Around and between the factory groups pressed the tide of suburban housing; and along the roads sprawled the brick semi-detached houses fastened, as it were, by drive and garage to the concrete highway.

The expansion of the built-up area was more rapid in some directions than others. It was especially forceful, for example, along the line of the Metropolitan Railway which gave direct access to the West End and City from points as far away as Harrow and Rickmansworth. At the same time, towns and villages just beyond the edge of the main built-up area grew individually, though as a result of the same generating forces that were prompting the expansion of the main urban mass. Processes generally similar to those at work in Middlesex were also active in changing the face of the urban fringe in the other counties adjacent to London. Essex, Kent and Surrey showed individual characteristics, but they held in common with Middlesex the outward spread of people and houses from London itself, and the dependence of a significant section of the new population upon employment in the central areas of London. In Essex, the concentration of industry on the riverside and in the Lea Valley gave to the county a higher proportion of local employment in manufacturing industry than in Kent or Surrey. In Hertfordshire, the direct impact of the outward expansion was felt later, but after 1931 the development of the southern fringe of the county as a residential area began. The opening of the Piccadilly Line to Cockfosters in 1933 permitted, for example, a considerable increase in the population of the East Barnet area.

In view of the spectre of the tragedies that were being enacted in the 'depressed areas', in view of the growing fears of war and of disaster to massed populations from aerial bombing, and with the unhappy spectacle of housing-estates of uniform and uninspired design sprawling across the countryside, it was not surprising that public opinion came to consider the growth of London a national evil. Fears, probably groundless it is true, but widely held, were expressed about the health of the urban population; the scandals of congestion and of the long journey to work were loudly condemned. In the public mind the need for national action to prevent further growth gradually crystallized. The Royal Commission on the Distribution of the Industrial Population, 1940 (the Barlow Commission) focused this general view and in its report called for action to halt further growth. 'It is not only that the mere size, spread and growth of the great conurbation tend to accentuate the various disadvantages present . . . in other conurbations, but also the trend of migration to London is on so large a scale and of so serious a character that it can hardly fail to increase in the future the disadvantages already shown to exist . . . the continued drift of the industrial population to London and the Home Counties constitutes a social, economic and strategical problem which demands immediate attention.'

Public opinion, the Barlow Commission, the war itself, thus brought about a new interest by the government in the use of land and, eventually, the taking of powers by the government, the exercise of which, directly and indirectly, were to lead to great geographical changes.

The essential plan, on which post-war development and change has rested, was prepared by Sir Patrick Abercrombie between 1942 and 1944. The plan covered a wide area, from Royston to Haslemere and from High Wycombe to Thames Haven, recognizing the inadequacy of the existing local government structure when faced with regional problems of industrial location and population movement. Events have proved even this wide area to have been too small; but the Abercrombie Plan must be recognized nevertheless as a masterly conception of a London region of the future. Its details have been changed and ignored, the pace of change has outrun it, but its main recommendations have been followed in principle. Abercrombie felt entitled to assume, in view of the Barlow recommendations, that the total population of the region would be stabilized, even reduced a little. He viewed the region as consisting of four concentric 'rings': the close buildings of the Inner Urban Ring; the dormitory suburbs of the Suburban Ring peopled by 'strap-hangers', and responsible for the 'waste of daily travel'; the open land of the Green Belt Ring threatened by engulfment; and, where rose the Chilterns and the North Downs, the Outer Country Ring, land of great possibility if well used and planned.

From the congested areas of the Inner Urban Ring he designed a mass regrouping of population, necessary if the closely packed areas were to be redeveloped to provide new environments for mid-century life. More than 1 m. people were to move, three-quarters of them to other parts of the region, the rest outside it. In the Outer Country Ring new towns were to rise, other towns to be enlarged, to provide homes and workplaces near together and so abolish long journeys to work. Abercrombie had a vision of a new London; the war and the evacuation of population provided the opportunity and the challenge. 'All things are ready if our minds be so,' he quoted. He drew accordingly a plan for the future geography of London in which the city and its dwellers could settle down in a life of peace. How far has the vision of a new city been achieved?

Ever since the reign of Elizabeth I prophets in every age have foretold that the press of population, the congestion of traffic, the increasing distances imposed by outward expansion, would bring to a halt the growth of London. But they have always been confounded: in spite of fears, in spite of plans, London has continued to grow. And so it has been in the post-war years. The prospect of unpleasantly hot and crowded journeys to work has not yet deterred men from enjoying London's opportunities for advancement. The advantages of central locations still outweigh the disadvantages of slow progress through the central streets. So long as the problem of ensuring that the user of central land pays the full cost to the community for his use of that land eludes solution, so long will there be an irrationally high demand for sites in the streets that attract most traffic.

The forces at work promoting increases in population and intensifying economic and social activities in London and its region are still very strong. Even a brief examination of the industrial structure reveals that in this area nationally declining industries are hardly present. Even the old staple trades of London's East End, the clothing and the furniture industries, have not suffered in the new age. Industries that have expanded nationally in the last thirty years, such as the electrical and general engineering groups, vehicle production, certain chemical and pharmaceutical industries, paper and printing, and food and drink, are strongly represented. Of major importance has been London's eminence in the building trades, wholesale and retail distribution, insurance and banking, public administration, the professions. This group of occupations has been increasing dramatically throughout the country; and in London six workers out of every ten find employment in it. Its total expansion has been a factor of the greatest importance in confirming the momentum of London's growth.

In the new industrial age no longer rooted in coal, London has many attractions for manufacturing industries. Thames-side sites have striking advantages for industries processing bulk materials; and the premier seaport, the premier airport, and the hub of radiating roads and railways is a convenient place for the assembly and processing of the many raw materials that make up modern manufactured goods, and for their distribution to markets at home and overseas. London itself offers the attractions of a great market: it is by far the largest of the regional markets in the country, both in number of population and, even more, in size of purchasing power. In an age in which the real value of wages and salaries increased by 60% in twenty years (1938–57) the Londoner has been even more fortunate than his cousin in the industrial regions of the north and west. Unemployment has been unknown, and London has continued to offer, to men of enterprise and ambition from all parts of the country, golden opportunities for advancement to economic, social and political power.

London is not only the greatest market: it possesses the largest supply of labour. The London labour market is a potent source of strength: manufacturers are able readily to hire skilled personnel and trained managers; here, too, large supplies of relatively unskilled labour are available, capable of profiting from the short training now necessary in many industries.

London's manufacturers are able to rely on the presence of a wide range of essential industrial services, from banking and insurance to merchanting and warehousing. They find at hand middlemen engaged in the procuring and distribution of raw materials and in breaking bulk. They derive prestige, and advertising power, from their position in the capital and, in an age in which the influence of government upon industry is increasing, they appreciate the advantages of nearness to the centre of national administration.

No student of urban growth can easily ignore the strength of the economic forces arising in and making for the concentration of industries in one place. The advantages that manufacturing and servicing industries alike obtain from proximity each to each and to one another are powerful agents in city-building.

If the strength derived from these conditions were not sufficient, the reorientation of the national economy to hide and recover from the scars of the war years has given a special impetus to the industries producing for export, a group in which London is strong, and to the trade of London's port. Around London, too, needing access to its libraries, scientific societies, universities and departments of state appear the governmental and industrial research institutes and laboratories, the seedlings of the industrial forest yet to be. London grows as an international city, as a main port-of-call of travellers on the world air-networks for whom it provides hotel, tourist and conference facilities. Each year this traffic increases in size and value.

London's economy continues to grow. How far have the checks on growth, envisaged twenty years ago, operated? What of the effects of public control of industrial location? The principal weapon of public control of industry and employment has been the Board of Trade industrial development certificate, applicable to all projected new industrial building of 5,000 sq. ft or more in area. Industrialists wishing to establish or expand factories in Greater London and in the surrounding region have been pressed to consider locations elsewhere. This policy has been applied fairly, and at times stringently, by the Board of Trade. But in practice, although it has been possible to restrict, it has not been possible entirely to prevent industrial growth and a consequent rise in employment opportunities. The need for modernization and re-equipment of thousands of factories, already within London, with almost inevitable expansion, has had to be recognized. Many of London's industries are bound so closely together that to insist on the removal of a firm to a Development Area might well have meant its death. The advancement of existing industries, the electronics group for example, can hardly be denied without severe cost to the nation's technological and economic progress. In fact, the application of locational controls has given no more than a moderate braking effect to the region's industrial growth. The control system has not impinged at all upon the main mass of the servicing industries and this group, employing more than one-half of the labour force, has been free to expand.

Not surprisingly, therefore, the post-war years, an age of national full employment and of an annual 3% increase in industrial production, have been for Greater London a period of overall increase in employment. The trends of the inter-war years have continued in only moderately abated form. During the 1950's, the London Region, defined as a zone of approximately 35 miles radius from Charing Cross increased in employment by over 7% compared with a national increase of only 5%. With one-quarter of the population of England and Wales, the region received two-fifths of the additional employment. During the 1950's the region's labour force increased by half a million workers. The increases in employment have not been uniformly distributed over the region as a whole. Important increases have taken place in the central office and business districts but the older manufacturing areas of inner London, in areas of characteristic nineteenth-century development, have not increased. Here planning controls have, in fact, caused a slight decrease, though the loopholes in the control systems are such that the decrease has been smaller than was anticipated. The industries in suburban areas show no such decrease. By contrast, the virility of the electrical, engineering, metal, food and drink, and chemical industries has resulted in substantial gains in employment in Middlesex and metropolitan Surrey and Essex. Even more dramatic have been the increases in employment in the outer parts of the London Region, beyond the main mass of the built-up area. Including the 'new towns', which have had little difficulty in attracting suitable concerns, the outer parts of the region recorded between 1950 and 1955 a growth of no less than 26%, no less than four times the national rate, in industrial employment. The new towns, it is true, had special attractions; good accommodation both in factory and in homes, the availability of new, well-planned factory buildings with improved plant layout, higher labour productivity, low rates of absenteeism. But the older towns of the outer region, such as Watford, Luton, Southend and High Wycombe have also shared in the high rate of growth. Also partly responsible for the expansion of employment in the outer region has been the growth of industry along the Thames estuary, with its special advantages for the chemicals, oil-refining and pulp and paper groups. In almost every town and village north of the

Thames employment opportunities have expanded greatly since the end of the war.

It is scarcely to be wondered that, once the war-time period of evacuation and re-adjustment of population had passed, the total population of the region began again to increase in total. Exact estimates of the rate of growth during the 1950's must await the results of the 1961 census: meanwhile, calculations from the General Register Office statistics show a total increase for the region of over 400,000 persons between 1951 and 1958. Even more significant than the total increase is the geographical pattern of change. Almost all areas within the main built-up area have registered a decline since 1951. Thus the pre-war trend for the decline of the population of inner London has continued and spread its influence geographically outwards. The overall decrease for the conurbation proper has been of the order of 176,000. On the other hand, surrounding the London conurbation, a wide zone may be distinguished in which the rate of increase since 1951 has been at least 10% or three times the national average, a zone which now contains (1959) 700,000 more persons than in 1951. The zone encircles London, embracing and passing beyond the new towns and extending its tentacles along the main routes from the south-east to the north and north-west.

This vast redistribution of population has taken place, in part, as a result of the plans to move people from congested inner London. Between 1952 and 1958 about 45,000 persons were moved to estates built and operated by the London County Council beyond the conurbation, 120,000 to the new towns and a further 10,000 to towns which have entered into 'expansion' schemes with the London County Council to receive 'overspill' population. On at least an equal, and currently growing, scale there has been again, as in the 1930's, a voluntary movement of people to the outer area in search of more attractive housing and living conditions, a movement made even easier now by the greater mobility granted by possession of a motor-car. There are signs that in 1958 and 1959 the rate of increase of population in the region as a whole was increasing. It is clearly true that the volume of voluntary and planned 'overspill' has been under-estimated and that the redistribution of population will continue.

So that, in seeking the Greater London region of 1960, the significance of the great peripheral zone of increase must not be missed. The great zone of interlinked workers and factories, tied to central

London by the daily ebb and flow of commuters extends outwards to Reading, Luton, Southend. Romford, Watford, Slough, Gillingham and Guildford are part of the region. It may be expected that, defined in these terms, the outward margins of the region will continue to expand, presenting new problems of regional administration and planning.

The strength of the daily movement from home to work is undoubtedly a force making for the coherence of the region. It is not the only way, of course, in which the suburbs and outer fringe depend upon the centre: central London provides for the region shops and entertainment, educational, medical and professional services. The difficulty of separating London's regional from its national function as 'Town' makes, in the present absence of detailed research, a geographical analysis of London's urban field an imprecise task. More simply it may be revealed by a brief analysis of the pattern of journeys to work. The journey to work from suburb to centre has become a dominant characteristic of life in London. Mackinder's 'beating heart' of London pulses ever more strongly, drawing to itself, each morning, over 1 m. workers and disgorging them each evening, between five and six o'clock, into the overcrowded tube-trains and omnibuses. Of the 1,250,000 commuters to the central areas recorded in November 1955, two-thirds were to the West End, the remainder to the City. Since 1955 an increase in the total number of at least 100,000 has been recorded and annual statistics continue to rise. The average Londoner makes a journey from home to work of about 45 minutes each way. The London suburbs, such as Finchley and Hendon, Beckenham and Bromley, form great crescents north and south of the river in which the majority of daily commuters dwell. As the outward movement of population proceeds, as the railway services are improved by the introduction of diesel and electric trains, and as more and more people buy motor-cars and can thus live farther from public transport stations, so the distance travelled from home to work may be expected to rise.

The main zone within which the journey to central London is a significant element in the human geography of the region is shown on figure 7. From the conurbation itself tongues extend outwards along the main and suburban railways to Reading in the west, Luton in the north, Chelmsford in the north-east. On either side of the Thames estuary, at Southend-on-Sea, in the Medway towns and,

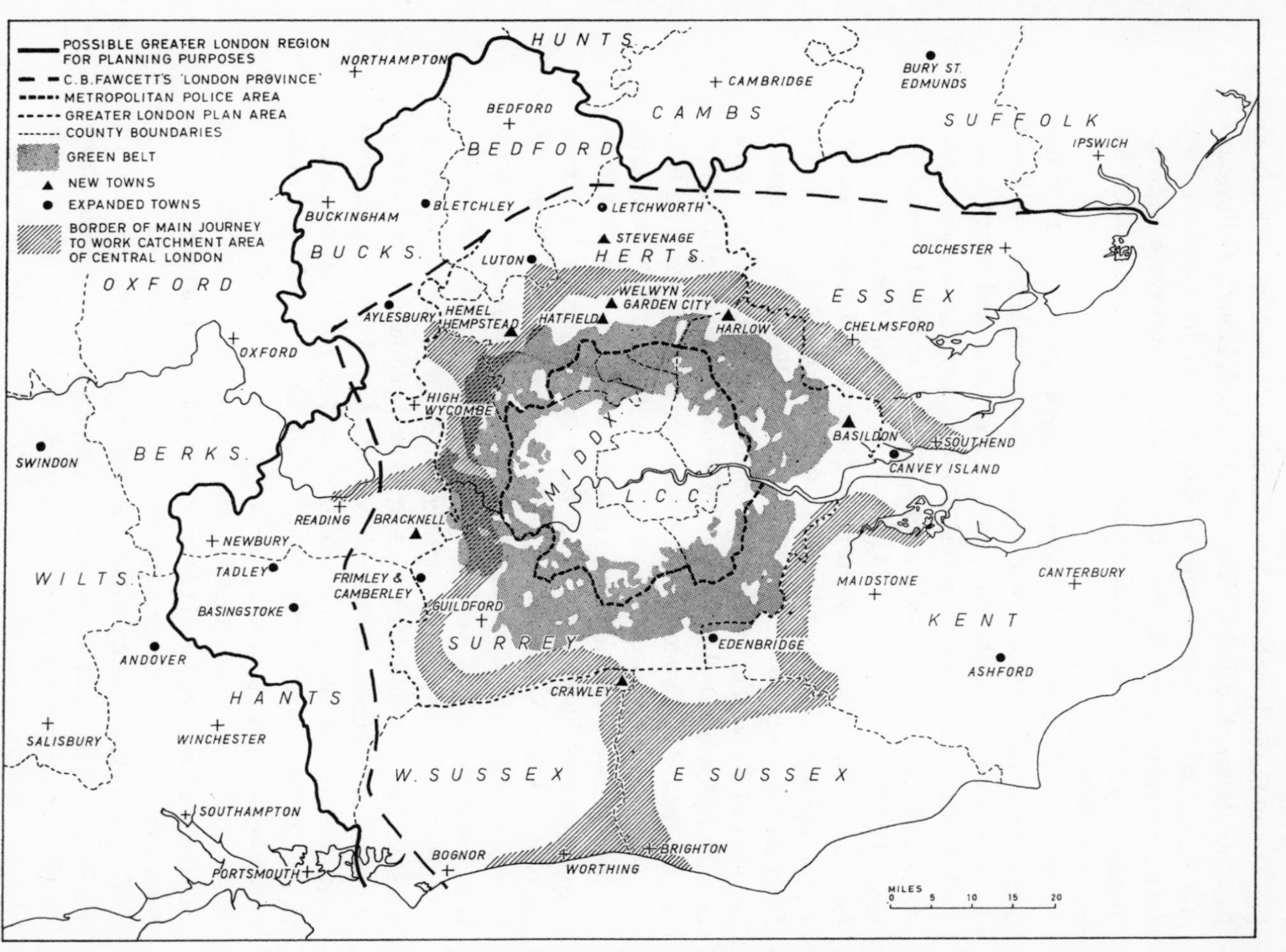

Fig. 7. The London Region in relation to south-east England.

increasingly, in Thanet are the homes of London workers, while from Brighton on the south coast the frequent electric trains carry London's business men back and forth from home to office.

The importance of the daily surge of population to the City and West End, and the problems of transport and congestion that are associated with it, should not obscure interest in the smaller-scale movements to the important industrial areas in the East End and along the riverside, in the Acton and Brentford areas and in industrial pockets farther from the centre, as at Hayes and Kingston. Nor must be forgotten the locally significant power of Watford, Luton and the Medway towns in drawing to themselves currents of movement. The pattern of journeys to work is not a simple one; but it is the tidal flow to and from the centre that grips the attention and provides so many of London's problems of organization and planning.

It was an axiom of C. B. Fawcett that regional boundaries should not cut across daily movements of population. Viewed in terms of the daily journey to and from employment, the region of London is seen to extend across the physical boundaries of the London Basin, especially southwards into the Weald and as far even as the towns of the Sussex coast.

At the centre of this great gathering-ground lies a vast expanse of building from Enfield in the north to Coulsdon in the south, from Hounslow in the west as far east as Dagenham and Thurrock, housing about $8\frac{3}{4}$ m. people. Whereas, for example, in the West Midlands a number of towns expanded to join with the greater city of Birmingham, London has been built almost entirely by outward expansion from the centre absorbing in its path only small towns and villages. Within this grouping of population there has occurred a well-marked specialization of human activity by area and locality to a greater extent than in any other British conurbation.

A geographical pattern in London's economic activities is clear. Near the centre lie the two principal office districts: to the east lies the financial heart in the City of London separated by the valley of the Fleet and by an area of wholesale trade (Covent Garden) and of light industry from the more recent office district in the West End. Of the two, the general office district, as it may be termed, is much the larger, possessing more than twice the area of office floor space of the City, though the disparity in the numbers employed is not quite so great.

The interests of the City are both more concentrated and more specialized than those of the West End. For centuries it has been the centre of the nation's financial life and the presence of the Bank of England, the Stock Exchange and the headquarters of the Joint Stock and Merchant Banks symbolize this ancient role. Around Fenchurch and Leadenhall streets are to be found the offices of the big shipping companies, while in the eastern part of the City are the wholesale markets in commodities as diverse as cereals and furs, wool and diamonds. It is a main centre of the insurance business of the country.

Though its functions range more widely than that of the City, the general office district of the West End also presents a strongly marked geographical pattern. Offices of similar kinds are often found together, thus obtaining advantages not dissimilar from those obtained by small or medium-scale manufacturing concerns gathered in industrial 'quarters'. It may not therefore be easy, as town-planners now wish, to persuade concerns to move their offices outward to the fringes of London thereby forfeiting these advantages, as well as those of a central site. For the head offices of many major national manufacturing concerns the prestige of a West End location and a situation near to government offices is considered essential. The concentration of government offices in the Whitehall area is well known: not so well appreciated has been the location in St James's of clubs and learned societies nor the relation to Whitehall of the distribution of the offices of representatives of foreign governments. In the pattern of the West End the shopping districts of Oxford Street and Regent Street and the Soho quarter form dominant elements. Within these districts smaller specialized areas may be distinguished; the music-publishing of Denmark Street, the tailors of Savile Row are famous examples. Around them other areas of specialized function are to be discovered: the doctors of Harley Street, small-scale clothing firms to the north of Oxford Street, newspaper publishing in Fleet Street, the engineering consultants and draughtsmen of Victoria Street, and Mayfair into which the office district now presses fiercely.

The present fashion is to build upward, though at high cost. The Shell building on the South Bank illustrates this new trend in the central districts. On the outer edge of the central area, in Kensington and Paddington, for example, the tall office buildings rise, replacing older houses and reflecting the keen pressure on scarce land in the heart of the capital. Thus the office districts of central London present distinct geographical patterns: their activities are of vital importance in national social, economic and political life. The further development of these areas seems at present to be inevitable, and the appearance of the City and West End changes rapidly as the high, rectangular blocks of steel, concrete and glass replace their smaller, though more ornate, Victorian predecessors.

A boundary for these office districts is difficult to trace. On figure 8 a wide boundary for the central area has been allowed, taking into account current developments in the south at the Elephant and Castle, in the west at Kensington and in the north-west in Paddington and St Marylebone. But, broadly speaking, lines joining the main railway terminal stations, lines that as recently as 125 years ago marked the edge of built-up London, contain the essential central districts into which the main daily movement of people and motor traffic is directed.

Linking London, the financial centre, and London, the port, is the zone of the commodity-markets between the Tower and Spitalfields. In them the nation's dealing is done in wool, tea, furs, rubber, cocoa, sugar, non-ferrous metals and grains. In the Baltic Exchange the world's tramp-shipping is on charter.

The vital importance of the port of London in the commercial and industrial life of the region has already been suggested. London sees 1,000 arrivals and departures of shipping each week, a fifth of the shipping using United Kingdom ports, carrying a third by value of the country's imports and exports. Down-river from the Tower to Woolwich and then discontinuously to Tilbury and Gravesend extends London's dockland. This is a strange, half-foreign world of high walls, brick warehouses and stores, and enclosed water-basins, lying within the fence imposed by H.M. Customs. Here an outstanding variety of goods is received and stored—specialized cargoes such as Persian carpets, furs, wines, silks, drugs; bulk imports such as timber, paper, pulp, grain, fruit and meat—the list is endless. The traditional trade in sea coal to London continues; almost 20 m. tons are received each year for consumption by gas-works, power-stations and the manufacturing industries of Thames-side. And a third of the

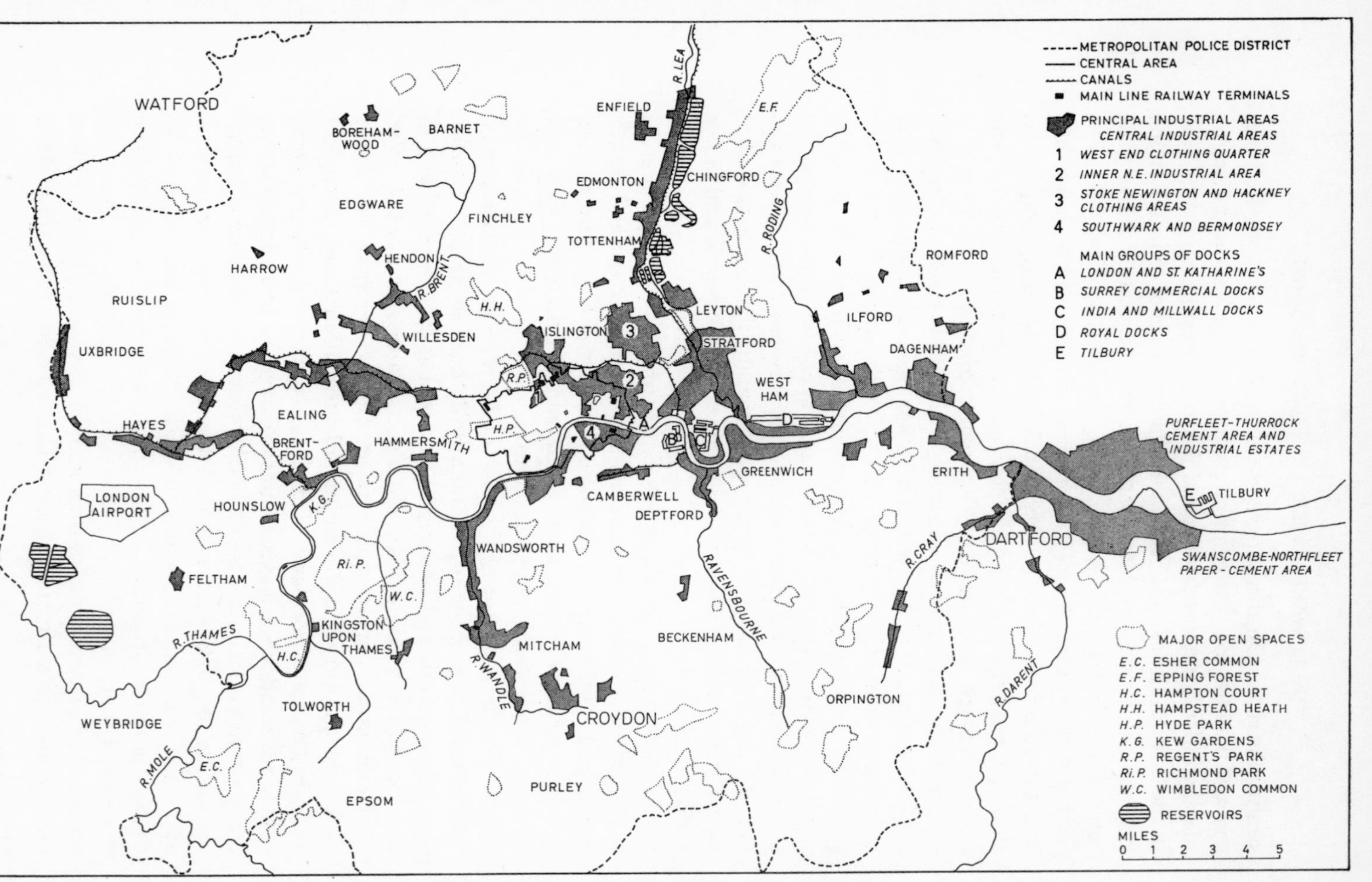

Fig. 8. London: distinctive areas.

nation's crude oil enters through the Port of London. Outward go the cars and vehicles, aeroplane engines, machine-tools, electrical equipment, chemicals, paper, textile yarns and fabrics, the staples of the nation's export trade.

There are five main groups of docks (fig. 8). Nearest to the city are the London and St Katharine's Docks, cramped in site and divided by public roads. Farther downstream, 3 to 4 miles from the City's eastern boundary, the India and Millwall Docks lie across the former marshes of the Isle of Dogs. Between these two groups but on the river's right bank are the Surrey Commercial Docks in which the timber-trade has traditionally flourished. The Royal Docks at Silvertown and North Woolwich are on a grander scale with long quays that accommodate vessels of up to 30,000 gross tons. Far away downstream at Tilbury, 26 miles from London, at the gateway of the port stand the Tilbury Docks, a system that has survived its early failure and seems destined to play an increasing part in the capital's trade.

Extensive as they are, the dock systems are but a part of the Port of London for less than a third of the ships entering the port find a berth in them. London's fleet of tug-drawn barges (there are about 7,000 of them) carries freight from ships in the river to the docks and to the premises of riverside wharfingers and industrial concerns.

Sixty years ago most of London's industry could be found in a rectangle extending for 7 miles from Camden Town in the north-west to Stratford in the north-east, and for 5 miles southward to Vauxhall and Greenwich. Today an area sixteen times as great, from Enfield in the north to Croydon in the south and from Hayes in the west to the motor-manufacturing area at Dagenham in the east, must be included. This expansion of the industrial map has been due in part to the outward movement of factories from inner London, often along radial lines, but also to the entry of new firms, many of whom have expanded quickly from small beginnings.

The present pattern (fig. 8) is at first sight complex but its main constituents may be indicated. Approaching London up the Thames estuary the traveller on board ship glides through a great zone of industry lining both banks of the river, many of the plants built on recently reclaimed marshland. About half the British production of cement is made on Thames-side. The Chalk of the Purfleet anticline north of the river, and of the river bluffs at Erith, Northfleet and

Gravesend, and the adjacent London Clay yield the main raw materials, while coal is imported by tidewater. Like the cement works, the paper mills depend upon water-borne materials, in this case pulp bales. The riverside gives excellent sites for the public utility industries as, for example, Beckton gasworks and Barking generating station on the north side of the river, on opposite sides of Barking Creek. Shellhaven illustrates the advantage of Thames-side for oil storage and refining, and downstream there is also the large Isle of Grain refinery. Here are fine sites for those industries that process heavy materials and use large quantities of fuel, which can be imported up the water highway of the Thames, and for which the London market is a principal attraction. The Ford Motor Works at Dagenham is one of the largest employers of labour in the region. Sugar-refining, oilseed-crushing, timber storage and processing are extensions of activities formerly carried on in the dock area. The capital costs of preparing sites are high and much land reclamation and site improvement have been carried on by industrial estate companies. James Bird has shown that for most industries the locational advantages of riverside sites begin rapidly to disappear beyond Tilbury and Gravesend, 26 miles downstream from the City centre, because of increases in distribution costs. But there are signs that this point will shift eastwards, especially when the long-awaited Dartford-Purfleet tunnel is completed.

The inner zones of London industry are to be found in a crescent-shaped area extending to the east and north of the City of London with a prolongation north-westwards to Camden Town. On the south bank of the river a corresponding zone is present in Southwark and Bermondsey. In the areas north of the river in the East End boroughs, notably in Stepney, Poplar, Bethnal Green and Hackney, lay the nineteenth-century industrial heart of London with its traditional associations in the clothing, furniture and allied trades. In a study of 10,000 firms employing about 240,000 workers in inner north-east London, Martin has shown the remarkable persistence of localized industrial quarters already well established a century ago. The clothing trade, especially the making of women's dresses, suits and coats, is clustered in adjacent streets in premises varying in character from modern blocks, which may be divided into flatted factories or used by one larger firm, to converted houses, tenements and back rooms. In Fleet Street and in Shoreditch printing dominates, in Clerkenwell the precision trades, including jewellery, scientific instruments, optical goods and glass and metal finishing are characteristic. In the furniture quarter the trade is carried on in small back-street workshops and many of the larger, more prosperous firms have moved away to the north-east. Few parts of inner north-east London contain no manufacturing: the environment is a distinctive one of closely packed industry and housing, much of it no better than a 'slum'. The attempts in planning to move industries away to improved environments and to new towns have met with some success but the volume of manufacturing is great and the concentrations are virile and not easily broken.

Moving outwards to the north the working-class and factory areas are continued from Hackney almost without a break up the valley of the Lea to Tottenham, Walthamstow, Edmonton and Enfield. The Lea valley is one of London's great industrial zones, exhibiting a remarkable intensity of economic activity as far north as Hoddesdon. Its manufacturing interests are varied but include the clothing and furniture industries characteristic of inner London, though here the size of the industrial plants is much greater. Still farther out, at Enfield, there has been much recent factory building, some of it since 1945, and here factories making electrical, light engineering, and automobile components are found. The Lea Navigation has been extensively used for timber and coal transport. Gas and electricity works rear to the sky. The valley has been for centuries a channel for London's water-supply and an almost unbroken chain of reservoirs extends from Walthamstow to Enfield. Old gravel pits remain. The celebrated glasshouse industry, founded on the terraces of the Lea and rapidly developed in the last two decades of the nineteenth century, has continued to retreat northwards before the advance of the urban area. During the last thirty years many glasshouses in the south of the old zone, around Edmonton and Enfield, have been demolished, and now the industry is leaving the valley bottom and expanding up the valley slopes on to the higher gravels and the London Clay. North of Waltham Cross a new distribution pattern has emerged with Cheshunt as a main centre. But despite its changing distribution the Lea valley glasshouse industry shows no signs of serious decline.

To the west and north-west of London, closely related in site to the main rail and road routes joining London with Bristol and the

Midlands, stand the great industrial estates at Park Royal, Willesden, Wembley, Perivale and Southall. Developed for the most part in the late 1920's and 1930's, often as speculations by building and estate companies, the original infant firms have in many cases expanded to become concerns of national importance. Not surprisingly, they reflect, in their industrial character, the industrial groups that have expanded since about 1930, and the electrical apparatus, vehicle, precision instrument, food and drink and pharmaceutical industries are strongly represented. Far to the west the London Airport area, with its auxiliary industries, has become a new major centre of employment.

In London south of the Thames, manufacturing industry is not so strongly marked. In the inner zone, at Southwark and Bermondsey, are the food and drink and printing trades. Elsewhere a continuous industrial zone is found only in the Wandle valley where the public utility industries, radio and electrical engineering and light engineering are characteristic. Important industrial pockets exist at Croydon and along the Kingston by-pass, while to the south-west there are the aircraft industries at Weybridge.

But offices, docks and factories, all the acres of bricks and mortar, do not cover all London. London's open spaces, inadequate in size though they are, form a distinctive feature of the urban geography. From the Horse Guards Parade in Whitehall one may walk on grass through the heart of London for 2 miles through the Royal Parks—St James's Park, Green Park, Hyde Park and Kensington Gardens; the green enclave, with its planned waters, offers welcome relief from the press of buildings and, in summer, refreshment of spirit to thousands of office workers. To north and south of the general office district lie Regent's Park and Battersea Park while still farther out is a series of large open spaces, all dear to Londoners, which have only been engulfed in the tide of houses within the last forty years. Hampstead Heath and Ken Wood, Highgate Woods, Hackney Marshes, Wanstead Flats, Greenwich and Dulwich Parks and, in Surrey, Wimbledon Common and Richmond Park are all favoured playgrounds and retain a surprising variety of natural life. Still farther out there is Epping Forest, while the motor-car has given to the south Londoner quick access to the North Downs and the Weald.

Areas in which people live have become increasingly distinct from the areas in which they work, and in its social geography, as in its economic geography, London exhibits many strong contrasts and a well-marked differentiation of districts. This may be observed down to a very fine scale involving, in some cases, a grouping of only a few streets. Broadly, however, each of the areas of distinct economic character shows a similarity in the types of houses and in the occupations of its population (fig. 9).

In Westminster, Holborn, St Marylebone, St Pancras and the City, the metropolitan centre of commerce and government, the majority of the houses have been converted into shops and offices, and the population has moved away. But houses and blocks of flats may still be discovered in quiet islands behind the great offices of the main streets. Perhaps 225,000 people dwell in the central areas of them: the majority of them follow occupations that require their presence in the centre—night-porters, caretakers, doctors, actors and Members of Parliament. Thus at night the central areas are deserted by all but a few: it is worth considering, in replanning, that more people might be encouraged to live in London's centre.

Around the centre lie the high-density, old residential and dock-side areas. In the East End and immediately south of the river this is close-packed mixed industrial and residential building, often poor and mean. Many areas, in Bethnal Green, in Poplar and, to the west, in Notting Hill are changing rapidly as the squat nineteenth-century terrace houses give place, under London County Council comprehensive redevelopment schemes, to high blocks of flats. The sense of achievement is best realized if the new, warm environment of Lansbury is compared with the dingy jumble of workshops and houses that remain for replacement. In Kensington and Chelsea, shops, administrative offices, centres of learning and culture have taken over large numbers of high Victorian buildings and replaced them in modern design. Here, too, much valuable residential property remains in an area conveniently near the West End.

Farther out again, in Kentish Town, Islington, Leyton, Hackney, Fulham and Camberwell, stretch mile upon mile of early twentieth-century terrace houses. With their yellow bricks stained by smoke, they give to London's inner suburbs a special character not found elsewhere in the country. They form, on the whole, residential areas of high density and, with the press of people into London, the larger houses have been split into flats. Broadly, residence is separated from

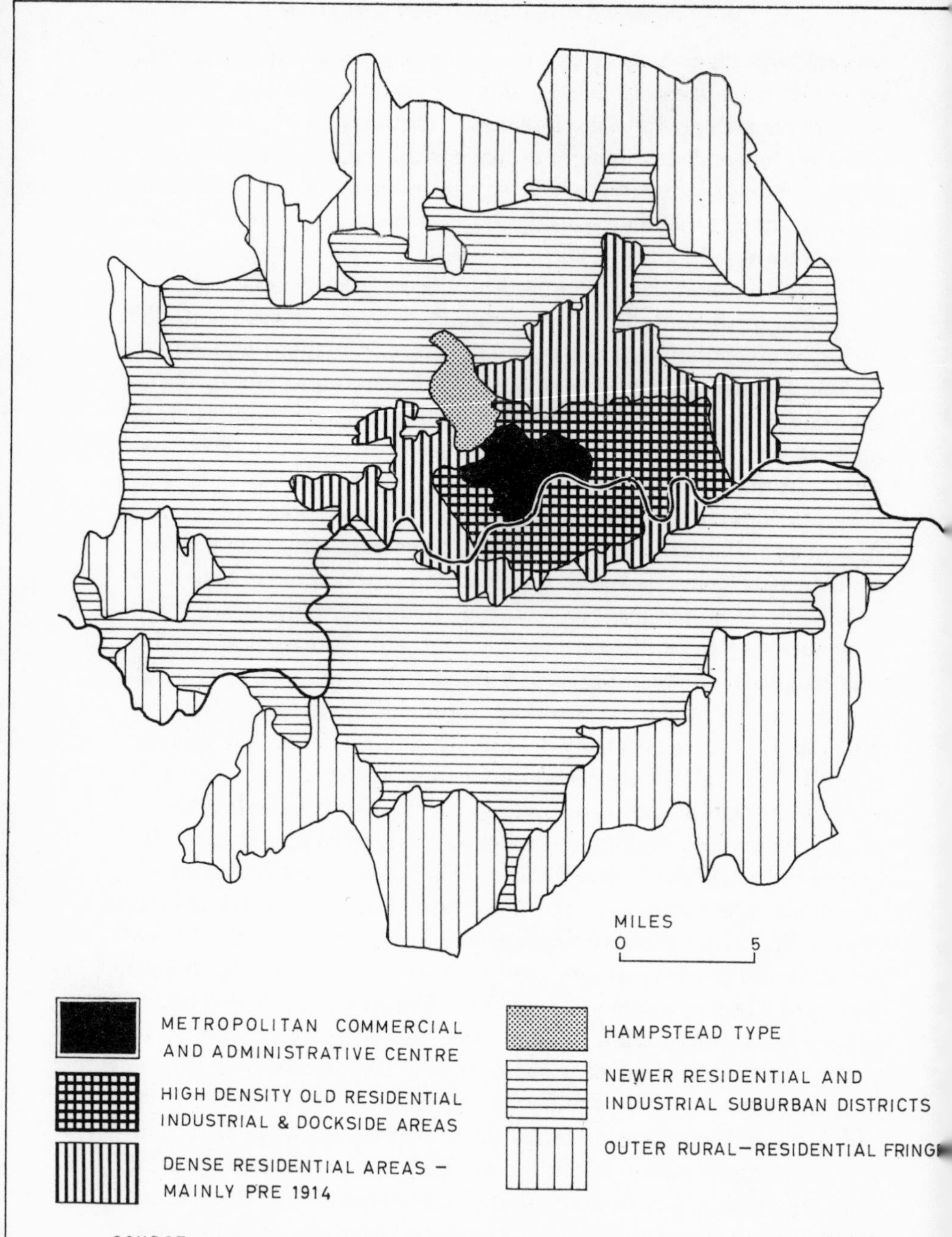

Fig. 9. Social divisions of Greater London.

industry, though occasional industrial pockets, as at Hammersmith, may be found. There is little doubt that in the next decade or so the forces of redevelopment will switch to this zone.

For professional people who can afford the high cost of living in large houses in pleasant surroundings relatively near to central London, the Hampstead and Highgate areas have great attraction. In these areas substantial houses with large gardens were built in the late nineteenth and early twentieth centuries. Many houses are now sub-divided, but Hampstead and Highgate villages preserve a special character and the high prices of houses in these limited zones indicate the competition for them. The element of foreign-born population is high. To the north-west of Hampstead Heath the same social character is continued in Hampstead Garden Suburb, a pioneer of its type, which has been a national example.

Beyond the terrace houses lie the apparently endless avenues and crescents of suburban development. In the main they lie between the industrial zones. Thus Friern Barnet-Finchley-Hendon-Wembley cover with inter-war suburban housing the high ground between the Lea valley and the industrial estates of north-west London. A division into sub-areas is not easily made since the changes in house-type, and in the occupational ranges of the residents, are gradual, though areas of council houses built by local boroughs or by the London County Council to house 'overspill' are soon picked out.

Still farther outwards from the centre lie the low-density residential areas of the outer fringe. On the one hand these include the out-county estates of the London County Council at Borehamwood, for example; on the other the highly desirable residences of the Surrey, Hertfordshire and Kentish fringes. Areas of special character exist: not the most pleasing among them the squat bungalows of Potters Bar. At present the demand for housing in the fringe zones is great: greater in fact than the supply of land, for here is the land of the green belt designed and applied to restrict the outward growth of the conurbation.

Thus the character of housing is one method of examining London's social geography. Equally revealing studies have been made of the variations by occupation and by the demographic structure of population in the different parts of the conurbation. For example, the London district as a whole is an area of the country in which the population structure exhibits a rather higher proportion of young people than the national average. Similarly there is a high sex ratio: females are more numerous in relation to males than is the general situation in the country. No doubt the social attractions of the area, coupled with the wide variety of employment opportunities in the entertainment industry, in light industries and in personal and public services must be accounted responsible. But there are marked variations within London. To take one or two representative areas: Finchley, a north London residential area, has a high sex ratio at 119·6 (the national average in 1961 was 106·6) while Kensington is even higher at 124·2. By contrast, Hayes and Harlington, a west London industrial district, has a much lower ratio at 99·9 and Stepney, as might be expected, is amongst the lowest in the area. The suburban fringe areas have high ratios. Such differences are of interest not only for themselves but for the wide variations that they imply of living conditions and employment opportunities in the different parts of London.

To the casual eye London is an amorphous mass of houses and shops, streets and factories. But in fact its social and economic life is highly organized and the geographical patterns that emerge are recognizable and open to understanding and explanation. London's shopping centres, and the areas that they serve, may be classified and mapped. Within the great conurbation, dominated by the shopping-streets of the West End, major sub-centres exist at Croydon, Ilford, Kingston and Wood Green. Subordinate to them, serving smaller areas, are at least 100 important groupings of shops, many of them situated at junctions of public transport routes. Each of these provides a local focus of activity and social interest. It used to be said of London that it was a collection of villages; certainly, local loyalties and community relationships are of great significance in its social life. They are strongest, perhaps, in the East End, but are present everywhere, be it in Finchley or Beckenham, Woolwich or Ealing. Some, at least, of the suburbs, derided as they have been, have come to possess a life of their own.

There have been great changes during the past thirty years not only in London, but also in the surrounding area. A major feature of the new regional geography that has emerged has been the green belt, a zone of varying width wrapped tightly around the main built-up area. In this zone strict control has been exercised over new building, with the result that it has been limited largely to the in-filling of spaces in existing settlements and to compact areas around them. Population growth has not been entirely restrained, for part of the area of over 10% increase of population since 1951 falls within the green belt, and its total population is higher by some hundreds of thousands than Abercrombie would have wished. But, despite the pressure on green-belt land, stemming from recent increases in employment and population, the policy has been successful in achieving its main object—the prevention of urban sprawl across the countryside. Around the conurbation fringe public action has succeeded in maintaining green land that, in a completely free society, would have fallen to piecemeal suburban housing. It is perhaps true that, had the green belt been delimited solely on geographical considerations, taking into account the need to satisfy some of the demand for housing land on the conurbation fringe, as well as the quality of the land itself for amenity or for agriculture, it would have been delimited differently, and placed farther away from London. But the motive was simple—to stop urban sprawl, and the negative consequence is a fact of great significance in the region's geography. County councils now press for additions to the green belt. But it may be argued that the first task is to delimit areas in which houses and factories may be built to serve current increases of population and to provide the open environments now sought by Londoners.

The achievements in housing overspill population—for even high blocks of flats cannot give room for all the people displaced by slum clearance and redevelopment programmes in the inner areas—mark a high point in British town-planning and have had important geographical effects. Prime amongst them must come the successful building of eight new towns (fig. 7). In each case the objective has been to construct a living community. Factories have been erected to attract firms from inner London; workers have come with their work. Broadly, employment has kept pace with the swelling numbers of dwellers. By 1956 the rate of housing in the eight new towns, Basildon, Bracknell, Hemel Hempstead, Welwyn Garden City, Hatfield, Stevenage, Harlow and Crawley, had reached 9,000 houses a year. More significant than the rate, has been the design of house and town lay-out, which have brought the world's experts to admire. It cannot be pretended that all problems of town birth and fostering have been solved; that, for instance, the provision of services has always kept pace with the need. Nevertheless, no other region in Britain can present such an achievement as eight new towns, housing now 300,000 persons and designed ultimately to provide homes for 500,000 including their original complement of 100,000.

The new towns were designed as self-contained towns. By this was meant that they would grow as communities rather than as suburbs, providing substantially for the work and play of their own inhabitants. The success in attaining this aim has been great but the ideal was unattainable. No town exists independently of the region in which it lives and new towns retain strong personal, economic and social links with London. Though not satellites they are part of the London region, showing the advantages for industry and for daily life of that region and dependent upon the centre of London for regional services.

By themselves they have been insufficient to house more than a part of London's overspill population and negotiations have proceeded between the London County Council and many existing towns for their planned expansion. Most of these towns lie far beyond the natural limits of the London Basin—Swindon, Bletchley, Thetford, Haverhill, Aylesbury, Huntingdon, Edenbridge, Luton, Letchworth, Ashford, Basingstoke and Bury St Edmunds. The provisions of the Town Development Act of 1952 have not been easy to apply; but the first moves from London—to Bletchley—took place in 1952 and already over 5,000 houses have been built under the scheme at Swindon. The present target (1960) for the twelve schemes is to provide homes and jobs for 28,500 London families (100,000 people), and the movement will increase. The size of the planned overspill movement must not obscure the fact that at least an equal number of Londoners have moved out of the conurbation into the wider region voluntarily. A country town is now becoming the home of thousands of Londoners who still travel each day to town. London has become 'a colonizing power . . . pouring out the treasure and labour of her citizens in order to make homes for them in distant lands'.

It is not merely the movement outwards that has caused a growth of the region's smaller towns. The prosperity of the region's industries, its supposed attractions in climate, social services and education relative to other parts of the country, have brought a new expansion. Beyond London, but still within its region, exist five main urban groups each with a population of ¼ m. or more, Reading, Luton,

Southend-on-Sea, the Medway towns, and Brighton and Hove. Each has a large local market, each (despite the decline in employment in the naval dockyards of the Medway) has a wide basis of employment and attracts labour from the smaller towns near by. Each is a local centre of communications for a considerable area, and is closely linked by rail with London. The expansion of these towns, together with that of some smaller towns within the orbit of London, such as Chelmsford and High Wycombe, is to be expected as the London Region continues its history of expansion and growth.

The achievements of the post-war years have been great and London's expansion has been braked, if not halted; new development has taken place in a controlled, not in a haphazard, fashion; re-development has created and is creating inner areas of which Londoners need no longer be ashamed; the new towns are built. Yet the circumstances of 1960 are very different from those envisaged by Sir Thomas Barlow or Sir Patrick Abercrombie, and new forces have risen to promote further growth. The increase in international air traffic widens London's market area; new industries are born yearly, the motor-car widens the radius of personal life and experience. It may well be true that new policies are now needed to provide a more positive approach to the problems of future development not of a conurbation but of a city region. The Report of the Royal Commission on Local Government in Greater London (1960) demonstrates convincingly that, for existing problems alone, the present structure of local government is obsolete; it proposes in its place a Greater London Council with powers to plan and act for the easement and solution of the great problems that have been sketched in this essay. The council would serve an area that is geographically essentially the built-up area. 'Lower-tier' authorities, reorganized from the present obsolescent pattern, would administer local services. Welcome though this report is, its conclusions may well prove inadequate in the face of the geographical changes now taking place. Recent reviews of the size of the overspill problem remaining have put a minimum figure of one-and-a-third million on those who will move from London to new houses in towns outside the conurbation in the next twenty years. In addition the population of those districts that lie beyond the conurbation will, from natural increase alone, rise by 700,000 in the same period. So that, by 1981, over two million more people will be living in the outer parts of the region, an additional population equal to that of Britain's second largest conurbation. Who is to plan? Where are they to go?

Faced with problems of such magnitude, one is drawn back to C. B. Fawcett's view of the size of the London Region (fig. 7). Almost certainly a region no smaller in size is the true London Region of today, for which overall study and planning is necessary. Certainly, in human geography, the London Region can no longer be equated with the London Basin. Both Fawcett and S. W. Wooldridge have shown that the London Region extends far to the south over the southern rim of the London Basin to include the Weald and the coastal towns. For Wooldridge, writing at the end of the Second World War, the Chiltern Edge still marked the northern limit of the region. But the changes of fifteen years have brought this equation into doubt. E. G. R. Taylor considered the London Region to have a radius of 50 miles from Charing Cross. Considering the wide area that London serves as a regional centre, the changing pattern of population and employment distribution, the widening zone of the journey to work to London, the needs of town and country-planning and the estimates of informed observers of the current scene, we find a London Region extending across nine counties and into parts of three others. These counties, Bedfordshire, Buckinghamshire, Essex, Hertfordshire, Kent, London, Middlesex, Surrey and Sussex, together with parts of Hampshire, Berkshire and Oxfordshire, contain a population of 14,450,000 (1960). With its vitality unimpaired, London continues to draw men to itself and to grow outwards. The city region that is now coming into being is a social and economic unit with a well-marked geographical pattern. At the heart lies London, its different parts becoming increasingly specialized in appearance and function; around it, and linked with it, grows its constellation of towns and villages.

Stand at the heart of it, on Waterloo Bridge, and behold the sweep of the River Thames from Westminster to the Tower lined with the palaces of government and commerce dominated by the dome of St Paul's; and remember that 'amongst the most noble and celebrated cities of the world, that of London, the capital of the kingdom of England, is one of the most renowned, possessing above all others abundant wealth, extensive commerce, great grandeur and magnificence'.

Selected Bibliography

Sir Patrick Abercrombie. *Greater London Plan, 1944.* London, 1945.
J. H. Bird. *The Geography of the Port of London.* London, 1957.
R. S. R. Fitter. *London's Natural History.* London, 1945.
Ll. Rodwell Jones. *The Geography of London River.* London, 1931.
Hilda Ormsby. *London on the Thames.* London, 1931.
S. E. Rasmussen. *London: The Unique City.* London, 1934.
Michael Robbins. *Middlesex.* London, 1953.
Royal Commission on Local Government in Greater London. *Report*, cmd. 1164, 1960.
D. H. Smith. *The Industries of Greater London.* London, 1933.
Ministry of Housing and Local Government. *The South East Study, 1961–1981.* H.M.S.O., 1964.

CHAPTER 5

EAST ANGLIA

J. A. Steers
J. B. Mitchell

East Anglia lies south and east of Fenland and north of the London Basin, an area in an accessible yet isolated position. It is a region with long-established traditions. By Anglo–Saxon times it was a clear-cut political unit, frowning at Mercia across the Fenland frontier and isolated by dense forests from the Saxon kingdoms in the Thames valley. By the Norman Conquest it was one of the most densely populated parts of England. From late medieval to early Victorian days it was famous for its woollen and worsted industries, and although the industrial revolution of the nineteenth century almost passed East Anglian industries by, the new industrial revolution of the twentieth century bids fair to catch them within its toils. This may bring great changes, but these if already foreshadowed are not as yet overwhelming. East Anglia is, and has always been, an agricultural area essentially dependent on its farmers. Fields and pastures dominate the scene; cultivated land stretches to the horizon in almost every landward view. Farms and villages and country towns house the people and even the largest towns, with modern industries, have corn exchanges and cattle markets and retain a country air.

Many think East Anglia a flat and dull country. But this is not so: at its centre it is true, lies a somewhat featureless plateau but it has lakes, heaths, brecks, fens, river valleys and wide estuaries, and a coast unrivalled for its marshes, its boulder-clay cliffs, and sand and shingle spits. Its geological structure certainly is simple; the rocks are all of Mesozoic and later ages, and none of them offers any serious resistance to denudation. Yet in colour, consistency and surface form there is the greatest contrast between the Lower Greensand, the Chalk, and the Crags. Over nearly all is a mantle of drift: it is sometimes a thick, sticky boulder clay, sometimes pure sand and stones; mostly an even blanket but occasionally built into ridges and small hills.

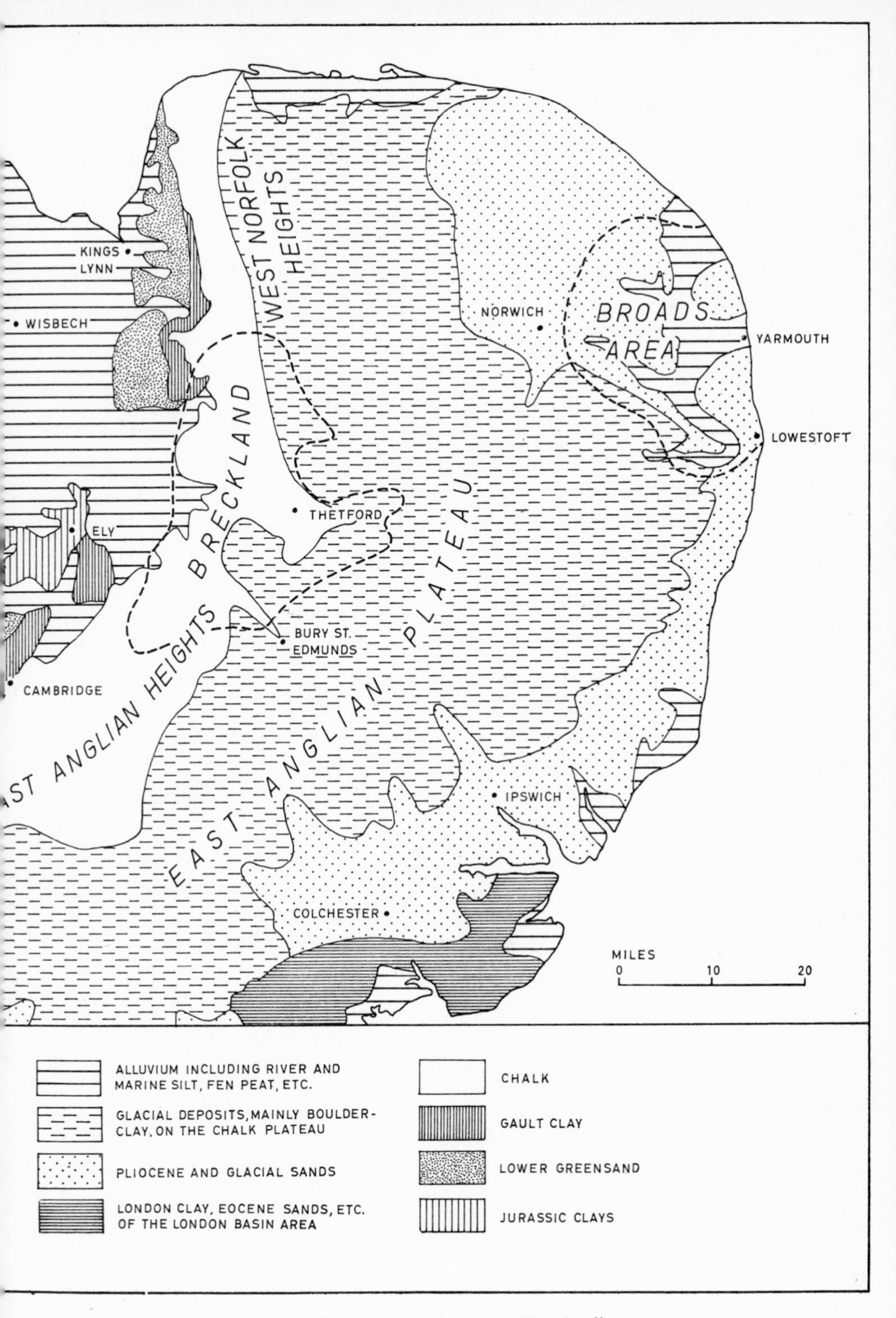

Fig. 10. The physical features of East Anglia.

The Chalk is the backbone of the region (fig. 10). The dip is easterly, and steepens beyond a north–south line through, approximately, Norwich. The maximum height of the Chalk in Norfolk is 250 ft north-east of Swaffham; in Suffolk it reaches about 270 ft between Newmarket and Haverhill. At Lowestoft it is 463 ft, and at Felixstowe 100 ft below sea-level. The Chalk scarp forms a belt a few miles wide from Hunstanton to Brandon where it is interrupted by Breckland and reappears in the Gog Magog hills in Cambridgeshire. Most of the Chalk is low plateau country, boulder clay covered and intersected by wide valleys. North of the Wissey there is a narrow tract of Lower Cretaceous rocks, where the Lower Greensand forms a low escarpment facing the Kimmeridge Clay and Fenland, and near the Wash passes into low cliffs. The Sandringham Sands provide heath country, and between them and the Chalk there is locally a narrow belt of Gault Clay which has little effect on the scenery. East of the heath country is the 'good sands' region, consisting largely of very chalky boulder clay and glacial sands resting on the Chalk, and passing southwards into the true boulder clay of High Norfolk and Suffolk.

There are extensive outcrops of London Clay in north-east Essex, and the Reading and Thanet beds appear in the valleys of the Stour, near Sudbury, and the Gipping (Orwell) near Ipswich. Eastern Norfolk and Suffolk and north-eastern Essex is the classic country of the crag deposits, mainly shelly sands, laid down as sand-banks, and by their fossils indicating a change of climate during the deposition. The Coralline Crag was deposited in a warm period, but the succeeding Red Crag—the base of which is now taken as the beginning of the Pleistocene—implies colder conditions, which continued into the Norwich Crag and culminated in the Ice Age. Much of the Crag area is covered by glacial gravels, rather than boulder clay, and the Westleton Beds, extending from Mundesley to Dunwich, were probably formed as pebbly shore deposits of an encroaching sea in front of an ice cap.

The relief of East Anglia owes much to the nature of the drifts. Detailed work is still necessary before a complete interpretation of the glacial beds is possible. On the old hand-coloured geological maps, some of which are still current, two boulder clays separated by 'mid-glacial' sands and gravels were distinguished. Between the wars four glaciations were recognized, each identified by the particular heavy minerals found characteristically in their deposits. They were differentiated only in the cliff sections of the Norfolk coast; inland their limits were unknown.

In 1956 the drifts were re-examined on the basis of stone orientation. The longer axes of the stones, it is assumed, lie parallel to the direction of the movement of the ice. The earliest advance of the ice, the Cromer advance, came from the north-west and left behind the North Sea Drift, the Norwich Brick Earth, and the Cromer Till. On the east coast there is an extensive bed of stratified sands and gravels, the Corton Beds, originally regarded as interglacial. This, however, is unproven, and the North Sea Drift, the Corton Beds, and the overlying Lowestoft Till are now regarded as the products of one glaciation named the Lowestoft (=Elster on the Continent).

At Hoxne, Ipswich and other places there are deposits resting on the Lowestoft Till which were probably laid down in lakes, and imply an interglacial period. Above them is another till, the Gipping, deposited by ice moving southwards from north of the Wash. It corresponds with the main chalky boulder clay of the Midlands. The final glaciation to reach East Anglia was the Hessle. This ice just reached the Norfolk coast, and left behind the Hunstanton Brown Boulder Clay.

It is not possible to date accurately the many interesting sand and gravel ridges that are common in parts of the region. The largest is the Cromer moraine. It consists of coarse (Cannon shot) gravels, carrying thin soils supporting bracken, heather, and woodlands. It makes the 'hills and holes' country around Sheringham, and the heaths of Kelling, Salthouse, and Weybourne. Landwards it gradually gives place to the flat-lying region of north-east Norfolk. Seawards it ends in a steep slope—which may be an ice-contact slope—standing a little way inland from the cliff edge. Of the smaller ridges the best known is the Blakeney 'Esker'. Similar features are found in several parts of Norfolk and Suffolk, and especially near Great Massingham and Castle Acre. They distinguish East Anglia from the monotonous boulder-clay plains of the Midlands. In the east of the region the glacial deposits and the crags may be regarded as one from the physiographical point of view. The heath country (the Sandlings) cut up by minor valleys around Orford, Dunwich, and Southwold, is most attractive.

The rivers at first sight seem simple in their arrangement. The

Chalk is the main watershed; the secondary shed of the Cromer ridge is very young. Rivers flowing to the Ouse and Wash are obsequents, and of Pliocene or earlier date. The Cam is not a strike stream, but superimposed since it separates a mass of Chalk to the west of the main outcrop. The age of the rivers flowing down the dip of the Chalk is equivocal since details are obscured by drift. Some valleys are cut in the oldest boulder clay, but contain newer drifts within them: in this sense they are interglacial in age. Most valleys contain terraces formed partly by erosion and partly by deposition of gravels and sands under more rigorous climatic conditions. Some have been investigated stratigraphically (Yare, Stour), but only in the Cam have the terraces been mapped in such a way as to bring out their significance in the landscape. Several follow earlier courses, some of which are well established by bores. These buried channels drained into what are now coastal estuaries and were formed at a time when the land stood higher relative to the sea. In south-eastern Suffolk and north-eastern Essex the pattern of the streams suggests captures which took place when the London Clay formed a low scarp facing north-west.

The Little Ouse and Waveney, the one flowing west and the other east, rise near Lopham Ford, but the valley there is broad and open and continuous from the Fens to the coast. It may have been formed when ice existed in Fenland and held up a lake which overflowed eastward, and cut the flat trough-like valley seen today. The shape of this valley is not unlike the valleys in which lie the Broads. Breydon Water, behind Yarmouth, is the only unfilled part of the joint estuary, and is natural tidal lake.

The broads in the upper Bure and Yare (rivers originally tidal to about Norwich and Wroxham) have been variously explained on the assumption that they were natural hollows. Recent physiographical and ecological work, necessitating bores about one metre apart, has shown that the peat in which they lie was cut in medieval times; in short they are all that remains of old peat cuttings. In recent decades vegetation has encroached rapidly in most of them, and if not kept open the Broads will gradually disappear. Some of the smallest are already completely overgrown.

Breckland, traversed by the Wissey and Little Ouse, is underlain by Chalk, which there lies at a level not much above Fenland. The chalk floor is covered with glacial sands and gravels, and even today sand-dunes occur in places; the blowing of sand and of sandy soils is common, especially in spring. Agriculturally, Breckland is marginal; ecologically, it is the most interesting part of East Anglia since the flora contains relics of a post-glacial steppe. Nowadays only a few parts remain in a natural state, including some small meres, for example Fowlmere and Ringmere. The meres probably originated as swallow-holes in the chalk, and the rise and fall of the water in them is related to the water-table, and indirectly to rainfall.

The coast is of great interest. The Wash is a breach in the Chalk formed like the Humber gap by river action followed by change of level. From Hunstanton, where the cliff is formed of Lower Greensand, Red and White Chalk, right round to Bawdsey where London Clay outcrops, there is no 'solid' rock in the cliffs. Between Hunstanton and Sheringham is the finest marshland coast in these islands. Seaward of the marshes there are barrier beaches; Scolt Head Island and Blakeney Point (technically a spit) are the best known. From Sheringham to beyond Happisburgh the cliffs, sometimes reaching 100 ft high, are formed of glacial deposits and are undergoing rapid erosion, partly by land-water and atmospheric agents, partly by marine action. Between Happisburgh and Yarmouth there are lines of low cliff, again glacial, separated by stretches of dune, through which the Thurne river once flowed out at Horsey. Winterton Ness is an anomaly; it is a local area of accretion on an eroding coast. Yarmouth stands on a sand-bank which dams the combined estuary of the Bure, Yare and Waveney. Near Lowestoft, the denes (sand flats) lie in front of old cliffs, and the harbour is artificial. Along the Suffolk coast there alternate stretches of low and easily eroded cliffs, mainly formed of glacial deposits, and sand and shingle beaches damming back or deflecting small streams. Near Aldeburgh the deflexion of the Alde begins. Orford Ness is formed of shingle, and since the time of Henry II its evolution in relation to the history of Orford can be traced with some accuracy. There is a small spit at the mouth of the Deben, and Landguard Point is a shingle foreland on the north side of the Orwell–Stour estuary. Hamford Water is a drowned lowland now filling with marsh. The coast from Walton to Clacton is particularly vulnerable.

The movement of beach material is southward from near Sheringham, although along the north coast of Norfolk the beach drift is mainly to the west. The southward movement is responsible for the spits and river deflexions, although it is not a simple process, nor one that always acts in a constant direction. Parts of the coast are protected by sea-walls, many of which were extended and strengthened after the great surge of 1953. The whole coast is slowly sinking relative to sea-level, partly as a result of isostatic adjustment of the land, and partly (at present) because of a slight rise of sea-level produced by the melting of the polar ice. In the Thames the movement is of the order of 8–12 in. a century.

The soils and climate of East Anglia on the whole favour the farmer. The drift deposits largely determine the quality of the soils. As a group they are potentially fertile and easily worked; gentle slopes and pervious subsoils allow good drainage, low rainfall prevents excessive leaching. Broad distinctions can be seen between the light loams of the chalks and chalky drifts of the west and north, the medium to heavy loams of the Jurassic boulder clays of the central plateau, and the sandy loams of the fringe of gravels along the east coast. However, as the drift varies locally so much in character and thickness there may be marked differences in soil within any one district, even in any one parish, farm or field. Rainfall though low, is in most seasons sufficient to nourish the crops and there is usually sunshine to ripen them. Continuing cold late into the spring, drought in the growing season, and heavy thunderstorms with wind-lashed rain near harvest, are perhaps the greatest hazards of the cereal farmers. Early and late frosts are a danger to the growers of fruit and vegetables, and the positions of frost-pockets are thus carefully studied and avoided. The yearly variations in the farmers' profits, in spite of government subsidies, reflect sensitively variations in the weather.

East Anglian agriculture has always been essentially arable. It was badly hit by the increasing import of foreign grain, after the repeal of the Corn Laws, and in the long period of agricultural depression from 1870 to1939 many arable acres were laid down to grass. However, even in the 1930's, when the plight of the arable farmer was at its worst, more than half the cultivated land remained in grain. Under the stimulus of war conditions the acreage under permanent grass fell markedly, and, since war-time restrictions were removed, it has not risen but has fallen further. There has been in recent years an increase in the number of farms where livestock is important, but this change is not the result of a decrease in the arable acreage but of an increase in livestock kept.

Barley and wheat are still the principal cereal crops; the greater part of the increase in arable acreage is, however, an increase in the acreage under barley. Sugar-beet is now the root-crop of choice on all but the heaviest soils; the acreage planted is controlled by the British Sugar Corporation and thus varies little from year to year, but the yield varies markedly with the weather. Other crops occupy only about 5% of the total arable acreage and give only 9% of the total farm income. Small acreages of main-crop potatoes are grown on many farms, but in contrast to the neighbouring Fenland, potatoes are nowhere important. Fruit and vegetables are grown on some farms as field crops; seed crops, especially clover seed, are important on others. A wide variety of less usual crops, varying from mustard for condiment and maize for canning to lavender for essential oil and poppies to provide capsules for the lipstick manufacturers, are found.

Arable farming throughout the region is highly mechanized and thoroughly scientific. The tractor has replaced almost every working horse; combine harvesters are in general use; most farmers use some mechanical aid in harvesting root-crops, and some are experimenting to bring efficient singling machines into the field. Mechanical means to lift and spread muck have increased the popularity of animal manure and encouraged the keeping of store cattle. There has been a fourfold increase in the last twenty years in the amount of fertilizer used per acre, an increase that has been accompanied by the use of a greater variety of fertilizers to meet special needs. Seed varieties are carefully selected to suit the soil, the climate and the market, and crops are sprayed against pests and diseases. Crop yields are thus high, and the increase in yield is one of the most marked changes of recent years.

Livestock provide about half the income of most East Anglian farmers. It now pays to feed part of the increased yield in crops to animals on the farm, and to use some of the labour freed by mechanization to tend them. Dairy herds, carefully bred and well managed, are very profitable in many districts and their numbers are growing. Milk yields in 1957 averaged 810 gallons per cow, an increase of about 200 gallons on the pre-war average. Pigs, much reduced in numbers during the war, now contribute more to farmers' incomes than before 1939; poultry is also a major source of income on many farms. On the other hand there are fewer sheep; the traditional folding of sheep fed on roots on arable farms to consolidate and manure the land is now, except in Breckland, uncommon. Horses, too, are becoming rare animals; only around Newmarket where stud farms replace grain farms are they seen in numbers, and here it is race-horses, not working horses, that fill the stables.

There are regional variations of these general characteristics related to soil, to water supplies, and to the demands and accessibility of the markets. These are, however, not so strong as they were even a few decades ago. Intensive cultivation and the lavish use of fertilizers are creating a uniform top soil on chalk and boulder clay alike; piped water-supplies are reaching every area and the irrigation of crops and watering of stock no longer depend on the abundance and reliability of local wells and springs; with modern motor transport, nearness to railways or the existence of through main roads no longer gives any great advantage, for lorries and trucks can penetrate a maze of country lanes. The interest, the judgements of advantage, and the capital resources of individual farmers, are fast becoming of greater importance in determining farming practices than considerations of soil, water and accessibility of markets. Nevertheless, these still have some effects if only because an earlier pattern is not yet wholly obliterated.

The chalklands of the west, particularly of the south-west, are the barley lands *par excellence*. Some 85% of the land is arable and most of it is sown with barley; the yield is high and the malting quality good. On the sticky, cold, intractable soils of the western borders of Suffolk and Essex, as much wheat as barley is grown, but even here barley is now gaining ground. On the extensive areas of loam soils of the high plateau, though barley, wheat and sugar-beet are still the principal cash-crops, livestock are more important than in the west; the chief animals are cows and store cattle in central Norfolk, pigs in central Suffolk. Along the east coast from north Norfolk to the Blackwater estuary there is great variety of farming practice because there is great variety of soil and great variety in the nature, extent and season of the markets. On the reclaimed marshes of the coast and Broadland, dairy cows and store cattle are fed. On the thin sandy soils of the coastal heaths and of Breckland, intensive farming does not pay; and much land is owned by the Forestry Commission and planted with conifers.

Manufacturing industries in East Anglia may be fairly described as of subsidiary importance; they are less important in the economy of the area than agriculture, and less important in the economy of Britain than those of many other regions. There is no single large-scale industry; there are no purely industrial towns or districts. The industries are diverse and widely distributed, and are either successors to those of the past or linked directly or indirectly to agriculture past or present. Many were begun by local men using local resources to supply local demands, and were carried on in small units. But this pattern is changing. Increasing specialization is concentrating manufactures more and more in the larger towns, and in the smaller country towns near to, and easily accessible from, the large towns. In recent years most of the new industries have been established, and many of the older ones taken over, as subsidiaries of big companies, often London firms, who can draw for management, technical training, and buying and selling facilities on the strength and organization of the parent concern.

The manufacture of textiles and clothing, the successors of the old woollen and worsted industries, are the most strongly localized; they still remain in the traditional areas, in and around Norwich and in the villages and towns of the Suffolk–Essex border. As the old industries lost ground to their more fully mechanized competitors in Yorkshire, the manufacture of silk, early introduced, continued in East Anglia and new crafts—the drawing and weaving of horsehair, the making of coconut-matting, and the manufacture of clothing—were taken up. These in their turn languished as fashion changed and foreign competition increased, but new life has once more been given to some of the old centres by the opening of new silk, rayon and nylon factories.

Of purely agricultural industries the traditional ones of milling and malting are still strong. The largest and most up-to-date plants are now in the towns, but attractive mills and maltings, once powered by water-wheels and now equipped with modern machines, often with Georgian or earlier dwelling-houses near by, may still be found by the bridge of many a town and village. Sugar-beet factories, and canning and freezing plants, more recently established, are necessarily also closely linked to farming in their neighbourhood.

Agriculture has also fostered engineering, now the most important industry in East Anglia. In its modern form, it dates from 1789 when Robert Ransome, an ironmaster of Norwich, set up a little foundry in a disused malting in Ipswich to make ploughshares for East Anglian farmers. In the first half of the nineteenth century agricultural prosperity fostered agricultural engineering, and, though the long years of agricultural depression reduced the home market, they at the same time encouraged migration from rural districts to country towns and many engineering firms supplied with cheap labour expanded, diversifying their products to suit a wider market. The return of prosperity to agriculture has increased again the demand for machines on farms. General engineering, some of it to meet local needs, such as machinery for maltings, engines for ships, dock gear for ports, grew alongside agricultural engineering, and has now in its many branches outstripped the latter in importance.

Of the great variety of newer industries, the majority have been attracted to the area not primarily by local demands or local raw material but by sites more spacious and labour more plentiful than can be found in the metropolitan area; thus they too are indirectly a reflexion of the rural environment. Some of these industries, such as the manufacture of brushes, barrels and baskets, saw-milling, woodworking and the making of furniture are traditional country industries in modern guise, though few depend on local raw materials now. Others have no particular links with this or any other country district: printing and bookbinding at Beccles and Bungay are old established, but the manufacture of metal goods, paints and varnishes, electrical goods, radio and television sets and their accessories, and electronic apparatus, are more recent examples. Local water supplies and communications are almost everywhere sufficient for their needs; but electricity supplies and sewage systems are in many places inadequate and this fosters the concentration of these newer industries in the bigger towns and ports, and in those smaller towns where sufficient power is or can be made available, and disposal of waste can be arranged. The newest comers of all, the atomic power-stations at Bradwell and Sizewell, with their need for water, are attracted to the area not by the amenities of East Anglian towns but by isolated stretches of coast without buildings, yet relatively near to potential users of power.

The modern world has not yet fundamentally altered the long-established settlement pattern of East Anglia. Growing industries have added to it but not yet destroyed it; Londoners are not yet numerous enough in any one district to dominate, though with new fast-train services some villages in the south may soon be overwhelmed. In Haverhill and Thetford, the new trading estates are not yet transformed into 'new towns'. Only along the coasts and in Broadland have many villages and towns changed wholly in character to satisfy the demands of a seasonal influx of strangers.

The position and character of the villages, the size of the parishes and fields, the run of the roads still reflect in surprising detail the conditions of the early settlement. East Anglia, easily accessible from the Continent, was entered from the coast on the east and from Fenland on the west. In the east settlers avoiding the coastal marshes colonized along the river valleys, building farms and villages on the river terraces; the settlements that grew most were often those at the lowest fords of the rivers, as at Colchester, Ipswich and Woodbridge, or near the head of navigation for small sea-going vessels as at Sudbury, Stowmarket and Norwich. In the east villages are now thick upon the ground. Often several church towers lie within one view; but rarely signposts on the by-roads give directions to places more than two or three miles away. Here, where farming for market was stimulated by trade with the Continent and by a flourishing local textile industry, and where there was considerable enclosure by Tudor times, farms and fields are often even now smaller than in the west. The villages are of many shapes and sizes; the church dominates most, and the number of great churches of many periods, but especially of the fifteenth century, bear witness to the long-established prosperity of the area and its one-time industrial pre-eminence. If an early house survives it is usually timber-framed; if later, of brick or perhaps of flint with stone or brick to make the corners. In East Anglia brick is early used in domestic buildings and there are some splendid examples of fifteenth- and sixteenth-century brickwork. The village streets are not always distinguished; in most villages the houses are still the working homes of the working population and not the country playthings of 'foreign' town-dwellers, but most present the neat and cared-for appearance that gives real attractiveness. In villages once important in the cloth trade are to be found not only an unusual number of bigger houses of Georgian or earlier date, but also cottages with large and well-lighted upper rooms that once housed the looms.

In the west, settlers approaching from the Wash and Fenland found a drier land with fewer suitable sites for villages, and the villages are still today more widely spaced and the parishes larger than in the east. The villages are often strung out in a line along the outcrop of water-bearing strata in the Chalk, along the margin of the fen alluvium at the scarp-foot, or along the edge of the boulder clay overlapping the Chalk at the crest. Positions on the edge of the fen or on the edge of the boulder clay had the added attraction that different soils lay near to hand, fen and chalk, or chalk and clay, each with their special advantages. Many of the fen-edge settlements were active ports reached by Fenland rivers and lodes that, before the coming of the railway, carried much of the heavy and bulky trade. The settlement pattern in the west has been remarkably stable. Enclosure came late in the Parliamentary period; the land was often then divided into a few big farms, fields remained large and often unfenced, new roads were made straight and wide; the pattern suits modern farming practice.

Lying between the east and west on the high plateau is a frontier area where colonization was probably later and less thorough. Into this area it would seem groups already settled in the east and in the west advanced, cultivating here an isolated pocket of specially attractive soil, pasturing there a group of animals in an open woodland glade; outlying farms were built sometimes in a scattered group, sometimes moated and alone. However this may be, hamlets not villages are today common in central Norfolk, and still more markedly in central Suffolk and north-west Essex; farms are scattered, fields are often very small, and present-day roads, even main roads, wind and curl around them—the modern versions of ancient farm tracks. Parishes are large, some, like Finchingfield, the largest parish in the region, containing several hamlets. In these central areas a church may stand alone in a field or near by a farm, so placed to serve several hamlets. Many of these churches have round towers built of flint and rubble, a reminder of the difficulties of builders with no good local building-stone when they were too poor and too far from navigable water to import coign stones.

Of the small country towns a few are resplendent in past glory, some have an air of decadence, but most are busy and bustling, though many are smaller now than they were at the beginning of the nineteenth century. Declining industry caused decay in some. Lavenham gives perhaps the best idea in England of a medieval industrial town; its old function is lost and it is too near to Bury St Edmunds and to Sudbury to be useful as an agricultural market. It is a living fossil. Haverhill and Leiston, with streets of nineteenth-century houses, have something of the air of industrial villages so common in the north of England, so rare in East Anglia. Both expanded insecurely with a single industry, clothing in Haverhill, engineering in Leiston, and have declined recently; but Haverhill is destined to take some of London's emigrants. Most of the little towns were once busy markets and their form reflects their functions: a market square with perhaps a medieval market hall or market-cross is at the centre of many; others have a very wide main street in which there was once room in plenty and now only just room, to erect market stalls and animal pens. With the coming of the railway and the motor-car some little towns lost their weekly markets and annual fairs to neighbours better placed. Industry in its modern form is differentiating the little towns still farther. On the outskirts of those best placed to serve its needs are now the new factories and the new housing-estates that are bringing new interests and new prosperity: North Walsham, East Dereham, and Wymondham within easy reach of Norwich; Stowmarket and Woodbridge near Ipswich; and Halstead, Braintree, and Bocking and Witham near Colchester—all these towns are growing fast.

The holiday towns stand apart. A few of the coastal towns, Wells, Cromer, Southwold and Aldeburgh for example, have not been completely overwhelmed by the tourists and retain at their core the architectural structure, if no longer the function, of small ports. Wells has declined in population in the last century and Southwold has only increased a little, but Cromer and Aldeburgh have grown considerably, and New Hunstanton, Felixstowe, Walton, Frinton and Clacton have been almost entirely made by the fashion for seaside holidays. Some of the little creeks, Blakeney, Walberswick and West Mersea, for example, now earn more by their attractiveness to yachtsmen than by their fishing boats; and along the east coast many villages, Hemsby, Caistor-by-Yarmouth, and St Osyth among others, have been fundamentally altered by holiday camps and caravan sites. Wroxham is the biggest centre for sailing on the Broads. Newmarket, with its race-course and the headquarters of the Jockey Club, has a character all its own.

Harwich, Lowestoft and Great Yarmouth are ports. Harwich (13,569) has the finest harbour on the coast and is the most northerly of the channel ports. Within easy reach of London much passenger and goods trade use Parkeston Quay both winter and summer. Lowestoft (45,687) is the most important fishing-port in East Anglia and one of the most important in Britain. With Yarmouth it handles catches of turbot, sole, plaice and cod landed by the trawlers, and, in the autumn season from late September to Christmas, the catch of herring and mackerel brought in by the drifters. Scottish boats follow the herrings south and large numbers of Scottish fishermen and fisher-girls come in to deal with the catch. Many fish-curers and merchants with businesses in Lowestoft have also Scottish addresses fringing the coast from Stornoway to Peterhead and Aberdeen. The prosperity of East Anglican fisheries, with those of Britain as a whole, is waning as a result of the competition of Continental fishers and loss of Continental markets. Decline of fishing has led to the growth of other industries in the town: to fish-curing has been added fish-canning and the canning of fruit and vegetables too. To marine engineering has been added electrical and precision engineering. Summer visitors also bring trade and money. Great Yarmouth (52,860), important once as a fishing port, is even less dependent on fishing now than is Lowestoft. Yarmouth has a threefold structure: the immense Church Plain testifies to the importance of the local market; the fish-quays, the timber-yards, the shipbuilding and repairing-docks along the estuary, and the oil and petrol stores and factories that fill the long spit between the estuary and the sea, witness to a busy port; and the hotels and cinemas, the piers and esplanades that front the open sea, belong to the holiday town.

Bury St Edmunds and Colchester, Ipswich and Norwich are the county towns of East Anglia, but King's Lynn and Cambridge on its western edge that serve both East Anglia and Fenland are also important. King's Lynn (27,554) is both port and market-town, and its structure emphasizes the importance of sea-borne trade both in the past and today. Between the great Tuesday Market with its hotels in the north and the Saturday Market dominated by a magnificent church and a fifteenth-century Guildhall in the south run the quays, and the streets parallel to the quays, with their warehouses, merchant houses, and the attractive seventeenth-century Customs House. Modern docks north of the Tuesday Market handle a considerable trade with

North Sea and Baltic ports. The modern town spreads out eastwards; commerce and trade are more important than industry but the industries—agricultural engineering, chemical works, a sugar-beet factory and the preserving and canning of fruit and vegetables—occupy about 3,500 workers. A great part of western Norfolk looks to King's Lynn as its centre. Cambridge (95,358) in one sense lies outside East Anglia yet in another sense belongs to it in a special way; it is the market town of the southern chalklands, it is the railway junction and the link with London for most of the western part of the area, and in some ways it is the regional capital of the whole.

Of the true East Anglian towns, Bury St Edmunds (21,114) is the most centrally placed. It is an important market; it has old-established and growing industries closely connected with the agricultural neighbourhood; it is a garrison town; it is an ancient abbey town and has recently become a cathedral town; and it is the county town of west Suffolk. It is a town of distinction and character; its spacious market place, its magnificent churches, its Georgian houses, the precincts of its abbey, give it grace and dignity.

Colchester (65,072) is the social and economic centre of north-east Essex; the length of its established tradition as a local capital is constantly emphasized to citizens and visitors by the impressive remains of the Roman walls that still ring the core of the modern town and the Norman castle, built largely of Roman brick, that dominates the High Street. It is an important shopping centre, a lively agricultural market and a garrison town. The river, though navigable only for vessels up to 10-ft draught, carries a considerable trade; timber-yards, petrol stores and gravel dumps line the quays, and corn mills, breweries and engineering shops are near by. The town is well laid out and the modern suburbs, though large, do not overwhelm or obscure the historic centre.

Ipswich (117,325) is the major town of the south-east; it is a market town, an industrial town and a port. In Ipswich as nowhere else in East Anglia the importance of the agricultural market town is at first glance obscured by the other commercial and industrial activities. Ipswich has the appearance and the feel of a busy modern town; with one or two carefully preserved exceptions the streets have few early buildings, though some in their narrow width and their names recall to mind the form of the older town; the power-station and the petrol stores seem to dominate the estuary and the quays; industrial estates are large and the suburbs spread far and wide. A closer look reveals even here the agricultural element: it is corn mills and breweries that crowd the Customs House on the quays; agricultural engineering works are large enough to create an industrial estate of their own at Nacton; the tanning, the sack and bag making, the processing of fertilizers, though many are concerns of national or international stature, have their local links and importance too. The hinterland of Ipswich overlaps that of Bury St Edmunds and Colchester; indeed for certain purposes the hinterland of Ipswich includes both these towns and their spheres, though London competes with Ipswich for the wider allegiance of Colchester.

Norwich (119,904) is not only in size but also in standing the leading city of East Anglia. Its visible structure helps to foster the consciousness of its importance: the cathedral by the bridge so clearly the heir of the great abbey that it succeeded, and the present centre of the religious life of the district; the castle with its Norman keep towering above abbey and market place, a witness to early wealth and military strength and, with its splendid museum and art gallery, to present wealth and strength of cultural interests; the market-place packed with stalls and seething with customers, all these lie at the centre of the town, almost within one view. It has all the characteristic industries of East Anglia with, in addition, a big share in the British manufacture of boots and shoes. Norwich is a flourishing city.

The East Anglian countryside, and life within it, is thus for all its diversity but variations upon a single theme: arable land and the men who farm it give cohesion and colour to the region. East Anglia is a part of lowland Britain but a distinctive part, less industrialized than Lincolnshire or the Midlands, more strongly an arable land than the Upper Thames basin or south central England, less closely linked with London than south-east England. It profits by close contact with London: its society can use the amenities that the capital offers; its economy shares the stimulus of the metropolitan market; but as yet it lies distinct from and beyond the area tributary to London. It lacks perhaps a sufficiently strong regional capital. Colchester and Ipswich are too far south and east, Norwich too far north and east, to attract every part of the region. Bury St Edmunds has an old tradition of leadership and is well placed but is so much smaller than the other three that it would take much conscious planning, artificial forcing even, to give it the size to become a regional capital. In so far as there is central administration for East Anglia, it is carried on from Cambridge. Cambridge, by virtue of the very fact that it stands apart from the region and yet is accessible to all of it, has advantages. But Cambridge cannot add a third function to its own two ancient ones of market-town and university without detrimental overgrowth; East Anglia needs a capital of its own where purely East Anglian interests dominate. Norwich has the present lead and may be this lead will strengthen. Colchester, and perhaps Ipswich too, may well fall increasingly within London's orbit and thus lose regional strength; Norwich at a greater distance is in less danger, and, the University of East Anglia, when established there, will increase her pull. A recognized capital would do much to preserve and further the regional identity of East Anglia.

Selected Bibliography

C. P. Chatwin. *East Anglia and Adjoining Areas.* British Regional Geology. London (H.M.S.O.), 3rd ed., 1954.

R. Rainbird Clarke. *East Anglia.* Ancient Peoples and Places. London, 1960.

W. J. Clarke. *In Breckland Wilds* (2nd ed., revised R. Rainbird Clarke.) Cambridge, 1937.

J. J. Donner and R. G. West. 'The glaciations of East Anglia and the East Midlands: a differentiation based on stone orientation measurements.' *Quart. Journ. Geol. Soc.*, **112**, 1956, pp. 69–92.

T. Eastwood. *Industry in the Country Towns of Norfolk and Suffolk.* Oxford, 1951.

J. M. Lambert and J. N. Jennings; and C. T. Smith, Charles Green and J. N. Hutchinson. *The Making of the Broads.* Royal Geographical Society Research Series, No. 3. London, 1960.

J. A. Steers. *The Coastline of England and Wales.* Cambridge, 1946.

University of Cambridge, School of Agriculture, Farm Economics Branch. *Reports on Farming.* Cambridge, 1931–59.

CHAPTER 6

FENLAND

A. T. GROVE

When the spring tides flood into the Wash and run up the embanked lower courses of the Witham, Welland, Nene and Great Ouse, more than 1,200 square miles of Fenland lie below the level of the water. Along the eastern shore of the Wash a narrow strip of farmland, nowhere more than 2 miles wide, has been won from the sea and is overlooked by an old cliff, cut into the sandy hills of Sandringham and Snettisham, and running south beyond King's Lynn as far as Denver. On the opposite side of the Wash, Lindsey's coastal marshes merge into a belt of drained Fen that widens from about 7 miles at the Steeping river to 10 miles near Boston, and stretches up the broad valley-floor of the Witham as far as Lincoln. South of the Witham and Wash, almost the whole of Holland and the Isle of Ely lie within Fenland, as well as a large part of western Norfolk and fragments of Suffolk, Cambridgeshire, Huntingdonshire, the Soke of Peterborough and Kesteven.

This is one of the most distinctive regions in the British Isles, even today when only small portions of it remain undrained and uncultivated. In no other part of the country of comparable size is so large a proportion of the land arable, and nowhere else is the productivity of farmland so high over so large an area. Eighty per cent of Fenland is under the plough and more than a third of the people living there are engaged in agriculture. Industry and mining are in comparison of negligible importance; this is the most completely agricultural region in lowland Britain.

Variations in the landscape and economy within the region are not striking. The most important contrasts, those between the siltlands near the Wash, and the peat fens farther inland, derive mainly from differences of soil and differences of a few feet in height above sea-level. The explanation for both these lies in the history of the accumulation of sediments and peat in the embayment, and the draining of the Fenland.

In early post-Glacial times, when the sea stood more than a 100 ft lower than now, the Fenland basin and the surrounding country were drained by a system of rivers that reached a shrunken North Sea by a broad shallow valley running north-east between the Chalk uplands of Lincolnshire and Norfolk. On the floor of the basin Jurassic clays, Greensand, Gault and Chalk Marl outcropped through spreads of glacial sands and boulder clays, as they still do on the islands rising above the general level of southern Fenland. The basin was not clearly distinguished from the country nearby, but its borders had already been sketched into the landscape by the marine and river gravels, now up to 50 ft above sea-level, which are today the sites of many fen-edge villages. As the climate ameliorated and meltwater from the ice caps caused ocean levels to rise, the North Sea spread south and west over woodland and peat bogs, and submerged the valley of the Wash. The gentle slopes of the surrounding country were covered with forests of pine and oak, but as the sea continued to rise and the climate became more humid, peat accumulated on the floor of the forests; the trees died and their trunks were buried in the thickening peat. These are the bog oaks that are revealed today in Fenland ditches, or are struck by the plough and dug up and left by the roadside. The surface of the peat emerged above the water-level and was colonized by alder, willow and birch; dry woodland extended over an increasingly wide area and rivers cut deeply into the peat. Then, in Neolithic times, the sea broke through some coastal barrier to flood again the woodland, and a layer of clay several feet thick accumulated quite rapidly over most of the peats except those at the Fen margins. The upper surface of this fen clay, often known from its consistency as buttery clay, lies a few feet below present sea-level (fig. 11). Once more sedge peat began to form above the fen

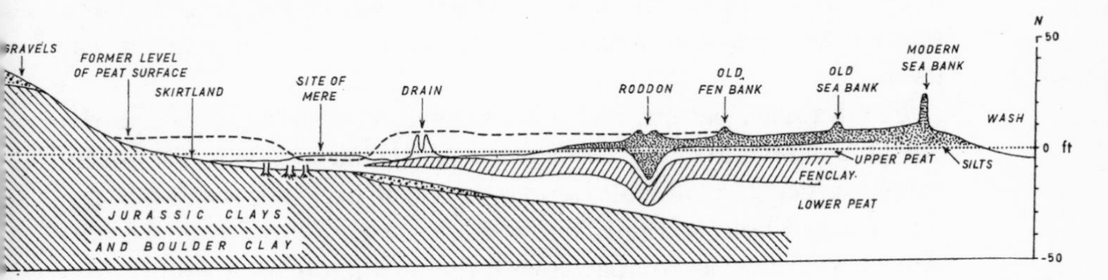

Fig. 11. Diagrammatic section north–south across Fenland.

clays and the older peats at the landward margins of the Fens. They were alkaline for the most part on account of the high calcium content of the streams from the Chalk uplands, but in the secluded embayments south of Whittlesey raised bog developed, giving peats of a somewhat acid type. Open stretches of water remained near the landward margins of southern Fenland, resembling in some respects the Broads of East Anglia, and various shelly marls accumulated in them. By Bronze Age times, judging by the archaeological finds there, Fenland was relatively dry and the higher parts inhabited, but the climate and drainage conditions again deteriorated during some hundreds of years, and the next signs of settlement date from the Roman period.

Ancient ditches and droveways show up on air photographs over a broken strip of country some tens of square miles stretching from near the edge of Fenland north-west of Spalding to an area north-east of March (fig. 12) They are believed to date from Roman times when, judging by the distribution of archaeological finds, the shoreline may have run not far to the north. Alluvium was accumulating in Fenland rivers and drainage channels in the Roman period, and since that time accretion has continued at the head of the Wash. Thus the earliest Anglo–Saxon settlements were sited on a chain of low sandy islets, few of them more than a mile or two across, stretching close alongside the present north-west shore of the Wash to Boston, south to Spalding, from Spalding east and then south to Wisbech and from Wisbech north-east to Lynn. From these centres the village lands were extended in two directions, seaward by embanking salt marshes as marine accretion permitted and fenward by protecting peatland from flooding by freshwater. The marine alluvium building the silt fens now stands about 10 ft above sea-level and rises slightly towards the newest reclaimed land at the coast, while the peat fens farther inland have been lowered by the shrinkage and wastage that has followed drainage and lie at or below mean sea-level.

The drainage pattern of Fenland is very complicated, for it is the outcome of several centuries of large- and small-scale efforts to overcome the peculiar difficulties presented first by the low relief of the region as a whole, secondly by the high levels of the silts at the outfalls of the rivers as compared with the peat fen behind, and thirdly by the high flood-levels of rivers draining a catchment that stretches

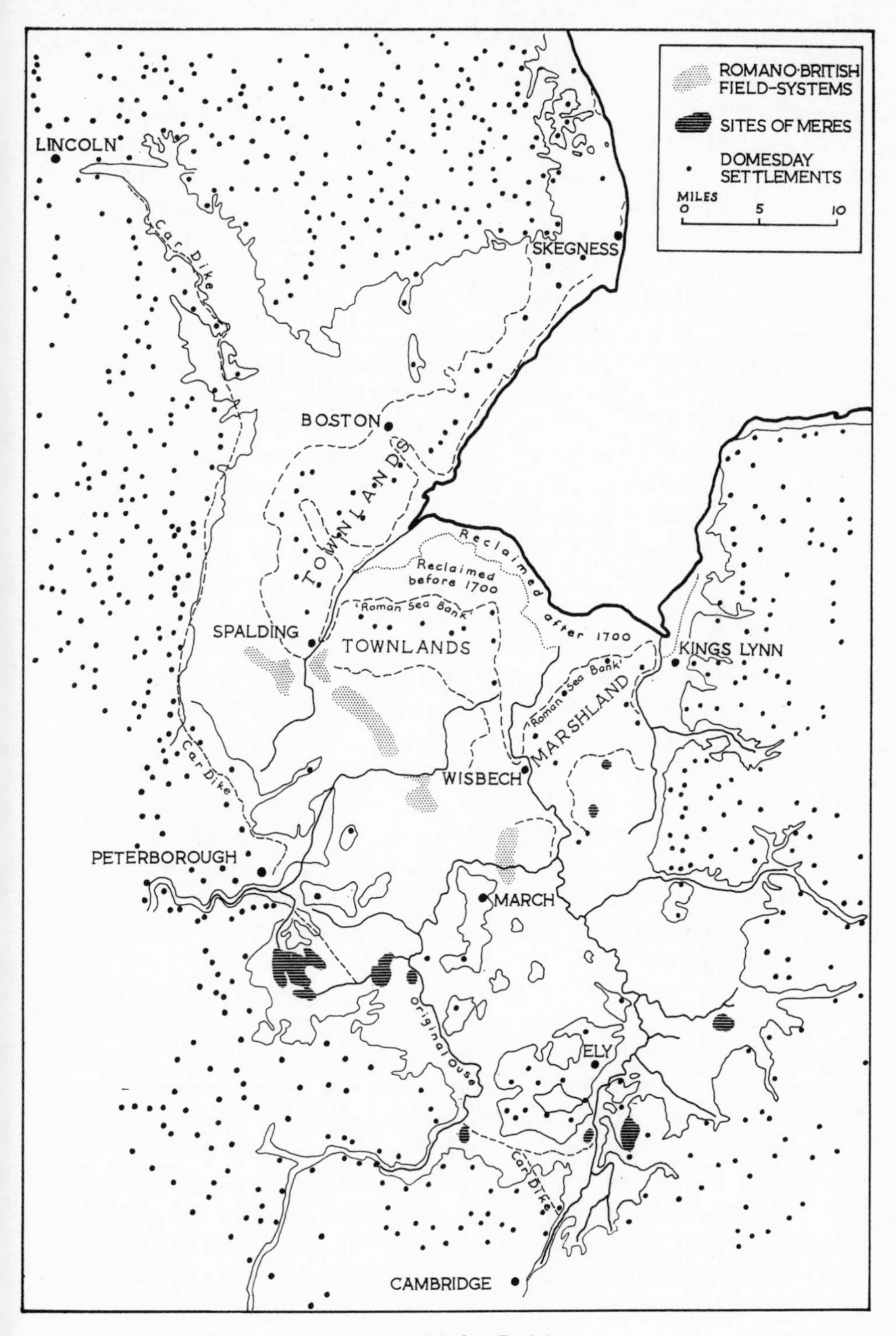

Fig. 12. Fenland before Draining.

as far as Lincoln, Leicester and Luton. Almost without exception the river channels of the present day have been artificially modified. The Car dyke, which can be traced westwards from Waterbeach on the Granta towards an early course of the Ouse, and its continuation from Peterborough along the western edge of the fen to Lincoln, was constructed by the Romans and may have formed part of an ancient drainage-scheme. However, its primary purpose was, probably, to allow boats to travel from Cambridge to Lincoln and then via the Foss Dyke and the Trent to York. An old course of the Nene is followed by the Cat's Water and by the county boundary between the Isle of Ely and Holland, and the medieval river crossed the fen Island of March by an artificial cut and joined the old course of the Granta–Ouse near Upwell. Low silty ridges called roddons, winding across the black peat fens of southern Fenland, have been shown by Gordon Fowler to preserve the traces of an old river-system that converged on an estuary at Wisbech (Ousebeach). This was modified when the Bedford Ouse changed its course to flow south of the Isle of Ely instead of west of it, and was transformed in late medieval times when the combined waters of the Nene and Granta–Ouse were diverted along Well Creek to join what had been a minor stream reaching the Wash at Lynn. The Old Croft River marks a reach of the old river-system abandoned when a cut was made from Littleport to Salter's Lode. On the east side of the Ouse–Granta, the boundary of Norfolk with Suffolk and the Isle of Ely follows a roddon representing the course of the Little Ouse three centuries ago.

The most important arteries of the existing drainage-system date from the seventeenth century. Until that time the fens had been drained and reclaimed piecemeal, but the reclaimed lands were frequently flooded, channels were blocked by debris, and banks collapsed in time of flood. Responsibility for upkeep of the works had always been divided and the dissolution of the monasteries added uncertainty to the situation. Then in the seventeenth century London businessmen and the new landowners began to visualize what might be gained from large-scale reclamation, and when the Duke of Bedford proposed to drain the Great Level he had the support of other local landowners and the financial assistance of the Adventurers. The engineer Vermuyden who was engaged to plan and carry out the project recommended, like earlier advisors, the replacement of the old winding rivers by straight cuts. The Old Bedford River was excavated in 1637 to take the water of the Ouse directly north-east from Earith to Salter's Lode. It attained its purpose, at least in part, freeing much land of water so that it could be used for grazing in summer if not in winter. The New Bedford River was cut parallel to the old one fourteen years later in an attempt to drain the Great Level more thoroughly, the space between them, called the Washlands, being left as a flood-reservoir. Tong's Drain was cut at this time; St John's Eau was dug to shorten the course of the Ouse from Denver to Stowbridge; Popham's Eau was resuscitated to lead the Nene into the Ouse below Denver; and the Forty Foot Drain was linked to the Bedford River (fig. 13).

Partly because of outfall difficulties, mainly because of the shrinkage of the peat as it dried, these works failed to drain southern Fenland adequately until they were supplemented by numerous minor pumping schemes. These were organized by local landowners, who formed independent Commissions of Drainage. Though now subject to the overriding authority of the River Boards, they remain responsible for draining the land; the River Boards ensure that the water gets away to the sea.

Until the eighteenth century large areas around Ely, and between March and Whittlesey, remained drowned and useless for much of the year, and the peatlands stretching north of the Glen and Witham were normally flooded in winter. Kindersley's Cut of 1773 greatly improved the drainage of the North Level, by taking the waters of the Nene directly north from Wisbech and thereby lowering the levels in the drains to the south and west. On the north side of the Wash great tracts of East and West Wildmore Fens lay under water until the high prices of corn during the Napoleonic Wars stimulated reclamation. The last considerable stretch of water in Fenland, Whittlesey Mere, persisted until 1853, a date that marks the completion of the drainage of Fenland.

Before draining, the surface of the peat is believed to have stood some 10 ft above mean sea-level; after draining it sank at a rate of a foot or two every ten years, at first as a result of the removal of water from the peat and later because of the wasting of the organic matter as a result of bacterial action. The rate of sinking is well demonstrated by Holme Post which stands a little to the south of the site of Whittlesey Mere. This was an iron post from the Crystal Palace which was driven into the peat in 1851 until its top was level with

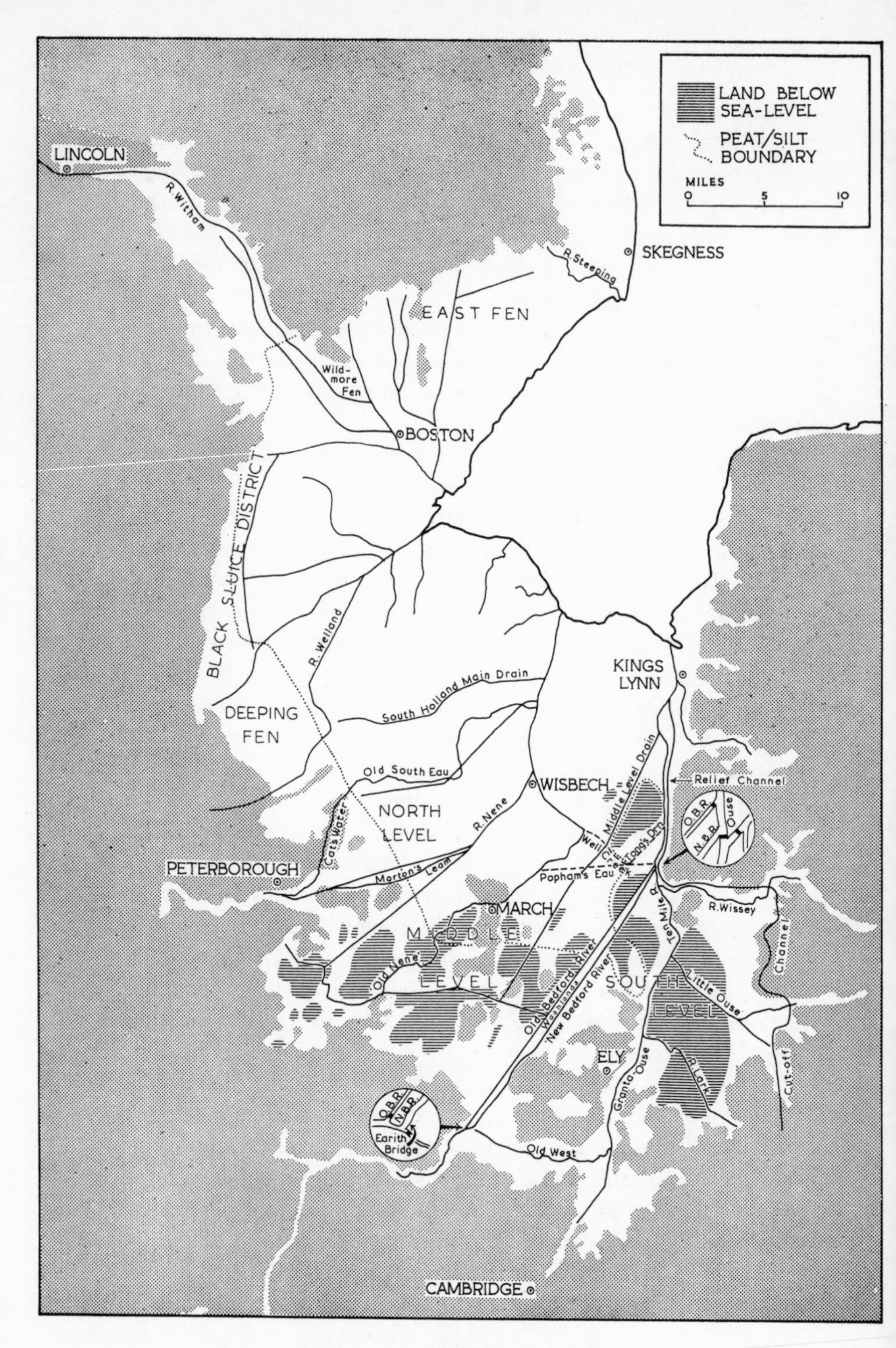

Fig. 13. The Draining of Fenland.

the ground. Whittlesey Fen was drained; by 1860 nearly 5 ft of the post protruded and by 1892 the surface of the peat had been lowered a total of 10 ft. Thereafter the rate of wastage diminished, because the land near by was abandoned for cultivation; it was colonized by a fine birchwood that now forms Holme Fen Nature Reserve. The peat continues to disappear from the cultivated fields in Fenland, exposing the uneven surface of sands and clays beneath it. The rate of wasting is less than in the early days of reclamation when paring and burning were a normal prelude to cultivation, but in many places it still approaches an inch every year. In southern Holland the peats were thin and had largely wasted away more than a century ago, exposing quite heavy clays. Skertchley noted the disappearance of peat formations from the neighbourhoods of Bourne, Spalding and Crowland in the score of years preceding 1870, and since then much greater areas have gone in the vicinity of Thorney and March.

In the early years of the nineteenth century it was noticed that the texture of the soils was much improved and crops benefited in places where the plough had penetrated the peats and turned up calcareous marly clays. Clay was therefore applied to the peat soils deliberately, by digging parallel trenches and spreading the clay over the peat at a rate of about a hundred tons to the acre. The effects lasted for several decades. Claying is one of the most effective ways of reducing the damage caused by wind erosion, which is liable to occur after spring droughts when the soils have been thoroughly cultivated and are still unprotected by growing crops. Strong winds can then lift the light soil particles, together with seed and fertilizer, and deposit them in drains and ditches, thereby causing considerable damage within a few hours. The results are spectacular and costly but steady wastage of the peat by oxidation is of more consequence.

The deepest peat remains at the southern margins of Fenland, but even here many feet have been lost in the last century or two, and the chalky floors of meres once covered by 10 ft of water now stand above the general level of the surrounding fields. Near the Isle of Ely it is still possible to drive for miles across black fen over concrete roads with fractured surfaces that tell of the peculiar difficulties the underlying material presents to the civil engineer and builder. Roads and railways have subsided while their bridges resting on deep piles have remained stationary. The permanent way sags under the weight of a passing train. Houses lean sideways and their walls crack because the peat beneath them has shrunk unevenly. But the most significant feature of the landscape is the high level of the rivers and main drains compared with that of the fields. As the fields continue to sink, their drains must be deepened, the river-banks have to be strengthened, and more powerful pumps installed to lift water from the field drains into the rivers and main drains.

The problem of draining the peat fens was largely solved by the introduction of steam pumps early in the nineteenth century, and these have been replaced by still more efficient machines driven by diesel oil and electricity. But sporadic floods have continued until recently. The last was in 1953 when the storm-surge of 31 January caused the waters of the Ouse to overflow and break the river-banks between Denver and the sea. This was an exceptional kind of flood; the main threat is from spring floods reaching high levels farther south. The North Level benefited from a new cut made for the Nene in 1832; and the Middle Level was freed from the danger of floods by a powerful pumping-station at Wiggenhall St Germans, installed in 1934. But the South Level, most remote from the sea, was seriously affected by flooding in 1937 and 1939. Again in 1947, after a snowy winter the rivers, swollen with meltwater and heavy rain, were held up by high tides and overflowed their banks, flooding 37,000 acres east of Earith. It was clear that similar floods would recur unless the rate of discharge of the Ouse–Granta at Denver could be accelerated. Work began in 1956 on a scheme that involved excavating a straight channel from immediately above Denver sluice direct to an outfall on the tidal river a little above King's Lynn. Its function is similar to that Vermuyden had in mind for St John's Eau, namely to lower the flood-level of the river at Denver. To take advantage of this lowering, the Ouse for several miles above Denver has been widened to allow the waters from upstream of Ely to move rapidly seawards at low tide. Finally a cut-off channel is being excavated (1960) along the eastern margin of the fens to carry the flood-waters of the Lark, Little Ouse and Wissey to the flood-relief channel at Denver. It is interesting to note that Vermuyden intended such a channel should form part of his works, but for lack of funds it was never completed. The current works as a whole will cost in the neighbourhood of £12 m., of which some 90% will be provided by the central government. With their completion it is believed that the danger of serious flooding in southern Fenland will have been removed.

In Anglo–Saxon times the number of sites in Fenland dry enough to allow settlement was restricted to the silt islets near the Wash and the islands of gravel and clay rising above the peatland. Most of the fen and marsh that has since been reclaimed has been incorporated within the boundaries of the villages that grew upon these sites and at the Fenland margins, and although some of them have been subdivided in recent decades, parishes are larger than in most other parts of England. In the silt-lands, settlement is now dispersed. Some houses are grouped around great churches, built in late medieval times on land rising above the spring tides and in some cases, as at Kirton and Holbeach, above the level of the highest tides; the majority of the houses straggle inelegantly along the highways for several miles from the village centres. The peat, on the other hand, remains to a large extent deserted of habitations, because of the difficulties presented by wastage and the depth to firm ground. Many isolated farmsteads lie on silt roddons or on patches of gravel, and all the hamlets and villages have a solid basis of some kind. The hamlet of Prickwillow is largely built on silts, laid down by the Ouse when its course ran farther east than now, and at Benwick the foundations of the houses rest on gravel lying close to the surface of the peat. The larger villages are all situated on larger islands of gravel or boulder clay resting on Jurassic clays or greensand. Several of them are centred on monastic buildings that were erected by religious originally attracted to the fen islands by their remoteness. The cathedral at Ely remains a landmark for miles around; most of the other abbeys are in ruins. At Crowland, the houses fronting the main streets have steadings and yards at the rear and recall the days when stock and crops from outlying farms could be accommodated there in time of flood.

Settlement between the Wolds and the silt fen north of the Wash is mainly confined to a boulder-clay ridge between Stitchford and Sibsey, though there are small hamlets on the lower ground with dull names like Midville and Eastville. On the eastern side of Kesteven a long narrow strip of peat remains practically barren of settlements, and the land is included in the long eastward extensions of numerous villages spaced at intervals of about a mile along the 50-ft contour at the margin of Fenland.

While great tracts of peat fens are interrupted only by straight drains, with farmhouses scattered sparsely here and there on rising ground and sheltered from the wind by clumps of poplars and willows, the landscape of the silt fens is much more intricate and congenial. The surface of the ground is gently undulating instead of being dead flat, with light soils on the crests of ridges originally thrown up by the sea, and heavier soils in the intervening depressions. The density of settlement is highest on the light soils first occupied by the Anglo–Saxons, which in Holland are sometimes called the Townlands. The roads in the areas of early occupation are winding and irregular, many of them running along ancient banks built in medieval times. In the Elloe district Hurdletree Bank, Raven Bank and others were built in turn farther south as fen was reclaimed for grazing and later used for cultivation. Nearer the Wash, the Roman Bank, which acquired its misleading name from Dugdale in 1660, is probably a composite feature of several different ages, and its position seems to correspond roughly with the stage that had been reached in reclaiming salt marsh from the sea at about the time of Domesday. Running from south of Lynn to Wainfleet over a distance of more than 50 miles, it is about 10 ft high and in two stretches north of the Witham, where accretion has been negligible, it still forms part of the existing sea-defences. The names of several villages in the marshlands of Norfolk, like Walsoken and Walpole, refer to these banks; and other place-names such as Gedney Dyke and Tilney Fen End tell of the extension of the siltland villages towards the sea and the fen. They are amongst the most elongated parishes in the country, so long that the southern ends of some of those in Elloe have been cut off to form separate parishes (fig. 14). On both sides of the siltlands of early settlement soils are heavy, houses are dispersed along roads that were once droveways to salt marsh or fen pasture, and villages are small and often strung along the sides of banks and drains. Minor place-names indicate that Danish settlement was important on the Townlands, and the customs and character of these colonizers may explain why the villages there have always been strongholds of small freeholders, and why in turn farmsteads are so dispersed over the countryside. Near the southern limit of the siltfens, Wisbech stands on silts laid down near the former estuary of the Ouse, and farther south still, Outwell and Upwell stretch for several miles along a low winding ridge marking the old course of the river.

In both peat and silt fens, ale-houses on the banks of the older

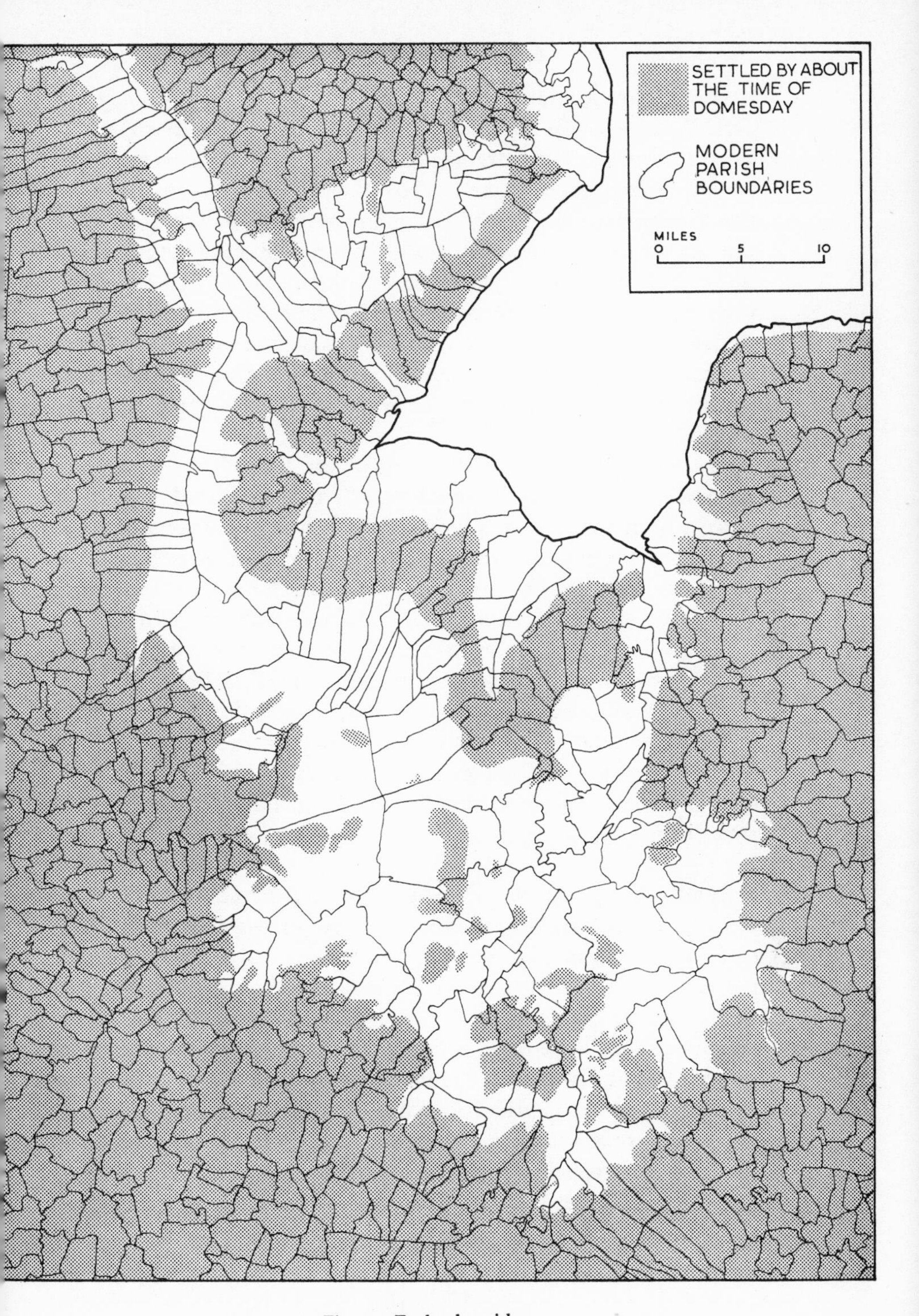

Fig. 14. Fenland parishes.

waterways are a reminder of the former importance of water-transport. Navigation in the Fenland had already suffered from the silting-up of river mouths, drainage-sluices interrupting movement, and the efforts made to keep the level of water in the rivers as low as possible for drainage purposes. The coming of the railways brought to an end almost all the remaining trade. Barges still carry beet to sugar factories, and clay for repairing river-banks, but the days have gone when they were towed up to Cambridge laden with coal, or carried cargoes of grain to Lynn from the uplands behind Burwell, Commercial End, and Reach. Now the ale-houses cater for sailing clubs and for the men working on drainage schemes instead of for thirsty bargees.

In the last few decades, the ports at the head of the Wash have enjoyed a revival. Boston and King's Lynn both handle more than ½ m. tons of cargo annually and Wisbech about 80,000 tons. In every case the inward traffic is in excess of that outward bound; petroleum is of growing importance, and such bulky goods as timber, grain, seed potatoes and fertilizers are commonly handled. Boston and King's Lynn are amongst the more important minor ports of the country; but the lack of industry in their immediate hinterland must prevent any significant increase in their status. The fairways of all the ports are inherently unstable and King's Lynn, which has always been opposed to fen drainage schemes for fear of their effects on the tidal channel, views the latest flood-protection scheme with more than suspicion.

Agriculture is the dominant activity in Fenland and its characteristics are remarkably uniform in view of the variations in soil conditions, and in particular the differences between the silts and peats. This is true of both farm size and productivity, and of the distribution of crops.

Fenland farms are smaller than farms in the nearby uplands and the rest of East Anglia. In Holland only nine estates exceed 1,000 acres, and only five in the Isle of Ely. About half the holdings in the Isle of Ely are less than 15 acres, and in 1948 of all the holdings larger than 20 acres, a half were less than 50 acres in size. The fertility of the soil is such that there has usually been a good demand for land, and in the past farms were commonly sold to settle questions of inheritance. It was possible for a man to build up his holding by purchase and this helps to explain why the plots constituting a single farm are often scattered over several square miles.

Although Fenland farms are small, the incomes they provide are higher than those from upland farms. The income from an average farm in the Isle of Ely in 1951–52 was of the order of £1,500 to £1,800, as compared with £900 to £1,200 from a farm of equivalent status on the uplands. Farming in Fenland is usually intensive. Outputs per acre of £50 or £60 or more compare with £20 or £30 on upland farms. Of course, costs per acre in Fenland are higher: rents are nearly twice as great, drainage rates have to be paid, and labour is employed more intensively. But the output from every £100 worth of labour invested is as great from Fenland farms as from upland ones and sometimes greater, and the income per acre may be three or four times larger.

The crops usually grown—corn, potatoes, sugar-beet and vegetables—show far higher profits than grass; animal products are consequently of relatively little importance. In the Isle of Ely there is an average of only two cows per farm, and three or four in Holland, as compared with between eight and twenty in East Anglia. Sheep are of still less importance, and while the number of cattle has remained fairly steady over the last twenty years, the sheep population throughout Fenland declined sharply during the last war and has never regained its pre-war figure. Horses have declined in number with the coming of the tractor, as they have elsewhere, though some fine teams have been retained. Pigs more than doubled in number between 1948 and 1958 and are back to their pre-war level, but neither pigs nor poultry, though economically successful, have shown the rapid post-war increases in numbers typical of most parts of the country. The high income that can be obtained from an acre of Fenland under the plough goes far to explain the agricultural landscape. No land is allowed to lie under bare fallow. There are fewer playing-fields than in most rural areas; and woodland is sparsely distributed, with a total in Holland of less than 10 acres. Permanent grass is restricted to river-banks and dykes, the Washlands, coastal marshes, and the heavy soils of south Holland and the fen islands.

A rough comparison of cropping on the peat-fen and silt-fen can be made by referring to the agricultural returns from Holland and the Isle of Ely. Holland is mainly siltland, with soils varying in texture from fine sands to heavy loams, and some peats in the extreme south-west and north-east. About one-half of the Isle of Ely is peat covered; in addition silts, varying widely in texture, occur in the north between March and Wisbech, and mineral soils have been derived from a variety of parent materials on the fen islands.

In both Holland and the Isle of Ely about 10% of the land is under permanent grass and rough grazing, as compared with nearly 20% before the Second World War; temporary pastures occupy a much smaller area. In both districts, about a quarter of all agricultural land is under wheat; this is about the same proportion as in pre-war years, but the acreage has increased considerably since then, partly because of the ploughing-up of pasture land and the decline in potato acreage. In both, the area under barley and oats together has remained about one-tenth of the total under crops and grass, but the reduced local demand for animal feeding-stuffs and the introduction of improved varieties of barley has resulted in the proportion under oats declining and of barley increasing.

Root-crops are next in importance to grain. Between the wars potatoes occupied a larger acreage than wheat but the area under main-crop potatoes has since declined on both peats and silts, largely on account of damage caused by eelworm infestation. In the Isle of Ely and Holland together, there are about 80,000 acres, about one-eighth of the total potato acreage of the United Kingdom; yields are above average and the two districts together provide about one-sixth of the country's supplies. The decline in main-crop potatoes has been accompanied by an expansion in the area under sugar-beet, which now occupies about one-tenth of the fen farmland. It grows well on deep friable soils and is now as important on the silts as on the peat-fens.

The acreage of vegetables in the Isle of Ely increased rapidly during the last war, but has since dropped back very markedly from 23,800 in 1948 to 13,500 in 1958. Production of vegetables in the Isle, which is mainly concentrated in the area around the town of Ely, is far less important than in Holland where about one-sixth of the total farming area, some 40,000 acres altogether, is under vegetables. Nearly half is under peas, increasing quantities of which are quick-frozen. The acreages of cauliflowers (about 4,500), cabbage and savoy (5,300), and broad beans (2,500, about one-third for cattle feed) are also impressive, and it must be borne in mind that the figures given for vegetables are under-estimates, for they are based only on returns from growers with holdings of more than one acre. In many cases, more than one crop is grown every year, and some are not recorded in the returns. The value of output per acre is greater for vegetables than for most crops. In both areas farmers are growing more and more of their vegetables under contract to the processors.

More detailed study indicates greater local specialization than the crude figures for Holland and the Isle of Ely would suggest. Early potatoes, for example, are a feature of the area north-east of the Witham, near Boston, and mustard of southern Holland. Most of the sugar-beet is grown within 10 miles of each of the big extraction factories at Spalding, King's Lynn, Wissington, Peterborough and Ely. Ideal conditions for celery are provided by deep moist peats, and most of the crop is grown in peat-filled depressions near Ely, Littleport, Stretham and Whittlesey. Carrots are mainly on peaty soils near Chatteris, Doddington, Welches Dam and Wimblington. Fruit trees are unable to root firmly in peat and most of the orchards are on silts near Wisbech and on the fen islands of Haddenham and Ely, with some plantations near Fordham, Wimblington, Cottenham, Chatteris and Thorney. Small fruits, of which strawberries are the most important, are grown in the same areas as orchards and also in the vicinity of Spalding.

To the visitor, the most striking and certainly the most colourful feature of Fenland in spring-time is the concentration of bulbs and flower-growing on the densely settled, light soils near Spalding, and in the area stretching east to Sutton Bridge and north to Kirton. Little or no wheat is to be seen there, but mainly small holdings with flowers, fruit and vegetables, and with glasshouses, used in the winter months for chitting potatoes and in summer for growing tomatoes.

A characteristic feature of Fenland farming is the high proportion of crops grown for sale rather than for use on the farm. Much of the produce is taken directly from the producer, by road transport, to markets in all the industrial areas of the country. In spring and summer the railways provide special afternoon and evening services from Spalding and Wisbech to deal with perishable flowers and fruit. Potatoes are sold to wholesalers, often through local merchants, and Boston is the chief centre for trading them. Wheat goes to London and Hull for milling; barley and oats are milled locally. Local industry based on the vegetable crops provides work, some of it seasonal, for several hundreds of people. Peas and other vegetables are canned at Boston, Long Sutton and King's Lynn, and both canning and quick-freezing are important at Spalding and Wisbech.

Except for the processing of agricultural produce, local industries are undistinguished and on a small scale. Chemicals for use in agriculture are now being prepared at King's Lynn, cans and baskets are manufactured at Wisbech, bricks at Ely and Whittlesea, and engineering is well established at Boston. But the lack of stimulus to industrial growth is reflected in the population figures for the region. Since the middle of the last century, emigration to the major industrial areas of England has been marked, and the populations of the smaller towns and villages have shrunk. Ely, important as a railway junction, has grown slowly as have the other large towns in Fenland with good rail and road communications. The marshalling-yards built at Whitemoor about 1930 were at that time amongst the largest in the country, and March near by has grown mainly because of the 2,000 people employed on the railway. Since 1921 there has been a persistent loss of population from Fenland to London and the Midlands, but over the last four decades natural increase has more than balanced emigration and so the population in most parts of the region has risen slightly. Wisbech (17,512), Boston (24,903), Spalding (14,821) and King's Lynn (27,554) are increasing in size at the expense of the rural areas, while the total numbers of workers in agriculture have gradually declined since the war as machines have replaced men and horses on the farms. Probably the most powerful centres of attraction to the people living in Fenland villages are the cities on the outskirts of the region, Lincoln, Peterborough and Cambridge, each with thriving engineering, electrical and manufacturing industries.

Agriculture continues to be the most important employer of labour, and the intensive cash cropping typical of the region will for long require a large labour force. In Holland the number of workers per 100 acres is about five, somewhat more on the light silts, less on the heavier land, as compared with three on the uplands. The demand for labour is probably greatest in May and June for singling beet, but in some areas more hands are required for harvesting in the following months. In this century, the Fenland farmer has been remarkably independent of imported casual labour; some Irishmen are employed at peak seasons, but reliance is placed mainly on the wives and families of regular employees who are skilled in assisting with the seasonal operations. Near Ely a large number of holdings of less than 20 acres are worked as allotments by people engaged in some other occupation as well as farming. In Holland, too, much of the land owned by the County Council and Ministry of Agriculture, some 20,000 acres, is farmed as small-holdings, and allotments of about half an acre can be rented from parish councils. As more equipment, fertilizers and insecticides are used, the small grower finds himself at a disadvantage, and in many areas there is a movement towards greater co-operation and amalgamation of holdings.

Little true fen now remains with the exception of certain patches that are deliberately conserved, such as Wicken Fen, where ecological research has thrown light on the succession of events in the accumulation of the peat, and its colonization by woody plants. Holme Fen and Woodwalton Nature Reserves are less representative of former conditions in Fenland, but they include much of interest to naturalists—bird-life, insects, and plants peculiar to the region—and offer a welcome contrast to the surrounding country where Nature has been so successfully tamed.

The character of Fenland has been transformed. Some trace of its ancient wildness can be recognized at Wicken; salt marshes outside the latest sea-banks preserve a simple freshness; and some say the sky is wider in Fenland than in hillier lands. But the old Fenland has faded, and as the remainder of the peat disappears, and farmers cultivate soils derived from clays and sands beneath, the distinctiveness of the region further diminishes. Pump-houses on the banks of perched rivers, and intricate systems of long straight drains, will continue to demonstrate Fenland's hollow form. The discerning traveller will recognize from the air white patches of shell marl, marking the beds of meres long since dry, and will note the geometrical patterns of fields tilled in Roman times when the sea had hardly begun to build the silt-lands. He will be reminded, perhaps, by the upstanding churches of Marshland of the people who won that country from the sea. The ruined abbeys and the Cathedral of Ely will still tell of the early efforts to drain the fen, and the Methodist chapels north of the Wash may hint at the late completion of the drainage works there.

The process of reclamation continues, and since the last war five more square miles of tidal lands have been embanked. But the expense is great. The Wash was never like the Zuyder Zee, a lake; its channels are deep, and it is difficult to envisage any wholesale reclamation that would eliminate that familiar bite out of the east coast of England and bring the evolution of the Fenland region to a fitting conclusion.

Selected Bibliography

A. K. Astbury. *The Black Fens.* Cambridge, 1958.

H. C. Darby: *The Medieval Fenland*, Cambridge, 1940; *The Draining of the Fens* (2nd ed.), Cambridge, 1956.

W. A. Doran. 'The Great Ouse Flood Protection Scheme.' *Dock and Harbour Authority*, 1956, pp. 368–73.

G. Fowler. 'The Extinct Waterways of the Fens.' *Geog. Jour.*, **83**, 1934, pp. 30–9.

H. Godwin and M. H. Clifford. 'Studies of the Post-Glacial History of British Vegetation.' Parts I–IV. *Phil. Trans of the Roy. Soc.*, Series B, **229**, 1939, pp. 323–406 and **230**, 1941, pp. 239–304.

CHAPTER 7

THE SOUTH-EAST MIDLANDS

C. T. Smith

The south-east Midlands is not very distinctive with respect to any one criterion, or even, perhaps, to any criteria at all. Unlike East Anglia it has no recognition in traditional usage, nor does it constitute any modern administrative division, for it is an area shared among its neighbours by such bodies as the gas and electricity boards, and the Ministries of Labour and Agriculture. The south Midlands has a shadowy recognition in common parlance or in nineteenth-century organization of census returns, but the south-east Midlands is not normally an accepted division.

The boundaries by which the area is here defined are either completely arbitrary, and decided upon for reasons of convenience, or they are boundaries which mark off other regions of greater individuality. The Welland valley to the north is a convenient boundary with the north-east Midlands. The margin of the reclaimed Fenland is of self-evident significance, and the march with East Anglia may be determined by its claim to consist of Norfolk, Suffolk and the northern part of Essex. To the south the London area, defined in terms of urban spread, is a clear limit. Finally the western boundary, parallel to Watling Street and the Euston–Rugby line, is the most arbitrary of all, cutting the south Midlands in two, and putting Oxford and Cambridge one without and one within an academic Danelaw.

The absence of definable boundaries is not in itself significant or important to the identity of the south-east Midlands, but it is also true that many of the characteristic features of the area are shared with neighbouring zones where they are often better seen. Like much of lowland England, the south-east Midlands share a relief which is developed on gently dipping Cretaceous and Jurassic rocks. But the Northamptonshire uplands are not so imposing as other sections of the great belt of Jurassic scarplands. There are few areas over 600 ft and the escarpments are sinuous, low and discontinuous. The outcrop of the Oxford Clay, which coincides with broad, low-lying plains in Oxfordshire and south-west Bedfordshire, and floors much of Fenland, is a low plateau capped with boulder clay in most of this area. The Lower Greensand produces a relief feature of variable significance, but to the south the chalk lacks the boldness of the Chilterns and the monotony of the boulder-clay-covered plateau of East Anglia.

The lack of emphasis which is characteristic of the physical features of the area is to be seen also in land use, settlement and industry. As a whole the area is one of transition from predominantly mixed arable and pastoral use in the north-west to arable farming in the east. Village settlement is characteristic in most of the area, for this was one of the strongholds of the Midland open-field system and one which yielded late to final enclosure. There is a wealth of stone-built villages, but they are by no means so widespread as those of the more famous Cotswolds. Apart from a few industries which spring from the use of local mineral resources, such as cement, bricks, and steel, manufacturing industry is scattered in and near the towns in a way which has not much altered the predominantly rural character of the area, except in the south towards London and in the north towards the East Midlands. Communications pass through rather than to the area. It is, in short, the subdued and transitional character of the area which very largely sets the keynote for more detailed examination.

The Northamptonshire uplands are essentially an undulating plateau tilted gently to the east and south-east and dissected discordantly by the broad valleys of the Welland and the Nene and their tributaries. There is no clear break in this pattern of relief until the escarpment of the Lower Greensand or the Fen margins. To the north-west a discontinuous, irregular escarpment, much dissected by obsequent streams flowing to the Avon or the Welland, may be traced to the south-west of Market Harborough as far as Long Buckby. It is held up by the Marlstone Rock Bed of the Upper Middle Lias, an oolitic, ferruginous limestone resting on softer clays and marls and overlain by the stiff blue clays of the Upper Lias. Farther east a second, Inferior Oolite escarpment is straight and continuous where it overlooks the Welland valley, but to the south of Market Harborough it becomes less clearly marked and very irregular. To the west of the main outcrop, outliers of Northampton Sand cap high plateaux rising abruptly to over 600 ft, notably at Cold Ashby and Naseby. It is only in this area that the escarpments can be said to be at all impressive (fig. 15).

Several features help to account for the relatively low relief and the poor development of the escarpments. The scarplands overlook the present watershed zone between the Warwickshire Avon and the Soar, which itself rises to as much as 460 ft. Thick beds resistant to erosion and weathering are poorly developed. The facies and thickness of Jurassic rocks vary greatly because they were laid down in a series of basins during a long period of intermittent downward movement. The Northampton area lay towards the western margin of a Midlands basin of deposition bounded by the axis of uplift of the Vale of Moreton to the west. In Northamptonshire the Inferior Oolite is represented by the Northampton Sand, consisting normally of 15–25 ft of ironstones and ferruginous and calcareous sandstones, but with a maximum thickness of 70 ft north and west of Northampton; and by the Lower Estuarine Series, consisting of 10–25 ft of sands, silts and clays. Near Cheltenham, however, the equivalent beds, mainly limestones, are 300 ft thick. The Lincolnshire Limestone of the Inferior Oolite, which forms Lincoln Edge farther north, and which helps to form the escarpment overlooking the Welland valley from Stoke Albany to Harringworth, thins out to the south and is completely absent in southern and western Northamptonshire. The Great Oolite Limestone is no more than 15–25 ft thick in this area and is nowhere of much importance in the making of the scenery.

The regional dip is not more than half a degree to the south-east, but the plateau surface cuts across successive strata from Lias to Oxford Clay. A substantial accordance of sub-drift summit levels suggests the existence of a pre-glacial erosion surface, perhaps of subaerial origin trimmed by marine action, which is thought to have been uplifted and tilted from west to east so that its remnants range from 750 ft in the west, in Leicestershire, to 200–250 ft in the east. The assumption of a tilted surface helps to explain many of the anomalous features of the drainage pattern of the Welland and the Nene systems. Predominantly west to east drainage was initiated on the uplifted surface, though what are now subsequent sections of the Welland and the Nene systems can best be explained by assuming that they have followed vestiges of earlier lines of drainage adjusted to structure. Alteration of the drainage to its present pattern was

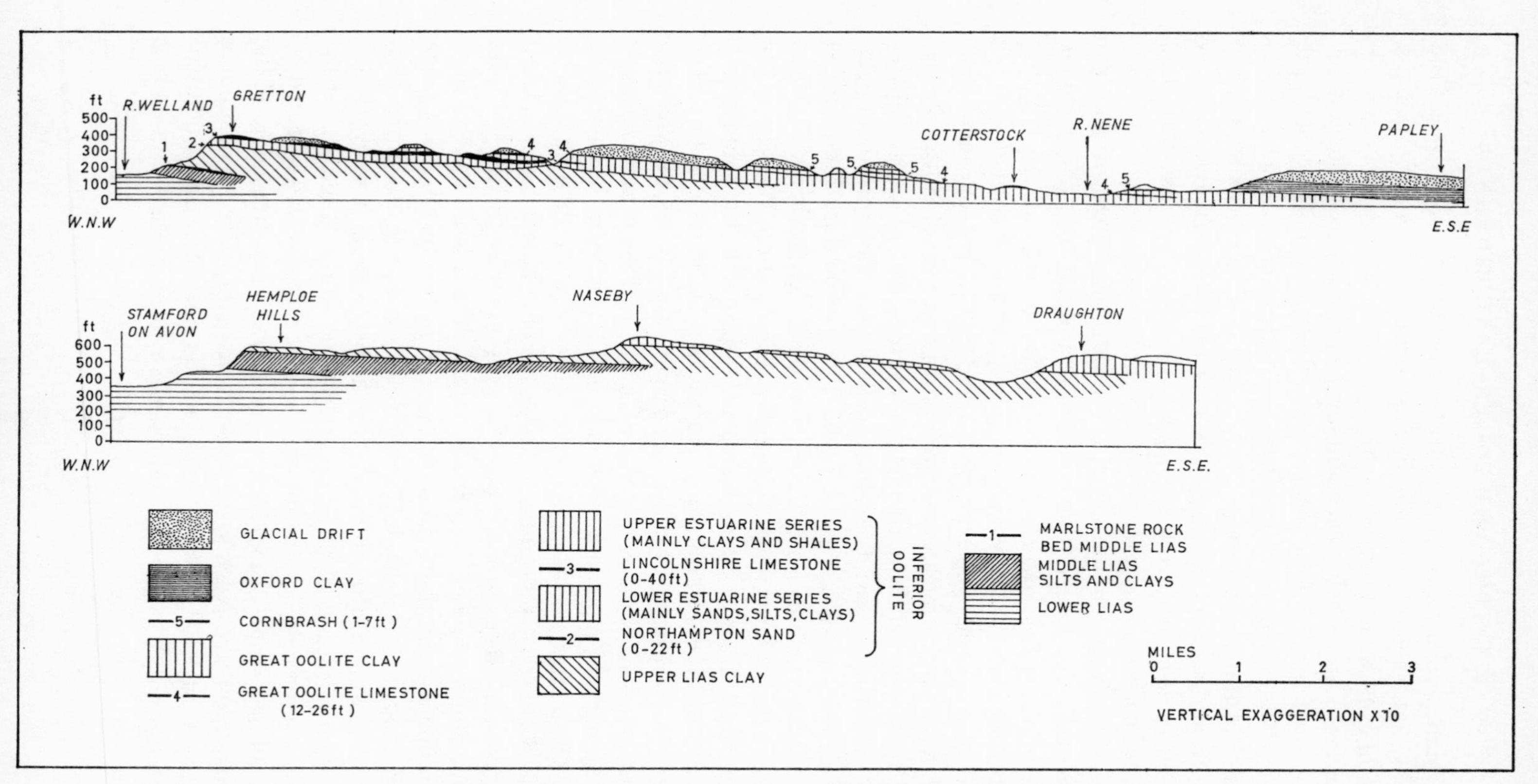

Fig. 15. Section across the scarplands of the south-east Midlands.

accomplished by river capture and glacial diversion, leaving some of the wind-gaps to be plugged by glacial drift.

In the neighbourhood of pre-glacial valleys, various superficial features have modified the original disposition of strata, and may have helped to bring about a rounding of relief forms. Under periglacial conditions, soft clays have bulged outwards towards the valleys under the weight of overlying rocks, producing local, superficial anticlinal structures in the valley bottoms. Overlying strata have cambered down towards the valleys, leading to the opening of superficial crevices or gulls, often filled with boulder clay, and to superficial dip-and-fault structures. Outcrops may thus dip locally towards the valley bottoms rather than in the direction of the regional dip. Glacial deposition has contributed much to the evolution of the present subdued relief. Boulder clays of different ages are plastered over much of the plateau surface except in the north-east, and have contributed to the smoothing of relief through the plugging of old wind-gaps, for example, north-east of Gretton—between Rockingham and Cottingham, or the filling in of small valleys, as for example in western Cambridgeshire.

South of the Oxford Clay and the small and undistinguished outcrops of the Ampthill and Kimmeridge Clays, the outcrop of the Lower Greensand usually forms a marked relief feature, chiefly because of the permeability of the current-bedded or incoherent sands of which it is largely composed. The Lower Greensand escarpment is at its most impressive near Woburn Sands, where the sands are some 220 ft thick. It may be traced north-east across Bedfordshire, declining in height and broken by the Ivel valley, until it disappears as a significant feature near Waresley, where the sands are no more than 70 ft thick, and they continue to thin out to the north-east to 10–12 ft. Variations in the thickness of the deposits, so important for relief, are related, like those of the Jurassic rocks, to movements of the Palaeozoic floor, for the Greensand was laid down in a shallow basin limited by an axis of uplift trending south from Nuneaton and another trending south-east from Charnwood.

Beyond the Gault Clay vale there are striking contrasts in the form and character of the Chalk escarpment between east and west. The scarp of the Chilterns is dramatic, rising from the Vale of Aylesbury to over 800 ft. Between Dunstable and Hitchin it is steep and fresh-looking, though much fretted by short dry valleys in the evolution of which spring sapping has played an important part. East of Hitchin the scarp is subdued, with slopes shallow enough for arable farming, and irregular in plan. Differences in the lithology, thickness and dip of the Chalk are slight and do not suggest why the contrast should exist. Ice moved across the eastern area, but if ice is responsible for the difference between east and west, it seems more likely to have been a result of the protection of an old landscape than of intensive glacial erosion, since terraces cut into the Chalk before the deposition of boulder clay farther south are still preserved at levels of 330 ft, 230 ft, 180 ft and 130 ft. The Totternhoe Stone, a harder bed in the Lower Chalk above the Chalk Marl, and the Melbourn Rock at the base of the Middle Chalk, have often been regarded as more resistant beds holding up benches at the foot of the main escarpment. But this appears to be true only where the outcrops of Totternhoe Stone or Melbourn Rock coincide with terraces at 180 ft and 130 ft which have been shown to have a wider distribution. The 130 ft terrace is almost certainly of marine origin.

Contrasts between the glaciated east and unglaciated west extend to the dip-slope of the Chalk, the former capped with drift, the latter with Clay-with-flints. In the west, areas over 700 ft have been interpreted as an old land surface. Below it lies an indefinite, degraded cliff-line and, at 600–650 ft, a marine-cut bench which has been identified as Pliocene in view of the existence of traces of Pliocene marine sediments overlying it and modifying the Clay-with-flints. Below this facet, the dip slope is held to represent the exhumation of a sub-Eocene surface. Deposits of Clay-with-flints, up to 50 ft thick, cap the watersheds on the dip slope, and represent not only the products of solution of the Chalk, but also the weathered and distributed remnants of Eocene rocks, modified under periglacial conditions and containing brick-earth deposits in addition. The drainage pattern is partly related to this series of events. Streams dissecting the older land surface extended across the newly exposed marine bench with the retreat of the Pliocene sea, and new streams, now represented by closely spaced shallow dry valleys, were initiated on this freshly exposed surface. A few valleys, however, cut more deeply into the face of the escarpment itself, providing gaps through which pass main railways from London to the north-west. The form of the Tring gap and the composition of the gravel spreads contained within it and at its mouth suggest that it was cut or at least greatly deepened by meltwater ponded up between the face of the Chilterns and ice to the north. The Tring gap is followed by the Grand Union Canal, and the main Euston line, completed in 1838. Formidable embanking and cutting made possible a ruling gradient of 1:330, and no tunnelling was necessary. The Luton gap, used very much later by the Midland line from St Pancras (1868), has a similar origin.

To the east of Hitchin ice overrode the Chalk, which is covered by boulder clays and glacial sands and gravels; relief is more subdued and the drainage pattern less regular than in the west. Fluctuations of the most southerly lobes of ice, reaching as far south as Finchley, have been held responsible for the diversion of the Thames southwards from an early course in the Vale of St Albans and later a course along the Finchley depression. Drainage has been modified in detail by the creation of marginal overflow channels and the plugging of old valleys by drift. It is partly by these processes that the convergence of streams in the area of Hertford is to be explained. The gaps followed by the King's Cross main line and the Liverpool Street line to Cambridge have an origin quite different from the gaps to the west. In both cases the watershed between streams flowing north and south lies well to the south of the escarpment. In the upper valley of the Saffron Walden branch of the Cam, terraces common to other parts of the Cam system are absent; tributaries form fish-hook bends on joining the Cam; and the existence of a deep, buried channel near Newport, probably of sub-glacial origin, and sloping generally southwards, suggests that reversal to the north took place on a surface of drift left after the decay of the ice. A similar origin is suggested for the Hitchin-Stevenage gap.

The gaps cutting through the Chalk have been of real importance to communications only in the nineteenth century, for of the major roads, only A11 using the Cam-Stort gap follows a low-level route consistently. The crossing of the Jurassic uplands by rail involved the solution of greater engineering problems, especially where the Midland coal traffic to London necessitated minimum gradients. In the west a group of routes (Watling Street, M1, the Grand Union Canal, the Euston and Northampton–Rugby lines) use the indefinite Watford–Kilsby gap, an early overflow channel from the great glacial Lake Harrison to the north, but even here tunnelling was necessary. The Kilsby tunnel was a major achievement of the early railway age, as indeed was the canal tunnel before it. Elsewhere the main lines tend to follow valley sides, crossing narrow watersheds in deep cuttings and tunnels and cutting obliquely across escarpments where necessary, as the Kettering–Manton line does above its magnificent viaduct across the Welland valley. Many of the roads follow high valley-side routes or watershed routes on the plateau surface (for example, A50 and A428 north-west of Northampton).

The landscapes of the Cretaceous country to the south follow an orderly sequence and scarp and vale relief in such a way that there is occasionally to be found a systematic and repeated relationship of relief, soils, land use and even settlement that permits of useful division and generalization on the scale of this essay. To the north, the patchiness and abundance of drift, the variations of lithology in plateau country dissected by drainage not too well adjusted to structure, combine to produce differences on a very local or very general scale.

From mid-Befordshire to the Welland, villages are distributed at quite regular intervals over plateau-top and valley sides alike, and there is little dispersed settlement anywhere. In general, the hill-top villages tend to be smaller and perhaps poorer than those of the valleys, especially on the clays, and this seems to have been commonly so in the past. The chronology of settlement is not reflected as it is to the south and south-east in the form and character of settlement. Pagan Anglo-Saxon burials and place-names indicate early settlement in the Nene basin before the end of the fifth century, long before the main penetration into the middle Ouse valley, though Gidding, Yelling and Earith are early names. Settlement of the watershed zones probably came later, and Rockingham Forest, which has left its traces in the parkland and woodland of large estates and a few large parishes, was densely wooded in Domesday times. The position of the eastern boundary of Northamptonshire may reflect the existence of a wooded frontier zone between Anglian settlement in the Ouse and Nene basins, which has, perhaps, left its mark in Salcey and Whittlewood Forests, now under the Forestry Commission. Yet penetration into the higher north-west was in places early. Weedon Bec and Naseby, for example, both contain pagan Anglo-Saxon elements. The siting of villages shows a preference for river terraces, or for places where water supply is assured at the contact of different outcrops, notably the junction of Northampton Sand and Upper

Lias, or the base of the Great Oolite Limestone. There are, however, many parts of the area where such favourable sites were not to be found, and settlement was on Oxford Clay or boulder clay, relying until very late on ponds and brooks for water supply.

The use of local building materials gives character to the villages and small towns, particularly in Northamptonshire. The creamy Lincolnshire Limestone, weathering to a soft grey or honey colour, has been worked for local building over much of the north-east of the county where many villages have a unity that comes from the reconciliation of different building styles by the use of a common material. Ketton stone, Barnack rag and Weldon stone have also been used farther afield, in Ely Cathedral, Cambridge colleges, and the Houses of Parliament, for example. Water transport by the Welland or Nene made possible an easy distribution to the coast or to the Ouse system at an early date. The easy cleavage of the sandy limestones at the base of the Lincolnshire Limestone, the so-called Collyweston Slates, once provided a roofing material as highly valued as the freestone above it. Other less famous building materials are no more than locally used. The warm brown Marlstone and the darker Northampton Sand are widely used in the west of the county, and in combination with other stones give a rich variety which is all the more attractive since the area is little affected by either the tourist trade or residential settlement.

Although heavy loams and clays are on the whole characteristic of the Jurassic area, there are many distinctive zones of lighter soils. Light, easily worked brashy soils occur on limestones in the drift-free areas of the north-east, and in small areas on the Cornbrash in Bedfordshire where the name indicates its reputation for arable farming. The redlands on the Northampton Sand and the Marlstone in the centre and west of the county were often heathland before the enclosure movement of the eighteenth century, and later highly valued as good turnip land. They are still largely arable. The alluvial meadows and pastures of the Nene and Ouse have had a great reputation for their fattening qualities, but better control of the Nene valley floods has recently encouraged farmers to grow crops on former meadows. The fatting pastures of the Welland valley have been famous since the seventeenth century for the quality of their grazing, which is classed among the best rye-grass pastures of the country, but it is possible that the skill of the graziers in the buying and selling of stock now counts for more than the natural capacity of the land to produce good grass.

The clays and heavy loams are by no means uniform. There is much variation in the texture, acidity and drainage characteristics of the boulder clays, depending on their provenance, thickness and the underlying strata. The Oxford, Great Oolite and Upper Lias Clays yield heavy, cold, tenacious clays. There is, in general, a transition on the clays from predominantly pastoral farming in the slightly wetter north-west to arable farming in the slightly drier east, and it is here that a history of convertible land use is written into the landscape. Old permanent pastures still betray a former arable use in their patterns of ridge and furrow, but more frequently in the west, for in the east continued arable farming has destroyed most of these vestiges of an earlier field-system. Deserted villages, depopulated in the great movement to enclosure and conversion to pasture in the fifteenth and sixteenth centuries, are more frequent too in the north-west of the area than in Cambridgeshire. Small and slightly irregular fields still sometimes betray early enclosure and conversion to pasture. In Northamptonshire enclosures which were considered old at the time of the Napoleonic wars covered a third of the county and had long been in grass, many of them on the cold clays of the north-west. In north Buckinghamshire seventeenth-century enclosure was frequent, and much of it must have been for conversion to pasture. In both of these areas the emphasis is still usually on stock, mainly for beef, though areas which were almost wholly in permanent pasture before the war, much of it of poor quality, are now diversified with leys, fodder crops, and with wheat and even barley as cash crops. Recent immigration of Welsh farmers to the neighbourhood of Daventry recalls the traditional association of the graziers with the Welsh drovers of the eighteenth century.

Clays in the east of the area have a much more doubtful reputation both for the poor quality of their grass and the difficulties of arable farming. In times of prosperity they have yielded good crops of wheat and beans, and are now productive of barley, temporary grasses and fodder crops as well. In times of depression the difficulties and expense of good cultivation have often proved too great and land has reverted to inferior grassland. As early as 1341, some Bedfordshire clay-lands were being left out of cultivation. Indifferent land in the Oxford Clay vale south-west of Bedford has been identified with, and given the reputation of, Bunyan's 'Slough of Despond'. In the 1930's the reversion of arable land to grass and of grassland to scrub, invasion by rushes and poor tussock grass, and frequent understocking were reported from west Cambridgeshire, Huntingdonshire and north Buckinghamshire. The major obstacles to good farming have been the poor drainage, the heaviness of the soil, and the delicate timing needed to maintain the soil in good condition. Attention paid in recent years to mole and tile drainage recalls an earlier emphasis on subsoil drainage in the nineteenth century in years of prosperous farming. Up to six horses were once used for ploughing for fallow in Huntingdonshire, but the rapidity with which crawler tractors can deal with the heaviest soils has been of enormous importance in recent years. Old scrub and heath have been grubbed up and fields are being amalgamated into units of 100 acres or more in parts of western Cambridgeshire. The mechanization of other farming operations, the improvement of seed varieties and price structures have also favoured cereal production rather than stock, so that the eastern arable area has now extended well into the Midlands.

Parkland and plantations on fairly large estates and tiny remnants of formerly extensive heaths, such as Rowney Warren or Gamlingay Great Heath, are characteristic of the sandy soils of the Lower Greensand plateau, often leached and acid, dry, hungry and subject to blowing. The sandy soils of the slopes are less leached, and downwash from the Greensand has lightened the heavier soils in the valleys. These warm, free-draining soils were sought out for vegetables near Potton and Sandy in the eighteenth century, but it was only after the building of the railway in 1848 that market gardening expanded greatly through a close connexion with London. Carrots, onions, parsley, beans, peas, cabbages and brussels sprouts were produced for Covent Garden, and in return the railway carried sweepings of the London streets, yard manure, soot and rags to maintain and improve the fertility of a hard-worked soil. Market gardening has spread to the sands and gravels of the Ivel valley with Biggleswade as the market centre, to the Ouse valley, and to southern Cambridgeshire. The area produces more than a quarter of the country's production of brussels sprouts, which have spread more widely than other crops since the sprout flourishes on heavier soils and can be easily integrated into the rotations of general arable farms. On the light soils, however, intensive market gardening on small holdings is normal. Fields are large and often unfenced, and holdings are frequently parcelled into small scattered plots. Farm buildings are few and concentrated near the modest houses in and near the villages, where the extent of Victorian and twentieth-century building gives a character quite different from that of most Bedfordshire villages. Farming is finely balanced, needing a liberal use of fertilizers, pest and weed-killers to maintain the production of two or even three crops a year. Since London's supply of manure has ended, the maintenance of organic content in the soil presents a problem, in spite of the continued use of rags and shoddy. Farms are too small to permit the land to lie 'unproductive' in green crops for manure, and farm buildings and fences are too few to permit mixed farming with livestock. Production would benefit from irrigation, but the possibilities of its expansion are very limited away from the farms with access to the Ivel or the Ouse, where it is already practised.

In farming and settlement the Chalk escarpment is distinctive, with some contrasts between east and west. In the west, steep slopes are associated with National Trust property, woodland and a pastoral land use, though town dogs have made parts of southern Bedfordshire an area in which sheep may no longer safely graze. The gentler slopes to the east are highly valued cereal land. Large, highly capitalized and highly mechanized farms laid out in vast fields with vestigial hedges, and often with new farm buildings, concentrate on wheat and barley, the latter often of malting quality and thus able to command a high price. Stock are few and rotations very flexible, for heavy fertilization, mechanization, pest and weed-killers have brought about a rejection of old values and of the tenets of the four-course rotation. Much of the Chalk land consists of the outlying parts of long, narrow parishes in which the villages may lie up to a couple of miles away near the spring lines at the outcrop of Totternhoe Stone or of Melbourn Rock, or on the boulder clay above the scarp, so that many of the remoter farms have been chronically short of labour. Early and rapid mechanization has thus been encouraged and even now the production of sugar-beet and other root crops is low because their heavy labour needs for singling and harvesting cannot be met. Large arable farms growing wheat, barley and clovers are characteristic, too, of the area below the escarpment where downwash from the Chalk, and in the west from the thin outcrop of the Upper Greensand, and the lime content of the Chalk Marl have contributed to the making of very fertile loams. These merge gradually into the heavy Gault clays, difficult to work as arable or pasture land.

Soils on the dip slope of the Chalk are very varied. Clays and loams on the Clay-with-flints are sometimes acid, but there is much mixture with brick-earths and pebbly gravels, and there is a long history of chalking and liming, so that generalization is impossible. The boulder clay plateau to the east had a reputation as the granary of Hertfordshire and for the fertility of its soils. Corn crops are now combined with substantial amounts of temporary grass, and farming is often oriented to dairying or fattening. Sheep are few. The proximity of London is to be seen in a variety of ways. New suburban building and the new towns are swallowing up agricultural land, and farming is sometimes disorganized by the threat of it. There are no less than 660 acres under glass in the Lea valley. Long-continued exports of meadow hay to London for horse fodder have left some of the alluvial soils of the area unusually poor in phosphates and potash. Large estates and imposing houses, built by men for whom accessibility to London was important before the opening of the railways, are perhaps less frequent in this area than to the west and south of London, but they have contributed greatly to the embellishment of a pleasantly undulating countryside. They have a modest counterpart in the post-war spread of part-time farming by business and professional men. Intensive farming and an undulating relief, frequent patches of woodland, often preserved mainly for game, and in the west, especially, a scattered settlement pattern reflecting late settlement, all combine to make the dip slope of the Chalk an area of great variety and charm, adequately equipped with amenities for the tourist and the semi-rural commuter because of its nearness to London.

London is of vital importance to the communications, industry and towns of the south-east Midlands as a whole, though the proximity of 'the Great Wen' may be partly responsible for the fact that no town within it has a population of more than 130,000. Roads and railways fanning out towards the industrial Midlands and the North dominate the small-scale map. What Morton said of Northamptonshire in 1712 still applies, and applies to the area as a whole. 'The County is the great Thorow-fare from the North: It lies in the Trade-way: and the furthest part of it is not seventy miles [*sic*] from the Grand Empory, the City of London.' So striking is the concentration of roads and railways, and their relationship to the growth of industry in a few places in the area, that some have been tempted to include it in a grand, generalized axial belt of industry from London to Lancashire and the West Riding. The industries of this area, though expanding, perhaps do not warrant such an exaggerated view of their importance. Three zones, however, may be recognized. The rapid industrial and urban growth of south-west Hertfordshire, Stevenage and Letchworth, belongs to the tale of London's expansion. Boots and shoes, clothing, light engineering and iron and steel are the major concerns of a second zone: the red-brick towns and overgrown villages along A6 from Rushden to Desborough, with Kettering and Wellingborough as the main centres, flanked by Northampton and Corby to west and east of this main axis. In the third, and intervening zone, industries have swollen the population of old market towns in a very uneven way. Bedford is growing rapidly; Cambridge has its flour-mills, cement, radio and television, precision instruments, and printing, and the villages around it are expanding with a new growth of industry, but Huntingdon and St Ives have been relatively untouched. The expansion of railway, brick and engineering industries at Peterborough, compared with Stamford's stability, is a familiar example of the accumulation of industry round the railway rather than the road in the nineteenth and early twentieth centuries. The contrast is equally striking between Stony Stratford, a linear settlement on Watling Street of old inns and new filling-stations, and Wolverton, the creation of the London and Birmingham Railway and once not only the centre of locomotive and carriage building, but also the 'grand intermediate station' of the line, with a Victorian notoriety for its refreshment-rooms. The carriage works survive.

Many of the industries of the area are concerned with engineering in its various forms, but some are related to older industrial traditions, farming or mineral resources. Flour-milling, malting and brewing were once more important occupations than they now are in the towns on the Ouse and its tributaries, and in east Hertfordshire on the roads to London, but many old mills have been converted to the production of other commodities, ranging from potato crisps to scientific instruments. Domestic industries were widespread over much of Northamptonshire, Buckingham and Bedfordshire from the seventeenth to the nineteenth centuries. Where straw-plaiting faded out to the north, the making of bonelace began. Luton, Dunstable and Hitchin were marketing and manufacturing centres for the former; Newport Pagnell and Wellingborough were famous for the latter. Northampton was well known for its leather and footwear in the eighteenth century, when Kettering and neighbouring centres still had a cloth industry of sorts. Straw and lace have left few vestiges, but Northamptonshire's boots and shoes and clothing industries are indirect legacies of the former domestic industries; and so are Luton's hatters, though they are far outnumbered now by workers making Vauxhall cars and other vehicles, manufacturing ball-bearings, and engaged in electrical engineering, chemical and other industries. Luton's population, now 122,880, has soared ahead of its former rivals, and it is the largest town of the whole area.

The mineral industries are perhaps of greater importance to the national economy or to London's needs than to the geography of the Midlands countryside as a whole. Most of the quarrying in the Chalk from Cherryhinton to Totternhoe is for the manufacture of cement, so that the marly horizons of the Lower Chalk provide optimum locations. Bedfordshire alone produces more than 10% of the national production. Some 13% of national production of industrial sands (for glass-making, moulding sands, refractory silica sand for open-hearth furnaces) is derived from a few hundred acres of pure 'silver sand', very free from iron-staining, on the Lower Greensand near Leighton Buzzard. The clays are no less valuable. Yellow or grey Gault bricks are characteristic of much of the pre-1914 housing to the south-east, but few pits are now open, for the Gault has yielded to the better brick-making qualities of the Oxford Clay since the end of the nineteenth century. For the railway traveller, indeed, the exposure of the Oxford Clay is marked by the forests of chimneys and acres of derelict land near Bletchley, south of Bedford, and at Peterborough, for these three areas make 35% of the national output of bricks. The 'knot' clays require little preliminary drying and much less fuel in firing than other brick clays because of a high proportion of combustible material in the clay. Large reserves, large-scale plant and mechanical methods, low costs and the proximity of main-line transport have enabled Fletton bricks to serve a national market as well as much of London's demands.

With an output of 5¼ m. tons in 1957, the iron ores of the Northampton Sand are by far the most important mineral product of the south-east Midlands. In spite of the low content of the ore (28–35%) a good deal of ore is still exported from the area to Lincolnshire and the north-east. Since the beginning of mining in 1853 at Blisworth, near canal transport, mining has spread north and north-east with the building of the railways, which provided new exposures as well as transport. The Midland Railway stimulated work in the Desborough and Wellingborough areas, and work began near Corby soon after the building of the Kettering–Manton line in 1879. Much of the older part of the field and the surface deposits are now worked out. In the new ore-fields of the north-east, it is necessary to remove up to 80 ft of overburden by specialized mammoth walking drag-line excavator. Since the Act of 1947 attention has been paid to the recovery of land for agriculture and there is a great contrast between the land left in 'hill-and-dale' by older operations, often derelict or thinly disguised by struggling plantations, and the barbed-wire fences, temporary leys and new farm buildings on newly reclaimed land. Although blast furnaces were established from 1857, notably at Wellingborough and Kettering, it was not until 1932 that Stewarts and Lloyd planned the integrated production of steel at Corby, using the basic Bessemer process since the ore is phosphoric (0·7%). Corby, now a scheduled 'new town' with 36,322 people has almost overwhelmed the village from which it takes its name. It retains a Scottish flavour from the origins of many of its early immigrant steel-workers, and is attracting light industry (food-processing, clothing, shoes, for example) to provide occupations for women.

The industrial structure of Northamptonshire has much in common with the East Midlands, just as many of Hertfordshire's towns are satellites of London. In the intervening zone industry is expanding, but the atmosphere of the market town is never quite lost, except at Luton and the railway settlements, and no single town dominates. Strongly influenced by the proximity of London, but not yet quite so metropolitan as Mackinder made out; varied in its farming, but not quite so uncompromisingly agricultural as East Anglia, the south-east Midlands seems fated, with its pleasing but unexciting landscapes to be an area through which many pass on their way to somewhere else. Mackinder's comment on a rather larger area: 'What gives the fundamental unity to the quadrant under consideration is the radial spread of the routes leading from the metropolis,' applies to the south-east Midlands, but is a thesis which, in detail, is singularly elusive to demonstrate.

Selected Bibliography

H. C. Darby (ed.). *The Cambridge Region.* Cambridge, 1938.

B. W. Sparks. 'Evolution of the Relief of the Cam Valley.' *Geog. Jour.*, **123**, 1957, pp. 188–208.

G. A. Kellaway and J. H. Taylor. 'Early Stages in the Physical Evolution of a portion of the East Midlands.' *Quart. Journ. Geol. Soc.*, **108**, 1952, pp. 343–66.

CHAPTER 8

THE UPPER THAMES BASIN

A. F. Martin

From White Horse Hill (856 ft), the highest point on the Berkshire Downs, the attentive walker may obtain a view over a large part of the central, most complete and most typical section of the English scarplands. To the south and south-east he looks over broad, bare ribs of Chalk, dipping insensibly beneath wooded Eocene deposits and down into the Kennet vale, until on the southern horizon, some 20 miles away, the Chalk anticlines of Kingsclere and Inkpen (975 ft) stand up abruptly on the northern rim of the Hampshire Basin. Westwards and eastwards he looks along the prehistoric Ridgeway and the steep scarp edge of the Chalk from the Marlborough Downs beyond Swindon to the Chilterns, visible behind the tiny buttes of the Sinodun Hills placed like sentries in the approach to the Goring Gap. Northwards and north-westwards the view is widest, plunging down the scarp, over the narrow Greensand bench at its foot, across the broad tree-studded Vale of White Horse 600 ft below, to the low ridge of the Oxford Heights picked out by the distinctive pine-clump of Faringdon Folly and capped farther east by the larger lump of Boars Hill (525 ft) which hides the spires (and gas-holders) of Oxford itself; behind the Oxford Heights he must divine the wide Oxford Clay vale of the upper Thames, but beyond that again he can see the lower Cotswold slopes rising towards an ill-defined horizon, near which on a clear day the Leafield wireless masts (600 ft) serve as a landmark 18 miles to the north.

The simple pattern of parallel scarped ridges alternating with clay vales, observed from this or many similar belvederes along the Chalk crest, may be traced, now with more, now with less distinctness, the whole way across England from Dorset to the North York Moors. Nowhere, however, is it more beautifully exemplified than in the relatively small sector forming the subject of this essay, drained mainly by the upper Thames above Reading, and extending from the foot of the eastern Cotswolds and of Edge Hill in the north to the

Kennet valley in the south, and from the Wiltshire boundary in the west to the neighbourhood of Henley, Aylesbury and Buckingham in the east.

This upper Thames country lies slightly west of the centre of the English Plain, and Oxford, its focus, is almost equidistant from the great cities of London (52 miles), Birmingham (58 miles), Southampton (60 miles) and Bristol (60 miles). In spite of its centrality and partly, indeed, because of its position midway between these busy commercial and industrial centres, it was until quite recently avoided by the main streams of modern economic life and remained remote and backward.

'The region, in short, is a typical scarpland,' and 'the keynote of the South-Eastern Midlands is its essentially rural character.' In these two sentences H. O. Beckit, in 1928, summed up the area, and his verdict, though still commanding respect today, should be re-examined, and qualified if necessary, in the light both of altered facts and of modern theories.

This scarpland (fig. 16) is, as has been said, the most complete and typical in these islands, though the cuestas do not equal in regularity or abundance those of eastern France. Four persistent escarpments, those of the Marlstone (Middle Lias), the Oolites, the Corallian, and the Chalk, dominate the whole area, wherein a gentle south-easterly dip prevails until, at the very southern limit, in the steep southern limb of the Kennet valley syncline the Chalk comes sharply up again. Other minor and shorter cuestas intervene and complicate the simple plan; such are the Malmstone (Upper Greensand) bench at the foot of the Berkshire Downs, the Lower Chalk bench at the foot of the Chilterns, and the fragmentary Portland scarp near Nuneham Courtenay. Of the main English scarpformers only the Great Oolite, although resistant and thick, fails, in the sector between the Evenlode and the Cherwell, to give rise to an obvious escarpment.

The scarpfaces of these cuestas are usually abrupt and straight; fretted scarps and outliers are few, though Blewburton Hill and Brailes Hill are perfect examples. Most of the dip slopes are deeply dissected but, especially in the Jurassic cuestas, their seemingly horizontal surfaces dipping perceptibly but uniformly and gently south-east can be strikingly seen in many places, for example on the A34 road near Chipping Norton, on A40 near Burford, or on A41 north of Banbury. These flat plateaux have a slope of around

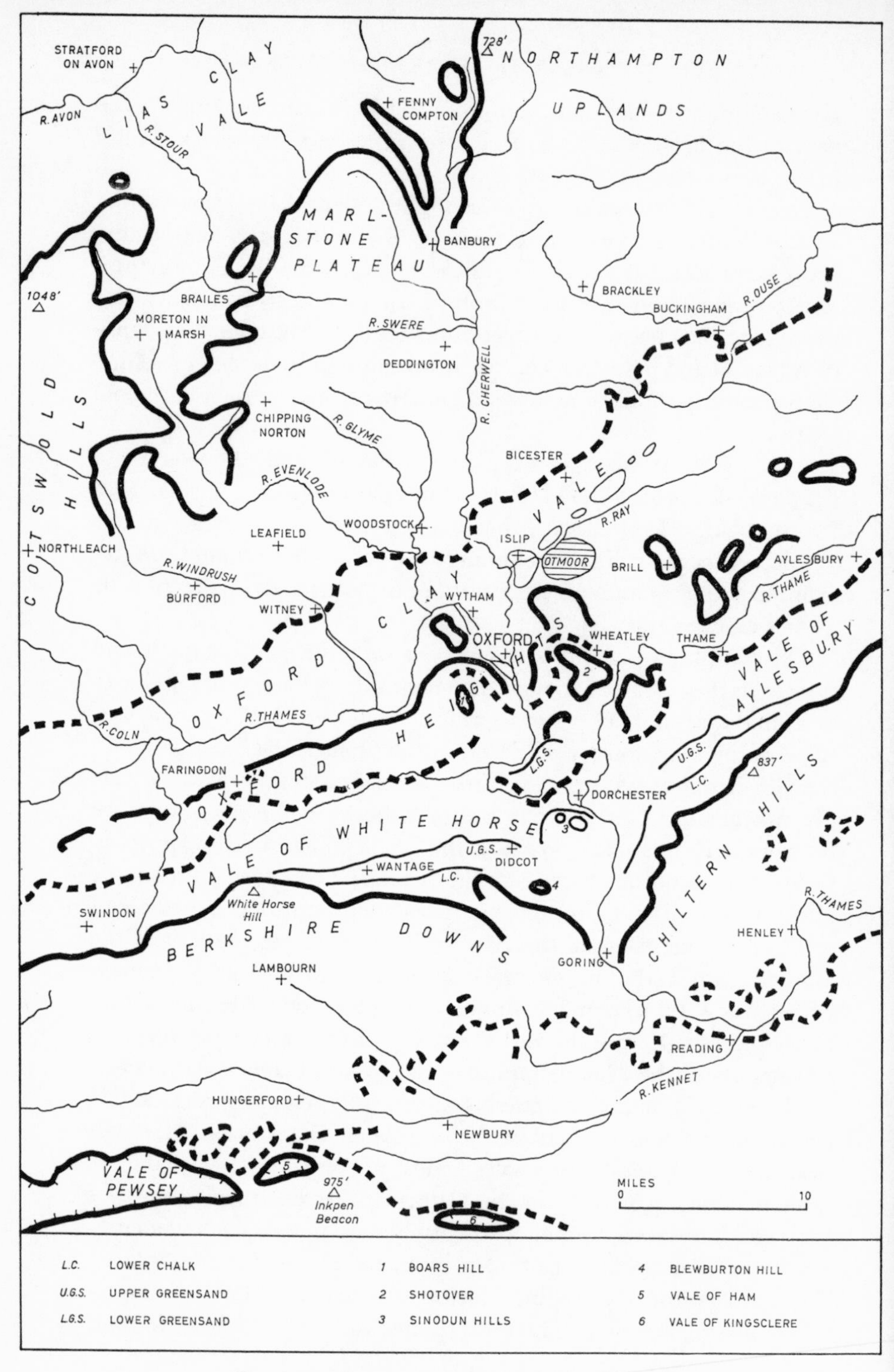

Fig. 16. The scarplands of the upper Thames basin.

40 ft to the mile, slightly less than the general dip, but it has not been possible, outside the London Basin, to relate the altitudes of their various remnants to each other in any significant way. All the plateau surfaces, however, are close to the inclined beds of resistant limestones, and they are found at almost all altitudes from over 900 ft on the road to Broadway in the north-west down to 300 ft or even less near the Thames. Only on the Chilterns and the Berkshire Downs is the supposed Pliocene marine bench visible at about 600 ft.

The pattern of rivers draining this corrugated surface shows many of the traits traditionally associated with a scarp and vale relief, and indeed W. M. Davis and later S. S. Buckman drew largely on examples from the upper Thames basin in expounding their theories of the significance of a roughly rectangular river-pattern. 'No explanation of the physiographic history', wrote Beckit, 'is as adequate or comprehensive as that which regards the Cherwell–Thames as the master consequent stream which has collected the waters of a number of other consequents, such as those which drain the long back slope of the Cotswolds, by a process of successive captures effected by the retrogressive erosion of longitudinal subsequents along the strike of weak rocks.'

The passage of over thirty years since Beckit wrote has not cast doubt on the broad lines of his account of the relief and river pattern, but geographers, taking its basic truths perhaps too much for granted, have turned their attention more to the complexities which make his by now traditional descriptions seem a sufficient account of neither relief nor rivers. 'The term "simple scarpland" [is] peculiarly inapt,' says R. P. Beckinsale; while of the rivers, K. S. Sandford writes: 'It is easy to recognize "consequent" dip-slope rivers and "subsequent" strike-streams. It is much more difficult to decipher how these conditions were created.' Today one would wish certainly to supplement and perhaps to revise Beckit's account by stressing two principal considerations: first, complications of the stratigraphy, lateral changes in thickness and lithological variations, flexuring, faulting, cambering and mass-movements; secondly, the effects of climatic changes on the processes at work on the scenery, and in particular the unerased imprint of glacial, periglacial and interglacial conditions in the Pleistocene period.

Nothing is more noteworthy, in this as in other scarplands, than the way in which the details as well as the broader lineaments of the geology are reflected both in the scenery and in the geography of man. Minor structures and changes of rock type along the strike affect practically all the strata and produce differences of scenery between neighbouring sectors of their outcrops. Thus the Oolite limestones of the Cotswolds proper, south-west of the Evenlode, with their well-developed dry valley systems, brown soils, dry stone walls, and pale buff-coloured buildings, differ entirely from the Marlstone plateau of north Oxfordshire, which takes their place in the escarpment east of Chipping Norton and north of Deddington, with no dry valleys, bright red corn-growing soils, hedgerows, ironstone workings, and warm almost orange building stone. The southern part of this plateau and the neighbouring edge of the Inferior Oolite near Deddington and Rollright are cut by a swarm of faults into strips, separated by large east–west valleys such as those of the Swere and upper Stour. Less than 5 miles north of Oxford, the small Charlton on Otmoor anticline trends north-east from the Cherwell; here the Oxford Clay and alluvium are punctured by a line of elongated Cornbrash islands which make the northern arable rim of the Otmoor bowl and the sites of the Otmoor villages. Nearby the transverse Wheatley–Islip fault zone not only shows wonderful small-scale examples of 'subsequent' hollowing out of clay-filled synclines through short limestone canyons, but also encourages a covey of villages (including Elsfield with its unsurpassed view of Oxford) and the once important London–Islip–Worcester coach road to venture northwestwards across the Kimmeridge and Oxford Clay vales, here reduced to a fraction of their normal width.

Many of these geological complications and their consequences are matters of local interest only. There are, however, others of regional import. As a broad generalization each of the great belts of rock may be said to show remarkable differences to either side of a north–south line close to the Cherwell–Thames. The Kennet valley, 'that most unmetropolitan western extremity of the London Basin', widens rapidly east of Reading into the Basin proper. The Chalk, which in the Berkshire Downs gives bare open country of wide arable horizons and but scattered windbreaks, is, in the Chilterns, plastered with superficial deposits and clothed with beechwoods to the crest. The Gault and Kimmeridge Clays overlap west of the Thames to form the broad Vale of White Horse, but to the east in the doubtfully subsequent valley of the Thame they are separated by the fragmentary ruin of a cuesta of Portlandian and Lower Greensand hills around Great Hazeley and Nuneham Courtenay with outliers well to the north in Brill and Shotover Hills. The Corallian Oxford Heights, on their low but persistent limestone crown, carry the main Oxford–Swindon road westwards past the crest-line villages from Cumnor to Faringdon, but not far east of the city the limestone thins out and the ridge dissolves into isolated hillocks, capped by patches of Wealden and Portland beds, rising amidst a curiously remote and deserted clayey tract. The Oxford Clay, by far the thickest of the formations, extends its vale more or less uniformly across the whole width of this district; but whereas to the west of the Cherwell it is largely veiled by spreads of oolitic gravels occupying up to a quarter of its outcrop, to the east these gravels, together with their light soils, dependable wells and large villages, are almost totally absent, their place being in part taken by the intractable marshy enigma, Otmoor, and the limestone islands of the Charlton anticline. Finally, a mile or so east of the Cherwell the main Jurassic cuesta itself suffers a radical change: the Marlstone Rock, which crowns the scarp at Edge Hill, disappears and the cuesta disintegrates into a gently dissected upland only 400–500 ft high. The Oolite plateau farther south is continued east of the Cherwell; indeed, between Brackley and Bicester it is better preserved between 400 and 250 ft than anywhere else, but it has neither the deep dissection nor the clear scarp of the Cotswolds. As though to emphasize these changes a patchy mantle of boulder clay begins to appear along roughly the same line or a little farther east, transforming the landscape into one of undulations and shallow drift-encumbered valleys.

It is not only, however, in this small drift-covered eastern strip but throughout the whole area that the events and conditions of the Pleistocene have clearly exercised a more decisive influence than seems to have been appreciated thirty or more years ago (fig. 17). Elsewhere (except for the boulder clay of Moreton in Marsh and some of the 'plateau gravels' farther south) Pleistocene deposits are limited to more or less water-sorted gravels. None the less the effect of the glacial and interglacial phases not only on the superficial deposits and soils, but more particularly on the incision of the drainage system, has been profound. The modern rivers do not appear to deposit anything coarser than silt. The terraced gravels which fringe the lowest Cotswold slopes and mask much of the Oxford

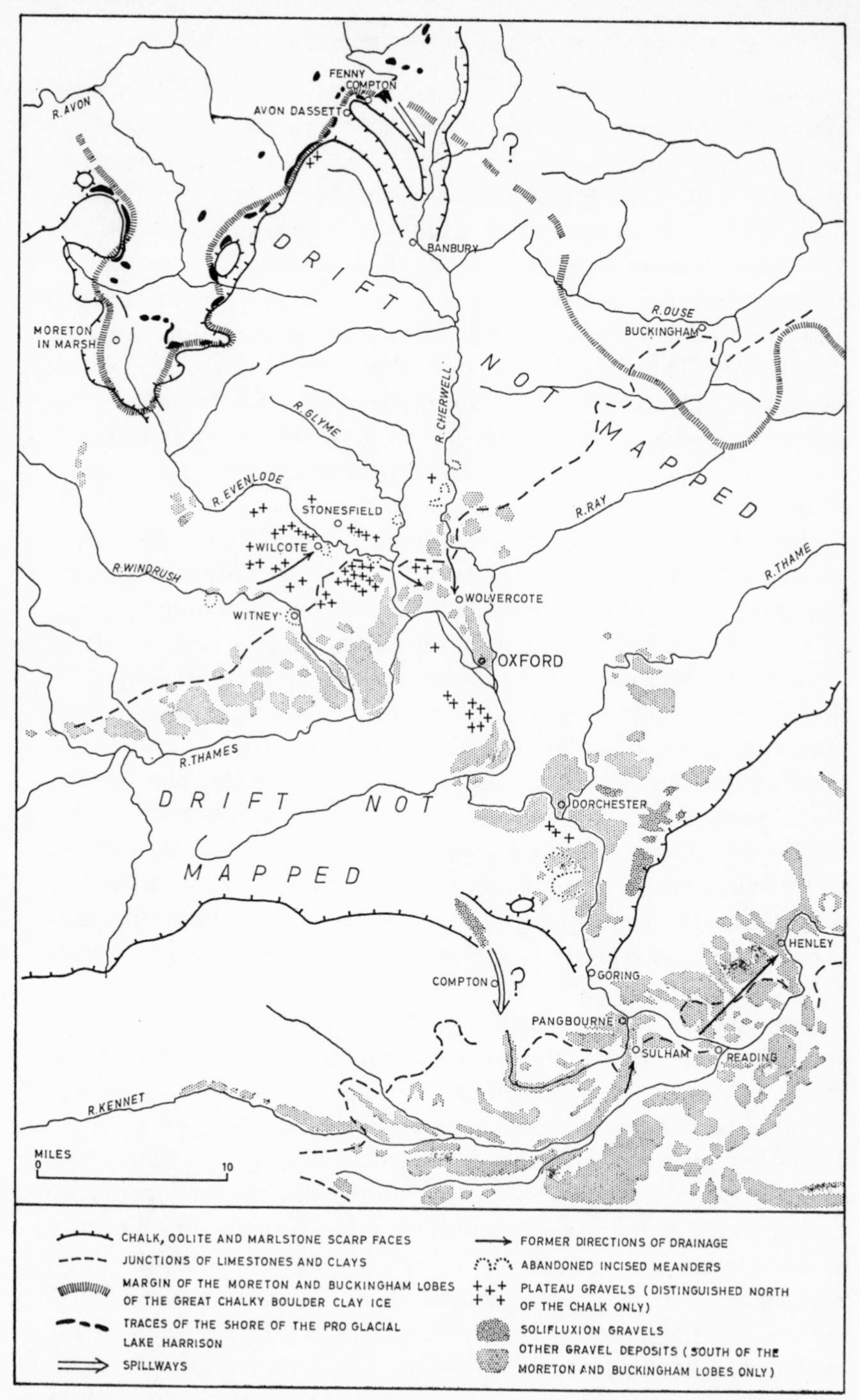

Fig. 17. Some Pleistocene landscape features of the upper Thames basin. No attempt has been made either to include any drift north or east of the Great Chalky Boulder Clay Ice margin, or to distinguish between 'plateau' and other gravels in the London Basin.

Clay are the remains of extensive fans deposited by the Pleistocene Cherwell, Evenlode and Windrush which were overloaded with debris formed under tundra conditions and transported and resorted by copious streams from Cotswold snowfields. Likewise in the Kimmeridge–Gault vale there are wide gravel spreads along the Thames itself and particularly around the lower Thame near Dorchester; in the Kennet valley, again, great flint-pebble terraces fringe the Chalk hills. These gravels are arranged in distinct terraces corresponding probably as much to phases of overloading and unloading of the contemporary rivers as to possible movements of base-level. All the gravels immensely affect not only the relief but the vegetation and settlement of the vales in which they lie, the oolitic gravels giving dry barley-growing soils, the flints heath and scrub. Finally, some of the varied surface deposits which give character to the Chilterns as well as the fan (tjaele) gravels of the Chalk slopes are attributable to colder climates than the present.

Of equal interest to the physiographer (though perhaps less obviously to the human geographer) are several features of the drainage system, in the understanding of which also change of climate may play a part. Earlier writers, though stressing the evolution of consequent and subsequent streams, devoted relatively little attention to one aspect of this evolution well illustrated in the Oxford area, namely the down-dip migration of subsequents and the recession of scarps. This recession is well exemplified all along the Chalk escarpment where magnificent cirque-like coombes (of which the Manger at the foot of White Horse Hill is an impressive example) bite into the scarp face, the work probably of nivation and certainly of headward sapping by springs. Such sapping has been and still is active in many other places, particularly where a relatively thin limestone capping protects upstanding clays. Cambering of the limestone, slumping and, less commonly, short underground streams also occur. These processes may readily be seen at work in the Corallian-capped hills east and west of Oxford or in the Marlstone plateau near Banbury; there are good examples close to Deddington Castle. It is interesting to note that in these conditions the scarp-line settlements are often found near the crest, as in these two areas. Where the scarpformer is thick, as in the Downs or the Cotswolds, the villages are usually at the foot.

These scarp attacking processes are relatively independent of the drainage pattern, but may in the right conditions be assisted by the scarp sharpening action of down-dip lateral shifting of subsequent rivers. This can be demonstrated near the only two sizeable subsequent portions of the large streams, the Thames above Oxford and below Reading. Above Oxford the shifting has been hastened by the building out of gravel spreads from the Cotswold foot, which has effectively pushed the Thames up against the insignificant Corallian scarp. Landscape evidence for such down-dip migration is provided by abandoned river courses such as the former channel of the Thames direct from Mapledurham to Henley, that of the lower Cherwell via Yarnton to the Thames near Wolvercote, or the presumed ancient high-level course of the Windrush from Crawley to the Evenlode valley near Stonesfield betrayed by a very fine dry suspended meander at Wilcote. More striking still are occasional examples of the prevention of this process where, by some accident of stratigraphy, a river, trapped in a resistant stratum, has become suspended, as it were, higher up the dip-slope than it should be, in a course which has now all the characteristics of superimposition. The almost west–east 'gorge' sections of the Windrush and Evenlode, with their well-known incised and abandoned meanders, and the anomalous Henley loop of the Thames curling back into the Chalk which it had practically quitted at Reading are surely examples of this.

The most far reaching of the effects of Pleistocene climatic conditions is the excavation of the vales to their present level below the scarps and the incision of the tributary valleys into the plateaux, the work of the more effective streams of that period. The earlier Pleistocene water-laid gravels, near the mouth of the Evenlode gorge, lie up to 175 ft above the present rivers; the steep incision of most of the valleys below the plateau surfaces reaches almost the same figure. The latest Pleistocene terraces lie but little above the modern alluvium. The landscape first invaded by glacial conditions must therefore have been much more subdued, even allowing the scarps to have retreated slightly south-eastwards since that time. The preservation at Edge Hill both of early Pleistocene gravels on the brink and of a Chalky Boulder Clay pro-glacial lake bench at the foot of the Edge leads one to doubt whether this scarp has been appreciably either lowered or cut back. By the end of the Pleistocene the relief of the valleys was almost as it is now. The modern rivers are doing little physiographic work. Likewise, even after allowing full credit to the efficacy of sapping by springs and underground watercourses, and of the lowering of the water table, it is likely that most of the dry valleys both in the Chalk and the Cotswolds were excavated under conditions in which frost and meltwater floods played a large part. Finally, former more copious run-off may explain the misfit character of the principal streams, while over-spill from a Midland pro-glacial lake through the Fenny Compton gap may account for the rapid down-cutting of the Cherwell valley, leaving the nearby tributaries of the upper Ouse perched high above it only a mile or so to the east as at Somerton or Farthinghoe. The same agent could have cut the gorge which forms the lower 200 ft of the Goring Gap, the base level for the whole basin, though A. Austin Miller revives the suggestion that the pre-Glacial upper Thames flowed north-east to the Wash, and that the Goring Gap is also a spillway from another lake, Lake Oxford, impounded by Eastern Ice advancing up the Oxford Clay vale. The nearby Compton Gap, followed by the Didcot–Newbury railway, and the abandoned course of the Kennet via Sulham to Pangbourne remain difficult to understand unless one may suppose ice at some stage to have blocked the Goring Gap.

The general effect of Pleistocene conditions on the landscape has thus been largely to reinforce the aforementioned west–east contrast: deep dissection of the pre-Glacial landscape west of the Thames–Cherwell line with wide spreads of oolitic gravels over the clay, and, to the east, preservation of the ancient subdued surface, accentuated by the deposition of till.

Re-examination of the physiography, even if it does not confirm in detail the traditional account of the evolution of the drainage pattern, does little to destroy the belief that this is a typical, though not perhaps a simple, scarpland. There does, however, emerge from it a pervading contrast between east and west: sharper relief to the west, greater prevalence of clays to the east; light arable soils, long cultivated and thickly settled in close-set villages and market towns built of stone and stone-tile in the west; to the east stiffer soils, shunned ever since Roman times, still heavily wooded and remarkably isolated, with rarer villages of timber, brick and thatch.

These contrasts between east and west provide a framework for a sub-division of each of the great parallel belts of differing rock and scenery. Of the dozen or so resulting smaller districts, none is without internal diversity attributable more often than not to geology, for in few parts of Britain is the influence of rock on landscape, history, settlement and occupations more remarkable. Almost everywhere this influence is still exerted, as it always has been, through the processes and needs of rural life, on the essential dominance of which Beckit was still insisting in 1928. 'The greatest changes of recent date', he continued, 'have arisen out of the growth of lines of through communication forming, as it were, so many bridges across a still rural, non-metropolitan, non-industrial zone which in itself seems like a surviving fragment of an England of the past.'

These words have since been both confirmed and falsified. The rural landscape indeed, still prevails over by far the greater part of the area. Oxford (106,124), Reading (119,870), and Banbury (20,996) are insignificant both in area and in population in relation to the whole. Remote and isolated areas there are, north of Chipping Norton, on the western Berkshire Downs, and even, interestingly enough, immediately to the east of Oxford around and to the east of Otmoor, but the remoteness and isolation are relative only and the effect of the motor car, rural electricity and television, measured in absolute terms, has been enormous.

On the other hand the importance of through communications has in the event exceeded Beckit's expectations for he was then referring to the main-line railways crossing the grain of the country. Railways, however, have affected the greater part of the area remarkably little compared with the country farther to the east. Only in the south, at Reading, in the Kennet valley, at Didcot and at Swindon, and in the north at Banbury did they have a marked effect. The growth of road transport rather than of rail has transformed the district even more thoroughly than most other parts of the country.

The effects of this growth are far from simple. The most direct has been to convert Oxford from a minor (rail) into a major (road) nodal centre (fig. 18). The crossing of routes that the railway makes at Reading, on the very margin of the upper Thames basin, is shifted by the road network to Oxford, at its centre. 'The city', says Thomas Sharp, 'is probably the most important nodal-point in the road system of southern England outside London. Here the road between London and the lowest bridging-point of the Severn at Gloucester crosses the fork of the Y made by the roads from the east Midlands and west Midlands where they join to proceed to Southampton.' The recent growth of Midlands–South Coast traffic, stimu-

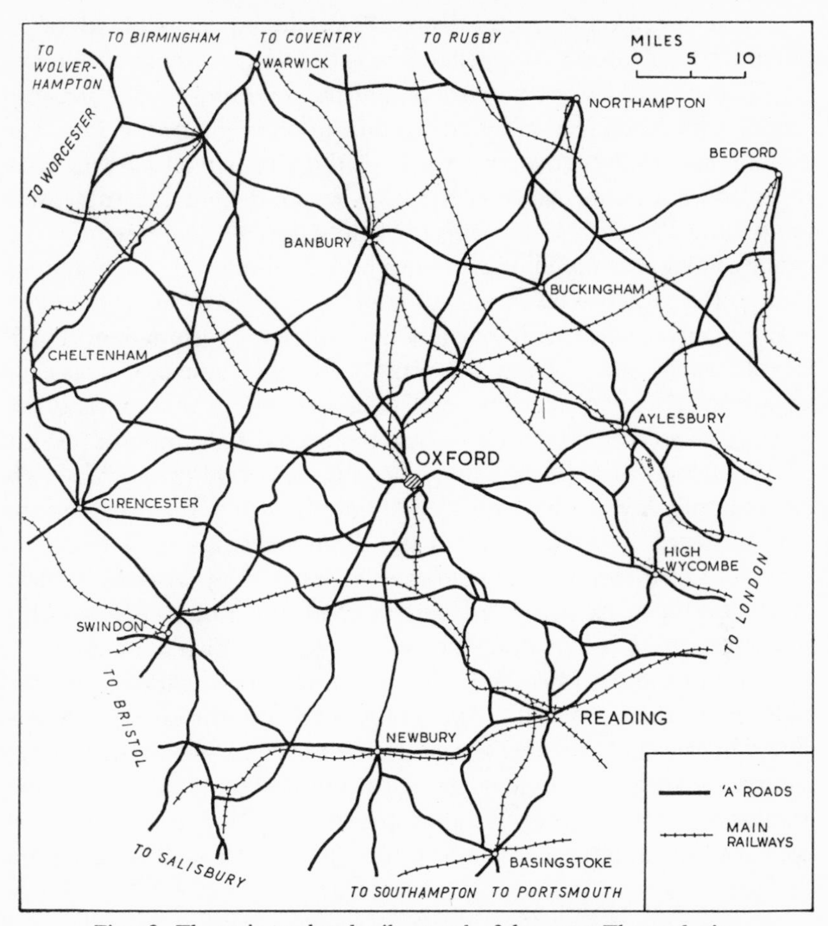

Fig. 18. The main road and rail network of the upper Thames basin.

lated by and stimulating the growth of Southampton, is of especial importance. Radial routes from London are much less closely spaced, and less busy than in the south-east Midlands, crossed by A1, A6, A5, A41 and now M1, as well as by all but one of the main railways to the North or Birmingham. Only five radial arteries from London cross the upper Thames basin and only three of these, A40 to South Wales and the Worcester and Bristol main railways, pass squarely through it. For north–south movement, however, Oxford is of crucial importance. The network is as though designed to concentrate vehicles on a central trunk (A34) between Oxford and Newbury. Traffic bound for Bournemouth, Southampton or Portsmouth, from a sector of quite ninety degrees (stretching from Wolverhampton almost to Bedford) is funnelled through Oxford by the need to cross the Thames and by the absence of alternative routes to east or west.

The congestion cause by this influx of through traffic into the already crowded central crossroads of the city is well known; it is repeated on a smaller scale at Newbury where the north–south trunk crosses the Kennet. Other effects of the suddenly increased significance of Oxford's nodality are of less parochial interest. First among these is the growth and stabilization of her motor-car and attendant manufactures, with an important dependency in Abingdon. Their establishment here, away from the accepted industrial areas, is usually assigned to accident, but their expansion and survival for nearly half a century (as well as the almost parallel example of Luton) may well be evidence of a more than coincidental connexion with an unusually favourable web of road links by which the products are mainly distributed and many of the components assembled.

Other industries, too, have arisen to spatter the prevailingly rural landscape in spots mostly related to road and rail facilities. Aluminium fabrication at Banbury emphasizes the town's superior rail service as do the service and commercial depots at Didcot, and the older biscuit-making and brewing of Reading, and printing at Aylesbury; but most of the other recent growths whether of manufacturing, quarrying for gravel, lime or stone, or commerce and administration (including atomic energy establishments at Aldermaston and Harwell, and a market research agency and publishers at Oxford) would

TABLE 2. *Percentage increase of population (1931–51)*

	Oxford-shire	*Berk-shire*	*England and Wales*	*No. of counties with higher %*
Administrative counties, including county boroughs	31·5	29·4	9·5	6*
Administrative counties, excluding county boroughs	37·0	34·7	21·0‖	7†
County boroughs (Oxford and Reading) ..	22·5	17·5	1·6‖	4‡
Municipal boroughs and urban districts ..	26·3	20·8	21·4	14
Rural districts	41·4	42·9	20·0	3§

* Hertfordshire, West Sussex, Buckinghamshire, Bedfordshire, Middlesex, Surrey.

† Hertfordshire, West Sussex, Surrey, Buckinghamshire, Bedfordshire, Warwickshire, Middlesex.

‡ Coventry, Blackpool, Doncaster, Bournemouth.

§ Hertfordshire, West Sussex, Lancashire.

‖ Excluding London.

be unthinkable without good road transport if only to allow the housing and assembling of their staffs.

Two further interrelated results of improved road transport are illustrated in Table 2. The two counties which make up the major part of the upper Thames basin have shown high rates of population growth surpassed by only one other county outside the penumbra of London. In particular they have experienced exceptionally rapid residential development in areas outside the towns. These high rates may be attributed partly to the general influence of superior nodality by road on opportunities for industrial and commercial employment, partly to the growing habit of living in pleasant rural surroundings, made possible by the motor-car and by fast suburban train services, affecting in particular east Berkshire and Henley.

This growth and spreading of population are not uniformly distributed, mainly for reasons of accessibility. There are parishes of exceptional growth, namely those near new or growing establishments (Harwell, Didcot, Brize Norton and Upper Heyford), those close to expanding towns (Oxford, Reading, Abingdon, Swindon), and those nearest to London. There are also parishes, less accessible from the main centres, in which population is decreasing or static, mainly in north Oxfordshire on or near the Marlstone plateau, and in the remoter parts of the Chalk cuesta and of the Kimmeridge and Gault Clay vales.

Finally road communications affect the relation between the towns and the surrounding country and between one town and another. Railways are important, especially as a determinant of the extent of London's predominance, but road transport has been the leading force in the latter day suppression of smaller by larger centres. The towns serving the district considered here, with their respective areas, are shown in fig. 19. They fall clearly into three categories: major, growing centres (Reading, Oxford), intermediate centres (Banbury, Newbury), and minor, relatively declining centres (Chipping Norton, Moreton in Marsh, Bicester, Thame, Wantage, Wallingford, and Henley). Oxford, with its subsidiaries Witney and Abingdon, serves the largest area, but Reading serves a larger population. A grouping round the two main centres naturally suggests itself, Reading with Newbury and Henley (and possibly Wallingford) in the south; Oxford with most of the minor centres farther north; but Banbury is left in an equivocal position perched between

Oxford and the Midland Plain. Other criteria, not so obviously related to road transport, seem to point to a somewhat similar area, widespread to the north but hemmed in on the south by the Downs and Chilterns as tributary to Oxford.

These inquiries and this map bring one back to the relief and geology with which this essay started, and in particular to the barrier of the Chalk cuesta. The same barrier has another function for which Oxonians are grateful, for it is a real obstacle to the creeping fringe of outermost London. Farther east, near Princes Risborough and Aylesbury, the barrier, weakened by wind gaps, has been breached, and commuters start from even as close as Thame. The time may come when suburban trains push beyond Reading and London coaches beyond High Wycombe, but at present the beechwoods and the bleak Downs still hold them back. Oxford and its environs remain insulated from and independent of the metropolis in a way in which Reading or Luton are not.

The scenic pattern of the upper Thames basin is one of geologically controlled south-west to north-east strips, each divisible into east and west sections along the Thames–Cherwell line. Even Nature has refused to be bound by this framework and has, hydrographically speaking, magnificently united a section of this scarpland by means of the Thames and its tributaries. Centuries ago Man, too, had begun to break down these geological boundaries by means of rows of villages and market towns established to exploit and combine the differences of soils and products on their either side. Thus, on the small scale, rural economy unites what geology clearly divides. On the larger scale, however, political, social and economic forces were insufficient to bind together the whole group of districts. The upper Thames basin has throughout most of its history been debatable ground between political or economic powers centred outside it, whether between Verulamium, Corinium and Calleva, between Mercia and Wessex, or between London and Birmingham. Oxford, although the ecclesiastical centre, the home of a university of nation-wide connexions, and the seat of several Parliaments, was unable to impose such economic or social unity. Only in quite recent years, since the advent of good road transport has endowed the city with an effective nodality it never had before, has Oxford been able so to extend its influence as to fill up the frame built for it by its physical setting.

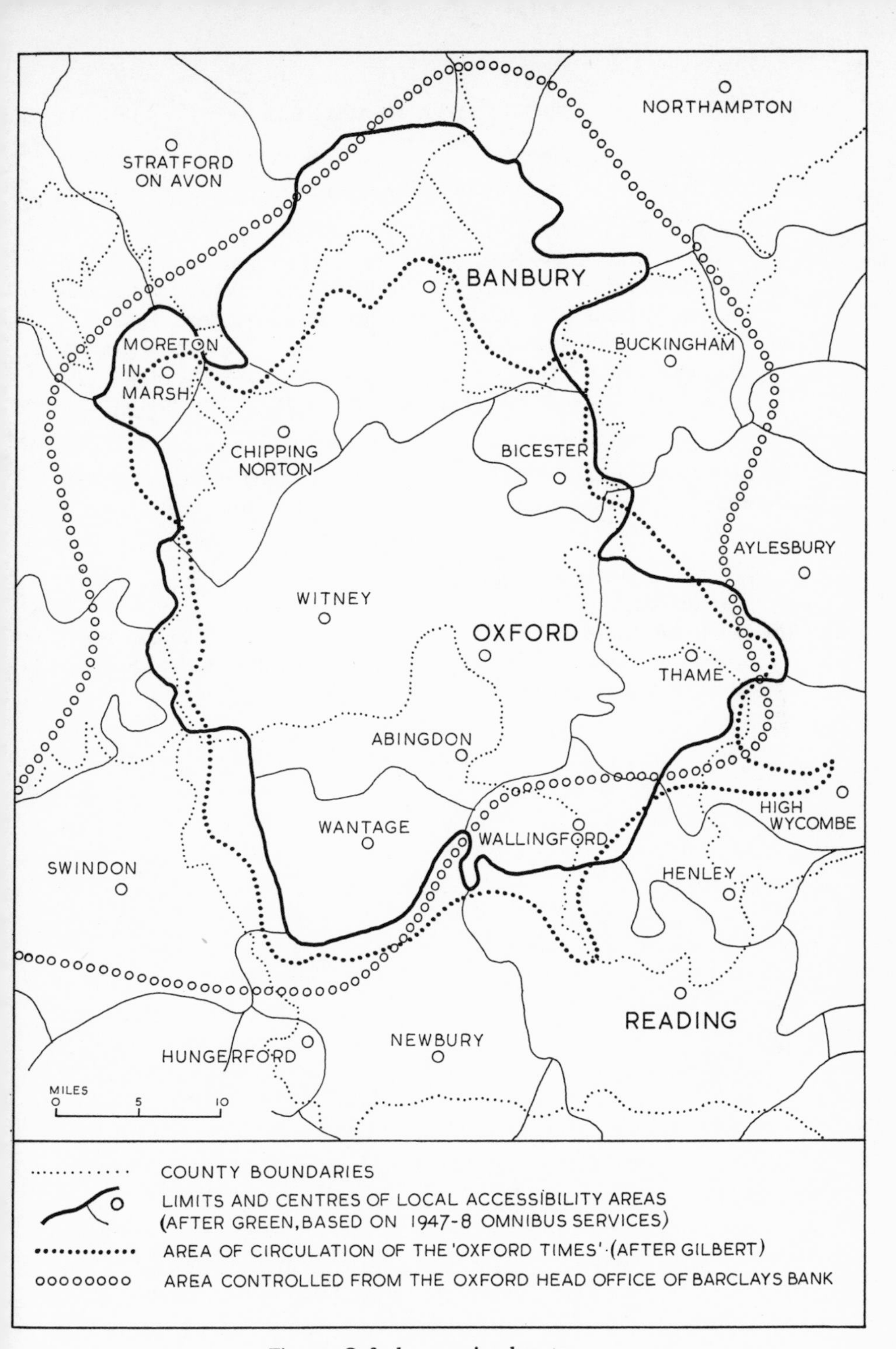

Fig. 19. Oxford as a regional centre.

Selected Bibliography

W. J. Arkell. *The Geology of Oxford.* Oxford: Clarendon Press, 1947.

A. F. Martin and R. W. Steel (ed.). *The Oxford Region: a Scientific and Historical Survey.* Published for the British Association for the Advancement of Science, Oxford: Oxford University Press, 1954.

E. W. Gilbert: 'The Industrialization of Oxford.' *Geographical Journal,* **109**, 1947, pp. 1–25; 'Reading: its Position and Growth.' *Trans. of the South-Eastern Union of Scientific Societies*, **50**, 1934, pp. 81–90.

The first two works include a large number of references to further literature. More exhaustive bibliographies may be found in:

(*a*) J. Sölch. *Die Landschaften der Britischen Inseln.* Wien: Springer, 1951, pp. 460–64.

(*b*) E. H. Cordeaux and D. H. Merry. *A Bibliography of Printed Works relating to Oxfordshire.* Published for the Oxford Historical Society, Oxford: Clarendon Press, 1955.

CHAPTER 9

SOUTH-EAST ENGLAND

B. W. Sparks

In many ways south-east England, as defined here, possesses a certain measure of uniformity and differs appreciably from surrounding areas in both its physical and human geography. To the east and the south lies the sea. On the west the Portsmouth Road, A3, makes an admirably distinct boundary in the south, while farther north the chalk escarpment of the western Weald makes a tolerably neat limit. On the northern side no precise limit may be set: it is obviously desirable to include the North Downs in south-east England as a physical unit and equally desirable to include the dip-slope in the London basin for it is a part of London's closer suburban area and parts of it are served by the denser part of London Transport's network.

To seek significant climatic differences is fruitless: the area is probably a little drier than the Hampshire Basin to the west, a little wetter than the London Basin to the north, a little more extreme in its temperatures than the areas west of it and a little less extreme than those to the north. But the differences are so small as to be negligible and are probably exceeded by internal differences in the region, so that climate does not differentiate this area from the rest of the country.

In structure and relief, however, the south-east has many claims to be classed as distinct. It shares with the whole of southern England certain structural features: it is a region of dominantly east–west structural trends which extend right through to the south-west and to Wales. More exactly, it is a zone of short, arcuate, asymmetric, periclinal folds arranged in echelon and of middle Tertiary (Alpine) Age. This structural characteristic it shares with the neighbouring region to the west, for zones of more intensive folding extend through both the Hampshire Basin and the south-east: for example, the zone of folds, which includes the Ham and Kingsclere periclines, continues into the Hog's Back region with its reversed fault and the important Peasemarsh anticline, while an important series of folds through Stockbridge and Winchester continues in the Weald through Hindhead and Fernhurst. These folds apparently disappear in the Weald Clay lowland largely because lithological division of this thick bed has only been attempted in recent years and folds cannot be mapped when the strata are not divisible and are not picked out by erosion unless they affect beds of differing lithology. They apparently reappear in the central Weald in the sandstones with subordinate clays of the Hastings Beds and are associated there with significant east–west faulting.

Yet, in spite of the continuity of these belts of folds, there is the world of difference between the updomed Weald and the downwarped Hampshire Basin, the one exposing in its eroded centre the sandstones and clays of the Lower Cretaceous and the other preserving in its lower parts the gravel-covered plateaux of the Tertiary beds. The differences between the south-east and the London Basin are even greater, for, in addition to the contrast between up-warping and down-warping, there is the absence in the London Basin of east–west folding. Even in the Weald such folds are not characteristic of the North Downs and, indeed, are not often found north of the centre of the Weald except in the west. These differences reflect differences in the thicknesses of post-Palaeozoic rocks: they are thick enough in the south to have been folded in the Tertiary, while to the north they are thinner and only reflect the movements of faults in the underlying Palaeozoics as slight monoclinal structures.

No other region of England has a development of Cretaceous rocks comparable with that of the south-east. The Hastings Beds and the Weald Clay are virtually absent elsewhere, though their equivalents in time if not in lithology may be found in Yorkshire. The Lower Greensand of East Anglia and the south Midlands is a thin, poor relation of the varied series of rocks in the Weald, while, around Weymouth, where the Lower Greensand might have been expected, it is missing because it was a period of folding locally.

The Hastings Beds in the centre of the Weald form a tract of uplands extending from the Channel coast near Hastings westwards almost as far as Horsham, a distance of some 45 miles, while they attain a maximum width of some 15 miles. They consist primarily of two important sandstone formations separated by a clay: at the base are the Ashdown Sands, in the middle the Wadhurst Clay, the source of the low-grade ironstones which formed with water power and charcoal the bases of the former Wealden iron industry, and at the top the Tunbridge Wells Sands. But the sandstones have clay horizons and the clays contain some sandstones. Folding and faulting ensure repetitions of the outcrops and these have been acted upon by erosion to give the dominant east–west trend found mainly in the east of the region. Here, longitudinal streams, such as the Brede and the East Sussex Rother, separate the east–west ridges. Farther west, the drainage is mainly effected by such streams as the Medway, the Ouse, the Adur and the Mole, themselves generally transverse streams though their headwaters may be longitudinal. The highest ground is in Ashdown Forest between East Grinstead and Uckfield, but it falls just short of 800 ft and is, thus, lower than both the Chalk and Hythe Beds escarpments. The region as a whole gives an impression of complexity, partly because it is not a simple escarpment and dip-slope, and partly because of the prevalence even to the present of a considerable percentage of forest and woodland, some of it ancient deciduous woodland as in Worth Forest, some of it a mixture of heath and conifers as in Ashdown Forest, and some of it merely coppice.

Round the central forest ridges developed on the Hastings Beds runs the horseshoe-shaped outcrop of Weald Clay, averaging perhaps 4 miles in width and terminated at one end by Romney Marsh and at the other by the Pevensey Levels. This and the Chalk are probably the most homogeneous of the beds responsible for the relief of south-east England, but the one is weak and the other resistant. The Weald Clay forms lowland, almost entirely below 400 ft and generally well below this, except on some watersheds, notably that of the Arun and Mole around Rusper, and where it is given greater resistance by the presence of sandstones and freshwater limestones. These are never very thick, a few feet at most, and they usually consist of a number of thin horizons interbedded with clay rather than a continuous bed. In places they affect the relief to form low cuestas within the Weald Clay, for example at Outwood a few miles south-east of Redhill, or low plateaux as between Charlwood and Leigh, a few miles south-west of Redhill, but elsewhere they have virtually no effect on the relief. Drainage is mainly across the Weald Clay outcrop rather than along it except for the Medway and its tributaries, the Eden and the Beult.

Outside the vale developed on the Weald Clay lies the complex escarpment of the Lower Greensand, which varies so much between its constituent beds and within these beds along the outcrop that a generalized description of the relief is hardly possible. The lowest bed, the Atherfield Clay, may, when relief is discussed, be included in the Weald Clay below, for its effect is the same. The succeeding Hythe Beds are the most prominent scarpforming unit and also the beds which show the greatest variation along the outcrop: where thin and uncemented they may have little effect on the relief, while, where thick and fortified with significant developments of cherts and cherty sandstones, they produce higher ground than any other stratum in the south-east. In eastern Kent the Hythe Beds are thin, 50–150 ft, and consist of alternating beds of limestone and sand, a facies known as rag and hassock. They are not very resistant and, east of Maidstone, the maximum elevation rarely exceeds 400 ft. West of Maidstone cherty sandstones become increasingly important and, immediately west of Sevenoaks, the Hythe Beds form a prominent escarpment reaching 800 ft. Between Sevenoaks and Dorking this becomes narrower and lower, but, west of Dorking, the beds thicken and again become more cherty, so that the best development of the Hythe Beds cuesta, reaching a maximum elevation just above 950 ft at Leith Hill, and of its vegetation of heather, bilberry, bracken, birch and pine, is found in the western Weald. The plan of the escarpment is complicated by folding, the Fernhurst anticline, for example, being followed by an attractive vale with inward-facing escarpments, and by a series of right-angled bends at Dorking, near Hascombe, and again at Blackdown. Although the latter could be explained by either faults or monoclines with downthrows to the west in each case, unfortunately no evidence of such structures has ever been found. The prevailing podsolized soils and heath vegetation are intimately connected and very little of the land, except low down the dip slope, is cultivated. Bare rock outcrops are very rare, as everywhere, in the south-east, except in the sunken lanes which are so characteristic of these beds. The height of the escarpment, which is broken by numerous gaps, declines eastwards along the southern outcrop in Sussex, gently at first and then abruptly as the River Arun is approached. East of the Arun, lithological changes ensure the almost complete elimination of the Lower Greensand cuesta, though it may be found as a minor break of slope in the plain which stretches from the

Chalk escarpment to the central Weald, for example, around Henfield.

The middle and upper divisions of the Lower Greensand, the Sandgate and Folkestone Beds, are less important in the relief than the Hythe Beds. The former are very variable, including Fuller's earth and concretions of sandy limestone in Surrey, but become less resistant in Sussex, where the southern outcrop has been carved out into one of the most continuous strike valleys in the Weald, that of the West Sussex Rother. The Folkestone Beds are not as variable as either the Sandgate Beds or the Hythe Beds. They are mainly coarse, unconsolidated sterile sands, readily leached to form very acid soils. They give rise to a series of low-lying heaths, used for military purposes in Woolmer Forest west of Hindhead and very well developed also in West Sussex, south of the Rother. Here, the outcrop of the Folkestone Beds is discernible on the 1 in. topographical map from the east–west band of heaths and commons. One of the few lithological variations in the Folkestone Beds is caused by irregular cementation with ironstone: this has been responsible for the Devil's Jumps a few miles north-west of Hindhead, a series of rocky little hills, which, in an age of unreason, were attributed to the feet of the devil and, in a later scientific age, to volcanic action.

The succeeding Gault and Upper Greensand are virtually facies of each other, the Upper Greensand replacing the Gault to a greater and greater degree westwards in the Weald. The Upper Greensand is important as a relief-forming bed in the western Weald only. It forms a ledge at the foot of the chalk escarpment of the South Downs west of the River Arun and a much more striking feature in the far west around Selborne. In this latter area the Chalk forms an escarpment, the crest of which reaches about 700 ft, while in front of it a second escarpment reaching 500 ft is developed on the Upper Greensand, which in turn dominates the Gault vale to the east by 200 ft or so. Along the foot of the North Downs the Gault, together with the uppermost parts of the Lower Greensand, forms a discontinuous strike vale. At the foot of the South Downs it has far less effect. West of the Arun, the strike vale is on the Sandgate Beds and the Gault is part, in places a lower part, of the low plateau between the South Downs and the River Rother. East of the Arun the Gault is one unit in the general lowland which stretches from the South Downs to the central Weald.

Although the Chalk is in a broad view a homogeneous bed, it is profitable to consider both its regional variation and variations from horizon to horizon within it. Even though it is characterized by the cuesta form, by the preponderance of dry valleys and by its smooth convexo-concave slopes, the landforms developed on it are not unique, as has sometimes been suggested or implied. It shares its smoothness with other soft limestones and with some sandstones, while its broad convexities have probably been stressed because they are revealed by its grassland and often concealed on other beds by woodland, for example on the Lower Greensand.

Even the North Downs are not satisfactorily treated as a unit, because variation is introduced by differences in dip, overlying deposits and geomorphological history. The effects of differences in dip are most apparent in the west between Dorking and Farnham, in which section the dip increases from a couple of degrees to about forty-five degrees in the Hog's Back. The outcrop not only narrows so that the form changes from a cuesta to a hog-back ridge, but the elevation decreases westwards by about 250 ft down to 500 ft in the Hog's Back, thus illustrating the control of the area of the Chalk on its height, a feature also shown by the Chalk outcrop in the Isle of Wight. Unlike the South Downs, the dip slope of the North Downs is covered with a variety of superficial deposits. Much of this cover is Clay-with-flints, a deposit tending to produce acid and infertile soils and hence to bear an increased proportion of woodland, but, along the lower part of the dip slope, notably around and east of Dartford, thin outliers of Thanet Sands improve the fertility of the chalk soils. The most continuously wooded area is between Dorking and Guildford, much of it on the Pliocene Netley Heath Beds, but it is possible that there are other reasons than purely physical ones for this distribution. Geomorphological history is responsible for the truncation of the summit of the escarpment east of the Medway by marine planation in the Pliocene and by the formation of a broad terrace elsewhere on the dip slope, especially south of Banstead in Surrey.

The Chalk plateaux west of the Weald belong more to the Hampshire Basin than to south-east England. The escarpment differs from that of the North Downs in several ways. It is much more irregular, in trend because of the way it crosses the structural grain, and in summit elevation because of the frequency with which the Upper Chalk is breached. The well-developed Upper Greensand escarpment in front detracts locally from the impressiveness of the Chalk escarpment, the appearance of which is enhanced by the frequency of beech hangers.

The South Downs differ from the other Chalk cuestas and also vary from part to part. Generally, they have the appearance of much greater dissection than the North Downs, a feature to be expected from their proximity, both at present and more significantly in the past, to the sea. The ridges are, therefore, narrower and the surviving portions of plateau smaller. In addition, the effect of minor folds on relief is marked in the South Downs. Abrupt bends in the escarpment at Duncton, Pyecombe and again near Glynde, are caused by the fact that anticlines and the complementary synclines north of them enter the Chalk outcrop very obliquely. Within the South Downs the synclinal valley of the Lewes Winterbourne and the anticlinal ridge south of it, and the anticlinal upper Lavant valley are examples of the effect exerted by structural control on the relief. Again, much higher zones of the Chalk are present in the South Downs than in the North Downs and within these high zones is developed a dissected escarpment half-way down the dip-slope between the Adur and the Arun, and again west of the Lavant. The same escarpment recurs in the Hampshire Basin. In the South Downs it occurs as a series of prominent hills on the ridges between the main dry valleys and usually rises 100–200 ft above the general level of these spurs. The lithological causes and the geomorphological processes responsible for it are matters for speculation, though the feature can be paralleled elsewhere, for example by the dissected Aymestry Limestone escarpment, View Edge, on the dip-slope of the undissected Wenlock Limestone escarpment, Wenlock Edge. The decreased importance of superficial deposits in the South Downs has led to the development of 'typical' bare chalk scenery here, especially east of the Arun. In this section, bareness, wide horizons and convexity are characteristic of the Chalk, for woodland is small in extent and mainly confined to lower slopes and valleys. West of the Arun, woodland is much more common, some of it splendid beech forest, recalling that of the Chalk plateaux of northern France. The causes of the differences between the eastern and western parts of the South Downs are probably to be found in the much higher frequency of large estates in the west and of grazing in the past in the east, for physical causes, including the higher rainfall which has been invoked, seem to be insufficient.

Finally, to the south of the South Downs west of Shoreham is the Sussex coastal plain, where planed-off folds involving Chalk, Reading Beds and London Clay, are largely concealed beneath a film of raised beach deposits, coombe rock and brick-earth. This most fertile region of Sussex expands westwards to a width of about 10 miles between Chichester and Selsey Bill. The landscape, much of which is built over, includes a number of very broad, ill-defined steps, for there are two raised beaches here, one at 15–30 ft and the other at 80–130 ft, and these are backed by one or two areas of wooded lowland, as distinct from plain, developed on Tertiary Beds in the Forest of Bere and locally around Clapham, near Worthing. At Portsdown and Highdown elongated ridges of Chalk mark the northern limbs of eroded anticlines. The coast in this part of Sussex is a fragile one, subject to rapid erosion when the groyne system fails and contrasts markedly with the eroded chalk coast to the east, typified by the Seven Sisters and Beachy Head.

While the dominant relief features are controlled, as they are through most of Great Britain, by major and minor variations in lithology, the peculiar nature of the drainage pattern of the Weald cannot be explained in such terms. Often cited as the type example of the development of a drainage pattern on a dome, the Weald is nothing of the sort. In the first place it is only in general form a dome and far more significant is the abundance of minor east–west folds on the southern side, for these mean that the original consequent drainage could not have flowed down to the English Channel except by devious and dog-legged routes. Secondly, the drainage has not the appearance of the pattern that might be expected to develop under long-continued subaerial erosion on a dome. The development of subsequent streams is far too small, and many of those which occur in the vale between the Chalk and the Hythe Beds escarpments are in peculiar positions on the Lower Greensand dip-slope and not on the Gault, for example the Tillingbourne, the Darent and the Len. On the other hand, the number of streams which flow across folds, the Wey across the Peasemarsh anticline south of Guildford and all four South Downs rivers, the Arun, the Adur, the Ouse and the Cuckmere, across one or more folds, clearly suggests regional superimposition of drainage, but this does not apply to all areas, for the central Weald is plainly an area of well-adapted drainage. It has been suggested for a century or more that the approximate accordance of summit levels throughout the Weald, even though the definition of accordance is largely subjective, was due to their once having formed an eroded lowland: to regard this plateau as marine makes it difficult to explain the well-adjusted drainage of the central Weald; to regard it as a subaerial peneplain, as Davis did, makes it impossible to explain the superimposition of drainage in other areas; to combine the two ideas, as Wooldridge and Linton did, and to regard the central areas as peneplained and hence characterized by well-developed subsequent drainage and the marginal areas as trimmed by the sea and therefore characterized by superimposed drainage, provides a very satisfactory solution. The presence of approximately datable Pliocene deposits on the marine terrace on the dip slope of the North Downs is very fortunate, for it does away with the necessity for speculation about which beds might once have extended across

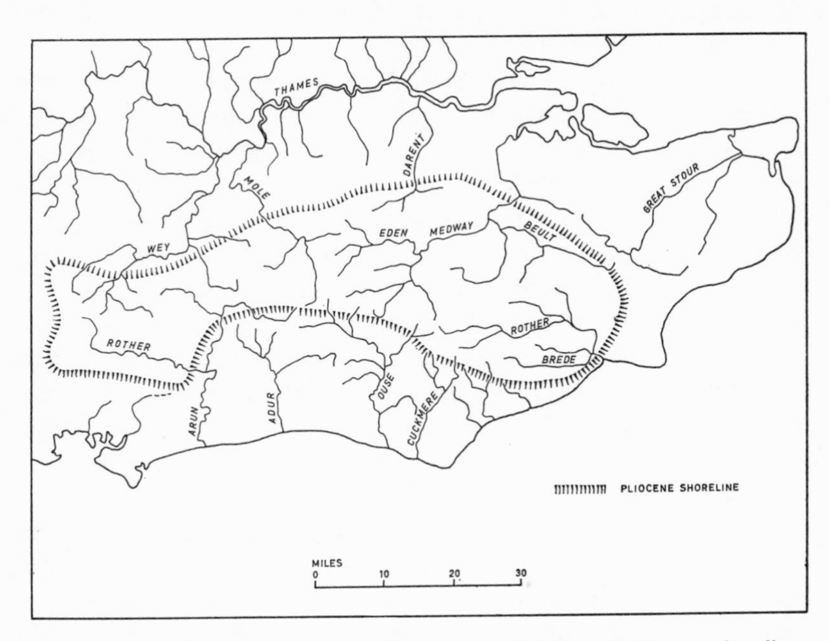

Fig. 20. The drainage of south-east England in relation to the Pliocene shoreline. (After Wooldridge and Linton).

the area and provided the mechanism for superimposition. The assumed extent of the Pliocene sea and the distribution of the two types of drainage pattern are shown in fig. 20.

Although the Weald has been the site of active geomorphological investigation into a number of problems, such as periglacial action, terrace sequences and river profiles, these features, however important to the geomorphologist, are usually of minor importance to the human geographer, for their effects on such natural resources as water supply and soils are, on a regional scale, small compared with those of the rocks.

The definition of natural resources is a vexed question. They have been extended, and quite rightly, to include for example bathing beaches, but some sort of division needs to be made between the natural, physical endowments, such as soils, mineral ores and building-stones, and values secondarily acquired, such as that of a plot of freehold building land within a quarter of a mile of the centre of a city. Such value can hardly be said to be intrinsic. Within the south-east of England the effects of both types of value may be clearly seen, although its natural wealth is probably not as great as that acquired by its position near London.

The soils, developed under temperate deciduous forest, are brown forest earths of, on the whole, middling quality. Although many facts must go to the assessment of the value of a soil, for example its general texture, its degree of acidity, the amount of humus and the subsoil drainage, the soils of the south-east may be grouped as mainly sands and clays, although there are many sandy clays and clayey sands among them. To a considerable, though not to a complete degree, they reflect the lithology of the parent material as might have been expected in an area which, although not glaciated, probably had its soil-forming processes largely upset by periglacial conditions. Thus, some of the poorest soils are found on the coarse sands of the Folkestone Beds. Unimpeded drainage ensures that the movement of soil water is predominantly downwards, leading to continued leaching and the vicious circle of greater acidity impoverishing the vegetation and a vegetation of acid type decaying to produce more acid humus further to leach the soil. The signs of podsolization, the surface humus and the bleached layer below, and the accompanying acid heath vegetation are well seen on Ambersham Common in western Sussex. But such podsolized soils are not confined to the Folkestone Beds, being found also on the Hythe Beds, for example at Hindhead. Elsewhere on both the Lower Greensand and the Hastings Beds the soils, though often sandy, are not sterile podsols. The presence in the Sandgate Beds of much silt and of clays in the Hastings Beds obviously tends to add body to the soil, while the calcareous nature of the rag and hassock facies of the Lower Greensand in east Kent alleviates the soil there. Elsewhere, a varying but usually thin deposit of wind-borne silt, the equivalent of the brick-earth of the Sussex coastal plain and the Thames terraces, helps to add body to light sandy soils and to lighten heavy clay soils. Such a deposit, similar in colour to the prevailing buff of the weathered rock, is very difficult to detect.

The clay formations, the Weald and Gault Clays, give rise to heavier soils though not to pure clays because there is a fair proportion of silt present. Adjacent to the Hythe Beds outcrop, the surface layers of these soils may be lightened by downwash and at the foot of the Chalk escarpments the Gault soils may benefit from calcareous downwash. But, as the rocks concerned are low-lying and clayey, drainage is one of the main soil problems. The light loams of the Upper Greensand in western Sussex, on the other hand, are good quality, lightish soils, helped again by downwash from the chalk.

Chalk soils are variable. Where the Chalk is covered with superficial deposits, the soils are not really chalk soils at all, though their drainage, an important soil feature, may be controlled by the underlying Chalk. The various superficial deposits give rise to a variety of heavier soils and, hence, woodland and heath may occur on the Chalk, while patches of brown loam, which might be a soil developed on Chalk or the last remnant of superficial deposits, on otherwise bare Chalk areas of the South Downs lead to a curious distribution of patches of acid vegetation surrounded by neutral chalk grassland or to a mixture of deep-rooting calcicolous plants and shallow-rooting calcifuges. Thus, practically any sort of vegetation may occur on the Chalk provided that there is a sufficient thickness of superficial deposits. Where the Chalk is really bare and provided that the slopes are sufficiently gentle for a soil to reach maturity, a humus soil, a rendzina, develops. This is a spongy black soil, up to about 9 in. or so in thickness, grading abruptly at its base into the underlying weathered Chalk.

On the brick-earth of the coastal plain of Sussex a rich loamy soil has developed. The brick-earth itself consists of 6 or 8 ft of stoneless silt, calcareous in nature and well-drained through the underlying gravel but not parched. Its very arid-looking surface in dry weather is no real indication of the amount of water retained near to the surface. This soil is the gardener's dream, apart from the fact that it grows weeds as well as it will grow anything else.

On the basis of its soils this south-eastern region is not destined to be the garden of England agriculturally. The Land Utilization Survey of the 1930–40 period showed English agriculture at its farthest swing towards pasture and at this period, apart from the woodland on the Lower Greensand and the Hastings Beds and the rough pasture on the Chalk, grassland prevailed in the Weald. Arable exceptions were provided by the light Upper Greensand soils of West Sussex, by the Sussex coastal plain and by the dip-slope of the Chalk in the Kent orchard zone.

Today the position has changed considerably. The Weald Clay lowland is no longer almost continuously mediocre pasture, but a mixture of pasture and arable, a considerable percentage having been ploughed up. The large percentage of woodland and common has limited the extent of the change on considerable areas of the sandstone districts, but the chalk downs have been drastically changed. The short springy turf of the barer eastern part of the South Downs is largely a thing of the past. After a phase of military use during the war, the South Downs were extensively ploughed and largely sown to wheat, thus virtually cashing the fertility of the rendzina soils. After a period under cereals many parts were converted to ley pastures, which were populated with dairy cattle controlled by electric fences and displacing the traditional sheep of the Downs to such an extent that the latter are now much more characteristic of Romney Marsh than of the South Downs. Some curious crops were tried under the pressure of subsidies, including a field or two of linseed. Relieved of the grazing pressure of sheep through arable cultivation and later of that of rabbits through myxomatosis, the small remaining unfarmed areas are changing from the traditional but not natural chalk turf. The typical South Downs view is now either one of cornfields and cows or, in winter, of thin poor-looking soils with great scabs of chalk showing through, for rendzinas do not last long under cultivation. The far west of the Downs was changed less, again because of the higher proportion of woodland, which, even if cut-over, is usually replanted, but even here the farmed area shows wheat and dairy cattle to a considerable extent. Thus, over most of the area today mixed farming is characteristic.

Specialized agriculture is not as evident in the south-east as might be expected from the proximity of London. The chief specialized areas are the orchard and hop regions of the Kent Chalk dip slope east of the Medway and of the area around and south of Maidstone. A variety of fruits is grown, especially apples and cherries, and this is the chief hop-producing area in England, its extent largely controlled by agreement. A second but minor hop-producing region is found on the fertile Upper Greensand dip slope around Selborne in the far western Weald. Worthing is well known for the production of tomatoes under glass, while nurseries producing vegetables and ornamental plants are fairly common on the Sussex coastal plain, though this is so large that it is not by any means entirely under market gardens. Much of it is built over and it reaches its highest value as building land; some considerable area of it is occupied by aerodromes; much of the rest is good-quality farmland. Generally this brick-earth region, which has been acclaimed as the most fertile land in Britain outside the Fens, hides its fertility well.

The south-east of England, the region which once provided the oak for England's ships and the iron for arming them, is still one of the most wooded areas in Britain, even though the forests are shreds of their former selves. While some good oak remains and some very good beech in parts of the South Downs, much of the woodland is coppice yielding its periodical crop of sticks and poles for a variety of uses. Conifer plantations, many under the Forestry Commission, are increasing as they are over much of Britain, not only on the obvious sandy soils but also on clay and even on chalk, for example between Seaford and Eastbourne.

Other natural resources, mineral supplies, are of limited importance, even though the area was an important producer of iron in the days when charcoal and water power were of more importance in siting the industry than coal and iron ore. Parts of the sandstones of the Hastings Beds, the Lower and Upper Greensand have been used for building stones in the past but their use is declining, as is that of flints. When it is far cheaper to build a house of bricks and tiles than of stone, even in such a traditional stone area as North Wales, it is not surprising that this is so in the Weald too. Horsham Stone, a flaggy bed in the Weald Clay outcropping near Horsham, provided a typical rustic roofing material comparable with the Stonesfield and Collyweston Slates of the Jurassic belt, but the material is rarely used today and is then probably salvaged from demolitions. Bricks and tiles have long been traditional building materials in the area: many of the older buildings are half-timbered and filled in with brickwork, while, in the west, rich red bricks, hung with shaped tiles and topped with heavy chimneys, are quite typical in the villages. Bricks and tiles are still made, the former, for example from Weald Clay at Warnham near Horsham, but the area cannot compare with the vast extent of modern brick-making near Peterborough and Bedford.

Chalk has long been worked for lime production and more recently for cement manufacture. The main cement-making area on the lower Thames around Gravesend is outside this region, but important cement works are found in the Medway gap. Formerly chalk was widely quarried in a great number of small pits, but its production has become increasingly concentrated into large pits usually in the face of the scarp and working Lower and Middle Chalk.

Gravel production is not very important within the Weald where there are few rivers large enough to have deposited extensive spreads of gravel, but its working becomes of importance on the Sussex coastal plain around Chichester. Sand production is much more important, for some of the solid rocks are sufficiently unconsolidated to be worked for this purpose: this applies very much to the Folkestone Beds.

Specialized mineral production includes gypsum, the basis of plaster of Paris, from the Purbeck Beds near Battle in the centre of the Weald, Fuller's earth from the Lower Greensand at Nutfield near Redhill in Surrey, and coal from a buried field in the Hercynian rocks underlying east Kent. Four collieries lying east of the Canterbury–Dover road produce slightly less than 1% of Britain's coal between them.

The industries of the south-east, apart from those based on local mineral resources and others originally using local agricultural products, such as brewing and flour-milling, show the influence of London. They are mostly light industries requiring labour rather than raw materials, with a few exceptions near the sea or on navigable water. They include paper manufacture (Sittingbourne and Aylesford), commercial vehicles (Guildford), pastes and potted meats (Chichester), cake mixtures and refrigerators (Brighton), and railway carriage work (Lancing). They reach their greatest concentration in the largest towns, notably Brighton (162,757) and the Medway towns of Rochester, Chatham and Gillingham (171,721), and, a new feature, in the satellite town of Crawley-Three Bridges (53,786).

Much of the settlement of the Weald is ancient, though the greatest change has been wrought by the spread of the influence of London largely in the present century. The main Roman entry into Britain via the Straits of Dover more or less ensured some Roman settlements in eastern Kent, notably Canterbury and Rochester, and the construction of Watling Street to the Thames crossing at London. Outside this area, Chichester was an important Roman settlement and a number of villas are known, as well as routes across the Weald. Much more comprehensive was the Anglo-Saxon settlement and Wooldridge has pointed to the coincidence of loamy soils and place-name endings of an early type, and also to the concentration of these early settlements on low-lying coasts well-indented with small harbours. Hence, there are concentrations of early settlements on the loamy brick-earths of the Sussex coastal plain, along the light scarp-foot soils on the Upper Greensand in western Sussex and again on the fertile Thanet Sand soils on the Kent dip slope, where, as on the Sussex coast, both harbours and soils occur close together. Colonization on the wooded lands of the interior was later, but often proceeded from parent settlements at the foot of the Chalk scarps either as a direct continuation or by the addition of an outlying section of territory. This emphasized the pattern of parish boundaries at right-angles to the outcrops. In the interior the villages, apart from being later, are often smaller and also more widely spaced.

Towns have not been very large in the south-east until recently, the most important being a series of small towns sited on the rivers, where these latter cross the ridges in gaps. This is the class of settlement loosely referred to as gap towns, but often probably functioning mainly as crossing points of the rivers by routes running along either the scarp-foot or dip-foot zones of settlement. The gaps are often poor routes through the uplands, especially in the South Downs, where they are thickly filled with alluvium: here the Arun gap is not passable by road, unless one is prepared for some rough going on minor roads and tracks that wind half-way up the eastern side, and the Cuckmere gap also lacks a through road.

The general scatter of villages and small towns, a reflexion to some extent of the moderate agricultural value of the land, has been greatly changed by the spread of London's influence across the area.

After the first royal interest in seaside holidays at Brighton some 150 years ago, the habit of holidays spread with the increase of wealth and leisure affecting the various groups of society at different times in the nineteenth and twentieth centuries. The coastal towns from Whitstable, through Margate and Ramsgate, Dover, Folkestone, Hastings, Eastbourne, Brighton, Worthing, Littlehampton, and Bognor Regis to Portsmouth and Southsea owe much of their growth and prosperity to the increasing habit of seaside holidays. Many had been established earlier for other reasons: Hastings, Romney, Hythe, Dover, Sandwich, Winchelsea and Rye were Cinque Ports in the Middle Ages, exchanging the furnishing of ships and men for the Navy for certain commercial privileges. Dover, Folkestone and Newhaven are cross-Channel ports, this function being of long-standing in the case of Dover and of doubtful future at Newhaven. Brighton was a fishing port before its great rise as a resort. With the rise of the coastal towns came the construction and emphasis of the radial pattern of routes from London. The Kent resorts, the farthest from London, could be reached along the Dover Road, but routes to the Sussex resorts caused increasing use to be made of the gaps through the North and South Downs. Although the river gaps are used, for example, the Worthing and Bognor roads through the Mole gap and the Bognor railway through the Arun gap, many of the most important routes use dry gaps, involving climbs of a couple of hundred feet or more, for example the Brighton road through the Merstham and Pyecombe gaps, and the Worthing road through the Findon gap in the South Downs.

The earliest coastal town to grow as a resort, Brighton, is the one which has spread most, right along the coast to Shoreham and also to the east, and inland to an increasing degree recently. But, as soon as the main towns became too popular—and Brighton is probably one of the three most popular seaside resorts in the country—people turned from the crowds to find quieter resorts in the sparsely settled areas between the towns. Week-end and holiday bungalows, whether of the less permanent type such as some of those of Peacehaven or of the more solid but equally land-consuming type common, for example, between Worthing and Littlehampton, rapidly encroached on valuable land. The more select areas became popular for retirement, and with the electrification and intensification of the Southern Railway services between London and the Sussex coast, it became feasible, with the decreasing length of the working day, for wealthier people to live by the sea and travel to London each day on business. Later on, with the recent general rise in prosperity, more and more people have become financially capable of doing this. The various types of settlement demanded services and so further increased their sizes that not only the area between Brighton and Shoreham but also that between Worthing and Littlehampton are now built up. The same result may occur in east Kent with the electrification of the railway, but the greater distance will probably always prove a handicap, or a blessing.

Although the growth of the coastal towns provides the more spectacular examples, the inland towns also have become increasingly important as London suburbs, for example Guildford, Dorking, Reigate and Tunbridge Wells. Further, land which was despised because of poor sandy soils, has acquired value as high, well-drained, healthy building land: its infertility no longer matters, for it will grow conifers and rhododendrons and these provide as good ornamental plants as anything else.

The expansion of London in search of homes and recreation has affected the area in other ways. Even in villages some of the more attractive and older houses are usually taken over by people in retirement or still working in London. Such houses acquire a neatness of garden and a freshness of paint not at all characteristic of, for example, real agricultural villages in Fenland and East Anglia or on the Chalk plateaux of northern France, all of which regions are agriculturally richer.

For recreation the high proportion of open spaces, hills, commons and woodlands attracts the Londoner southwards. One has only to think of such places as Box Hill, Ranmore Common, Leith Hill, Ashdown Forest, Newland's Corner to realize this. The position north of London is very different. True there are such places as Epping Forest, Burnham Beeches and Ivinghoe Beacon, but towards East Anglia the land is almost entirely given over to agriculture, far less interesting as natural scenery, however attractive the settlements, and it does not lead to such a series of seaside towns at a comparable distance. Hence, London goes south for its recreation, demands food, drink, and petrol from the villages and so changes their character, and at the same time congests the inadequate road network between itself and the sea.

Selected Bibliography

F. H. Edmunds. *The Wealden District.* British Regional Geology (3rd ed.). London (H.M.S.O.), 1954.

E. W. Gilbert. *Brighton: Old Ocean's Bauble.* London, 1954.

S. W. Wooldridge and F. Goldring. *The Weald.* London, 1953.

S. W. Wooldridge and D. L. Linton. *Structure, Surface, and Drainage in South-East England* (2nd ed.). London, 1955.

CHAPTER 10

CENTRAL SOUTH ENGLAND

R. J. Small

Central south England, comprising the Isle of Wight and parts of Hampshire, Wiltshire and Dorset, has for the most part a well-defined boundary. In the east the Chalk forms a sharp edge looking down on the wooded ridges of the western Weald; in the north-east it plunges suddenly and steeply beneath the newer rocks of the London Basin; in the north-west it again forms an escarpment bounding the Vale of Pewsey; and in the west its much indented face offers a striking contrast with the gentle clay-plain of the Vale of Blackmoor. Yet within central south England regional unity, apart from that provided by a centripetal drainage system, is not particularly apparent. There are major contrasts of physical geography, agricultural land use and rural settlement between the great girdle of the Chalk, broader in the north and enclosing small areas of older rock, and the core of the region, the Tertiary-filled Hampshire Basin. More important, the varied coastline of the region has been the scene of large-scale urban developments which appear to be quite out of harmony with the rural inland areas and indeed have very little economic dependence on them.

The Chalk uplands which cover so much of the region possess an intrinsic charm and character. Their smoothly rounded outlines, their wide vistas, their deep-cut coombes, their wide-spaced beech and yew woods, their grey-green downlands, their buff and creamy soils, their flinty tracks, their ancient lynchets—all add up to an instantly recognizable assemblage of features which renders them quite unlike other types of hill country. Yet this uniformity of aspect should not be overstressed, for within the Chalklands of central south England are both distinct and subtle landscape variations, the origin of which can be appreciated only in the light of an examination of the area's complex structural and denudational history.

The first uplift and flexuring of the Chalk occurred in pre-Tertiary times and was followed by a period of prolonged erosion involving the formation of an extensive peneplain, itself later 'fossilized' by burial beneath Eocene sediments. The late-Oligocene period saw the beginning of the Alpine earth-movements, which accentuated existing structures (notably the broad syncline of the Hampshire Basin) and

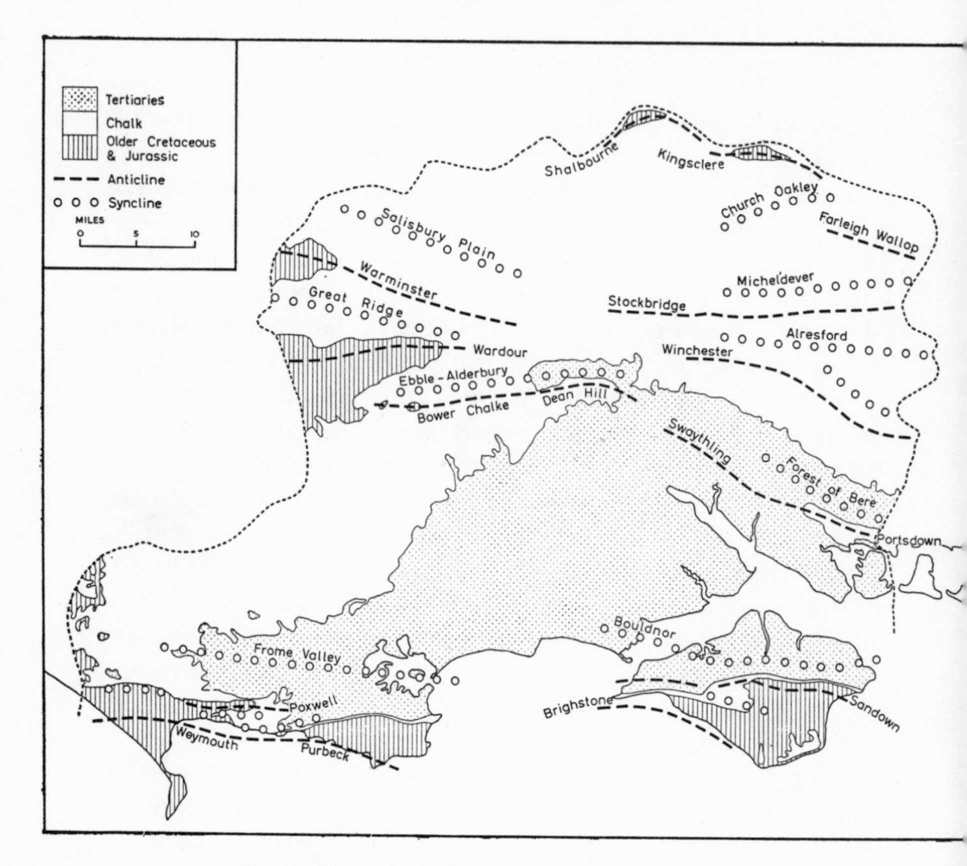

Fig. 21. The geological structure of central south England.

added new east–west folds (fig. 21). In the south, a belt of major movement affected Wight, Purbeck and Weyland, where large almost monoclinal flexures were formed. To the north of the Hampshire Basin the folds, with the exception of that of the Vale of Wardour, were smaller but still morphologically significant. Of especial interest are the periclines of Shalbourne and Kingsclere, and the fold-axes passing through Warminster, through Winchester, and through Bower Chalke and Dean Hill to Portsdown, where the upraised Chalk forms a striking hog's-back projecting from surrounding Tertiaries. Separating these anticlines are synclinal axes, notably those of Salisbury Plain, the Great Ridge, the finest example of inverted relief in southern England, the Ebble Valley, and Alderbury, in which Tertiaries are preserved.

The Chalk itself is a soft, pure limestone, except in its lowermost horizons where it becomes very marly. That it forms for the most part land over 500 ft, with culminating points in excess of 950 ft, is due to its permeability, for it is penetrated by a network of joints which readily allows the passage of rainwater to the water-table, lying at places at a depth of some hundreds of feet. The upland blocks separating the main river valleys are thus almost entirely waterless, and are at present subjected to the slowly acting processes of solution and creep, a fact which partly accounts for the dominantly convex forms of the Chalk slopes. However, the existence of numerous systems of dry valleys points to effective surface run-off at earlier periods, and even today ephemeral flows are common.

As indicated already, the outer margins of the Chalk are marked everywhere by an imposing escarpment face, usually topped by the *Micraster* fossil-zone. Within the confines of the Chalklands as defined here, equally fine scarps overlook vales formed by the unroofing of the larger Alpine folds (fig. 22). Perhaps the most stately is that bordering the Vale of Wardour, and the least impressive the hogbacks of Wight and Purbeck. In addition to these main scarps there is in Hampshire and Dorset a 'secondary' escarpment, formed by *quadratus* Chalk, which is represented by a series of low, isolated hills lying close to the Chalk-Eocene boundary and culminating in Pentridge Hill (600 ft). This resistant horizon also gives rise farther to the north, for example near North Tidworth, to steep-sided hills rising like islands above a lower undulating plateau of *Micraster* Chalk.

The anticlinal vales themselves form sub-regions of considerable individuality, for within their Chalk rims lie subsidiary cuestas and valleys developed in older Cretaceous and Jurassic rocks. Thus in Wardour the Upper Greensand gives rise near Shaftesbury to a magnificent escarpment, whose face is much pitted by land-slipping, whilst in sharp contrast the adjoining Kimmeridge Clay forms a featureless, ill-drained plain. In Purbeck—and to a lesser extent in Wardour—scarps of Portland and Purbeck Beds are found, and in

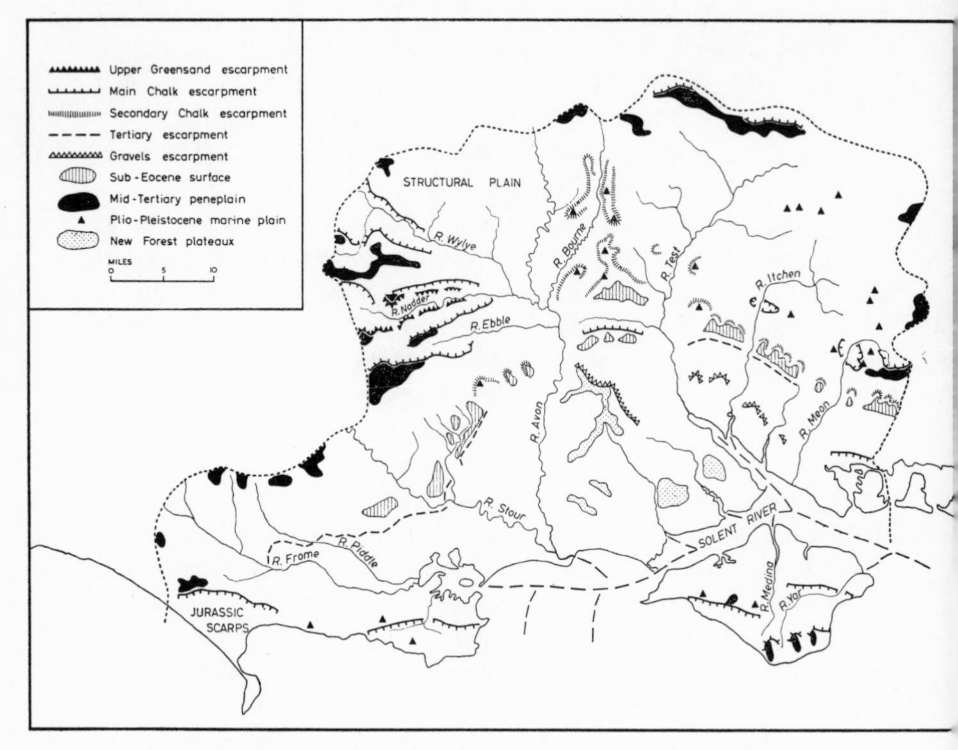

Fig. 22. The geomorphological features of central south England.

Weyland the picture is further complicated by the addition of cuestas and hog-backs of Corallian Beds, Cornbrash and Forest Marble.

The evenness of the Chalk crestline in many parts of central south England suggests a past history of peneplanation, but particular care in interpretation is needed here, for structural surfaces resulting from the stripping of softer Chalk layers are present. Salisbury Plain, a remarkably homogeneous area ranging for the most part between 450 ft and 550 ft in elevation, is the prime example, for it is everywhere underlain by *Micraster* Chalk from which a former, weaker *Marsupites* cover has been cleanly removed by past stream action. None the less, it can be shown that the highest Chalk summits, particularly in the west (fig. 22), form hill-top remnants of a once-extensive surface which truncates the Alpine folds and is demonstably of mid-Tertiary age. The Great Ridge lying above 700 ft, the Upper Greensand escarpment at Shaftesbury (801 ft), the crests of

Cranborne Chase (911 ft at Win Green) and the Dorset escarpment west of the Stour (902 ft at Bulbarrow Hill) are all important representatives of this peneplain, the subaerial origin of which has been deduced from its gentle slopes towards existing drainage lines. The mid-Tertiary can be readily distinguished from the pre-Tertiary peneplain; the latter, although distorted by earth-movements and reduced in extent by erosion, may be observed as the tilted 'sub-Eocene' surface of the northern margins of the Hampshire Basin (fig. 22). A more recent stage in the denudational history of the Chalklands has been partial inundation by the sea during late Pliocene–early Pleistocene times. Very little of the plain of marine erosion so formed may be traced today, though numerous plateau-like summits of the Chalk at 540–690 ft may contain significant vestiges.

The drainage pattern of the Chalk, and of the Tertiary area too, is in outline simple. On the mainland (excepting the Isle of Purbeck and Weyland) rivers such as the Stour, Avon, Test and Itchen drain southwards or south-eastwards, meandering through broad, alluvium-floored valleys towards the line of the Solent trunk-stream, whilst their diminutive southern counterparts, the Corfe River, Medina and Eastern Yar, run northwards. All cut across the Alpine folds in strikingly discordant fashion, and most probably were superimposed from a former cover of Plio-Pleistocene marine deposits. In the north-west, beyond the limits of the transgression, river development has proceeded more normally, and the Wylye, Nadder and Ebble run accordantly in structurally guided courses. Apparently similar accordant streams within the Plio-Pleistocene area, for example the Upper Itchen, are in reality recently formed subsequents which have etched out synclines once occupied by weak Tertiaries.

The Tertiary rocks forming the core of the region occupy a shallow synclinorium—the broadest of the structural features completed during the Alpine folding—and are relatively unconsolidated sands and clays, with some limestone and pebble-bands. Their formation began some time after the uplift of the Chalk, lasted through several marine and fluviatile episodes in the Eocene and Oligocene, and came to an end with the general uplift which followed the deposition of the Hamstead Beds. Formerly the Tertiary rocks were far more extensive, and spread beyond the Basin proper to occupy the synclinal troughs of the present Chalklands; but virtually uninterrupted erosion in Miocene, Pliocene and Pleistocene times has driven back their margins, which are now defined at many places by a feeble and discontinuous escarpment overlooking the freshly revealed 'sub-Eocene' surface of the Chalk (fig. 22).

The relief of the Tertiary area is much less complicated than that of the Chalk. Minor folds of Alpine date do exist (fig. 21), but have by comparison subdued topographical effects. Further, owing to their overall weakness, the Tertiaries fail to form land high enough to preserve the remains of the mid-Tertiary peneplain. On the other hand, they do maintain admirably a record of the final stages in the retreat of the Plio-Pleistocene sea in the shape of old marine benches, which are particularly prominent in the New Forest. These erosion residuals, the highest of which lies at 419 ft, form in effect a dissected staircase declining in height to 35 ft near the Hampshire coast. They are not simply of academic interest, for they are in themselves striking elements in the landscape. They are often capped by 10 ft or more of 'plateau-gravel', a coarse deposit of sub-angular flints in a sandy matrix. This protects the underlying weak rocks, with the result that the northern limits of the gravels are coincident with a pronounced scarp slope which may be traced from near Fordingbridge to Southampton and beyond (fig. 22).

In general, the Tertiaries form a landscape dominated by heath and woodland. Most local variety is to be found in the New Forest, where the gravel-topped plateaux support heath, bracken, hardy grasses, gorse and birch, and the valley slopes give rise to natural oak–beech forest or more heathland. In many valley bottoms narrow tongues of bog, with a plant association of alder, cross-leaved heath, sedge, bog-moss and cotton-grass, are developed. In marked contrast to the desiccated Chalk the Tertiaries form extremely ill-drained land; even long dry spells will not disperse entirely the pools of stagnant water lying on plateau top and valley floor alike. This impeded drainage is due both to the subdued relief of much of the area and to advanced podsolization—a process favoured by the initially porous gravels and sands—which has caused the accumulation in the subsoil of impervious clay-pan or even intractable iron-pan. To the west of the Avon the Dorset heaths form generally lower land, but although the gravel plateaux are not in evidence the soil and vegetation characteristics are if anything more forbidding than those of the New Forest. The podsols of the Bagshot Beds are particularly acid and deficient in mineral nutrients, and are widely used for army manœuvres and forestry. The landscape is perhaps most sombre and desolate in the Frome valley near Wareham and Poole; here lies the 'heathy, furzy, briary wilderness' of Thomas Hardy's Egdon Heath.

In terms of agricultural land use the contrasts between the Chalklands and Tertiaries are easy to point. Both are characterized by comparatively limited productive capacities, yet in each the agricultural emphasis has in many ways been strikingly different. Thus on the Chalk extensive farming has been the rule, whereas in the Tertiary area much undeveloped land alternates with pockets in which the land use has been of the most intensive kind.

The Chalklands as a whole are not well endowed with fertile soils, though a good deal of variation in character and quality exists in spite of the overall lithological uniformity of the parent rock. Thus thin, porous rendzina soils occupy most slopes, whilst heavier, more acid clays (supporting uncharacteristic hazel-oak woodland, as on the Great Ridge) are derived from the problematical Clay-with-flints which caps some uplands. Within the anticlinal vales the soils vary according to aspect and parent material. Probably the most favourable are those of the Upper Greensand, which are light in texture, usually well-drained and easily worked, and the least favourable the colder, heavier soils of the Kimmeridge and Oxford Clay outcrops.

These soil changes are naturally reflected in some degree by the agriculture of the Chalk; none the less there are many uniform features which must be emphasized. In general the farms are large, frequently exceeding 1,000 acres in area, and the fields are often bounded by the barbed-wire fences indicative of late enclosure; there is indeed an immediate contrast with the smaller holdings and close-set hedgerows of much of the Tertiary farmland. Open, short-grassed downland is still found in the higher or more deeply dissected areas, but its extent has been seriously reduced as external stimuli have prompted far-reaching changes in the farming economy of the Chalklands.

In 1928, O. H. T. Rishbeth wrote: 'The Chalklands seem to have arrived at a more stable and balanced farming, based upon their three natural major products, sheep, cattle and grain.' This statement would certainly not have been true a decade ago, for the catastrophic decline in sheep grazing which may be traced back to the beginning of the nineteenth century had not been arrested but rather accelerated. Conversely the numbers of cattle kept, particularly for dairy purposes, had shown a striking increase, whilst grain production was considerably in excess of pre-war days. Today the position is again somewhat altered, for in response to high world prices of mutton and particularly wool the numbers of sheep on the Chalk have shown a noteworthy increase, without in any way disturbing the dominance of the new cattle economy.

The most important single cause of the increase in cattle-keeping has been national farming policy. The 'monthly milk cheque' provided by the Milk Marketing Board, together with 'fat cattle subsidies', have until recently proved enticing substitutes for the more delayed and less remunerative returns to be had from sheep. Another telling consideration in the opening-up of the waterless downlands to cattle has been the large-scale extension of piped systems, from deep wells. The milch cows are no longer tied to the water-meadows of the main Chalk valleys, but can be transferred to the once unfavourable hill-top pastures, where they often graze in temporary enclosures of electrified wire and are milked there with the aid of mobile machines.

Hand in hand with this trend towards increased cattle-rearing have gone alterations in arable land use. First, there has been a general net increase over the past twenty years in the total arable acreage, though there has been a slight but understandable decline in the last decade. This represents to some extent an extension of barley, wheat and oats production, but is largely accounted for by a significant reduction in permanent grass at the expense of temporary varieties. The large-scale introduction of the latter has necessitated modifications of the classic rotation of Chalk farming (wheat-oats-roots-grass), and today the system of alternating long leys of four to five years of temporary grass with several years of grain and root production is widely practised. Secondly, there have been significant changes in the actual crops, other than grass, that are grown. Oats has always been a characteristic crop, especially on the poor Chalk soils of the higher west, but it is now increasingly supplanted by wheat, formerly the preserve of the better land, and to an even greater extent by barley. Roots such as turnips and swedes, admirably suited to winter sheep feeding, have now been largely replaced by kale and cabbage. It only remains to add that these changes in the arable of the Chalk may have been dictated by national needs, but have been rendered practicable only by technological advances. Much of the rolling Chalkland is highly suited to mechanized farming, and the introduction of the combine-drill, combine-harvester and grain-drier has played a prominent part in permitting the use of soils hitherto regarded as too poor to merit cultivation.

A somewhat dismal picture of natural conditions in the Tertiary area has already been painted, but it would be inaccurate to give the impression that favourable soils are entirely lacking. Besides the rich alluvial bottom-land of the river valleys, river terraces such as that displayed prominently on the left bank of the Avon below Fordingbridge give rise to warm and fertile loams. Even the plateau-gravels to the east of Southampton Water have been successfully worked into a good sandy loam, whilst in the west the London Clay, forming a low-lying zone between the Chalk and the sandy Bagshot area, provides a medium loam and generally good farming country.

On these more amenable soils mixed farming is typical, though there is a distinct emphasis on milk production for much the same reasons that apply in the Chalklands. There is, too, an important development of market gardening near Bournemouth, Southampton and Portsmouth; here numerous small-holders grow on the relatively warm soils potatoes, peas, beans, greenstuffs and tomatoes, which compete in the nearby large urban markets. But the two areas of greatest individuality are the New Forest and that to the east of Southampton around Swanwick.

The New Forest, that 'miraculous survival of pre-Norman England', covers an area of 144 square miles lying between the Avon valley and Southampton Water, and between the Chalk and the Solent shores (fig. 23). The 101 square miles of Crown land within the Perambulation (boundary) of the Forest are today administered by the Forestry Commission, and accordingly the most important aspect of forest life is no longer hunting or even grazing but timber production. Approximately 39% of this land is under woods (as opposed to 59% open heath and 2% cultivated land) which are themselves largely concentrated in 'Inclosures'. The latter were first legalized as far back as 1483, when they were introduced to enable protected regeneration, and they have gradually become more extensive, until in 1949 the total area of enclosed woodland was 27 square

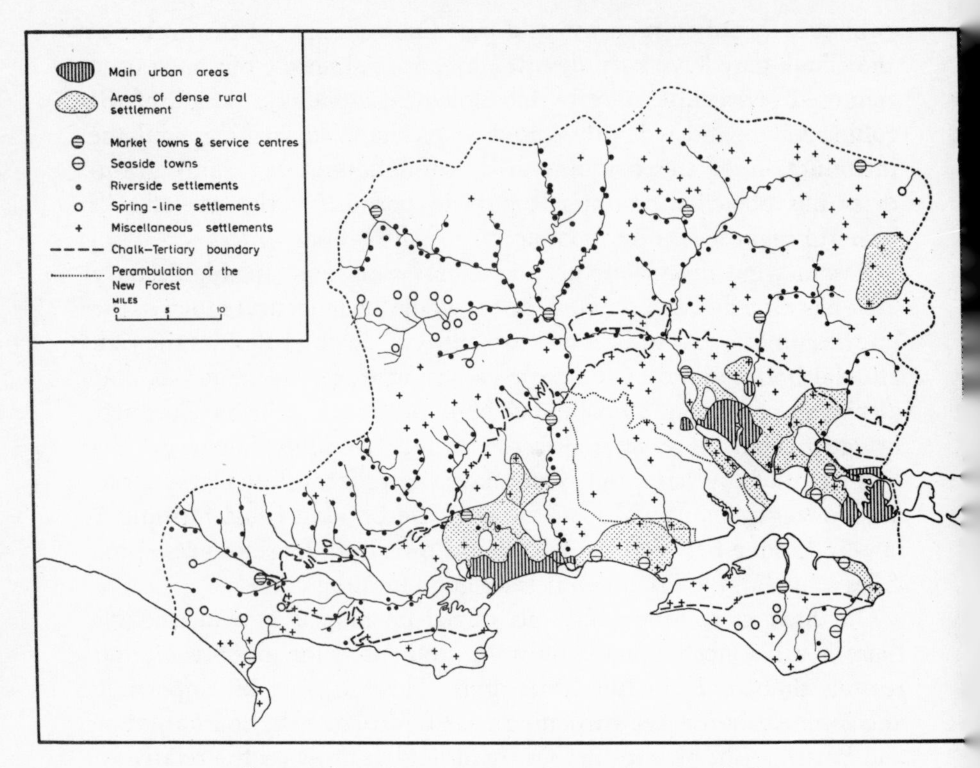

Fig. 23. The settlement pattern of central south England.

miles. Outside the Inclosures were a further 11 square miles of 'Ancient and Ornamental Woods', of amenity value only. By the New Forest Act of 1949 the Forestry Commission was given limited powers of enclosure of these, solely for preservation purposes; and the same Act also empowered the Forest Verderers (Administrators of Common Rights) to lease to the Commissioners up to 5,000 acres for further enclosure and afforestation for timber.

Thus forestry has grown in significance and the resultant changes in the Forest landscape are manifest. Before the eighteenth century the virgin woodlands were of oak and beech, but in 1776 the Scots pine was introduced and quickly spread, showing great powers of adaptation even to the adverse conditions of the plateaux. Large pure stands have since been developed and have been supplemented by other exotic species such as Douglas fir and the very productive Corsican pine.

Other important aspects of Forest life today are pastoralism and tourism. The former is of particularly long standing, for since early times the Commoners have had hereditary rights to pasture their animals on all land not enclosed for forestry or agriculture. Their ponies, which number approximately a thousand, add charm to the Forest scene, but of far greater economic value are the 2,500 or so milch cows. Tourism, on the other hand, is a phenomenon of the twentieth century, and has been promoted by the accessibility of the Forest, its wild and largely unspoiled beauty and its unique history.

It would be difficult to imagine a sharper contrast in land use than that between the agriculturally underdeveloped Forest and the Swanwick area. The latter has been the scene not only of market gardening in general, but also of a marked and famous specialization in soft-fruit production. This branch of horticulture has been favoured here by three main factors: the physical conditions of warm soils, southerly aspect, and relative freedom from frosts. Other important considerations have been the proximity of Southampton and Portsmouth, which offered in the early days a good market (though the fruit soon began to compete in a wider field, and is today taken as far as Edinburgh and Aberdeen), and the suitability of soft fruit as a crop for the 'squatters', who set up their small-holdings here in the nineteenth century and established a tradition which has lasted to the present day.

Strawberries have for several decades been the most important individual fruit-crop and bush fruits have not been widely cultivated. The 1959 acreages for the area were: strawberries 831; blackcurrants 141; gooseberries 62; and raspberries 51. Unfortunately the 'industry' has suffered a serious decline since pre-war days, when in some years the acreage of strawberries alone exceeded 3,000. The reasons are not hard to find: casual pickers are nowadays difficult to obtain; there has been much loss of land to speculative building as both Southampton and Portsmouth have overspilled; and, especially important in the light of these other problems, even here vagaries of climate are a source of considerable insecurity.

In 1961 the population of central south England numbered 1,600,000, with 400,000 inhabiting the Chalklands and 1,200,000 the Tertiary area. This surprising discrepancy is the result of the inclusion within the latter of the large coastal towns of Portsmouth (215,198), Southampton (204,707), Bournemouth (153,965) and Poole (88,088). These, together with their physical setting, will receive separate consideration.

As may be seen from fig. 23, the settlement pattern of the dominantly rural Chalklands essentially reflects problems of water supply. Most villages are strongly nucleated and are found close to scarp-foot springs, as in Wardour, Wight and Weyland, and to permanent streams, where they frequently receive a bonus from dip-slope springs; or they occupy dry valley floors and lower hill-slopes where the water-table may be easily tapped, as to the north of Andover. The larger market towns and service centres, showing today many signs of the prosperity which pervades the Chalklands as a whole, invariably stand on valley-floor sites. Of particular note are Dorchester (12,226), Andover (16,974), and the cathedral cities of Salisbury (35,471), with its remarkable focal position, and Winchester (28,643), which commands the gap cut by the Itchen through the ridge of the Winchester anticline.

In the rural Tertiary areas, on the other hand, nucleated settlements are unusual (though notable exceptions are Botley, Wickham and Titchfield), and villages such as Cadnam and Netley Marsh to the west of Southampton, North Baddesley to the north, and Hedge End and Bursledon to the east, straggle without form or focus. Numerous explanations of this feature may be put forward. For example, the relatively late settlement of much of the Tertiary land, together with the general abundance of water, have both permitted and favoured dispersal; indiscriminate 'squatting', especially on the many commons adjoining the New Forest, and the concentration of certain areas on horticulture have further fostered it. The pattern of settlement where horticulture rules the economy is one of numerous intersecting minor roads and many urban-type houses, each with its separate holding, but few shops and amenities. More recently there has been speculative building in the 'countryside' around Portsmouth, Southampton and Bournemouth; thus many of the empty spaces have been filled in, areas of unusually dense rural settlement have been created (fig. 23), and pre-existing settlements have lost what little identity they had. Among the inland towns of the Tertiary area, Eastleigh (36,577) is pre-eminent. This important railway engineering centre is made distinctive by its 'planned' grid-iron street pattern dating from the 1890's, though recent residential and industrial growths are steadily altering its character.

The main outline of the coast of central south England may be related to the marine erosion of a resistant southern barrier of Chalk and older rocks, and to the large sea-level rise associated with the post-Glacial transgression. In Wight and Purbeck this barrier remains effective and gives rise to much fine cliff-scenery, but in geological terms its future seems short-lived, for the sea is relentlessly seeking out points of weakness, for example at Arish Mell in Purbeck. Between Wight and Purbeck the barrier has been broken, and, deprived of their protection, the weak Tertiary rocks lying to the north have been rapidly hollowed out. It seems likely that the early breaches were established during inter-glacial periods, but that the final break-through took place in post-Glacial times. Thus occurred the formation of Poole Bay and the disruption of the 'Solent River', the great drainage artery once dominating the region. The results of the transgression are also manifest in the Solent itself, Poole, Christchurch and Portsmouth harbours, and Southampton Water, all formed by the inundation of shallow river valleys. An interesting result of the rapid erosion of the Tertiaries, which form low cliffs when open to wave attack, has been the release of much plateau-gravel; this has been refashioned by west–east longshore drift into numerous constructional shore-forms, for example Hurst Castle and Calshot Spits. Within the Solent and in the estuaries protected from wave attack there has been the widespread development of salt-marshes, formed largely by the plant *Spartina Townsendii*, with its great powers of trapping mud.

Two features of this coastline have favoured the remarkable growth of towns along it. First, there is the pleasant and sunny nature of many of its beaches, both to the west of the Solent and on the Isle of Wight. Secondly, there are numerous, well-protected inlets, the value of which as potential harbours has been enhanced by the famous double high-tide usually associated with Southampton Water but actually experienced over a far wider area. Thus of the four major towns, Bournemouth may be classed as a resort, whilst Poole, Southampton and Portsmouth are ports; additional smaller seaside towns include Weymouth, Swanage, Sandown-Shanklin and Ventnor (fig. 23). But there is more to the story than this, for both ports and resorts depend on their hinterlands. Central south England itself has little to offer except the products of agriculture, mining and quarrying (for example sand and gravel, building-stones from Portland and Purbeck, and kaolin clay from Corfe), and has a limited population. But nearby London, linked to the towns by excellent rail communications, can provide far more in the way of both goods and people; and indeed this access to the metropolis has undoubtedly been the prime reason for the growth of the coastal towns. Unfortunately, London's influence has been felt in another way, namely the serious undermining of central south England's regional consciousness.

The growth of Bournemouth has been astonishing, for the town is now spread over an area that was at the beginning of the nineteenth century virtually uninhabited heath. Today its development continues unchecked, and it is now contiguous with the adjoining towns of Poole and Christchurch. In 1961 the population of this conurbation totalled 268,000, representing a 42% increase over the 1931 figure of 188,000. This long-continued expansion cannot be related solely to geographical factors—though the unused heathy areas behind the town have facilitated it—but rather to Bournemouth's acquired reputation as a fashionable place in which to live and to its undoubted attractions for holiday-makers. Many of the inhabitants are, for instance, retired persons of the middle and upper classes who have migrated to the town from London. It is interesting that the maintenance of this 'class' character has been made possible by the distance from London (100 miles), which is not excessive but just sufficient to prevent an influx of day-trippers.

Of the three ports, the smallest, Poole, participates in coastal trade, exchanging kaolin clay and agricultural products for coal and timber, and has alone developed primarily to serve its immediate hinterland. The much larger Portsmouth has long been essentially a naval port, and its rapid growth to cover Portsea Island during the nineteenth and early twentieth centuries was a reflexion of Britain's role as a pre-eminent naval power. The harbour itself was excellently suited to its task, having an exceptionally narrow and defensible entrance, and further off-shore protection was afforded by the Isle of Wight. Today the Royal Dockyard continues to exert a dominating influence on the life of the town, for manufacturing industry (including corset-making, timber working, and the processing of food and drink) has been less important here than in other comparable towns.

Subsidiary employment is afforded by the resort of Southsea. Portsmouth itself is the only large town of central south England to show a recent decline in population (14·7% decrease 1931–61), but this is in large measure due to decentralization, the outcome of war damage and the severe overcrowding of Portsea. Conversely, the adjoining towns of Fareham and Gosport experienced startling increases of 167% and 62·5%.

Southampton, the country's third largest port, lays incontrovertible claim to be the commercial centre of the region, even though its true hinterland lies beyond the latter's bounds. The physical advantages of its situation include a 6-mile stretch of protected water from the Solent to the docks, the unusual tidal phenomena, an approach channel (minimum depth 35 ft) which is largely kept free of silt by tidal scour, and the practicability of dock expansion on the easily reclaimable mud-flats.

Southampton's rise to eminence as the country's premier passenger port dates from the formation of the Southampton Dock Company in 1836, and the subsequent establishment of railway connexion with London. Its facilities for handling the world's largest liners have since been steadily improved until they are now unrivalled. Handling of cargo has also grown in importance; indeed the port's functions are now changing somewhat, for the ocean passenger traffic has undergone decline in the face of competition from air traffic. The principal imports of the docks are fruit and vegetables (as much as 98% of the deciduous and 54% of the citrus fruit from South Africa have been handled here in one year), grain, flour and timber; a variety of manufactured goods (notably cars) is exported. The most recent and far-reaching changes have been prompted by the construction of the great Esso refinery at Fawley; thus of Southampton's total imports in 1958 of 12,800,000 tons, 11,500,000 were crude oil, and of her exports of 10,500,000 tons, 9,800,000 were oil products. The refinery, whose 1958 output totalled over 10,000,000 tons of fuel oils, gasoline, gas oils and special fuels, has had other repercussions; for instance, the hitherto undeveloped west side of Southampton Water is now the scene of allied petro-chemical industries, typified by the large synthetic rubber plant at Hythe.

The growing importance of Southampton, particularly as a cargo port serving a very large hinterland and as an industrial centre meeting nation-wide demands, seems to emphasize the role of central south England as a zone of transit between London, and also the towns of the Midlands, and the south coast. Within central south England regional unity is, as stated already, not particularly apparent. A closer examination of the physical and human geography of the area does not reveal it to be any more real.

Selected Bibliography

H. C. Darby. 'The Regional Geography of Thomas Hardy's Wessex.' *Geographical Review*, **38**, 1948, pp. 426–43.

C. E. Everard. 'The Solent River: a geomorphological study.' *Trans. Br. Inst. Geographers*, **20**, 1954, pp. 41–58.

L. E. Tavener. 'The Port of Southampton.' *Economic Geography*, **26**, 1950, pp. 260–73; 'Dorset Farming, 1900–1950.' *Proc. Dorset Nat. Hist. and Arch. Soc.*, **75**, 1955, pp. 91–114.

E. Thomas and G. B. Bisset. *The Strawberry Industry of South Hampshire.* Cambridge, 1933.

S. W. Wooldridge and D. L. Linton. *Structure, Surface and Drainage in South-East England.* London, 1955.

CHAPTER II

SOUTH-WEST ENGLAND

M. A. MORGAN

South-West England is a peninsula that has long been physically isolated from the rest of the country by the swampy Somerset Levels and by Exmoor and the Blackdown Hills. In a subtle fashion the individuality it possesses has been created out of this isolation which has been gradually but significantly broken down within the last century. The people of the region have enjoyed a long tradition of self-sufficiency and independence, and although in times of crisis they have more than once rallied loyally and effectively to support the national cause, there have been occasions when they have fought splendidly but unavailingly for their own special interests. The intimate association with the sea has modified the narrowness of vision which often goes hand in hand with isolation. West country sailors in West country ships have roamed the world's oceans in search sometimes of loot but more often of conventional trade. Although changing circumstances have sadly reduced the fishing fleets and eliminated the sailing vessels many cottages still bear witness, with faded photographs, bric-à-brac, and exotic plants to the contacts with foreign parts that have been made in the not too distant past.

Until the railway system was completed towards the end of the last century internal movement was difficult; less than a hundred years ago, for example, the quickest way to travel from west Cornwall to London was to go by sailing packet from Hayle to Bristol and by train from Bristol to Paddington. The upland areas are so disposed as to make travelling difficult, and the difficulty is increased in the west by the rias which penetrate far inland forming small peninsulas. Consequently several minor regions have grown up, each with a distinctive character. Even a rather insensitive ear can detect very marked variations of accent and dialect, many far removed from the soft, gentle speech popularly associated with the whole region. Their survival is adequate testimony to the strength of local sentiment and of the physical barriers which have made it possible.

The great improvement in accessibility that came with the railways and motor vehicles has changed but not destroyed the character of the region. Agriculture has become more specialized as distant markets could be reached and as the peculiar advantages could be exploited, but the slow unhurried tempo of rural life has not been greatly affected. Better communications have helped to make this area a great centre for holiday-makers. This has led to a rather self-conscious but profitable emphasis on the quaint and picturesque. It has also, unfortunately, brought in its train the speculative builders whose activities have ruined many miles of beautiful coastline.

The physical geography of the region has not yet received the attention that it merits. Certain fundamental problems remain unanswered. The drainage pattern, for example, cannot yet be adequately explained. It is clear that in general terms the water-parting runs, not as one might expect along the central axis of the peninsula, but close to the Atlantic and Bristol Channel coasts (fig. 24b). It is evident, too, that the present drainage pattern is a compound one, hinting at the relics of much older patterns, some of which have been superimposed and to a greater or lesser degree have adjusted themselves to the structure. Amongst the features that need to be related and explained are, for example: the strong indications of a north-west to south-east pattern across the heart of Devon; the alinement of some of the Exmoor streams which is oblique to the structural grain; the east–west elements in the rivers of east central and south Devon; the strong southerly component most clearly marked in south and east Devon; and the apparent reversal of drainage implied by the courses of the Taw, Torridge, and Camel rivers. No satisfactory explanation of these features will be possible without more detailed research and it therefore seems wisest in this essay to avoid speculation and to emphasize the broad outlines of geology and structure with particular reference to their effect on the human geography of the region (fig. 24a).

The Palaeozoic upland which forms the greater part of the peninsula consists mainly of Devonian rocks and the younger Culm Measures. The great earth movements of Armorican times, thrusting from the south, have crumpled these rocks into a series of roughly parallel east–west folds. The forces were strongest in the south, and in the Lizard peninsula and in the extreme south of Devon older Pre-

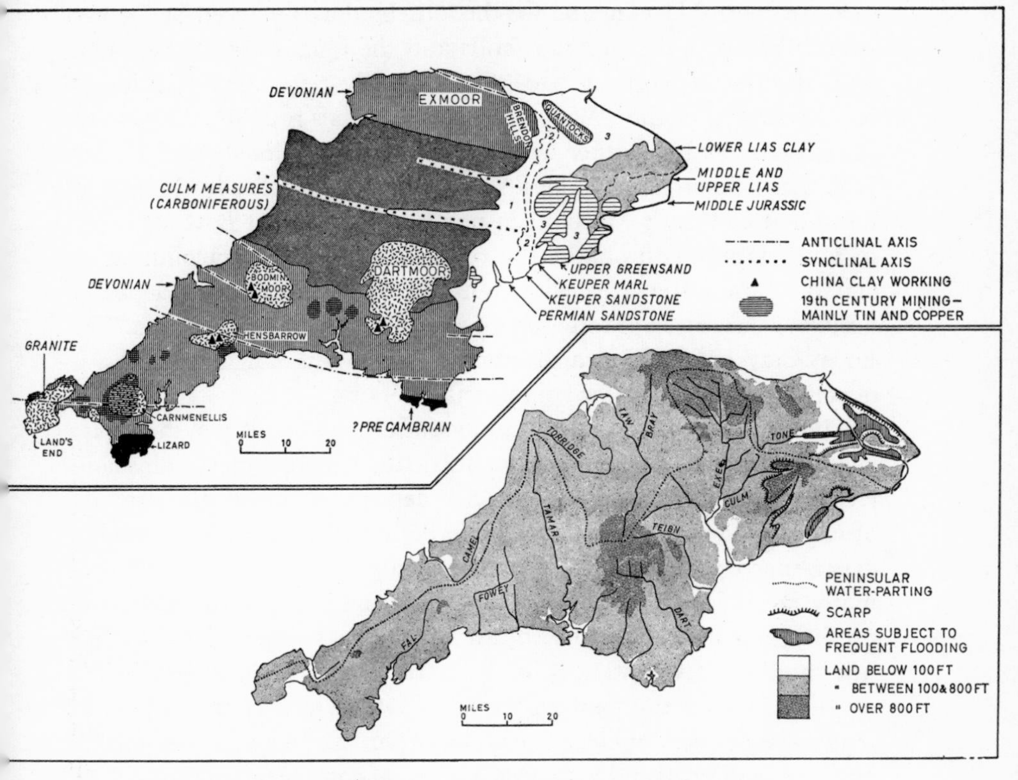

Fig. 24. *Upper*, geology and structure of South-West England; *lower*, landforms and drainage of South-West England.

Cambrian rocks from great depths were thrust with great violence over and amongst the splintered fragments of Devonian strata. A little farther north a weakness in the Devonian cover, preserved perhaps from Caledonian times, was opened up and extensive masses of granite were intruded. Later erosion has exposed the granites which now form the rounded uplands of Dartmoor, Hensbarrow, Bodmin Moor, Carn Menellis, Land's End and the Scilly Isles. The enormous heat and pressures generated within the molten granite baked the surrounding rocks into 'killas', permitted the formation of many different minerals of which tin and copper have been most important and also caused some of the granite to become kaolinized. This kaolin, or china clay, is worked in western Dartmoor at Lee Moor, and in the Hensbarrow area north-west of St Austell. Armorican folding is also responsible for the broad synclinorum which runs through central Devon, and for the anticlinal axis marked by Lundy Island, Exmoor, the Brendon Hills and the Quantocks. Very much later flexures of Tertiary age were probably responsible for downward warping of the Palaeozoic rocks: one axis roughly alined with the Tamar, and the other with the Lower Taw, Creedy and Lower Exe. The occurrence of deposits of ball- and pipe-clay in the neighbourhood of Peter's Marland in north Devon and at Bovey Tracey in south Devon is probably the result of local down-warping, associated in the latter area with faulting. These river-borne deposits, derived from rotted granite, are of considerable economic value. The Bovey Clays are found in association with deposits of lignite that have never proved suitable for commercial working.

The Palaeozoic rocks form mainly undulating country of moderate elevation much dissected by an intricate network of streams and rivers. Valleys are often floored with deposits of 'head', the product of solifluction during the Pleistocene Period when most of the region experienced periglacial conditions. Evidence of the rejuvenation of many rivers is provided not only by the sharp breaks in their long profiles, but more obviously by the precipitous slopes of the valley-sides and in a few instances by dramatic gorges, such as the one cut by the Lydd on the western flanks of Dartmoor. That the present landscape is a very ancient one is clear from the lack of a relationship between landforms and structure. The great complexity of the geology is only in evidence along the coast where the old, hard and often greatly contorted strata are displayed in magnificent cliffs.

The granite masses and Exmoor together with the fault-bounded Quantocks stand boldly above the general level. Dartmoor is the highest and most extensive of the granite intrusions which become smaller and lower towards the west. Exmoor, although formed of Devonian sandstones, slates and grits, with its rounded, smooth upper surfaces, its deep wooded valleys, its silence and its solitude is strikingly similar to Dartmoor. The main difference is that Exmoor lacks the tors which cap the summits of Dartmoor and the other granite areas.

Most soils on the Devonian rocks are inherently acid and because of the comparatively high rainfall they are subject to leaching. This means that they need regular applications of lime, nitrogen and potash if they are to be productive. Lime has been traditionally taken from the beaches, where the sand is rich in calcareous material, and today in Cornwall 70% of all lime comes from this source. In some parts of the north Cornwall coast it has become necessary to control the removal of the sand because of accelerated coastal erosion. The organic material lost through leaching is generally made good with farmyard manure, but in market-gardening areas, especially in west Cornwall, where few cattle are kept, the deficiency is remedied by using seaweed collected from the beaches and spread on the fields.

There is a considerable variation in the quality of the soil over these older rocks, but generally they consist of medium quality, dark red, clay loams, which drain poorly and tend to get waterlogged in winter. In the South Hams and in Combe Martin the presence of Devonian limestone has lightened the texture and increased the fertility of the soil. The Culm Measures weather to form a stiff, yellowish clay soil which is cold and heavy and often badly drained. The higher granite areas and Exmoor are covered with thin peaty sands, often with iron-pan layers beneath which give rise to patches of bog. On the lower slopes, however, the soils are rich in potassium and give good quality light loams.

In the east of the region the Palaeozoic basement is overlain by Mesozoic rocks dipping generally south or east, many of which once extended well westwards of their present position and probably covered the entire peninsula. The oldest of the Mesozoic rocks are the Permian Marls and Sandstones, which have weathered to produce the rich, deep, warm and red soils so often regarded as the typical soils of the whole of Devonshire. The Permian rocks are succeeded in the south by the narrow outcrop of the Budleigh Salterton Pebble Beds. This is a limited local exposure extending only some 15 miles north of Budleigh Salterton, but it forms a distinctive feature with a strong westward-facing scarp. The soil is excessively stony and supports only a heathland vegetation. The Keuper Marls are very widely exposed and form either a lowland or gently undulating country with red but rather heavy soils. They are overlain by the succession of clays and limestones which make up the Jurassic series. The base of the Lower Lias Clay forms a low but prominent escarpment running east-north-east from the Blackdown Hills and overlooking the southern part of the Somerset Levels. The broad dip-slope of rolling country is succeeded by escarpment of the Mid Lias Marlstone, outliers of this rock forming small 'islands' within the Levels. Liassic soils are generally heavy and best suited to pasture, but the underlying rocks are much more varied than the geological maps suggest and patches of shales, limestones and valley gravels support arable farming in a number of places. The Midford Sand exposures above the Marlstone can be made productive and are now almost entirely under the plough.

In the south-east the Jurassic and Triassic rocks have been planed off and covered with extensive deposits of Greensand capped with patches of Chalk. Remnants of this cover are found on the Haldon Hills south of Exeter. The fact that the upper beds of the Greensand consist of a hard nodular rock known as Calcareous Grit probably explains why the surface has been so well preserved. The plateau has been dissected by rivers with very deep, steep-sided valleys. Travelling across the divides is difficult; the roads climb steeply up one valley only to plunge sharply down another. In at least one instance a tunnel has been made through the resistant Greensand capping on a crest in order to reduce the gradients. The Calcareous Grit gives a hungry soil of little use except for afforestation or rough grazing, but the Chalk is more productive. The rivers have exposed the richer marls and clays in their valleys which form good quality pastures and give to the area variety which it would otherwise lack.

In the extreme north of the region, south of the Polden Hills, is part of the Somerset Levels. In origin these are remarkably similar to the Fens of East Anglia. They have been formed by the accumulation of silt and peat, deposited under conditions of changing climate and sea-level and most recently affected by marine transgression. The soil is rich and contains a high proportion of alluvium, but flooding is a constant hazard. After heavy rain the swollen rivers coming down from the surrounding highland carry a burden too great to be discharged into the sea, even with the help of modern pumping machinery and of artificial drains.

The present pattern of agriculture in the region results in large measure from the quality of the soils, but it has also been influenced by the climate which is mild, equable and wet, by relief, and by prevailing space relationships. Over almost the whole of the region the main crop is grass. The climate permits a growing season of about ten months in favourable circumstances and this means that cattle need spend only a short period under cover in the winter. However, it must be remembered that the acidity of the soils results in a rapid deterioration of pastures unless they are periodically ploughed, limed and fertilized. So it is normal for farmers to practise a system of ley-farming under which the land is ploughed, treated and sown with grass which is allowed to remain for periods ranging from three to eight years. It is then ploughed in and replaced by oats, barley or dredge corn (a mixture of the two) which are fed to the cattle, pigs, sheep and poultry. Turnips, swedes, mangolds and fodder cabbage also play their part in the rotation. It follows then that natural grassland or permanent pasture is rarely found, except in valleys whose slopes are too precipitous for ploughing or whose aspect is unfavourable for market gardening, and locally in individual fields which farmers for various reasons are unwilling to improve. An important exception is found in the rich water-meadows of the lower Exe valley, but although not ploughed the lush pastures are maintained only by very careful control of irrigation, and to this extent they are not entirely natural.

Nearly all farms carry some stock cattle but dairy farming is the most common agricultural activity. Improvements in communications permit milk produced in the eastern areas to be sent in liquid form by road and rail to markets in London and the Midlands. Farther westwards, distances to these markets are still too great and so most of the milk is first processed in a number of large factories and then sent out of the region in the form of butter, cream, cheese or dried products. The availability of waste products in the west explains the much higher number of pigs in this part of the region compared with eastern areas where sheep normally take second place to cattle.

Where the soils are richer, mixed farming is more common. It is found, for example, in the broad, curved belt of country around the southern margins of Dartmoor from Exeter to Plymouth and westwards to St Austell; in the Vale of Taunton; and in north Devon between Barnstaple and Hartland. But even in these areas of more diversified farming, dairying or stock rearing is still very important. In the Chagford–Moretonhampstead district extensive patches of gravel lying above the granite give rise to warm light soils which have proved particularly suitable for potatoes.

The growth of market gardening illustrates very well the close relationship between natural and economic factors which in a less obvious way is responsible for the agricultural pattern of the whole region. Market gardening requires suitable soil conditions, a mild climate (unless crops are to be grown under glass), favourable aspects providing shelter for tender young plants and a market for the produce. Fig. 25 shows the numbers of market gardeners in 1856 and 1939, and emphasizes the remarkable expansion of this industry. The period of greatest growth occurred between 1880 and 1914, when the railways were built which enabled the produce to be speedily despatched to markets in London, Bristol and the Midlands. Without the railways no amount of individual initiative would have been able to overcome the problem of distance, at least as far as perishable fruit and vegetables were concerned. With markets assured the industry flourished. The Scilly Isles, with an especially early season, have specialized in flowers which reach the market a week or ten days before those from other parts, although competition from early-flowering bulbs developed on the Continent is greater than it used to be. The alluvial deposits of the Mount's Bay–St Ives corridor are the basis of market gardening in that area which like the Fal valley produces not only flowers but also broccoli, spring cabbage and early vegetables. The small fields here, with their stone walls or thick hedges of decorative shrubs, are intensively cropped, and it is common to see the field gates draped with sacking to keep out the wind which is the only climatic hazard. The Tamar valley concentrates on top fruits, apples, plums and cherries, undercropped with bulbs, rhubarb, soft fruits and vegetables.

Around Torbay, Dawlish and Exeter, there is a greater emphasis on vegetables of all kinds which are sold in the large local markets. The Combe Martin market-gardening area is small but noteworthy, being highly concentrated on good soils along the bottom of a deep sheltered valley. As early as 1850 growers of Combe Martin monopolized the prizes given by the Ilfracombe Cottage Garden Society, and the stimulus to large-scale growing was provided by the expanding resorts of Ilfracombe and Lynton. The economic importance of market gardening is clear from the fact that Cornwall alone commonly exports produce every year worth over 4 m. pounds.

Both Somerset and Devonshire enjoy a long-established reputation for cider. In the past most farmers kept a few apple trees and made their own cider, but today most apples are sold off the farms to large factories. Orchards are still found in many sheltered areas, but they are becoming slightly more concentrated in the Clyst valley and in

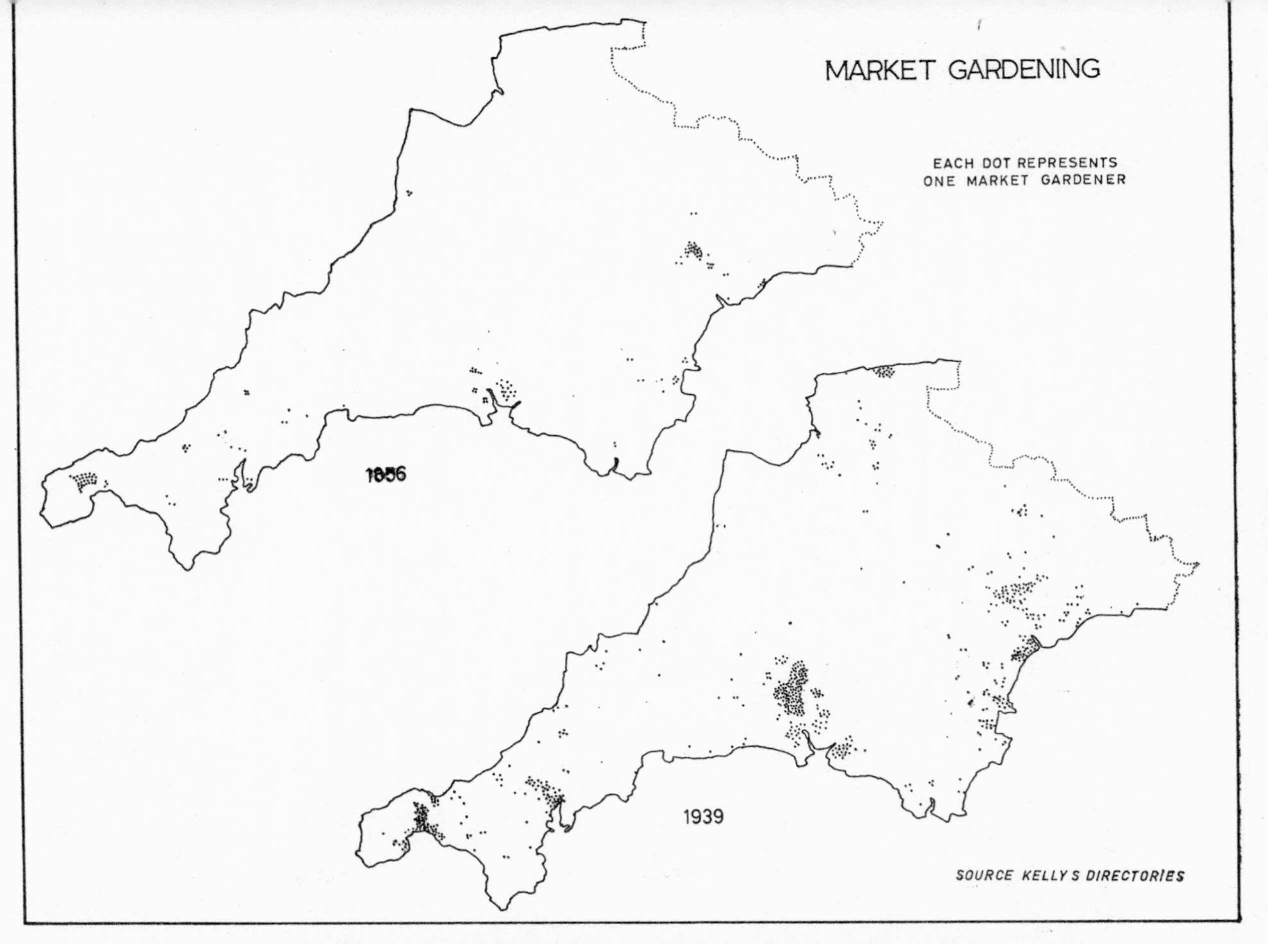

Fig. 25. Distribution of market gardeners in South-West England in 1856 and 1939.

the neighbourhood of Whimple in Devonshire, and in the lowland areas of south Somerset. Most orchards are grassed and serve as pasture for cattle, pigs and sheep.

The use of marginal land is another aspect of farming that deserves mention. On the flanks of Exmoor, Dartmoor and Bodmin Moor farmers have for years waged a dour struggle with advancing bracken and coarse grass to maintain a precarious foothold on the absolute upper limits of cultivation. Fluctuations in these limits reflect both the degree of individual initiative and the prevailing economic situation which determines whether financial rewards are commensurate with effort. Exmoor in particular has witnessed the failure of spectacular and costly schemes directed at introducing arable cultivation on the uplands. They failed because they ignored the effects of elevation and exposure, but now, after more than a century of experiment, the upper margins have been reclaimed. Natural grassland has been ploughed and resown with grasses better suited for pasture, with sheep and store cattle forming the basis of the economy. Hensbarrow, Carn Menellis and Land's End, the other granite uplands all show a fairly high proportion of reclaimed land. This is partly because, being lower, the climate is less severe and partly because they have all been settled by mining communities. In the past much of the reclamation has been undertaken by miners establishing small-holdings near their homes in order to supplement their wages. The decline in population following the closing of many mines at the turn of the century has resulted in the abandonment of some of the improved land, particularly in the Land's End area where the old allotments have been invaded by bracken and coarse grasses. On Carn Menellis and on Hensbarrow, where the mining of china clay still flourishes, the land is still well cultivated and the nineteenth-century pattern of small rectangular fields with stone walls has been preserved.

Especially difficult problems have been encountered in the Somerset Levels which, although similar in origin to the Fens, present a very different landscape. Some attempts were probably made in the Middle Ages to drain parts of the great marsh of Sedgemoor and more concerted efforts following the Enclosure Acts in the eighteenth and nineteenth centuries met with some success. However, there are peculiar difficulties in drainage which arise, not only from the great increase in the discharge of surrounding rivers after heavy rains, but also from the exceptionally high tides of the Bristol Channel which hold up the flood waters. The inclination of the underlying rocks is also significant in that they form a slight ridge of higher land near the coast through which the rivers have to cut their channels. It may also be noted that this area, unlike the Fens, has never benefited from the enterprise and capital of a few large landowners capable of integrating major drainage works over the whole district. In winter the cattle which feed on the lush pastures are moved to the surrounding highlands and the lower ground is often deliberately flooded in order to reduce the strain on the ditches and flood-banks. The cultivation of withy willows used in basket-making is locally important, although competition from abroad is causing some embarrassment to the smaller growers at present.

Nearly all the farms in the South-West are small or medium in size and they are usually worked by family labour. Small farms are characteristic even of the moorland margins, but their effective size is increased since their owners are able to exercise traditional grazing rights on the common lands which have not been enclosed.

The present agricultural pattern has clearly been very much influenced by the character of the soils and by the climate, but it has also been affected by improvements in communications which have allowed the land to be used in a more rational manner. Before the growth of railways and of markets outside the region, self-sufficiency was much more important. Land that was better suited for pasture was often made to produce cereals for human consumption, and failure of harvests was not uncommon. With the decision to import foreign grain the region, not without difficulties, has adjusted itself to a different, more specialized and in the long run more profitable economy based on the sound principle of comparative advantage.

The distribution of the population within the region can best be described in terms of the different kinds of settlements that have evolved. Many of the towns and villages have a special character which they have inherited from the past whilst others have been forced to alter their character to conform to changing conditions. Some of the earliest settlers were the Megalithic peoples who came to Cornwall from Brittany and they and their descendants evolved a way of life which was not seriously challenged for many centuries. The remoteness of Cornwall largely preserved it from the influence of the successive waves of immigrants from the Continent who settled in the rest of England. The Saxons colonized most of Devonshire and it is from them that Devonshire has inherited its characteristically compact villages, with houses, farmsteads and cottages clustered around the church. The River Tamar, however, marks the westernmost limit of Saxon influence and in Cornwall Celtic tradition is still strong. Here there are fewer nucleated villages but instead scattered, isolated farmsteads occasionally grouped together to form hamlets. The churches often stand apart and alone. Cornish farm-houses are usually built of granite with slate roofs and even when they are brightened with lichens they give an impression of sombre austerity. Devonshire builders could draw on a wider range of materials and here houses of brick and timber, limestone, sandstone and colour-washed cob, with roofs of thatch and tiles, add much to the charm of the countryside.

Over the whole of the South-West a network of more or less evenly spaced market towns has grown up. These little towns are usually quiet, unpretentious and attractive, and provide, besides the market itself, the services of, for example, chemists, doctors, auctioneers, solicitors, ironmongers, drapers and outfitters for people in the surrounding villages. Some of these smaller market towns have become in the course of time more important regional centres serving a wider area. North-west Devon, for example, looks to Barnstaple, south Devon to Newton Abbot, east Cornwall to Launceston, and west Cornwall to Truro for rather more specialized needs in terms of shopping, entertainment or medical attention. The functions of these regional centres are varied. Newton Abbot, for example, has railway workshops, Barnstaple factories making furniture, gloves, paints and pottery. Penzance is a resort, fishing port and depot for the collection and despatch of market-gardening produce. Although Bodmin is nominally the capital of Cornwall its role has been usurped by Truro. Truro used to be a port; it was closely associated with the mining industry and was much favoured by the wealthier classes who built many gracious town houses here in the eighteenth and early nineteenth centuries. The splendid Victorian cathedral which rises, like Chartres from the tight cluster of houses around it, confirms the ascendency gained by Truro over the more isolated Bodmin.

Even more important than these regional centres are Plymouth, Exeter and to a lesser extent Taunton. These cities, in addition to many other activities, are the great commercial, social and administrative centres; Plymouth serves the whole of Cornwall and part of south Devon, Exeter almost the whole of Devonshire, and Taunton most of south Somerset. Since the most densely peopled part of Cornwall lies in the south-west, Plymouth acts as a focus for most of the axial routes to and from that county. The presence of Dartmoor to the north and of Bodmin Moor to the west has helped to emphasize Plymouth's position by deflecting the major routes to their southern flanks. The submerged estuary of the Tamar forms a huge natural anchorage the value of which was early appreciated by the Board of Admiralty who established the Royal Dockyard at Devonport in 1688. Railway building led to a great increase in commercial activity in the nineteenth century. New harbours and port facilities were built, new industries mainly devoted to processing imported raw materials grew up, and the fishing industry flourished as the large markets became more accessible. The increase in population is shown in Table 3. New building was at first restricted to the area between the three towns (Plymouth, east Stonehouse and Devonport), and later peripheral growth was conditioned to a large extent by available transport facilities. The city has attracted a great deal of attention by the imaginative and comprehensive programme of rebuilding undertaken after grievous war-time damage. Despite doubts over the future of the dockyards (the largest single source of employment) Plymouth should be well able to maintain its traditional role as a regional centre.

TABLE 3. *Estimated population of Plymouth and Exeter*
(All figures relate to the area within the present municipal limits.)

	1856	*1883*	*1914*	*1961*
Plymouth ..	118,000	153,000	200,000	204,279
Exeter	36,000	42,500	59,000	80,215

Exeter is a fine example of an English provincial capital that has been a city of importance for almost a thousand years. First a cathedral city, it became a great centre for the West country woollen industry in the seventeenth century. It has had an influence extending far beyond the narrow limits of the medieval walls behind which it was long confined. Since the middle of the last century its population has increased steadily. Centralization of local government, the growing complexity of wholesale and retail trade, the establishment of a university and the attraction of the city and its environs as a place of retirement have all been responsible in differing degrees for the increase of population shown in Table 3.

In the region as a whole most of the villages and towns fit into an overall pattern which is characteristic of mainly agricultural areas anywhere in Britain. People in the villages and hamlets are occupied with farming and those in the towns with providing commercial and professional services for the farming communities. However, many of the settlements in the South-West cannot easily be fitted into this general framework. Many of them are, or have once been, engaged in more specialized activities. In broad terms they can be grouped into mining towns; ports and harbours; and holiday resorts.

With the exhaustion of easily worked alluvial deposits of tin the progress of mining operations in the nineteenth century became closely linked with the adoption of new techniques for deep mining. Copper production reached its peak between 1830 and 1860 and that of tin, normally found at greater depths, between 1865 and 1892. Ore-deposits are scattered throughout the granites and killas, but the main areas of exploitation were between Tavistock and Callington, Carn Menellis, and Land's End. The rapid growth of mining created a large demand for labour, and following immigration from the agricultural areas the population rose. Many new settlements, inhabited almost entirely by mining families, grew up and spread around and between the older villages. The long-established towns such as St Just, Penzance, Camborne, Redruth, Truro, Helston and Tavistock, shared in the increase, developing commercial, administrative or industrial activities. Mining of tin and copper virtually ceased about 1918, not because the mines were worked out, but because the cost of pumping them would have made them uneconomic in face of competition from foreign ores. The decline of mining brought great distress. Many Cornishmen emigrated and carried their skills to distant parts of the world. The full harshness of depression has been alleviated, but not yet eliminated, in the Camborne–Redruth area by the survival of factories making mining equipment and by the introduction of some light industries. The landscape still bears eloquent testimony to this short-lived but spectacular episode in Cornish history. Ruined engine-houses and chimney-stacks stand out on the upland skylines and in many small towns and villages the large numbers of Victorian artisan dwellings, the Mechanics Institutes, and the brick chapels strike an alien note, serving as a reminder that the South-West, too, has shared in the Industrial Revolution.

The china clay industry still flourishes and since working on a large scale began in the eighteenth century the major producing area has been on Hensbarrow. This upland is dominated by the ghostly white conical waste-heaps overlooking the deep pits from which the clay is extracted. Most of the villages grew up in the early period of mining and their appearance today is consistent with their origins. Some are small, others like Bugle and Roche have grown into small towns but all look to St Austell, the administrative centre of the industry, for their more specialized needs. Almost the entire output of china clay is exported, mainly from Par and Fowey, but in the past Pentewan, Charlestown and Newquay have also been used.

The long coastline and the character of the rivers help to explain the large number of ports and harbours that flourished in the days of poor overland communications. The bigger ports such as Plymouth, Truro, Barnstaple, Bideford and Exeter, acted as collecting and distributing centres. From them sailing vessels laden with timber, coal and general merchandise penetrated far up the winding creeks and wooded inlets of the Channel coast, and into the small harbours painstakingly created along the hostile Atlantic coast. New ports were built near the mining areas which needed facilities for exporting ores and for importing coal and timber.

The most famous shipbuilding yards were at Kingsbridge and Salcombe in south Devon; Barnstaple, Bideford and Appledore in north Devon; and Bridgwater in Somerset; but in the middle of the last century boats were being built in scores of other places all around the coast. Salcombe, in particular, enjoyed great prosperity between 1800 and 1870 through specializing both in building and operating fast merchant schooners for the fruit trade. Of the many fishing ports Brixham once ranked as the most important in the whole country. It was Brixham men who pioneered the technique of trawl-fishing, and Brixham trawlers dominated the London market through their success in fishing the North Sea grounds.

The decline of many of the small ports was brought about by improvements in communications, and by the introduction of new materials and techniques in ship construction and propulsion. The new iron and steel steamships were faster and cheaper to operate and they rapidly replaced the schooners on long-distance routes. The coastwise traffic, once so important, fell off with competition from road and rail services. Some fishing ports it is true derived a temporary access of prosperity from rail connexion, for their catches could be sent to larger markets farther afield. However, in the final analysis rail connexion only permitted a few fishing ports to maintain themselves in the face of conditions that were constantly changing for the worse. In fact, many of the little ports have only managed to cling to a degree of prosperity by catering for holiday-makers attracted by their picturesque qualities.

The South-West is fortunate in having a very beautiful coastline. The resorts which have grown up along it are varied in character and range from highly exclusive small yachting centres like Salcombe to large popular centres like Newquay and Paignton. One reason for the growing success of the tourist industry in the region is the wide range of tastes for which it can cater. The interest in the sea and in sea-bathing that began amongst the wealthier classes in the eighteenth century started a movement that has been gaining momentum ever since. As was often the case, a royal visit to a beauty spot usually guaranteed its future prosperity, and the Land's End district in particular found favour in this manner. Resorts like Westward Ho!, Salcombe, Fowey, Newlyn and St Ives, owe much to literary and artistic association. In the 1850's determined and affluent travellers reached the less accessible beauty spots by taking coaches from the nearest points on the embryonic railway systems. Summer coach services from Launceston, Barnstaple and Minehead carried many holiday-makers to the little coastal villages of north-west Cornwall and north Devon respectively. It is quite clear that in almost every case the resorts were either created, or grew most rapidly after they became linked to the growing railway network. This relationship is hinted at in fig. 26, which shows not only the growth but also the relative size of the resorts. Naturally not all resorts wished to become popular; some actively discouraged railways in the early period on the grounds that they wished to preserve their exclusiveness. Today it is generally true that exclusiveness and inaccessibility go hand in hand. The cars and motor coaches, which are such a problem now in the season, do permit a greater flexibility of travel, but even so an exclusive resort can retain its character by the simple expedient of not catering for the type of visitor it wishes to discourage.

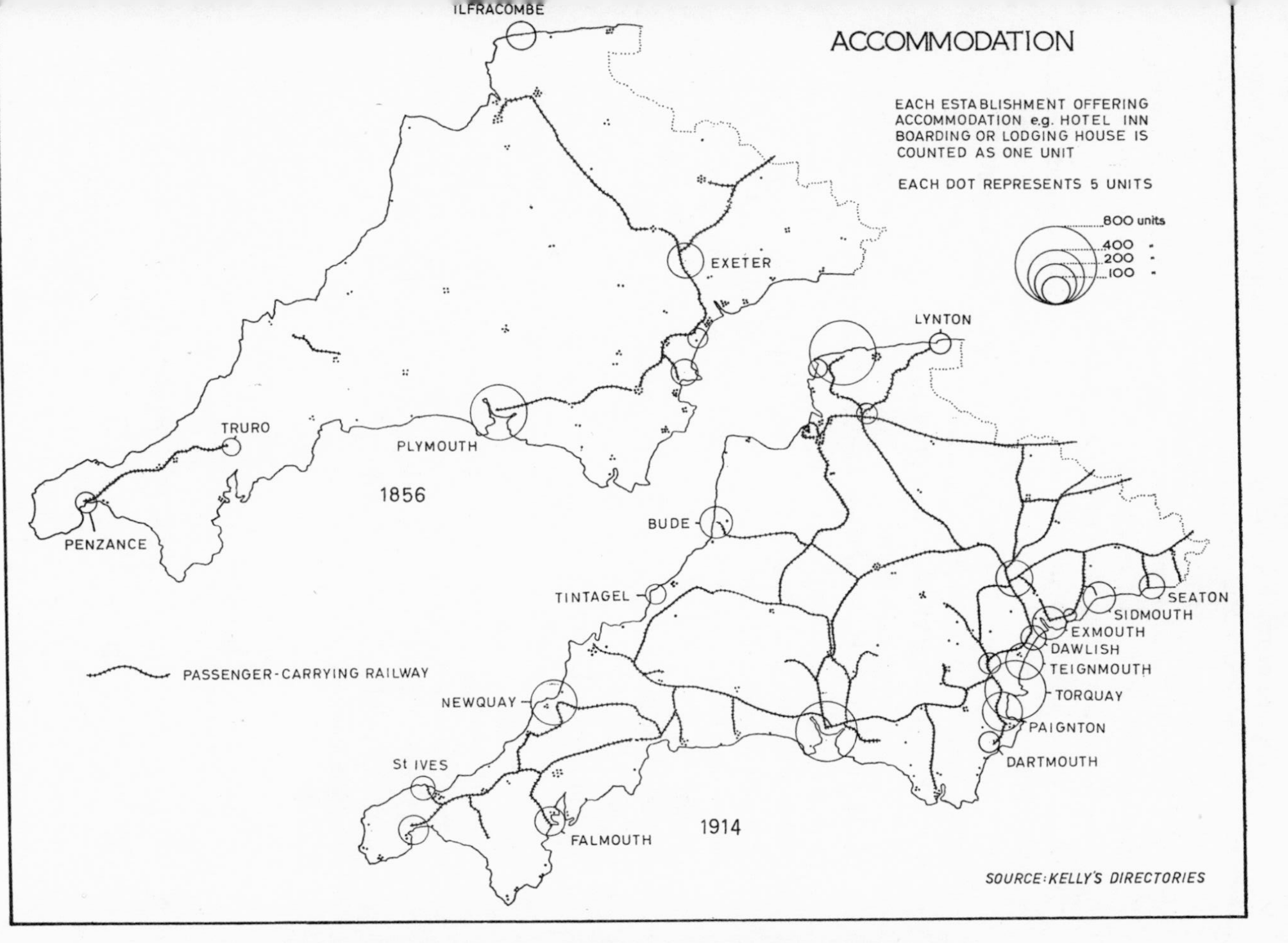

Fig. 26. The resorts and railways of South-West England in 1856 and 1914.

The growth of resorts, although closely related to improved communications, was also a reflexion of much more widespread social and economic changes within the country as a whole. In the nineteenth century the industrial population gradually acquired the taste for an annual holiday by the sea, the satisfaction of which had for a long time been the prerogative of the rich. The growth of this by now traditional social custom was the result of many factors: the general improvement in the standard of living; legislation in the form of successive Factory Acts that established the principles of annual holidays; and the introduction of 'Parliamentary Fares' on the railways making cheap travel available for the first time.

The same natural advantages of climate and scenery that attract holiday-makers also commend themselves to retired people, who, in settling especially along the south coast, have been responsible for a substantial increase in the resident population of this area. With few exceptions, however, the local influence of even the large resorts such as Torquay and Paignton rarely extends beyond their boundaries, and their growth has done little to disturb the long-established relationships between village and market town.

South-West England, then, stands out as a very distinctive region. The many facets of its character derive partly from the isolation which it enjoyed for such a long period; partly from the extent to which the varied potentialities of soil, climate and the sea have been exploited; and partly from its response to changing conditions. It has perhaps been fortunate in escaping the worst effects of the Industrial Revolution whilst benefiting from the social and economic changes which it brought in its train.

Selected Bibliography

V. C. Boyle and D. Payne. *Devon Harbours.* London, 1952.

A. A. L. Caesar. 'The South-West.' In *Studies in Regional Planning,* (G. H. J. Daysh, ed.) 1949.

A. Davies. 'The Personality of the South-West.' *Geography,* **39**, 1954, pp. 243–9.

Devon and Cornwall: A Preliminary Survey. Exeter, 1947.

B. Greenhill. *The Merchant Schooners*. London, 1951.
R. Roddis. *Cornish Harbours*. London, 1951.
J. Rowe. *Cornwall in the Age of the Industrial Revolution*. London, 1953.
J. A. Steers. *The Coastline of England and Wales*. Cambridge, 1948.
S. W. Wooldridge. 'The Physique of the South-West.' *Geography*, **39**, 1954, pp. 231–42.

CHAPTER 12

THE BRISTOL DISTRICT

B. H. Farmer

In 1751 Thomas Kitchin, 'Geographer', published his map of Wiltshire, which showed in the northern part of the county the contemporary east–west roads from Oxford to Bristol, through Highworth and Malmesbury, and from London to Bath and Bristol, through Chippenham. Perhaps because distances along these routes were better measured than those along less frequented north–south lanes, Kitchin's map, although said to be 'drawn from the best authorities and regulated by astronomical observations', is curiously distorted, north to south distances being too short compared with those from east to west.

Kitchin's map typifies a common and still persisting attitude to Wiltshire, and, for that matter, to the whole of the area that forms the subject of this essay (namely, Wiltshire north of the north-facing escarpment of Salisbury Plain, Somerset north of the Polden Hills, most of Gloucestershire, and southern Herefordshire). Many people have distorted ideas about this area, largely because they know little of it except that which they see fleetingly as they speed along the modern equivalents of Thomas Kitchin's east–west roads: for example, along Brunel's main line to the west, from Swindon through Bath and Bristol to Taunton and beyond; or along the newer cut-off through Westbury and Frome; along the South Wales main line, tunnelling under the Cotswolds and the Severn; or along great trunk roads like A4 (the Bath Road) or A46, which runs from Cheltenham to Bath just east of the summit of the southern Cotswolds.

The traveller on such routes as these cannot fail to be impressed by the existence and by the individuality of such cities as Bath and Bristol, or by the Mendips and other spectacular hills. But many, if they stopped to think, would be inclined to agree with Albert Demangeon, who wrote: 'Whilst a journey from Paris to the Central Highlands of France displays to the traveller a series of little districts each with its own special scenery and methods of land utilization, it is difficult to imagine a journey duller and less varied than that from London across the English Plain to the borders of Scotland or Wales.' It will be a main purpose of this essay to prove that, so far as the Bristol district is concerned, Demangeon was wrong. For, whatever may be true of the duller, drift-smothered regions to the north (of which, to be fair to Demangeon, he seems mainly to have been thinking), the Bristol district is emphatically not dull. It must be admitted that the district as a whole does not possess such a well-defined character as the Weald or Fenland, though it has unity of a sort. But its several parts are indeed 'a series of little districts each with its own scenery and methods of land utilization'. There are but few *noms de pays*, it is true, but local regional character has for a very long time been recognized by the inhabitants. One of the more illustrious of these, John Aubrey, wrote nearly three hundred years ago:

> In North Wiltshire, and like the vale of Gloucestershire (a dirty clayey country) the Indigenae, or Aborigines, speake drawling; they are phlegmatique, skins pale and livid, slow and dull, heavy of spirit: hereabout is but little tillage or hard labour, they only milk the cowes and make cheese; they feed chiefly on milke meates, which cooles their brains too much, and hurts their inventions. It is a woodsere country, abounding much with sowre and austere plants, as sorrel, etc. which makes their humours sowre and fixes their spirits. . . .
>
> On the downes, *sc.* the south part, where 'tis all upon tillage, and where the shepherds labour hard, their flesh is hard, their bodies strong: being weary after hard labour, they have not leisure to read and contemplate of religion, but goe to bed to their rest, to rise betime the next morning to their labour.

Even in John Aubrey's day, then, the contrast of scarp and vale was recognized, and its influence on land-use was noted, though few would carry this influence as far as Aubrey did. This contrast, in all its interest and subtlety, has much to do with the present geographical pattern of the Bristol district.

Scarps and vales, in fact, make up something like two-thirds of the area of the district; they are formed in sediments from Triassic to Tertiary in age, affected by a gentle regional dip eastwards or south-eastwards and complicated by gentle warping (fig. 27). Most of the land surface that is neither scarp nor vale is made up of what may for short be called 'old-masses': they are, areas of Palaeozoic

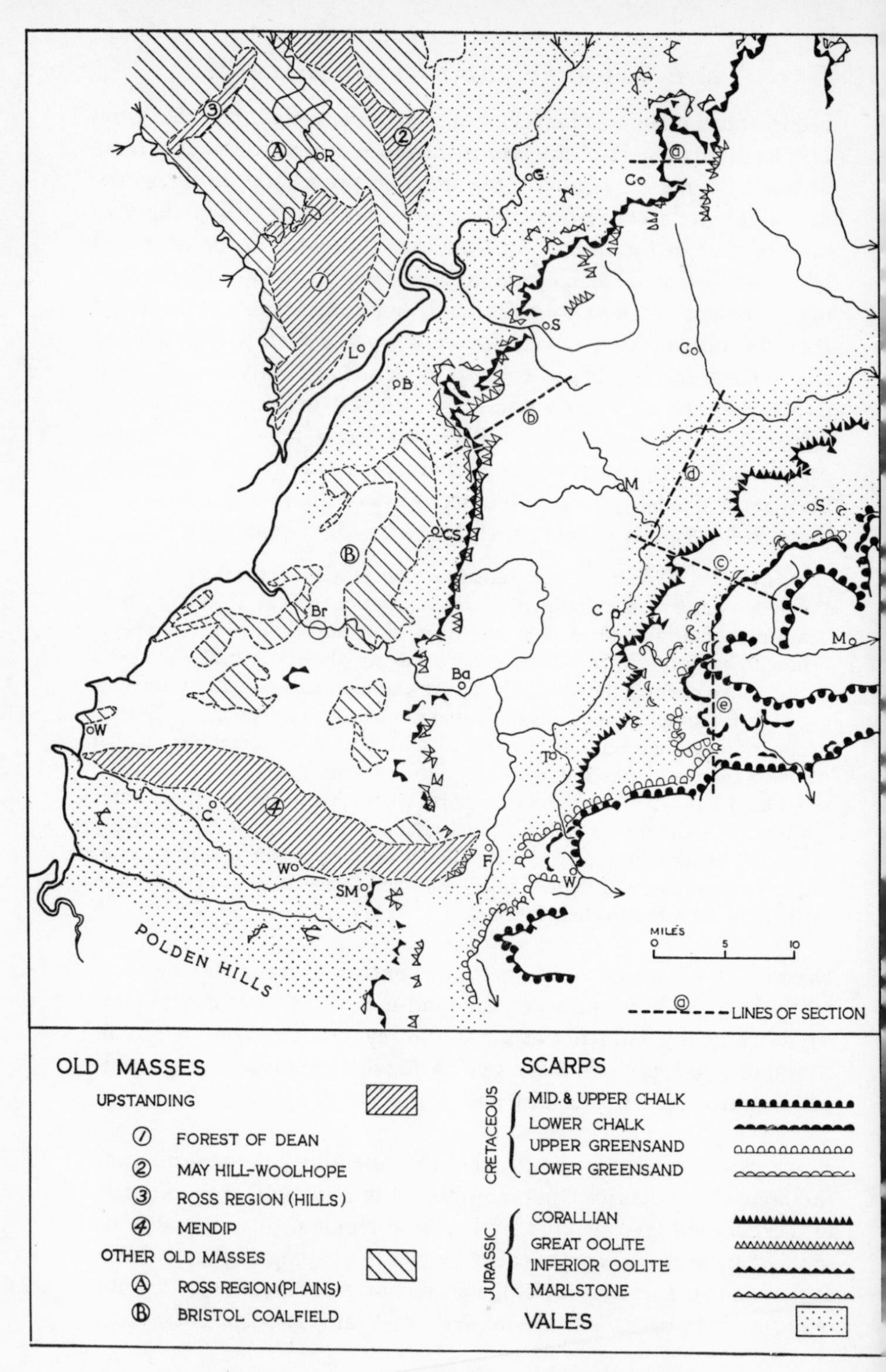

Fig. 27. Scarps, vales and old masses of the Bristol district.

rock, mainly though not universally up-standing, that were probably all once covered by Mesozoic rocks, but are now revealed by the erosion of the latter. Perhaps nowhere else in England is there such an evident interplay between the relief of scarp and vale on the one hand and of old-mass on the other; pairs of old-masses are severed by tracts of scarp, or vale, or both. The intermingling of Palaeozoic and Mesozoic outcrops and of associated relief forms is at its most complex north and south of Bristol itself, but is characteristic of the whole of the western part of the Bristol district.

The old-mass with the strongest character is, without a shadow of a doubt, the Forest of Dean. Here is an English *pays* indeed, lifted above its neighbours by steep wooded slopes, and strongly individual in its relief, vegetation, economy and scenic attractiveness, and in its historic survivals and ancient name. The bounding slopes are developed in Old Red Sandstone, Carboniferous Limestone and Millstone Grit, whose ring-like outcrops betray the basin-structure of the Forest. Within, the dissected Forest plateaux are developed in the sandstones and shales of the Coal Measures, which give loam soils that, under the moist conditions here prevalent, are highly favourable to oak and beech. The woodland became, and has remained for nearly a thousand years, a royal forest. As such it has passed into the care of the Forestry Commission, who husband its timber resources and have done much to develop the amenities of this, Britain's first 'National Forest Park' and most beautiful coalfield. The Commission is still bound by many ancient customs and enactments. For instance, the Deputy Gaveller, who allots portions of seams to the 'freeminers' who have the right to work coal in the Forest, is now an official of the Commission. A few freeminers are still at work, others long ago sold their rights to colliery companies, but by 1965 only one big colliery survived. The Court of Verderers, originally set up to try forest offences, is another historic survival: it may take on a new lease of life and control sheep pasture in the Forest. But for all its beauty and interest, the Forest is a problem area. Haematitic iron ore is no longer worked, the collieries will soon all be closed, Lydney's last tinplate mill shut in 1957; and forestry, sheep and cattle-keeping within the Forest boundary, and mixed farming on the slopes down to the Wye, provide an inadequate living for the people. There are a handful of new industries, but many persons now travel daily to work in Gloucester.

Separated from the eastern ramparts of the Forest by a marked depression lie two small old-masses, mainly of Silurian rocks: May Hill, 969 ft high and crowned by a clump of trees that is a famous landmark; and the Woolhope anticline, a miniature Weald with inward-facing scarps only a couple of miles apart. Here are tiny isolated villages, much woodland, and some orchard and pasture.

North of the Forest and west of Woolhope stretches the Ross region. The northern part of this is an upstanding old-mass made up of the north-west-facing scarps of such hills as Orcop and Dinedor; these are developed in harder beds of the Old Red Sandstone, and are covered largely with woodland and rough pasture. Dinedor Hill magnificently dominates Hereford from the south. The dissected dip-slope of these scarps passes into what is almost vale country. Fashioned in sandstones and marls of the Old Red, it sends a tongue of lowland down between the Forest of Dean and the South Wales coalfield and another past Ross to the Severn. But it is vale country with rolling relief, cut up by the grossly meandering Wye: a chequered red and green country, long recognized as distinctive because of soils lighter than those in most of Herefordshire. There is mixed arable farming, with a certain amount of dairying and much fruit-growing. Ross (5,641) is very much the regional centre.

South-east of the Severn the high limestone plateaux of Mendip rival the Forest of Dean in individuality; and, as in the Forest, steep bounding slopes lift the plateaux up from the surrounding lowland. Structurally Mendip is made up of a series of elongated domes, in the centre of which Old Red Sandstone is apt to appear and to give rough land (for example, Blackdown, 1,068 ft) that contrasts with the bare, pasture-covered, stone-walled plateaux on the Carboniferous Limestone. The effects of the limestone are there for all to see: caves, gorges, swallow-holes, grasses characteristic of calcareous soils. This is traditionally country for sheep, but dairy cattle have displaced many of them in recent years. The steep bounding slopes are often the exhumed and much reduced remnants of desert scarps first formed in Triassic times: with their ash woods and limestone flora they have a beauty of their own. At the foot of the south-facing slopes lie a notable line of settlements, from Axbridge through Cheddar to Wells (6,715). Cheddar is today more conspicuous for the vulgarity of its commercialized tourism and for its market gardens and strawberry growing than for the production of the cheese that bears its name.

The detached limestone hills round Winscombe, Bradfield Down towards Bristol, and other smaller uplands, may be regarded as outriders of the Mendip old-mass at somewhat lower altitude.

North to Berkeley from the Mendips and their outriders, and east to the Cotswold scarp from the Severn, there stretches an area of low but varied relief, of ridges, valleys and small flat-topped hills, with very little obvious pattern; a geological map of it shows outcrops of Palaeozoic and newer rocks scattered in apparent confusion. Yet for all the variety there is some unity, and for all the confusion there is a pattern. For, partly concealed by patches of Mesozoic rocks, the whole region is essentially underlain by a single old-mass, a Hercynian structural basin complicated by faults and subsidiary folds, but, broadly, folded on a north–south axis to the north of the Bristol Avon and on east–west axes to the south. Within the basin, partly concealed, lie the coal measures of the Bristol and Somerset coalfield. The coalfield is unique: if the Forest of Dean's was Britain's most beautiful coalfield, this is its most rural. At peak, the Somerset portion was producing 1,208,000 tons a year (1910) and the Bristol coalfield proper, north of the Avon, 532,000 tons (1870). The latter field declined steadily from 1870 and produced nothing at all from 1949 until the recent opening of a drift at Harry Stoke. The Somerset field declined between 1920 and 1950, but modernization has given it a new, but probably short, lease of life. Outside the towns and their sprawling outliers agriculture is almost as varied as the geology and the soils, though the demands of Bristol encourage market gardening, and there is much dairying and mixed farming.

The Bristol and Somerset coalfield forms an easy transition from the old-masses to the scarps and vales. For its Triassic and Liassic rocks form outliers of the Vales of Gloucester, Berkeley and Sodbury; and church-crowned Dundry Hill, which interposes between Bristol and Mendip, is a detached piece of the Cotswold escarpment.

The Vale of Sodbury follows the outcrop of the Lias clays between the Bristol coalfield and the Cotswold scarp, and consists of gently rolling country at 250 ft to 450 ft, covered mainly by pasture and woodland, and on the whole sparsely peopled away from the town of Chipping Sodbury which, with its district, is growing rapidly in population. At its northern end the Vale of Sodbury opens out into the Vale of Berkeley, which in turns runs northwards into the Vale of Gloucester. The two last-named vales together make up a lowland from 10 to 15 miles wide, seldom over 300 ft, but by no means flat away from the Severn alluvium; for valley gravels, and limestones in the Rhaetic and Lower Lias, give mild but pleasantly varied relief. But the real former of character is clay, either Liassic or Keuper. The heaviness of the soil has long been notorious; this is Aubrey's 'dirty clayey country', long under pasture and in other days the home of Double Gloucester cheese. Now it is a great producer of milk for Bristol, Gloucester, London, and the chocolate factories of the Midlands; a country of early enclosure, small fields and much hedgerow timber. But there are some orchards and market gardens on the gravels. Berkeley is historic but small, and the great market centre is Gloucester (69,733), which has grown tremendously and has engineering, aircraft and other industries. Cheltenham (72,154) has changed from a quiet residential spa into a busy industrial town.

South of the Mendips there stretch the plains of Somerset. Only the part north of the Polden Hills will be touched on here. It contains a number of components to be found in the vales just described: Keuper marl country just south of the Mendip edge, and in the Wedmore 'island'; Rhaetic limestones pleasantly capping the 'island' and forming miniature scarps on the south face of the Poldens; and heavy, pasture-bearing Lias clay country in the Poldens, in Wedmore and, on a larger scale, south-east of Glastonbury. But over a wide area the Mesozoic rocks are hidden by the peat and alluvium of the drained Middle Somerset Levels, along the Axe and Brue.

The Jurassic escarpments to the east of the regions so far described are complex and variable in character. Sometimes, as near Sodbury, they run in a simple unbroken curve; sometimes west-draining valleys penetrate to varying degrees and cut a complicated fretwork pattern (the Bristol Avon penetrates right through scarp and dip slope to the Oxford clay vale beyond).

In section the main scarp is usually formed by the Inferior Oolite, but varies from over 1,000 ft behind Cheltenham to just over 600 ft behind Sodbury and at Bath; while, south of Bath, the scarp, broken by the Mendips, is on a minor scale and is often no more than a series of little steep-sided hills (fig. 28). In places the Great Oolite forms a minor scarp behind the main one; while north of Wotton-under-Edge the Marlstone, an impure iron-stained limestone in the

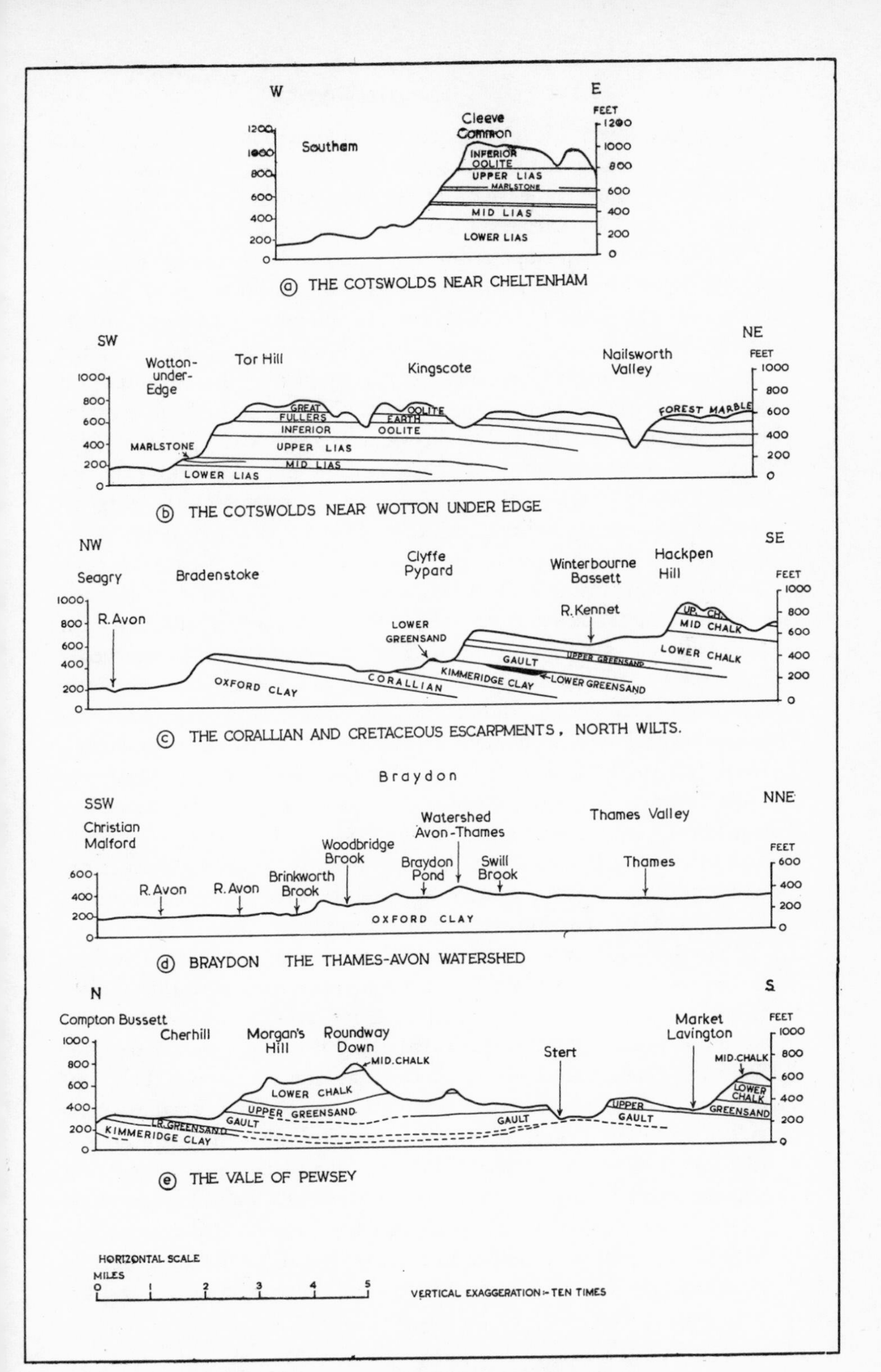

Fig. 28. Generalized sections across the Bristol district.

Middle Lias forms a ledge or even a minor scarp in front of it. There is much woodland on the steep slopes of the scarp edge and on the west-facing valleys; otherwise the land is in pasture, with some arable on flatter land, for instance on the Marlstone shelf.

The Cotswold dip-slope is developed in the limestones of the Great Oolite which give thin, stony soils, and in the variable limestones and clays of the so-called 'Forest Marble' which give deeper soils. At its eastern edge are the limestones and marls of the Cornbrash, so called by Wiltshiremen because of the suitability of its soils for cereals; on south-facing slopes these soils even bore vines in the Middle Ages, for example, at Malmesbury. The dip slope as a whole is a plateau, high, bleak and windswept in the north, lower and more hospitable farther south; this plateau is broken into separate fragments by the deep incised west-draining valleys, and diversified by dry valleys that lead down to the headstreams of the Thames and Bristol Avon. The high Cotswold was traditionally sheep country. In the last twenty years improved methods have increased the output of cereals; and beef stores as well as ewes are grazed. The Cotswold region is deservedly famous for its valley villages and small towns; nowhere does domestic architecture so modestly blend with the natural landscape. In many of the towns there was once a woollen industry. Now there is but little wool, but the Stroud Valley and Dursley have acquired newer engineering and other industries, which have recently also spread to some of the dip-slope towns.

South of the Bristol Avon, Cotswold dip-slope characteristics are only mildly developed, though they are often recognizable in miniature, especially near Frome, which also was a woollen town.

The Oxford Clay vale falls into four well-marked units. The first, the upper Thames valley, is wide, flat, open and pastoral. Gravel terraces follow the Thames and its tributaries, and there are very wide flood-plains. South-westwards along the clay outcrop past the Swindon–Gloucester railway there is suddenly a second belt, one that seems most unlike clay country, for a series of small valleys sever ridges up to 120 ft high. This is the area in which the tributaries of the Bristol Avon are cutting back into the former preserves of the Thames, and it forms a miniature *pays*, locally known as Braydon after a former royal forest. This is indeed Aubrey's 'woodsere country, abounding much with sowre and austere plants'; the clay is unmitigated by gravel or alluvium. Much is still in woodland, the rest but poor pasture. Past Dauntsey lies, thirdly, the valley of the Bristol Avon, which follows the strike of the Oxford Clay from this point until it takes its westward plunge through the Oolites at Bradford. This is a narrower vale than that of the Thames, with a floor some 100 ft lower, and less prominent terraces and flood-plains; a great dairy country, the ancient home of Little Wilts cheese. Chippenham (17,543) lies here, and its engineering and other industries have made it grow rapidly. Trowbridge (15,844) has a similar story. Beyond it the vale narrows farther, and in this, its fourth portion, runs over into Somerset.

In front of the chalk scarps there is a number of discontinuous minor scarps and vales. One, formed by the Corallian limestone, runs brokenly from east of Trowbridge to the Berkshire border, and on to form the Oxford Heights. Behind it is the Kimmeridge Clay vale, the eastern part of which, the Vale of White Horse, is followed by the Great Western Railway through Swindon (91,739), a town that has acquired industries besides its great railway works. The Lower Greensand is mainly obscured by the overstep of the Upper Cretaceous, but locally, as at Bromham, gives good market-garden soils. The Upper Greensand is much more continuous in its outcrop and usually gives rise to a line of villages. It has wide outcrops and marked scarps in the Vale of Pewsey, where it is brought up by an anticlinal axis and separates two inward-facing chalk scarps; and again south of Warminster. All of these minor scarps and vales add their own variety to the delightfulness of Wiltshire: woods on steep slopes, pasture and much hedgerow timber on the clays, a higher proportion of arable on Corallian limestone and on Greensands.

North of the Vale of Pewsey the Lower Chalk forms a steep, prominent, straight or gently curving scarp rising some 200 ft above the clay vale. Its dip slope is some 3 to 4 miles wide and is largely arable and dairy pasture; from it there rises the more irregular, more maturely rounded Upper Chalk downlands, the Marlborough Downs proper. These are more akin to textbook chalkland, traditional sheep-walks but now with much more arable. Marlborough (4,852) lies here. South of the Vale of Pewsey there runs the massive chalk scarp that forms the northern edge of Salisbury Plain.

Such, then, are the old-masses, scarps and vales whose presence and whose intimate association give the Bristol district much of its unique character and unrivalled variety. But a further source of unity and character must be explored. Apart from eastern Wiltshire and eastern Gloucestershire the whole of the Bristol district drains to the Lower Severn. How much, then, has the Severn influenced the nature of the district?

It is undoubtedly in physical geography that the character-forming influence of the Severn has been most felt. For the Lower Severn estuary and its extension, the Bristol Channel, between them provide a base-level at or very near sea-level, and only a very few miles distant from the highest hills of the district. This is, in a sense, only to say that relief here is steep. But there is also a further consequence of this state of affairs. The streams flowing west from the Jurassic escarpments, nibbling back into the scarp by a process of spring-sapping at their heads, have great power because of the low base-level to which they precipitously descend. It seems quite clear that the characteristic fretting of the Cotswold scarp by such streams as the Isbourne behind Cheltenham and the Frome and its tributaries in the Stroud valleys, and by a host of lesser streams, is caused by this process. Perhaps the Bristol Avon has been the most successful of them all; and perhaps it has sapped back through the Oolites to capture not only the subsequent drainage of the Oxford Clay vale, but also a number of Cotswold dip-slope consequent streams that once went to the Thames across what is now the strange country of Braydon Forest. Whatever the explanation of the Avon, the sharp relief of the scarplands overlooking its clay vale, together with the steep slopes of Braydon and the By Brook valley, are alike derived from the Severn base-level, and contrast with the mature aspect of the Upper Thames and Upper Kennet catchments.

But the 'high wild hills and rough uneven ways' of much of the Bristol district cannot merely be explained as the product of a sudden descent from hill-top to Severn plain, or of combes cut by Severn-aided spring-sapping. Why do such high hills survive so near to what is now a major river, and why have they not been reduced in height to become something like the poor apologies for hills that pass for scarplands in East Anglia? Why, too, is the relief so sharp and fresh? Hill-top plateau often meets valley slope in an angle that can almost be measured with a protractor, and valley slopes themselves subsist at a very steep angle in the most unpromising material. The conventional answer to the first two questions is that the dip of the scarpforming rocks carries them higher here as they approach highland Britain, so naturally the hills are higher. This is no doubt true, or partly so. But a fuller answer to all three questions comes from an understanding of the fact that the low Severn base-level is a comparatively recent phenomenon. Pre-Glacially the Lower Severn valley was 200 ft or more higher than it is today and correspondingly shallower in relation to the surrounding hills. The valley was, moreover, occupied by a much smaller and more puny river, for the Severn above the Ironbridge gorge then flowed out to the Dee and, according to recent workers, the Warwickshire Avon above Bredon Hill flowed north-east to the Trent. Quite possibly, too, the preglacial Severn entered the sea considerably to the west of the present mouth. During and since the Pleistocene the Lower Severn has by stages cut down to its present level, aided by the diversion into it at various stages of the Upper Severn and Upper Warwickshire Avon (and perhaps of the Upper Bristol Avon, too); aided also, possibly, by marine erosion giving the Bristol Channel its present shape and the Severn its present point of entry into the sea. Clearly, if the establishment of the Lower Severn as the estuary of a major river at or near present sea-level is a comparatively recent phenomenon, then the characteristic high scarps, fretted spring-sapped valleys, and sharp outlines must owe much to rejuvenation.

Rejuvenation has not only left its mark on scarp and vale; it has caused the most recent incision of the Wye meanders in the Forest of Dean and accelerated the exhumation of the old-masses of the Mendips and of the Bristol coalfield.

The steep relief thus produced by the Lower Severn has had a very great effect in the human geography of many parts of the Bristol district. The water-power of Cotswold streams; the predominance of woodland on steep valley-sides; the provision of defensive sites like that of Malmesbury, poised high between confluent incised meanders; all of these, in greater or lesser measure, owe their existence to steep, rejuvenated landforms and therefore indirectly to the work of the Severn.

But the Severn does not have the direct, integrating effect on the Bristol district that, say, the Rhine has on so much of its valley: the human geographer would hardly, in seeking an alternative name for 'the Bristol district', decide to call it the 'Lower Severn Valley'. For the Severn's own way of life is only lived along a narrow and curiously isolated ribbon following its shores. Here are to be found the declining number of those who sail on the river, who fish salmon by net or by trap or by specially constructed pools, and who catch shrimps and eels and wildfowl (sometimes in conflict with the Severn Wildfowl Trust). But the villages and hamlets and little decayed ports (like Framilode) are cut off from the hinterland by a belt of marsh, salt or fresh, which owes its existence to the extreme muddiness and very high tidal range of the Severn. The Severn, in fact, acted in history as a barrier rather than as a link. For many decades it was only crossed between Gloucester and the sea by the Severn Tunnel and the single-track railway that used the old Severn Bridge:[1] the road bridge is due to be opened in 1966 and linked with M4.

The considerable volume of traffic that uses the Sharpness–Gloucester Canal instead of the twists and turns of the river has its effect on Gloucester and, to some extent, on Sharpness; but otherwise it is but another example of the traffic that, as was said at the outset, regards the Bristol district as something to be crossed on the way from one place to another. The Severn, in fact, has done much to shape the physiography of the Bristol district; but man's use of the Severn does little to give the district its character.

The influence of the sea is another matter. Resorts like Weston-super-Mare (43,938) and Clevedon (10,658) obviously owe their existence to it, and, much more important, so does the westward-looking port of Bristol. But the influence of the sea on Bristol and of Bristol on the district around must be considered separately.

Bristol contains 437,048 people in the county borough but the conurbation is bigger, and is by far and away the largest town in the district; its nearest competitors clearly fall in another and much humbler category. Bath, also a county borough, has a population of but 80,901.

Bristol was already a fortified place and trading centre of importance in late Saxon and Norman times. It lay 8 miles up the Avon, protected from the sea by the heights on which Clifton now stands, and its original site was safe from attack by land, for it lay between the confluent Avon and Frome. By 1250 it was already showing enterprise and civic spirit, and was trading with Wales and Cornwall, and exporting produce from a wide hinterland. By the fifteenth

[1] Two spans and a pier of this bridge were destroyed by an accident in October 1960.

century Bristol merchants had become numerous and wealthy, and their city one of the most important in the kingdom; Bristol's trade had expanded to take in France, the Baltic and Iceland. But the city really came into its own and made the most of its western outlook between 1500 and 1750. From Bristol John Cabot and many other merchant venturers sailed to the New World and helped not only to build up the trade of the city but also to develop territories across the Atlantic: it is significant that there are no fewer than thirty-three 'Bristols' or 'Bristows' in the United States. By 1790, the commerce with America had made Bristol the third richest city in England, and given it such industries as sugar-refining and cocoa, chocolate and tobacco manufacture. By this time, too, the well-known triangular trade had established itself; cloth and trinkets from Bristol to West Africa, slaves thence to America and the West Indies, and sugar, rum and tobacco home again to Bristol. But the earlier part of the nineteenth century saw a decline. Bristol was not well-situated in relation to the new industrial areas on the coal-fields, and lost ground to other ports; the abolition of slavery was a setback to the sugar trade; and the new big ships found it hard to sail up the river to the ancient wharves. The tide turned to some extent with the construction of the river-mouth outports of Avonmouth (1877) and Portishead (1879), and revival was aided by the extension of railways and the cutting of the Gloucester–Sharpness Canal. But Bristol has never regained its former position relative to other British ports, and in particular continues to lack a really heavy export trade; this is in spite of the growth of diversified industries in its hinterland in the last few decades. 1,027,685 tons were exported in 1964, compared with total imports of 7,137,967 tons. There is a bulk trade in the import of grain, timber, oil, tobacco and ores, and a specialized trade in fruit, bananas, and meat. To the deep-rooted industries that sprang from the American trade there have been added paper-making, printing, aircraft construction, engineering of many kinds, zinc and lead smelting, chemicals and oil refining.

Bristol is one of the cities in England whose name springs readily to mind if one is thinking not only of places with ancient histories and a profound civic sense, but also of places which are truly capitals of provinces. It clearly cannot compare in size with many great industrial conurbations, yet, like Norwich, it has a hold on the area round it which is unrivalled by any city within 75 miles or more. It possesses a number of things which are usually associated with provincial capitals: a stock exchange, a branch of the Bank of England, wholesale markets of various kinds, the headquarters of all manner of bodies, and a great and renowned civic university and, since 1965, it has become the headquarters of a new economic planning region for the whole South West.

What relations does the 'Bristol district' that has been considered in this essay bear to the sphere of influence of Bristol as a provincial capital? This is a difficult question to answer, for it all depends on the criteria adopted. Bristol as a shopping centre, as the place to which the housewife goes for things she does not buy from her local or village shop, serves an area that is bigger than that served by any other centre in the 'district'. Bristol, moreover, probably serves in this respect nearly as many people as Swindon, Gloucester and Cheltenham put together.

But, dominant as Bristol's position is, it is by no means the shopping centre for all of the 'Bristol district'. Even apart from the three large towns just mentioned, many smaller centres, down to towns of some 3,000 people only, serve the needs of parts of the district. Moreover, very few people shop across the Severn, leaving autonomous centres like Chepstow and Ross to hold sway there untouched by Bristol's influence though the completion of the Severn road bridge is likely to alter this state of affairs.

In terms of other criteria, Bristol's sphere of influence extends considerably outside the 'Bristol district', notably into the whole of Somerset or even into the whole south-western peninsula; though once again the Severn tends to cut off the Forest of Dean and south Herefordshire and force them to look to Cardiff or some other city; and the length of the south-western peninsula, combined with regional feeling in Devon and Cornwall, means that for some purposes Exeter and Plymouth establish themselves as subsidiary provincial capitals.

But this much can safely be said. Though the 'Bristol district' of this essay is more than the Bristol shopping district and less than the full extent of Bristol's influence as a provincial capital; and though the areas to the north of the Severn are somewhat loosely attached; yet it remains broadly true that over nearly the whole of what has been here called the 'Bristol district', and to nothing like the same

[1] See *Handbook of Marketing Areas of Great Britain* (1955).

extent outside it, going to a city, whether for wholesale marketing, or to consult regional organs of government, or for cultural purposes, means going to Bristol.

Selected Bibliography

John Aubrey. *The Natural History of Wiltshire* (ed. J. Britton). London, 1847. (Aubrey's work was written 1656–86.)

Report of Forest of Dean Committee, 1958. London, 1959.

S. J. Jones. 'The Growth of Bristol.' *Trans. Inst. Br. Geographers*, **12**, 1946, pp. 55–85.

G. A. Kellaway and F. B. A. Welch. *Bristol and Gloucester District.* British Regional Geology (2nd ed.). London (H.M.S.O.), 1948.

F. W. Shotton. 'The Pleistocene Deposits of the Area between Coventry, Rugby and Leamington.' *Phil. Trans. Roy. Soc.*, Series B., **237**, 1953, pp. 209–60.

F. Walker.' 'Industries of the Hinterland of Bristol.' *Economic Geography*, **23**, 1947, pp. 261–82.

F. Walker. 'Economic Growth on Severnside.' *Trans. Inst. Br. Geographers*, **37**, 1965, pp. 1–13.

CHAPTER 13

INDUSTRIAL SOUTH WALES

D. TREVOR WILLIAMS

Industrial South Wales is rightly regarded as extending west to east from the Gwendraeth valley in south-east Carmarthenshire to the Usk valley in Monmouthshire, and north to south from the north crop of the coalfield between Amanford and Abergavenny to the Bristol Channel. Within this arcuate belt, 55 miles from west to east and of a varying width of 15 to 25 miles from north to south, are the coalfield, the metallurgical districts, and the ports. South Pembrokeshire and east Monmouthshire have made their contributions to, and been affected by, developments in industrial South Wales during the past two hundred years, and now seem destined in their own right to expand industrially even more in the future. Beyond the industrial area to the north are the Carmarthenshire Vans and the Brecon Beacons, and to the north-east the Black Mountains which form part of a National Park, sparsely peopled, and forbidding rather than beautiful in its scenery; but in the valleys are man-made reservoirs that supply the coalfield towns and the ports with water for domestic and industrial needs, and along the outcrops of Carboniferous Limestone and the quartzitic rocks of the Millstone Grit are the scars of many disused, and some still active, quarries.

The southern third of the industrial region includes the coastal areas of Carmarthen and Swansea bays, Gower, the vale of Glamorgan (bro Morgannwg), the water-meadows of the reen-drained Wentloog and Caldicot levels, and the plain of Gwent (bro Gwent). Much of it remains as agricultural land, or as cliff lands and sweeping sandy bays of great scenic beauty, as in Gower and the Vale of Glamorgan, attractive to the holiday-maker; but from Kidwelly to Swansea and Port Talbot in the west (where the southern edge of the coalfield cuts across Carmarthen bay, the Gower, and Swansea bay), from Aberthaw to Barry and Cardiff in east Glamorgan, and from Newport to Cwmbrân and Pontypool in the plain of Gwent, metallurgical and manufacturing industries and port activities dominate the landscapes.

Greater warmth, longer hours of sunshine and less rainfall distinguish the low plateaux and ridges of Gower and the Vale of Glamorgan from the coalfield uplands and Brecon Beacons. In the former, mixed farming for dairying and the growing of hay, roots and oats, and some horticulture, are the types of agriculture mainly practised. Their wheat and barley lands of the nineteenth century have become fields of permanent grass and rotation grass for hay, and, in the St Athan-Rhoose area of the Vale, airfields. The coasts of both Gower and the Vale attract many visitors in summer; Mumbles, Porthcawl and Barry Island are the most popular centres, visited by daily trippers from the coalfield valleys and the coastal towns. Under the Planning Acts, Gower is designated an area of outstanding natural beauty, wherein development must be seemly and unobtrusive. The low-lying Wentloog and Caldicot levels, protected by a sea-wall from the incursion of high tides, are under grass and hay for the grazing of dairy cows and the winter-feeding of sheep and cattle. Parts of the levels, immediately east of the Usk estuary and of the town of Newport, have become the sites of a number of large industrial works. The western half of the plain of Gwent, from Newport to beyond Pontypool, is almost entirely industrialized.

The northern two-thirds of industrial South Wales is the coalfield, traditionally called the 'hills' (blaenau) of Glamorgan and Monmouthshire, but, by the dwellers in the coastal towns and ports, the 'valleys'. Most of this area is a wind-swept plateau country of Pennant sandstone, between 800 ft and 1,200 ft above sea-level, where the annual rainfall is from 45 in. to 75 in., and the acidic soils support a natural vegetation of poor grass, peat bogs and marsh, suitable only for the pasturing of sheep, beef cattle and pit-ponies on the unfenced moorlands.

The narrow and deeply incised valleys of north-to-south-flowing rivers that cross the coalfield plateau from the high ground of the Vans and Brecon Beacons, and the wider valleys of some of their tributary streams, etched into the softer shales of the coal measures, give the relief a gridiron pattern. The present courses of the main rivers, with few exceptions, bear little relationship to underlying geological structure; their valleys are cut through successive outcrops of hard and soft rocks alike, and the rivers seldom seem to follow the north–south fault lines that lie in their paths to the sea. Largely for this reason, some geologists believe that the existing river valleys present a splendid example of superimposed drainage, originating on a former covering of Mesozoic rocks since eroded away. The valley floors and sides, once well-wooded and their rivers teeming with fish, are with one or two exceptions, notably the Neath, now masked by slag-tips and surface buildings of disused and active collieries and ironworks, by canals, railways and roads, and by the parallel terraces of single and two-storeyed houses, shops, chapels, pubs and cinemas of ribbon-type towns and villages that merge, unrelieved, into one another down each valley. Towering slag-heaps, linked to the valley pitheads by aerial ropeways, rise above the plateau divides.

The valley routeways from north to south are many: all have their railways and roads and, in the Tawe, Neath, Taff and lower Ebbw, are canals which, in the pre-railway period of 1790 to 1850, were the main means of transport and communication. Through-ways across the coalfield uplands from west to east are but two in number: the northern 'Heads of the Valleys' route from the upper Neath past Hirwaun, Merthyr Tudful, Tredegar and Brynmawr to Abergavenny, to which the north–south valley routes are linked, and the east-central route within the coalfield from Pontypridd via Maesycwmer, Pontllanfraith and Crumlin to Pontypool.

The Carboniferous rocks of the South Wales coalfield, accumulated in the fore-deep to the north of the Hercynian mountain folds of late Carboniferous–early Permian times, are preserved in an oval-shaped synclinal basin, hard up against the older Lower Palaeozoic rocks of central Wales. South of the coalfield, a few eroded stumps of the vast mountain chains of Hercynian–Variscan times remain in the 600 ft monadnocks of Cefn Bryn, Rhosili and Llanmadog downs in Gower, whilst up against the Carboniferous Limestone rim of the south crop of the coalfield in the Vale of Glamorgan are banked the almost horizontal beds of conglomerates, sandstones and marls of the Trias and the shales and limestones of the Lias.

Productive coal-seams occur, within the rocks of Upper Carboniferous age, in the Lower Coal Series and the Upper Coal Series, separated by several hundreds of feet of tough and barren Pennant Sandstone. In the lower half of the Lower Coal Series are some ten to twelve workable seams, 2 ft 9 in. to 9 ft in thickness, with associated bands of shales and fireclays and, in some seams, pins and nodules of clay ironstone. The Upper Coal Series occur in the top half of the Pennant and Supra Pennant; in it are many rich coal-seams such as the Mynyddislwyn, No. 3 Rhondda and Swansea 4 ft seams. The siliceous marls, brownstones and conglomerates of the Millstone Grit and the massive Main Limestone and black shales of Lower Carboniferous age encircle the coal measures, and stand out as high plateaux in the north, and scarp ridges in the west, south and east. As a consequence of the asymmetrical synclinal structure of the coalfield, with its deeper part along an east–west axis lying some distance to the south of the central line of the oval-shaped depression, and with the northern limb rising much less steeply than the southern limb, the surface outcrops of the strata in the geological succession are wider in the north than in the south. Structurally, the coal basin was itself affected by the intense mountain-building to its south during late Upper Carboniferous and Permian times.

East of the Neath valley disturbance are the Meiros, Maesteg and Pontypridd anticlines, arranged in echelon, which have brought the valuable coal-seams of the Lower Coal Series closer to the surface in the southern half of the coalfield, actually outcropping in the Maesteg district of mid-Glamorgan, consequently making them more easily mined. Indeed, in the case of the Pontypridd anticline, the Upper Coal Series above the barren lower part of the Pennant Sandstone have been preserved in the Gelligaer–Blackwood, and in the Caerphilly and Llantwit–Llantrisant synclinal basins to its north and south, respectively. Moreover, greater accessibility to the coal-seams of the Lower Coal Measures occurs along the valley floors of the many rivers that cross this southern part of the coalfield.

West of the Neath valley the principal structural feature is the major asymmetrical syncline, but with the east–west anticlines of the east missing. Along the trough of this syncline the Lower Coal Series are, at present, too deeply seated to be mined but, where they outcrop to the north and west at gentle dips of up to 10°, they have been extensively worked in levels, adits, and relatively shallow pits, in contrast with the far fewer coal workings along the south crop in north Gower where dips up to 40° are met with. In recent years, deep pits (Cefn Coed, Cynheidre and Abernant) have been sunk through the Pennant Sandstone, a few miles south of the north crop, in order to reach the rich anthracite coal-seams of the lower half of the Lower

Coal Series. Geological disturbance (overthrusting and faulting at right-angles to the normal north–south faults) has been severe along the western north crop, adding considerably to the difficulties and cost of mining operations. The strata of the Upper Coal Series are preserved in the Llanelli–Gowerton syncline where they form somewhat featureless low-lying country and, also, in an outlier along the deeply-faulted Dyffryn trough of the Seven Sisters–Crynant valley between the Tawe and Neath Rivers.

The South Wales coalfield is fortunate in having a wide variety of coals, including gas and house coals, coking coals, blended manufacturing coals, caking steam coals, dry steams and anthracites. The proportion of volatile matter in South Wales coals ranges between 5% and 36%. The coals with the highest volatile content are found along the south-eastern and southern outcrops and, when any one seam in the Lower Coal Series is followed to the north-west and north across the coalfield, its volatile content decreases so that the coals with the least amount of volatiles—the high-grade, smokeless and non-caking anthracites (5–9% volatiles) which give out great heat and need a strong blast of air for combustion—are along the north-west outcrop, west of Hirwaun and the upper Neath valley. The coals of the Pembrokeshire coalfield, now no longer worked, are also anthracites. The smokeless steam coals (dry steams 9–14% and caking steams 14–18% volatiles) extend in a south-west to north-east belt across the coalfield from Swansea Bay into the mid-Glamorgan valleys of the Maesteg district, the Rhondda valleys, and the upper valleys of the Cynon, Taff and Rhymni. South-east of this belt are the best coking and manufacturing coals (19–30% volatiles) of the mid-Taff and west Monmouthshire valleys. There occurs, too, in any one locality a less defined vertical increase in volatile content from the lowest seams of the Lower Coal Series to the topmost seams of the Pennant and Supra Pennant. The coals of the Upper Measures of the Gelligaer–Blackwood, Llantwit–Llantrisant and Caerphilly synclinal basins are, generally, good house and gas coals, but many of the most productive seams of these grades of coal, such as the Mynyddislwyn seam, have been worked out. The high-flame coals of the Swansea coastal district are excellent for use in the reverberatory furnaces of the non-ferrous smelting works of this area.

Associated with some of the seams of the Lower Coal Series, especially along the north and north-east crops of the coalfield from the Aman valley to Pontypool and, again, in the Maesteg district, where they come to the surface in the mid-Glamorgan valleys, are the argillaceous, blackband ironstones of low iron content (25–33%) which were extensively exploited in the second half of the eighteenth century and during the greater part of the nineteenth century. The fire-clays of some seams are made into refractory bricks for the lining of steel furnaces, and these clays are also used for 'plugging' in both ferrous and non-ferrous smelting furnaces. The Pennant sandstone is quarried for flagstones, kerbing, dam construction and general building; most of the older houses in the valley towns are built of this type of stone, and it has been greatly used in the construction of the quay-walls and harbour-protection works of South Wales docks and of many reservoir dams in South and Mid-Wales. The Carboniferous Limestone and the silica rocks of the basal division of the Millstone Grit, encircling the coalfield, have been, and are still, much quarried, the former for building purposes and for use as a fluxing agent in blast-furnaces, and the latter for the manufacture of silica bricks used in acid-steel furnaces. Between Pentyrch and Risca the Main Limestone has been metamorphosed into a dolomite, and it is quarried and used for the making of basic-steel in the furnaces of the integrated modern steel-works of South Wales. Haematite iron-ores, as well as lead ores, have been mined in the Main Limestone of this area during the nineteenth century, but the only productive haematite iron-ore mine at present is at Llanharry, and there is no lead-mining at all. The mining of bituminous coals of the coastal district from Llanelli to Margam and of the culm (shattered anthracite) of the Pembrokeshire coalfield dates from at least the early fourteenth century. An expanding coastwise coal-trade to the south-west counties of England and an overseas trade to Ireland, the Channel Islands and France existed in Tudor and Stuart times.

An early centre of metal-working in South Wales was around Pontypool on the eastern edge of the coalfield; here, it is claimed, began, in the late seventeenth century, the coating of iron sheets with tin and the lacquering of iron sheets for decorative ware. Another was the Swansea metallurgical district where the first copper works was built at Neath Abbey in 1584. More than a hundred years later, copper works were sited at Melincrythan, Neath (1695) and Swansea (1717). Copper ores were imported as return cargoes in coal-ships trading to Cornwall and Devon, and supplemented by Welsh and

Irish ores. Rapid expansion of the smelting of both non-ferrous ores of copper, lead and zinc, and of iron ores, took place during the eighteenth century, and the Swansea district became a thriving metallurgical area in the following century.

The wealth of its coal and iron ores, and its position on an inlet of the Atlantic Ocean, made South Wales a favoured industrial region; the decisive period in its history has most certainly been the past two hundred years. The Industrial Revolution in South Wales was founded upon two activities, coal-mining and the metallurgical industry: the former was, by its geology, restricted to the coalfield, but its indirect influences led to the establishment and growth of the exporting ports along the seaboard from Llanelli to Newport; the latter contributed to expansion in the older districts around Pontypool and Swansea, and to the formation of new industrial areas within the coalfield and along the coastal plain.

Coal mining spread, progressively, to all parts of the coalfield during the nineteenth century. The steam and bituminous coals of the north and east outcrops between Hirwaun and Pontypool, the anthracite coals west of the upper Neath, and the steam and bituminous coals of the Maesteg district were mined to provide fuel and coke for the blast-furnaces and re-heating furnaces of the ironworks. The house coals of the Upper Coal Series in the Gelligaer, Caerphilly and Llantrisant synclinal basins were mined from the early decades of the nineteenth century. From the 1840's onwards, with the introduction of hydraulic pumps and better ventilation-airways underground, the valuable steam and coking coals of the Cynon and Taff valleys in east Glamorgan, and the Rhymni, Sirhywi and Ebbw valleys in west Monmouthshire, were extracted to satisfy demands not only of the railways and industries at home but also of the merchant and naval fleets of the world, and of markets overseas.

Mining of the steam coals of the Rhondda valleys flourished after 1860, too late for the development there of a metallurgical industry, as the low-grade iron ores of the coalfield were already being displaced by the higher-content haematite ores from Spain and Scandinavia. Cliffe, writing in 1847, could then describe the two Rhonddas as beautiful streams, flowing through peaceful meadows and a well-wooded countryside, and Blaenrhondda, at the head of the Rhondda Fawr, as a pastoral district of sheep farms, its air aromatic with the smell of wild flowers 'where a Sabbath stillness reigns'. Wilkins mentions the practice, at this time, of the ironworkers of Hirwaun and Aberdâr of dynamiting the waters of the two Rhonddas on their poaching excursions.

Although an iron industry had existed in the anthracite field for a few decades after the introduction of Neilson's 'hot blast'—pre-heated air—in the 1830's, the mining of anthracite did not become of real importance until after 1880 when, in addition to its special uses in the brewing and baking industries, it found new markets on the continent of Europe and, in the 1920's, in Canada as a fuel for domestic stoves and the central-heating of blocks of flats and offices.

South Wales became, more and more, an exporting coalfield. In 1913, South Wales ports exported to overseas markets as cargoes and ships' bunkers nearly 35 m. tons of the coalfield's total production of 56·8 m. tons of all classes of coal; 3 m. tons of a saleable output of 4·8 m. tons of anthracite in that year were shipped abroad, practically all of it from the ports of Swansea and Port Talbot. Cardiff and Barry docks exported 26·5 m. tons of coal, coke and patent fuel in 1913, and Newport over 5·5 m. tons. The two docks at Barry, completed in 1890 and 1898, and linked to the valleys by a railway via Peterston and Pontypridd, were financed largely by coal owners of the Rhondda and Taff valleys in an attempt to break the monopoly possessed by the Bute docks, Cardiff, and by the Taff Vale and Rhymni railways in the shipment of coals, coastwise and foreign.

The iron industry in South Wales really began in the second half of the eighteenth century with the extraction, by scouring and shallow patchworkings, of the blackband and clay ironstones of the north and east outcrops of the coalfield. It gave rise to the belt of 'hill' towns from Pontypool to Hirwaun, of which Ebbw Vale and the dual town of Merthyr Tudful-Dowlais are, perhaps, the most famous. The heyday of their prosperity was from 1760 to 1860. The next forty years, from 1860 to 1900, when the change-over from iron to steel for rails and sheet-rolling occurred, were the most crucial in the history of the 'hill' ironworks. In 1860, there were 29 ironworks along the north and east outcrops of the coalfield between Hirwaun and Pontypool with 148 blast-furnaces built, of which 104 were 'in blast'. By 1880, the number of ironworks had dropped to 20, and only 52 furnaces were 'in blast'. Twenty years later, in 1900, there were but 9 active works, and 23 furnaces charged during the year. A few firms installed Bessemer converters or Siemens-Martin open-hearth furnaces, and became steel companies; the large majority simply closed their works; some became primarily coal-mining companies while retaining the foundry and engineering sections of their iron- and steelworks for the manufacture of rails, fastenings (chairs and sleepers) and mining gear for their collieries, and to carry out maintenance repairs.

The early iron towns of the hills were tied to their raw materials (iron ore, limestone, coal and coke), but when home and foreign high-grade ores were brought by sea to the South Wales ports there was a consequential shift of blast-furnace smelting and semi-manufacturing of iron and steel to dockside sites at Cardiff, Newport and Port Talbot. The manufacture of pig iron (blast-furnace smelting) and of iron and steel ingots (re-heating furnace) involved an appreciable loss in weight of as much as two-thirds in low-grade ores and of one-third in haematite ores, and they came to be associated with the importing ports of raw materials to save transport costs. The rolling of iron or steel ingots into bars in steelworks, and of bars into sheets in blackplate and tinplate works, meant a high weight-ratio of raw materials (steel ingots and tinplate bars) handled per man-unit; thus, tinplate mills and steelworks became closely linked in location. When steel tinplate bars could be imported cheaply by sea from Belgium and Germany during the twentieth century, new tinplate and sheet-works in South Wales were erected near to the importing ports of the Swansea district and, to a less degree, of Cardiff and Newport.

The iron industry of the anthracite coalfield and of the Maesteg district, as well as that of the localized ironworks in the Pentyrch-Trefforest area of the mid-Taff valley south of Pontypridd, did not outlive the crucial period of 1860–1900. However, the Swansea industrial district succeeded in maintaining a prosperous non-ferrous smelting and refining industry and, after 1870, in adding to it a flourishing ferrous industry. The iron and steel industry was based, in the main, on about twelve steelworks, with Siemens-Martin open-hearth furnaces and bar mills, and over fifty sheet and tin plate works in which the steel bars were cut and rolled into sheets. Steel sheets when uncoated are known as blackplates and are used in the manufacture of hollow-ware and enamel-ware and can be stamped, pressed and shaped for many purposes, including the making of car bodies, washing-machines and refrigerators. In tinplate works, steel sheets are coated either with a thin layer of tin to make tinplates, much used for the canning of food and drink, or with a mixture of tin and lead to make terneplates, used as drums for paints and varnishes. Galvanizing works make heavier sheets coated with spelter, almost pure zinc, which are used for roofing purposes, especially in tropical and undeveloped lands, and for storage containers.

Coal mining and the iron and steel industry led to a remarkable growth of population in South Wales between 1760 and 1913; an indication of this population growth is given by a comparison of the populations of some coalfield towns and ports in 1851 and 1921 (Table 4). People migrated from the rural and agricultural counties

TABLE 4

Town	*Population in*	
	1851	*1921*
Merthyr Tudful ..	46,000	80,000
Aberdâr	15,000	55,000
Rhondda	2,000	162,700
Ebbw Vale	14,400	35,400
Newport	19,320	92,350
Cardiff	18,350	200,180
Barry	346	39,000
Port Talbot	7,100	40,000
Swansea	31,460	157,550

of South and mid-Wales, attracted by the comparatively high wages paid in the mines and ironworks. During times of depression in the North Wales slate quarries, large groups of quarrymen moved south to seek employment and, eventually, to settle in South Wales. The Welsh immigrants from the rural areas and from North Wales spoke Welsh; they were of peasant stock, Nonconformist in religion, and with a great love of education, choral singing and the Eisteddfod. It was their culture that fashioned the ways of life in the industrial valleys of South Wales. Immigrants came, too, from the rural counties across the English border, more especially from Gloucestershire, Somerset, Devon and Cornwall in the early decades, and, after the mid-nineteenth century with the extension of railways and the growth of shipping, from the southern and eastern counties of England, and from Spain, Italy and continental Europe.

The fusion of people of diverse origins and heritages has resulted in virile, hard-working and talented communities, with a deep attachment to their new homes. As late as 1913, almost one-half of the inhabitants of the coalfield spoke both Welsh and English, and the proportion was much higher than this towards the north and north-west, but it decreased to the east and south. The linguistic trend in industrial South Wales has been, first, to an increased bilingualism and, later, to almost monoglot English-speaking communities.

The exporting coalfield of South Wales was severely hit by the loss of foreign markets, and increasing competition within them, at the end of the First World War. The introduction of diesel engines and internal-combustion engines, and the use of oil and petrol in transport on land, sea and air, and the electrification of railways, were some of the contributory causes. Amalgamation of ownership of mines rapidly took place; by 1938, the production of over 40% of the output of steam and bituminous coals was in the hands of two colliery companies, Powell Dyffryn Ltd., and the Ocean Coal Company, and nearly 90% of anthracite coal was mined by two firms, Amalgamated Anthracite and Messrs. Evans Bevan. Pits were ruthlessly closed and concentration of output centred on the larger, and more economic, mines.

Since 1920 there has been a continuous decline in output and employment in the South Wales coalfield; some parts suffered more seriously than others. The coalfield produced over 46 m. tons of saleable coal in 1920, and there were 225,800 wage-earners employed; in 1938 the output had fallen to 35·3 m. tons, and the number of wage-earners to 134,000. Since nationalization in 1948 the South-western Division of the National Coal Board has spent over £100 m. in improved lay-outs of underground workings and in mechanization at the most productive pits, and in sinking new mines in the anthracite coalfield. Uneconomic mines have been closed. The output in 1958 was 22 m. tons (19 m. tons of steam and bituminous coal and 3 m. tons of anthracite), and the man-power employed was about 90,000, but the average output per man-shift had risen since nationalization by over 25% to 20·85 cwt. The number of collieries, excluding small licensed mines, is now 120 (fig. 29) whereas in 1920 it was 467. Coal-mining in South Wales still employs more workers than any other single industry. Two mining areas stand out prominently, the middle coalfield belt of coking and manufacturing coals, east of the

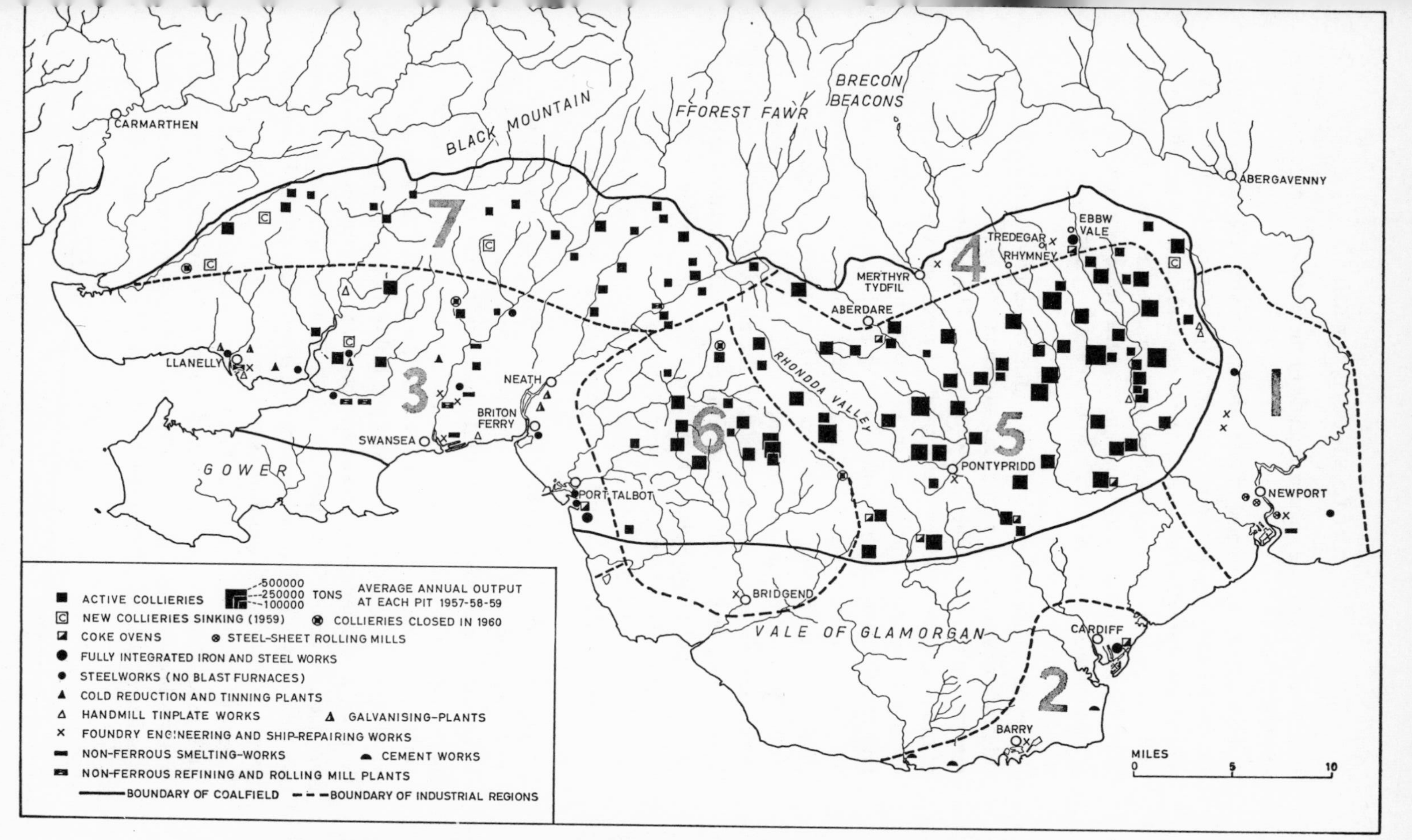

Fig. 29. Industrial South Wales. Coalmines, metallurgical works and industrial sub-regions (1–7).

Neath valley, and the anthracite field of the north-west outcrop, west of Hirwaun.

The output (19 m. tons) of bituminous coals in 1958 was consumed in coke-ovens (5·3 m. tons), industry (over 5 m. tons), thermal-electricity generating stations (3·5 m. tons), and railways (2·2 m. tons); only about ½ m. tons were exported. However, nearly 1 m. tons of the 3 m. tons of anthracite mined in that year were exported overseas; ½ m. tons were used in electricity power plants, and 1½ m. tons supplied the home market.

The metallurgical industries of South Wales have undergone considerable changes in location, organization and manufacturing processes since the end of the First World War. The non-ferrous industry of the Swansea district moved rapidly towards amalgamation of producing firms and concentration of production in a relatively few large works, such as the Hafod copper works, Swansea, the Llansamlet spelter works, and the Mond nickel works. New non-ferrous plants have been built in this area in the last twenty years; these include the duralumin and titanium smelting and rolling works at Waunarlwydd, west of Swansea, and the aluminium rolling mill at Resolven in the Neath valley. Moreover, substantial non-ferrous smelting and rolling plants have been sited in the Newport area since 1940. At Rogerstone, near Newport, is a large aluminium works with bar and rolling mills and, at Newport, a modern works, sited close to the Uskmouth electricity generating station, smelting imported bauxite to alumina.

Soon after the end of the First World War industrial depression again assailed the four iron and steelworks of the north crop. Cyfarthfa works, Merthyr Tudful, shut down entirely; Dowlais, in 1930, completed the transference of its blast-furnaces and steel-making furnaces to East Moors, Cardiff, retaining only the foundry and engineering branches at Dowlais; Ebbw Vale ceased production in 1935; and Blaenavon, whose two blast-furnaces were put out of blast in 1926, continued for some years the operation of the waggon-repair shop, steel-tyre section, coke-ovens and by-product plant. Blaenavon works is now dismantled and its site cleared.

Ebbw Vale was, somewhat fortuitously, saved from extinction as an iron and steel centre. The introduction of the American-type, continuous hot-strip mill for the rolling of steel ingots into coiled sheets was being actively considered by the steel firm of Messrs. Richard

Thomas in the 1930's. The process cut down the weight of material handled per man-unit to a remarkable degree as did, to a lesser extent, the earlier introduction of semi-continuous rolling mills at the Orb works, Newport, and the Albion steelworks, Briton Ferry, near Neath. The hot-strip mill could be integrated with blast-furnaces, coke-oven plants, and the cold-rolling and electrolytic-tinning of the coiled strip into blackplates and tinplates in one works.

There were at least three competing areas for the siting of such works: the old iron centres of the Welsh hills, the iron-ore importing ports and steel towns along the South Wales coast from Newport to Llanelli, and the Jurassic iron-ore belt of Lincolnshire, Northamptonshire and Oxfordshire, relatively close to the Yorkshire-Derbyshire-Nottinghamshire coalfield, and to the Midland industrial towns in which the sheets would be used for the manufacture of motor-car bodies, boilers and stampings of many kinds. Geographical inertia, social considerations, relative needs and, possibly, governmental pressure and strategic advantages turned the scales in favour of Ebbw Vale. Messrs. Richard Thomas, whose main activities hitherto had been in west South Wales, bought the Ebbw Vale steelworks, the associated Lancaster's steam collieries, and the iron-ore mines at Irthlingborough. The long and narrow site of the steel plant and mills was cleared, and two blast-furnaces, a battery of coke-ovens and three modern open-hearth furnaces were built, and an American-type, continuous hot-strip mill, with cold-reduction and electrolytic tinning-plants, laid down. The new developments revivified the town of Ebbw Vale and the nearby centres of Tredegar and Beaufort. The new works started producing in 1938 in time to cope with the armament demands of the Second World War. Its prosperity continues; additions and renovations to blast-furnaces, steel-furnaces, coke-ovens and finishing departments have been made since 1938, and its main output is in the form of sheets and plates for the automobile industry of the Midlands.

The selection of Port Talbot as the site of a second strip-mill (the Abbey works) after the Second World War was, at any rate, in an existing ferrous-smelting and semi-manufacturing district. The Port Talbot steelworks was long established and, during 1915–18, the adjoining Margam steelworks, with modern blast-furnaces and coke-oven plants, was built. Potential man-power was available on the consequential closure of out-moded steelworks and tinplate mills;

iron ores could be imported from West and North West Africa, Canada and Scandinavia, through both Port Talbot and the nearby Swansea docks, and the haematite ores of Llanharry in the vale of Glamorgan and the Oolitic ores of the Jurassic belt could be brought to it by rail. But it was not a complete integration of manufacturing processes at Port Talbot; the cold-reduction and electrolytic-tinning plants were sited at Trostre, Llanelly, and at Velindre, Swansea (20 miles and 12 miles by rail, and 15 and 9 miles by road, respectively, from the Abbey works), in the heart of the long-established Swansea steel and tinplate area. Employment and social factors undoubtedly were weighted against the full and strict application of economic laws.

A third strip-mill, integrated with blast-furnaces, coke-ovens and steel-furnaces, is, in 1961, being built at the Spencer steelworks, Llanwern, near Newport. The site is 45 miles nearer to the Jurassic iron-ore fields and to the consuming areas of the Midlands and the Home Counties than is the Abbey works, Port Talbot, and transport and power facilities, as well as port advantages, are equally available. The concentration of production at the integrated steelworks of Ebbw Vale, Port Talbot and Newport, has meant the closure of almost all the outmoded blackplate and tinplate mills in South Wales, and of many of the steelworks in west South Wales.

During the inter-war years re-rolling firms sought coastal location for their plants in order to take advantage of importing cheaper Continental steel bars, and of exporting finished sheets to overseas markets. In the 1920's Messrs Lysaghts built the Orb works at Newport and, in the 1930's, Messrs Whitehead of Tredegar transferred the sheet-rolling sections of their works to a new site at Newport and erected there a semi-continuous heavy sheet-rolling mill (Courtabella works). Newport and Cardiff had become important steel centres. By 1938, more than 6,000 workers in the town of Newport were employed in its ferrous industries, including re-rolling, engineering and ship-repairing, foundries, and the making of weldless steel tubes and railway waggons. To this total must soon be added the estimated employment figure of about 6,000 at the Spencer steelworks, Llanwern, on the eastern outskirts of the town. The integrated Cardiff-Dowlais works makes heavy structural steel as well as bars, and the adjoining Castle engineering works of an associated company manufactures wire-rod, nails, nuts, screws and bolts (previously produced in the company's works at Rogerstone, now closed); together they employ over 6,000 workers. In the area of Cardiff docks are structural steel and heavy engineering works, ship-repairing yards, foundries for castings, and factories making steel window-frames, enamel hollow-ware, boilers and chains.

In 1938 more than one-half of the output of South Wales tinplate was exported and most of the production of sheets and tinplates for the home market was sent to the Midlands and the Home Counties for final manufacture. Now, however, the semi-manufactured sheets are turned into car-radiators at Llanelli, into refrigerators and cold-storage equipment at Swansea, and into washing-machines at Merthyr Tudful and Aberdâr. The steel produced is also made into end-products in many light and heavy engineering factories throughout South Wales.

The industrial depression and unemployment of the inter-war period was accompanied by mass movements of population away from South Wales, more especially from the eastern coalfield valleys. It is estimated that the absolute loss in population during the years of *laissez-faire* between 1920 and 1939 was 400,000 persons. The unemployed workers were compelled to chase the new manufacturing whose favoured location was the belt that stretches across England from the Mersey–Humber line through the Black Country to London and the Home Counties.

Government action to ameliorate unemployment and to arrest the exodus of people from the depressed coalfield areas resulted in the greater part of the coalfield area, east of the Neath valley, becoming a Special Area. The chief result of this legislation was the development of the Trefforest Trading Estate to serve, primarily, the badly hit Rhondda, Taff and Caerphilly valleys. During the years 1937 to 1945 a number of Royal Ordnance factories and war 'shadow' factories in South Wales attracted both male and female labour. To maintain full employment in the post-war years, South Wales was declared a Development Area, and benefited from the government's industrial policy as laid down in the Distribution of Industry Act of 1945.

Since 1945 over 450 new factories and extensions of existing factories have been erected in South Wales which employ 120,000 persons in a variety of manufacturing industries. The sites of these factories range in type from the nodal industrial trading estates, with their 20 to 70 factory buildings, at Trefforest, Hirwaun, Bridgend and Fforestfach, near Swansea, to the nucleated group of 2–5 factories, usually in the larger centres of population, and to the single factory, which may be small in area or may extend over several hundred acres and employ 500 to 2,000 people.

Such factories have re-vivified many coalfield towns, created new industrial districts and expanded others, and brought diversity of occupations to most parts of industrial South Wales. Although the inherent locational advantages of the southern coastal plain and of the eastern fringe-belt from Newport to Pontypool surpassed those of the coalfield, there has been achieved, on the whole, a well-balanced regional pattern with a wide diversity of occupations. At the present time, about two-fifths of the insured workers in industrial South Wales are employed in coal mining (90,000) and metal industries (130,000), one-fifth in the light manufacturing trades, and the remaining two-fifths in the so-called tertiary occupations, including transport and the distributive trades, shops and offices, professions, personal services and central and local government. Directly related to this expansion of industry has been the concomitant development of the ancillary services of gas, electricity, water and transport (fig. 30).

The Wales Gas Board has linked industrial South Wales by two gas grids, one for the west and the other for the east, and thereby given to both domestic and industrial users a greater assurance of supply. The Board purchases great quantities of gas from the coke-oven plants of the National Coal Board and of the steelworks.

The demands of the new factories and of the expanding heavy industries for electric light and power have been satisfied by the building of new thermal electricity generating plants on carefully selected sites. In 1948 the chief power plants were at Swansea, Cardiff and Newport in the coastal towns, and at Upper Boat, near Trefforest, and Llynfi, near Maesteg, on the southern border of the coalfield; all were thermal generating stations using duff, or small coal, for firing boilers, and tremendous quantities of water. Since then, the South Wales Electricity Board has built new generating stations at Carmarthen Bay, near Llanelli, Uskmouth, near Newport, Rogerstone, and Aberthaw, near Barry. Sites had to be carefully chosen away from the coalfield, because of subsidence dangers, and near to the coast to obtain supplies of cooling water from the sea,

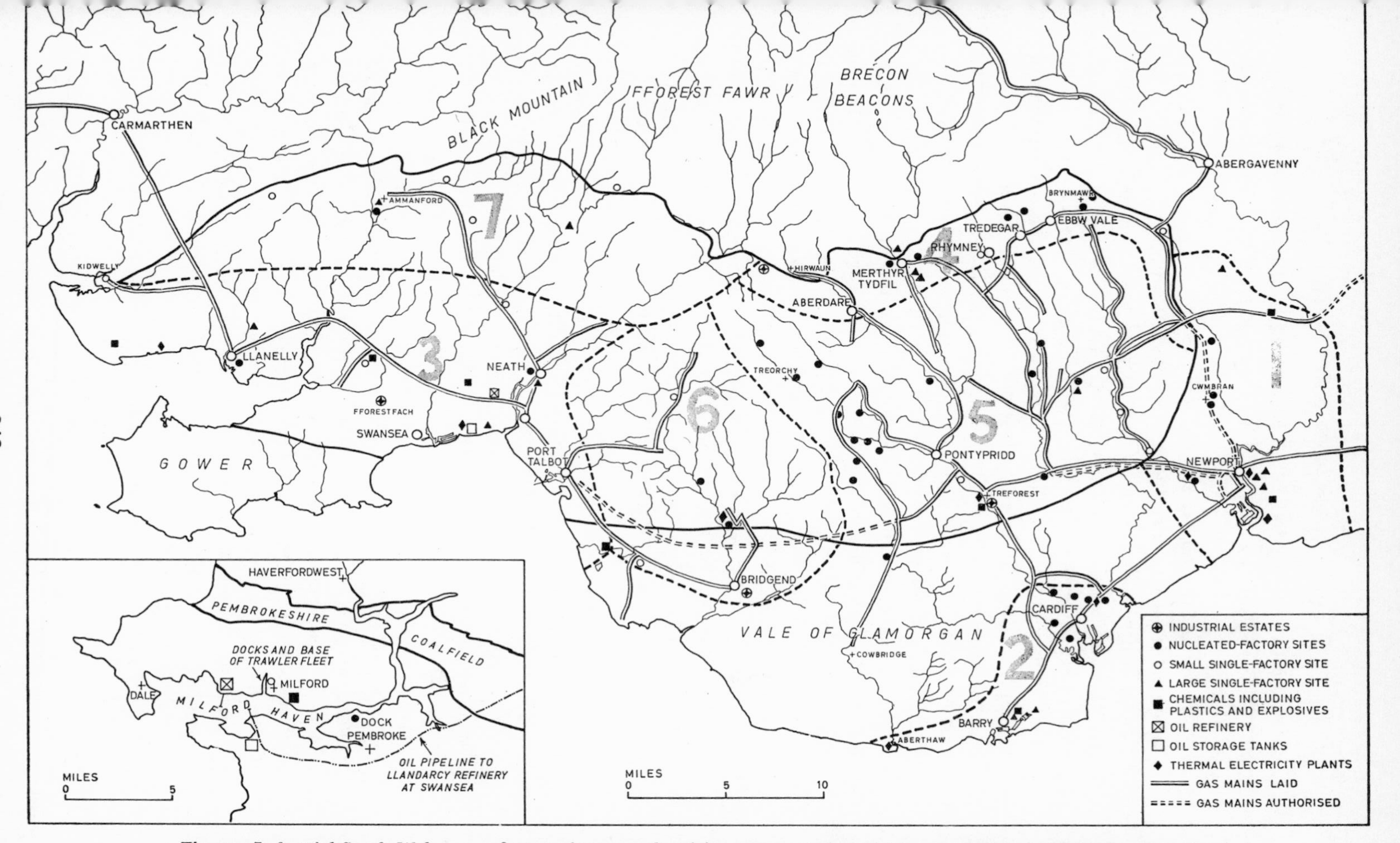

Fig. 30. Industrial South Wales: new factory sites, gas, electricity, water supply and transport and industrial sub-regions (1–7).

sites not always easy to find on account of the tidal rise and fall of as much as 45 ft, and of the need for firm and solid rock foundations in building.

There has been no extension of the existing railway network either for goods or passenger traffic, and no electrification of a single railway line in South Wales as yet. Great improvements have been carried out to many roads, more particularly to the 'Heads of the Valleys' road, to the Merthyr Tudful to Cardiff road, and to the main coastal road. The route to the Midlands is improved by the Ross spur road which links up with the main Bristol to Birmingham road. The lower Neath bridge has shortened the distance by road between Port Talbot and Swansea and west Wales by many miles, and has by-passed the bottle-neck of the narrow river bridge in the town of Neath.

Since 1920 there has been a large petroleum refinery at Llandarcy, Swansea; its productive capacity has been greatly increased since 1945, and the refinery is now linked by a 60-mile long pipe-line with an oil-discharging depot on the south side of Milford Haven as well as with a tank depot at Queen's Dock, Swansea, 3 miles from the Llandarcy refinery, where the oil-tankers at present discharge their crude oil cargoes. A mile or two west of the town of Milford, on the northern side of Milford Haven, a refinery has been built, together with a discharging jetty, to deal with the oil cargoes of super-tankers of the future.

Three important consequences of this modern industrial revolution in South Wales have been a greater stability in population numbers with less migration, employment opportunities locally for women and girls, and an increasing demand for technicians and scientists. Socially, the effects have been a numerical equality between the sexes (a change especially marked in the coal-mining valleys), an increase in the number of family units, but of smaller size, and a greater need for housing and the improvement of amenities and living conditions generally. Residential housing estates have multiplied since the last war; a new town is being built at Cwmbrân between Newport and Pontypool, but a proposal to build a new town at Church Village, near Trefforest, was ruled out because surface development there would have meant the sterilization of millions of tons of the best manufacturing coals which are to be extracted by horizon-mining methods.

The three coastal districts of Llanelli–Swansea–Port Talbot,

Cardiff–Barry, and Newport–Pontypool, can be said to be the most diversified in their industries, with a well-balanced distribution of occupations in metal industries, chemicals, light manufacturing, and ancillary services of many kinds. The ports are no longer largely dependent on coal exporting; they are increasing both the import of mineral ores, timber, oil and general cargo, and the export of manufactures, not only of the immediate hinterland, but also of the more distant hinterlands of the Midlands and southern England.

Coal mining remains the principal industry in the east–central bituminous field and in the anthracite field of the north-west. In the former area, additional employment is now provided in the many light engineering and manufacturing factories of the industrial estates at Bridgend and Trefforest, and in the grouped factories of such centres as Maesteg, Treorci, Porth, Ferndale, Pontypridd, Caerphilly, Bargoed and Pontllanfraith. On the other hand, owing to its inland, and more westerly, location, it has been difficult to attract new industries to the anthracite area. The most successful centre has been Ystalyfera in the upper Swansea valley, where two adjacent factories make clocks and watches and employ over 500 people.

Industry and occupation in the hill towns along the north crop of the coalfield have undergone a radical change. The coal reserves of the area are almost worked out. A few mines are still working around Hirwaun in the west and at Blaenavon in the east; and since 1945 there has been much open-cast coal working. Miners who live in the hill towns travel daily to work in pits farther down the valleys and in the anthracite mines of the upper Neath valley. Ebbw Vale has been given a new lease of industrial life. The foundries and engineering sections of the former ironworks at Dowlais, Rhymni and Tredegar continue in operation; but the rehabilitation of the hill towns has really been brought about by the location of light manufacturing industries in such centres as Hirwaun Industrial Estate, Merthyr Tudful, Tredegar and Brynmawr.

It is primarily on the foundations of coal mining and ferrous and non-ferrous industries that the physical, economic and social structure of industrial South Wales has been built during the past two hundred years. It is a story of long years of continuous expansion and development interrupted by short, violent periods of stagnation and unemployment which have been overcome in part by technological changes applied to the heavy industries themselves, by the introduction of light manufacturing industries in modern factories to provide a broader occupational basis, and, most recently, by the effects of government action in fostering the rehabilitation of industrial areas such as South Wales hit by depression in the inter-war years.

By letters patent, the city of Cardiff (256,270) is now the capital of Wales. In its peripheral position as a capital, Cardiff compares with London and Edinburgh; it is, however, centrally placed within industrial South Wales where two-thirds of the population of Wales live. The civic centre (Cathays Park) was laid out on land given to the Corporation by the Marquis of Bute in 1904. With its spacious rectilineal avenues and its flower gardens, it is an attractive and pleasing feature of the capital city. Around the civic centre are many fine architectural buildings of national status, and close by are the Norman castle and the ruins of a medieval priory. In some ways, perhaps, it may be said that the cosmopolitan city does not express the true character of the Welsh nation. Comparatively few of the citizens are Welsh-speaking, and the cultural traditions of the Welsh peasantry have been forgotten or were never assimilated. But, when Wales plays England in an international rugby match at Cardiff Arms Park, the vast crowd expresses the feelings and soul of the nation in the singing of the martial airs, folk-songs and hymn-tunes of Wales.

Selected Bibliography

E. G. Bowen (ed.). *Wales: A Physical, Historical and Regional Geography*. London, 1957.

British Association. *The Cardiff Region*. Published for the British Association for the Advancement of Science, Cardiff, 1960.

E. D. Lewis. *The Rhondda Valleys*. London, 1959.

L. R. Moore. 'The South Wales Coalfield.' *The Coalfields of Great Britain* (ed. Sir Arthur Trueman). London, 1954.

Ministry of Fuel and Power. *South Wales Coalfield (including Pembrokeshire)*. Regional Survey Report. London (H.M.S.O.), 1946.

J. Pringle and T. Neville George. *South Wales*. British Regional Geology. London (H.M.S.O.), 1937.

Trevor M. Thomas. *The Mineral Wealth of Wales and its Exploitation*. Edinburgh and London, 1961.

D. Trevor Williams. *The Economic Development of Swansea and of the Swansea District*. Cardiff, 1940.

CHAPTER 14

RURAL WALES

E. G. Bowen

The Welsh massif is usually considered to embrace the entire upland country lying west of the English lowlands between the Midland Gap and the Severn Sea. Thus the south Shropshire highlands together with the hills of north-western Herefordshire belong to Wales physically, if not politically. Rural Wales, therefore, is that portion of the Welsh massif so defined that lies outside the industrialized coal basins of the North and South. South Pembrokeshire alone appears in some ways anomalous. Geologically it is part of the South Wales coalfield, although its industrialization is no longer apparent and its landscape is thoroughly rural, yet, on the other hand, it has been drawn more closely into the orbit of the industrial world in recent years by new developments at Milford Haven. On balance, therefore, it was decided that it could best be treated with Industrial South Wales rather than form part of a chapter entitled Rural Wales.

The structural framework of the area thus defined as Rural Wales shows a simple gradation from very old rocks in the north-west to those of a much later age in the south-east. In the north-west, Anglesey is built of crystalline rocks which were severely folded in Pre-Cambrian times. This folding continued through Caledonian and post-Carboniferous times so that one finds narrow bands of Carboniferous Limestone pinched among the ancient strata. The Cambrian, Ordovician and Silurian rocks of North and Central Wales are also folded along the Caledonian axis, but much of their rugged mountainous character in Snowdonia is due to massive outpourings of contemporary lava, and to igneous masses which were intruded into the rocks before and during the folding. These have proved very resistant to subsequent erosion, and in part explain the mountainous character and scenic beauty of the Snowdon countryside. The complicated fan-folding and metamorphism resulting from this structural history is the foundation also of the slate industry in these parts. As one passes from the more mountainous country of the north-west, south-eastwards into mid- and south Central Wales, the igneous masses become less numerous and the rugged mountain country is replaced by the more rounded relief of the Hiraethog and Pumlumon (Plynlimon) moorlands. It would appear that although the ancient intrusive rocks are not so apparent in these parts, nevertheless, intrusions of another kind have impregnated these Lower Palaeozoic rocks with mineral veins. Copper, gold, lead and zinc have been mined extensively in past centuries. The Caledonian trend is clearly visible in the south Shropshire uplands, in the Longmynd, the Wrekin and especially in Wenlock Edge. These uplands, like those of Central Wales, are composed for the most part of Lower Palaeozoic rocks with Pre-Cambrian exposures in the upstanding ridges of the Longmynd and the Wrekin. Finally, in the south-east, there is a triangular area of Old Red Sandstone rocks lying between the north–south Malvernian folds to the east, the east–west trend of the Coal basin to the south, and the Lower Palaeozoic rocks to the north-west. When the Old Red Sandstone rocks are composed of hard sandstone and conglomerates they, too, form high moorland country like the Brecon Beacons, but when they are composed of soft marls, low-lying areas such as the Hereford plain result. This plain is an extension of the English lowlands and, thus, lies outside the area now under review.

These structural elements are by themselves only partly responsible for the present physique of Rural Wales. Geomorphologists have been able to identify the remnants of at least three major erosion surfaces which have been cut across the geological strata regardless of their age, or of their relative hardness or softness, or ol the degree of folding, faulting or metamorphism they may have suffered. It is thought that these surfaces are the remains of peneplains formed by subaerial denudation in Miocene or Pliocene times. E. H. Brown has called them the High, Middle and Low Plateaux. The High Plateau is a widespread plain between 1,700 ft and 2,000 ft in general elevation. It can be traced from the Snowdon country south-eastwards to the coalfield margins, while the fragments of the High Plateau found on the Longmynd, Stiperstones and Clee Hills in the Border Country are indicative of former extension eastwards. It has obviously cut across a great variety of structural features and geological outcrops, among them being the Snowdon anticline, the Harlech dome, the Central Wales syncline, the Bala cleft and the

Church Stretton disturbance. Rising above the High Plateau is a series of massive monadnock structures ranging from 2,100 ft to 3,500 ft. These include the mountains of Snowdonia, Cadair Idris, Pumlumon, Radnor Forest, the Black Mountains and the Brecon Beacons. The Middle Plateau forms a surface between 1,200 ft and 1,600 ft, and on the whole appears to be better developed than the High Plateau. It is particularly well shown on Mynydd Hiraethog in Denbighshire and east of Y Berwyn generally. These North Wales surfaces can be correlated with those of Central and South Wales where the Middle Plateau is well represented on Mynydd Llanybydder in Carmarthenshire and on Mynydd Epynt in Breconshire. The Low Plateau, or Low Peneplain, forms a surface lying between 800 ft and 1,000 ft above sea-level. It is well developed in north-eastern Wales around the head of the Vale of Clwyd and on the Halkyn mountain in Flintshire. In South Wales extensive stretches are found along the flanks of the Wye, Usk, Irfon, Ieithon and Towy valleys. It appears again between the Prescelly Hills and Mynydd Llanllwni and can be traced up the Teifi valley on to Mynydd Bach in central Cardiganshire, and beyond this to the northward, where it links with the remnants of the Low Plateau found along the flanks of the Dyfi valley. Around its seaward slopes the Low Plateau is fringed by a series of wave-cut platforms approximately 600 ft in general elevation. These are now considered to be of different origin from the three major surfaces discussed above, attributable not to sub-aerial denudation but to a marine transgression of the Low Plateau in Tertiary times to a height just below 700 ft. The resultant picture, therefore, of the physique of Rural Wales is a series of erosion surfaces superimposed upon each other and separated from each other by marked breaks of slope, the lower platform fronted around its seaward slopes by the remains of former wave-cut surfaces, and the highest platform topped by massive monadnock structures, culminating in the peak of Snowdon (3,560 ft)—the highest point in the Principality (fig. 31).

Most authorities think that the present rivers were actively engaged in cutting their way into the erosion surfaces just described in late Tertiary times, thereby helping to account for the severe dissection of these surfaces as seen today. The present streams can be seen flowing away in all directions from the central highland core made up of the major monadnocks and aligned from north-north-west to

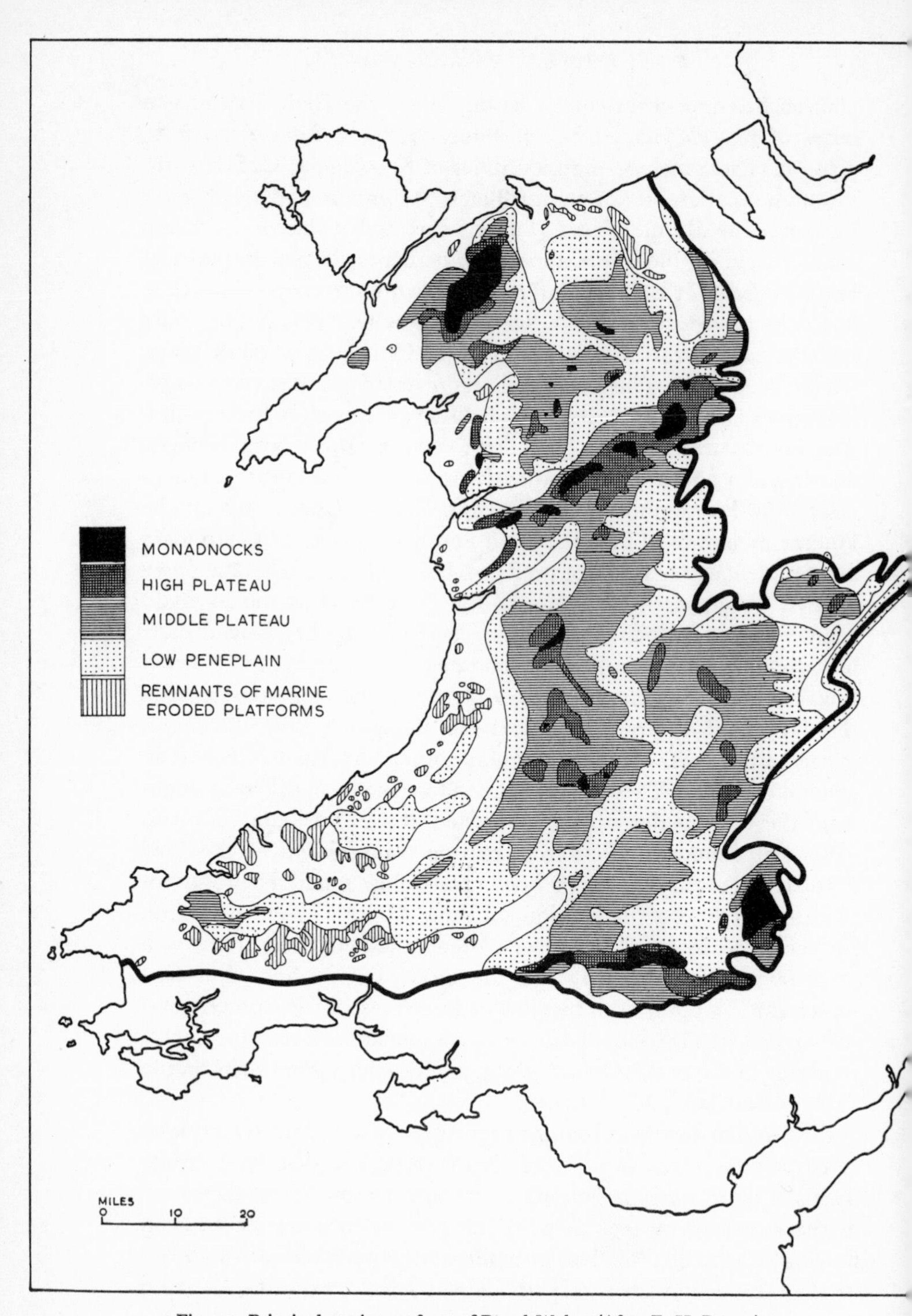

Fig. 31. Principal erosion surfaces of Rural Wales. (After E. H. Brown).

south-south-east. The eastern-facing valleys are those that have provided routeways for alien languages and cultures to penetrate the hill country and so make Rural Wales today an area of diverse cultures.

To the potent influences of subaerial weathering and running water on the evolution of the present physical landscape must be added the third—ice action. The destructive and depositional activities of the Pleistocene ice-sheets wrought profound changes upon the pre-glacial surfaces that had hitherto been exposed to subaerial and fluvial erosion only. Possibly more significant than the destructive work of the ice, important as that was in deepening and reshaping the valleys, was its depositional work. Much of Rural Wales is smeared with patches of boulder clay, while glacial drift and peaty accretions unquestionably mask considerable irregularities in the underlying rock surface. This mantle of cold, stony, acid soil, thick in places but very thin on the steeper slopes, is further heavily podsolized by the existing climatic conditions of high rainfall, little sunshine and weak evaporation. In a broad general sense these conditions are alleviated only by the alluvium of the narrow valley floors. Such a situation has had a profound effect not only on the natural vegetation but also on cultivation and agriculture.

In addition to the features typical of the oceanic climate of western margins in temperate latitudes, two special matters need underlining in Rural Wales. First, orographical rainfall is of supreme importance. The major erosion surfaces previously described are clearly mirrored in any map showing the average annual isohyets. The immediate coastal margins have between 35 in. and 50 in. per annum, while the Low Plateau records 50 in. to 60 in. The Middle Plateau ranges between 60 in. and 80 in., and the High Plateau over 80 in., reaching up to well over 100 in. on the monadnocks. Secondly, the shape of the coast and the trend of the mountain axis ensure that the coastal regions are more affected by oceanic conditions than the interior. The annual range of mean monthly temperatures at Holyhead is 16° F., while inland, beyond the mountain crest at Brecon, the corresponding figure is 20° F. The overall effects of relief and climate are well illustrated by a map of the average annual floral isophenes for a selected number of common wild plants. From such a map it is apparent that the onset of spring in north-west Wales is approximately three weeks earlier than it is at Brecon. This factor of 'earliness' is obviously important when considering man's use of the land.

The biogeography of Rural Wales is, indeed, significant and holds the key to a real appreciation of land use and associated problems at the present time. Among the latter are fundamental changes in the economy of the countryside and the fate of what has come to be known as the Welsh way of life. The marine-eroded platforms that form the seaward edge of the Low Plateau are, unquestionably, a zone of cultivation with woodland. The average total rainfall lies between 35 in. and 50 in. per annum. It is the area most altered by man, and where most of the present woodland is not in its natural state but represents mixed woodland of alder, ash and birch regenerating after felling. On the Low Plateau itself, sloping from 600 ft to 1,000 ft, there is a narrow belt of fescue pastures with or without gorse and bracken. Natural woodland composed for the most part of scrub oak (*Quercus sessiliflora*) and some birch, is found on the steeper valley-sides. In modern times much of the Low Plateau has proved suitable for tree growth and now carries great stands of conifers under afforestation schemes. The Middle and High Plateaux, and in most cases their monadnock superstructures, carry the *nardus-molinia* moorland so typical of highland Britain. All this territory lies above 1,000 ft and has an annual rainfall varying between 60 in. and 100 in. with many instances, especially in Snowdonia, of over 150 in. being recorded. The hollows on these moorlands are filled by vast peat bogs. The vegetation is determined by the depth of the peat; the thin peaty soils usually carry fescue and *nardus* grasses, and the deep peat areas *molinia* and *nardus*. Bilberry and heather moors also occur. *Nardus* grass is found in great masses well above 2,500 ft in areas that are useless for both afforestation and grazing. The monadnocks that reach beyond that height develop the bare slopes and huge boulders that characterize the true mountain scenery of Wales. Because of the general arrangement of the plateaux surfaces in the manner described earlier, the altitudinal vegetation zones of the western-facing slopes are repeated, but with slight changes of level owing to rain shadow effects, on the eastern slopes. The plateau remnants and monadnocks of east Central Wales and southern Shropshire, such as Radnor Forest and the Clun Forest, carry more heather and heathland than the moorlands farther west. This is due in the main to their sheltered position with reference to the incoming rainstorms resulting in a much lower aggregate of rainfall, approximately 40 in. to 45 in. per annum. It is apparent once again that the fundamental distinction in Rural Wales lies between the central high moorland core and the peripheral lowlands.

This distinction so apparent on the physical side is clearly reflected in a variety of human distributions. It is appropriate to look first of all at what must always be considered to be the most important distribution of all in this respect—the distribution of population. Fig. 32 shows the density of population throughout rural Wales. Apart from the very high density (over 640 persons per square mile) shown along the North Wales coast between Prestatyn and Penmaenmawr (a coastal strip that is an overspill of industrial Lancashire rather than a part of Rural Wales) there is a remarkably symmetrical pattern. The peripheral areas, with densities between 50 and 400 persons per square mile on the lower lands and between 32 and 50 on higher ground, stand out in sharp contrast to the core area with densities everywhere below 32 per square mile, and in many parts below 12 per square mile. Within the latter area there are territories as large as 100 square miles without a single inhabitant.

Closely allied is the study of the settlement pattern. Fig. 33 shows tracings of the pattern in two different sections of Rural Wales—one in the north-west, in Merionethshire, and the other, in the south-east, in Radnorshire. It is an easy matter to isolate four clearly marked superficial features in both areas. First and foremost the scattered nature of the individual farmsteads is unmistakable. The single farm (*Y Tyddyn*) is unquestionably the dominant element in the pattern. Secondly, there are clear examples of clustering around an ancient church. Sometimes, these clusters are very small, and at other times, they are of considerable significance. Thirdly, there exists a variety of nodal settlements, the majority of which are likely to be of recent growth, mainly at cross roads, bridge heads and near railway stations. Finally, there are the true towns such as Harlech and New Radnor. These towns are an interesting feature of the settlement pattern. Urban life is essentially intrusive and owes its origin to the Norman conquerors. 'Bastide' towns are numerous in the north-west where their establishment followed upon the Edwardian Conquest: such are Flint, Caernarfon, Beaumaris, Conway and Harlech. Many towns in South Wales had a similar origin: Carmarthen, Brecon, Haverfordwest, Cardigan and Aberystwyth are clear examples. The towns of this class that were well situated have served

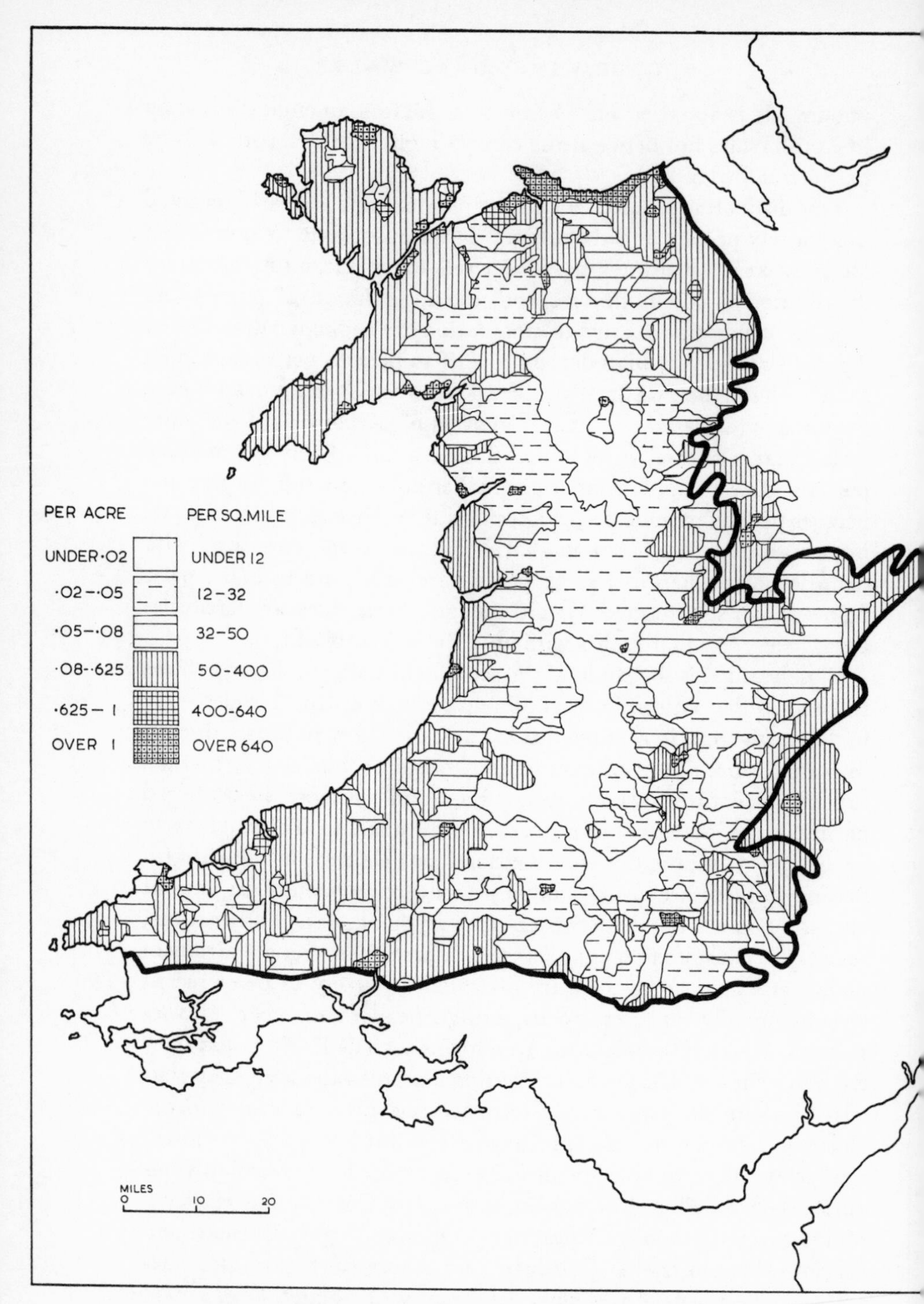

Fig. 32. Density of population of Rural Wales. (After H. Carter).

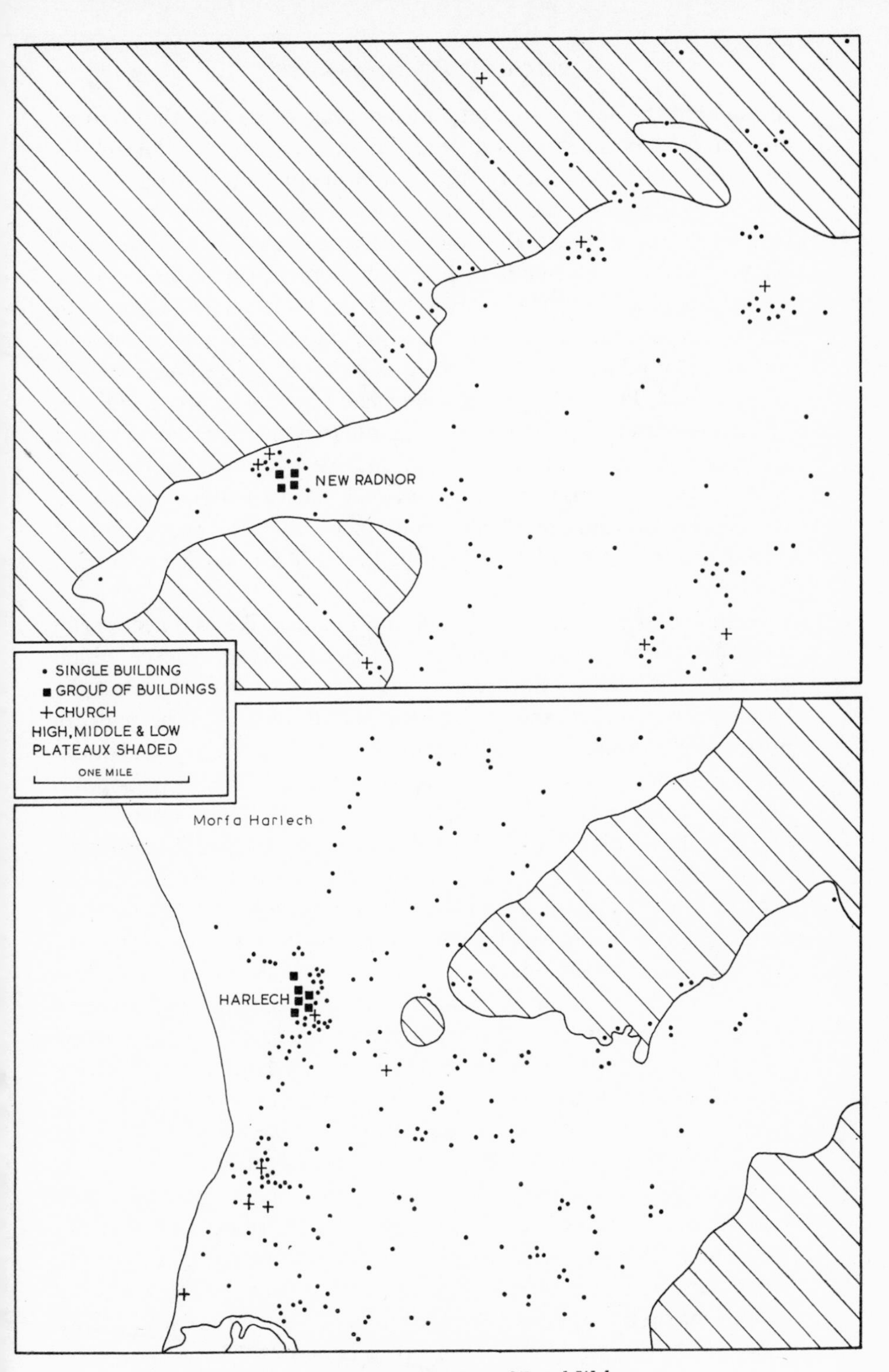

Fig. 33. Settlement patterns of Rural Wales.

throughout the ages as market and shopping centres for the country around. Communications since Norman times have followed the low-lying valleyways deeply eroded into the plateau surfaces. These routeways frequently focused on the older Anglo-Norman towns, but it must be remembered that movement through Wales is far more important than movement within Wales. In this way the westerly packet stations for Ireland, Holyhead and Fishguard, came into being. In addition to the above some native settlements such as Bangor, Llanidloes, Bala, Newtown, Tregaron, Dolgellau, Llangollen, Corwen, Pwllheli and Llangefni have grown to urban stature, but it must be remembered that size alone is no criterion of urban status for, sometimes, as at Harlech and New Radnor, the Norman bastides failed to grow to the size of such native settlements among the hills as Rhayader or Machynlleth, or to reach the proportions of recent tourist resorts like Barmouth, Llandrindod Wells or Church Stretton.

A study of the contemporary economy of Rural Wales is necessarily more complicated, but a number of salient features emerge which are closely related to the foregoing distributions and, thereby, enhance the regional synthesis. In terms of definition alone agriculture must be the dominant element in the economy of a rural area. The fundamental distinction brought out by the density of population map between core and periphery can largely be explained by the fact that the relatively low-lying areas have responded to the mechanization and national organization of the dairy industry in recent years, while the upland areas have remained sheep farming and store cattle regions, unable to respond in anything like as effective a manner to modern mechanization. The farms of Anglesey, Llŷn and Arfon and those on the coastal plateaux of central and southern Cardiganshire, north Pembrokeshire and Carmarthenshire, are all well suited to become efficient dairy units, while the lower-lying areas of eastern Denbighshire, the alluvial tracts of the Severn, Wye, and Usk, and other streams on the eastern side, show, on balance, an agricultural economy in favour of dairying because of the assured markets made readily available under modern organization.

Mechanization of the dairy farms in modern times has meant not only the establishment of electrical equipment for milking, but the abandonment by the farmer and his family of the making of butter and cheese on the premises, the giving up of many of his arable fields to pasture, and the elimination of all animals, except dairy cows, on the farm. A few poultry are kept, while many of the farmers take in sheep from the higher plateau areas for winter keep. Outside the farm itself there is the daily collection of milk and its transportation to the local factory. This involves the farmer taking the milk churns (drawn by his own tractor) to the wayside stand, where they are collected by motor lorry and taken to the factory. All this means employment for lorry drivers, milk-factory workers, mechanics and garage-hands in the countryside, a whole range of supplementary occupations which keep more and more people in rural areas. In this way the peripheral areas of the Welsh massif have not only been able to arrest the seepage of population from their rural and urban districts but, indeed, to record slight increases during the intercensal period 1931–51. Nowhere has this been more marked than in the small market towns which act as service centres for the countryside.

The recent demographic history of the central upland country stands out in sharp contrast. Here the difficulties of mechanization as applied to the raising of sheep and store cattle, and the physical difficulties of introducing dairy farming into the area on a large scale have meant that this type of country shows a marked decrease in population as compared with the peripheral areas, while migration out of the area is still a very important feature in the life of the community. In competition with the assured national market of the dairy farms and the up-grading of the standard of living generally, the small farmer of the uplands is unable to compete with his lowland neighbour. Those who remain on the upland farms have to rely heavily on government subsidies for hill-cows, hill-sheep, lime, the reclamation of marginal land and many similar items.

It should be noted in passing that apart from agriculture no other industry is now sufficiently prominent to leave its mark on the general distribution of population. Lead and zinc mining in Central Wales is now extinct; the woollen industry of the middle Teifi valley and of other parts of Rural Wales is no longer important, while even the one-time famous slate quarrying and mining of Llanberis, Bethesda and Blaenau Ffestiniog is rapidly declining. It is true that the coastal areas have the tourist industry, but this is on a scale very different from that on the North Wales coast previously mentioned. In west and north-west Wales it is mainly a domestic industry as at New Quay, Aberystwyth, Borth, Barmouth, Harlech, Cricieth, Pwllheli and Amlwch. There is nothing at any of these places to compare with the large hotels and mechanized entertainment characteristic of Rhyl and the north-east coast resorts. The reception of tourists on the farms, and the establishment of vast temporary châlet and caravan settlements are new developments, but their total effect on the population density in the off-season is slight indeed. Thus, agricultural pursuits are, for the time being, virtually unchallenged in the economy of the countryside, but everywhere one is made conscious of a sharp contrast between the relative prosperity of the peripheral lowland as compared with the general decay of upland farming.

A challenge of a very serious nature to the traditional economy of the upland areas is, however, already looming large on the horizon. In 1954 the Welsh Agricultural Land Sub-Commission under the Welsh Department of the Ministry of Agriculture, Fisheries and Food initiated a special investigation into the agricultural economy existing along a transect of the uplands in Central Wales. The reference area included most of eastern Cardiganshire together with most of Radnorshire and southern Montgomeryshire. In other words, the transect cut across the Lower, Middle and High Plateau areas of the Welsh massif. Some 1,404 farms within this area were examined in detail and the Report indicates that 57% of these were too small to be considered efficient economic units today, while another 31% of them were only marginal, leaving but 12% that could be considered as profitable enterprises. Of the farms examined, 570 possessed no tractors, while 786 had no motorized transport of any kind: 1,200 were still without electricity. More startling still was the fact that seven out of every ten had no piped water supply in the house, in spite of rainfall aggregates ranging from 50 in. to over 100 in.

The Report indicated that a rapid amalgamation of farm holdings was already taking place; a farmer living near to a secondary road among the hills would buy up a farm, or many farms, whose lands abutted on to the roadway, or possibly he would buy a larger farm higher up the valley. The homesteads of the newly bought farms are allowed to decay, or become cattle and sheep shelters after the family living in them has departed, while the farmer controls his newly acquired territory from his original homestead. In fact, one of the implications of the Mid-Wales Report was that this amalgamation of holdings should be allowed to proceed in spite of the depopulation that accompanies it, and that the best quality land (from the point of view of the sheep and store-cattle farmer) should be run more on a ranching system on the New Zealand model. Large units of more than a thousand acres in size are certainly replacing the smaller *tyddynnod* on the Welsh uplands. Such a change only partially solves the problem.

Geographical and other surveys have indicated very clearly that the land well suited to ranching of this type involves but one-third of the total. A second third of the core-lands of Rural Wales, known to be too limited in its natural endowments for profitable sheep and cattle pastoralism, could be given over to the Forestry Commission to develop. The Mid-Wales Investigation Report indicates that very considerable tracts are already held by the Commission in these parts, but points out that there are considerable portions that are suitable for further afforestation. It is obvious, also, that there must be limited areas where the land is equally suitable for both ranching and afforestation. In such areas the forecast is that afforestation will result, so that much of the landscape of the Low Plateau of Rural Wales will be completely changed in the next few decades. The remaining third of the upland area is known to be made up of such poor quality land, with very steep slopes, that it is suggested that it would be better to let it be developed by the water engineers and gather up in it the abundant rainfall of the hills. Rural Wales possesses great water resources that should be used for electrical power as well as for water conservation. It is from such sources that the large farms envisaged after amalgamation should be supplied with water, both for agricultural and domestic purposes. In the Mid-Wales reference area there is only one group of large-scale reservoirs at present—the Elan and Claerwen valley schemes of the Birmingham Corporation. Outside this area Liverpool receives much of its water supply from Lake Vyrnwy, and will receive more in due course from the Tryweryn scheme in Merionethshire. Much of the industrial area of South Wales obtains its water supply from the reservoirs of the Brecon Beacons. In addition, hydro-electric schemes are located at Dolgarrog in the Conwy valley, Cwm Dyli in the upper reaches of the Glaslyn in Snowdonia, and at Maentwrog, with its collecting ground near Trawsfynydd in Merionethshire. The maximum physical advantages for such power schemes are in North and West Wales where the rainfall aggregates are high and the slopes steep. Future schemes are to be located in this area, the six major ones being the

Nantffrancon, Snowdon, Ffestiniog, Upper Conwy, Mawddach and Rheidol schemes. Work is far advanced on the last named and production is expected to begin in 1962. The number of persons that will be employed permanently in works associated with afforestation and water conservation is not likely to be very considerable, so that the overall effect on the distribution of population will not be great.

Thus the general agricultural prosperity of the lowland margins and the decay of the small upland farms stand out clearly and there is an increasing interest in the attempts that are being made to re-evaluate the natural resources of the highlands in terms of present-day requirements. These changes should not obscure the fact that throughout rural Wales there remains a close relationship between the distribution of population, settlement pattern and rural economy on the one hand, and the physical build and natural resources of the area on the other.

If attention is now directed to elements of the non-material culture it is found that approximately 330,000 people out of a total of some 660,000 in the entire area can speak Welsh. Ability to speak Welsh is an index of the greatest importance since Welsh life and culture can claim descent from the Celtic invaders of highland Britain in Iron Age times. Their culture is known to archaeologists as Iron Age B, and its influence reached over most of highland Britain south of the Scottish Lowlands. These Iron Age peoples spoke a dialect of Celtic speech that was ancestral to modern Welsh and it was at this period that much else that was to characterize Welsh life and culture in later times took root. The emergence of Wales as a unit in highland Britain followed upon the Saxon victories at Dyrham (577) and Chester (616) when the Celts of Wales became separated from their fellow Celts in Cumbria on the one hand, and from those in the south-west peninsula on the other.

The leaders of early Welsh society inherited much along with their pastoral conditions. They were particularly rich in their non-material culture. Great emphasis was placed on oral tradition and the use of the spoken word. It is not surprising, therefore, that in medieval times the Welsh word for language *iaith* was used synonymously with *cenedl*, the word for nation. Welsh people think on similar lines to-day. From the point of view of inter-relationships on the ground it should be remembered that it was in the isolation of the *tyddyn*—the dominant unit in the later settlement pattern—that the characteristic features of Welsh life as we know them grew up, and not in the towns, or in the industrial settlements. In their scattered settlements the Welsh people developed a love of music; found leisure to read poetry and philosophy, and to think about religion; took time to prepare for the Eisteddfodau, the singing festivals and the chapel meetings; and all this through the medium of the Welsh language. The Welsh countryman rarely visited the Anglo-Norman towns except on market days. In this way the traditional Welsh way of life became indissolubly linked both with the Welsh language and with the fortunes of the scattered farm. It is because the *tyddyn* and the Welsh language will disappear when attempts are made to replan the upland area of Wales on lines described above, that these schemes still meet with strong local opposition. The census reports for the Welsh counties record the number of persons able to speak Welsh in each parish. This data is usually shown cartographically by a scale of shading indicating the percentage of the total population in each parish able to speak Welsh. Such a map is, however, misleading, for in one parish the percentage of Welsh speakers may be very high while the total population is very low, so that the actual number of Welsh speakers is, in fact, small; while in a second parish with a very dense population the percentage of Welsh speakers may be low though, in fact, their actual numbers may be far greater than those in the first parish where the percentage of Welsh speakers was so high. A closer approximation to reality is, therefore, obtained by attempting to show the percentage of Welsh speakers in relation to the density of population on one and the same map (fig. 34).

It is clear that the whole of the eastern, and particularly the south-eastern, part of Rural Wales has ceased to be Welsh-speaking in any significant sense of the term. On the other hand, the population of the Low, Middle and High Plateaux in north-eastern Wales and in western Breconshire and Radnorshire, and in eastern Cardiganshire and Carmarthenshire retain over 80% of their population able to speak Welsh, but in these parts the population density is so low as to make their total numbers very small. It is only in north-west Wales (in Anglesey, Arfon and the Llŷn peninsula) and again in the south-west (in southern Cardiganshire, north Pembrokeshire and western Carmarthenshire) that there are relatively high population densities with over 80% of the people able to speak Welsh. It is in these areas

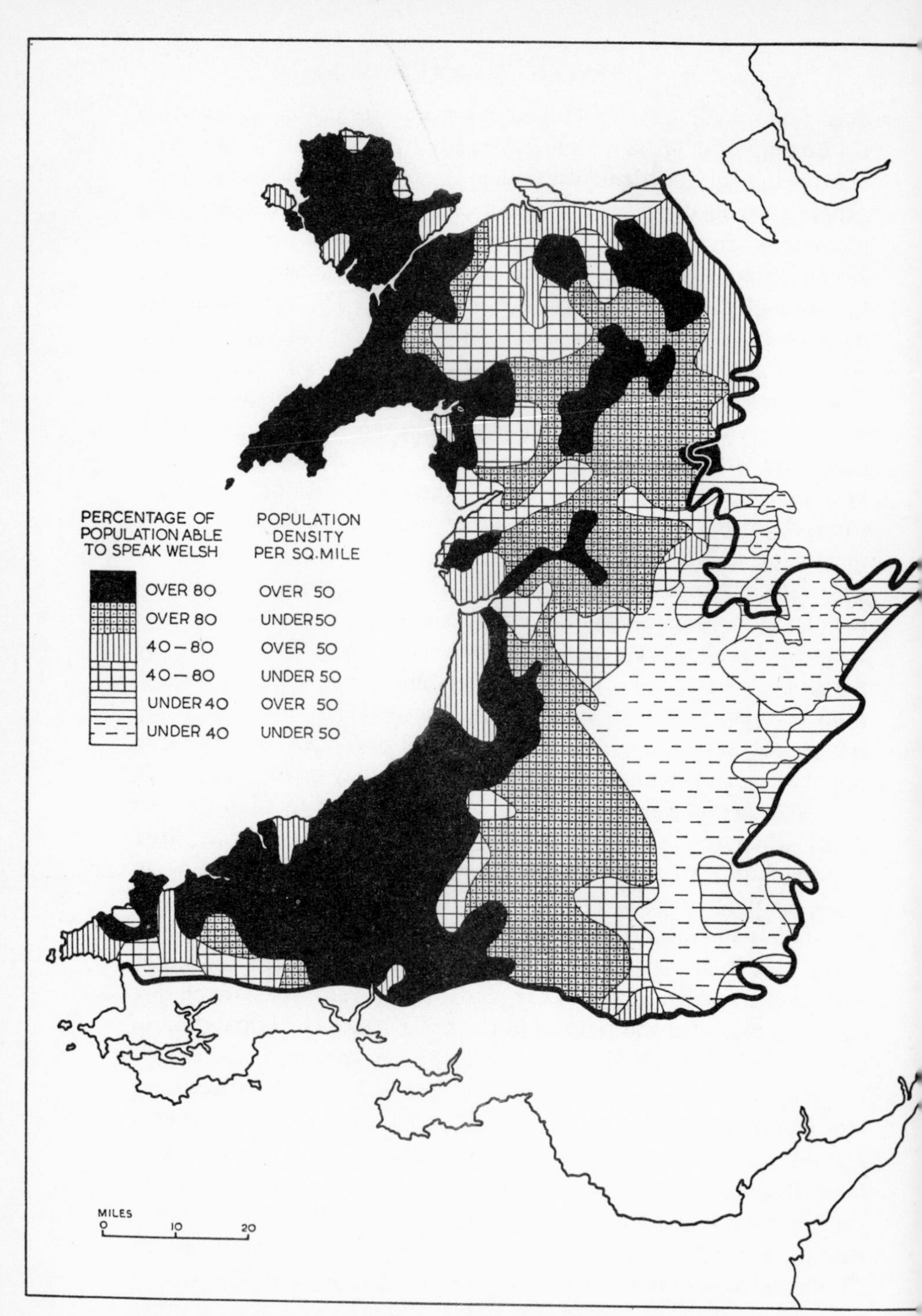

Fig. 34. Density of Welsh speakers in relation to density of total population, 1951.

that Welsh life and culture are most strongly represented at the present time.

The regional geographer may be permitted three comments. First of all, it is evident that the same physical background that gave a relatively symmetrical distribution pattern, relative to the lowland and the major plateaux structures, when population density, settlement pattern and rural economy were being discussed, produces a markedly asymmetical pattern when matters relative to non-material culture are considered. The explanation is obvious. The eastern-facing valleys, cut into the great plateaux and monadnocks, afford easy lines of entry for English speech and English ways of life, whether in Anglo-Saxon or in Anglo-Norman times, or even later with the coming of modern roads and railways. Plateau areas surrounded by lowlands such as Radnor Forest, Clun Forest, the Black Mountains and the highlands of southern Shropshire were easily engulfed, and soon overwhelmed. The industrialization of the coalfield restricted still further the Welsh-speaking area of Wales so that it fell to the main section of the High Plateau with its monadnocks to act as a bastion against incoming influences, and to shield the areas of cultural survival in the coastal lowlands of north-west and south-west Wales. Secondly, the resulting regional picture would appear to be but another example of a feature so strongly stressed by American geographers in recent years, namely, that the boundaries, or the margins, of physical regions rather than their centres, or cores, are likely to be the centres of culture areas. This is seen to be particularly true of Rural Wales, especially if the word centre is used to distinguish the more active and virile parts of a culture area with reference to present conditions. In the region under review the physical core has already been clearly defined, while it is obvious that the present centre of the culture area lies markedly to the westward of the mountain axis (fig. 31). Finally, it must be clear that the major conclusion to be drawn from this study is that Rural Wales, as defined at the beginning of this chapter, cannot be considered a *pays* in the full geographical sense of the term—if it were, the entire area would be Welsh-speaking, irrespective of population density, and similar in culture to those parts of the Principality where over 80% of the population can still speak the Welsh language.

Selected Bibliography

E. G. Bowen (ed.). *Wales: A Physical, Historical and Regional Geography.* London, 1957.

E. G. Bowen. 'Le Pays de Galles.' *Trans. Inst. Br. Geographers*, **26**, 1959, pp. 1–24.

E. H. Brown. *The Relief and Drainage of Wales.* Cardiff, 1960.

H. Carter. *The Towns of Wales, A Study in Urban Geography.* Cardiff, 1962.

G. M. Howe. *Wales from the Air: A Survey of the Physical and Cultural Landscape.* Cardiff, 1957.

G. R. Jones. 'Basic Patterns of Rural Settlement in North Wales.' *Trans. Inst. Br. Geographers*, **19**, 1953, pp. 51–72.

J. G. Thomas. 'The Geographical Distribution of the Welsh Language.' *Geographical Journal*, **122**, 1956, pp. 71–9.

Welsh Agricultural Land Sub-Commission. *The Mid-Wales Investigation Report.* London (H.M.S.O.), 1955.

CHAPTER 15

THE WEST MIDLANDS

R. H. KINVIG

The five counties Warwickshire, Worcestershire, Staffordshire, Shropshire and Herefordshire obviously do not form a natural region but they possess an intrinsic wholeness and at least a measure of economic and social coherence. The area thus forms a reasonably satisfactory administrative unit so that it has been adopted as a 'standard' region for a variety of governmental and other purposes. Some parts are, of course, marginal. North Staffordshire, in particular, with its Potteries and its own University of Keele is very different from the rest of the region, and its people consider themselves to be a separate group with some interests divided between Manchester and Birmingham. Similarly parts of Shropshire still tend in economic and social matters to look northwards to Liverpool or Manchester, while the more westerly areas of both Shropshire and Herefordshire have many links with Wales.

The heart of the West Midlands undoubtedly consists of Warwickshire, Worcestershire and south Staffordshire, within which the conurbation of Birmingham and the Black Country holds a dominant position. Birmingham is the undisputed regional capital. In earlier days this heavily industrialized area was one of the least significant economically and culturally, so that there has been a fundamental change in its role.

The landforms of the West Midlands range from featureless alluvial lowlands, less than 100 ft above sea-level, to high moorlands exceeding 2,000 ft. At the core is an alternation of uplands and lowlands very unlike the 'Midland plain' of popular imagination, composed mainly of Trias and Lias rocks, but also with the highly significant Palaeozoic inliers containing their vital stores of coal, iron ore and limestone now much depleted if not actually exhausted.

To the south-east lie the Jurassic scarplands, historically important; to the north-east the Pennine fringe with the adjoining Potteries coalfield; and to the west the Palaeozoic and Archaean areas of the Welsh border, largely of higher ground but also including the lowland of Hereford.

Dominating the relief of the West Midlands is the Birmingham Plateau, a pear-shaped mass, mainly above 400 ft and rising to 1,036 ft in Walton Hill, with the longer axis, about 45 miles, running north-north-west to south-south-east from near Stafford to near Stratford-on-Avon, while the shorter axis, some 34 miles, trends east-north-east to west-south-west from Nuneaton to Kinver Edge. Surrounding this unit, and indissolubly bound up with it, are the river valleys of the mid-Severn with the Stour from the western section of the Plateau, the Avon, joining the Severn at Tewkesbury, and the upper Trent, into which flows the Tame.

The Plateau is generally highest round its margin, especially in its south-western section where the bold ridge of the Clents reaches 1,000 ft. Moreover, on all sides except the north-east, there is a steep face to the surrounding river valleys so that transport has often been handicapped.

Internally the Birmingham Plateau consists of two major sub-units, the South Staffordshire and the East Warwickshire Plateaux, these being largely made up of several level tracts of different heights separated by bluffs of varying steepness and dissected by minor valleys (fig. 35). Thus in the former, the Cannock Chase High Plateau overlooking the Trent valley to the north and the Cannock coalfield to the south forms a very distinctive and thinly settled tract mainly of Bunter Pebble Beds, generally between 600 ft and 700 ft high.

Along the south-western flank of the Upper Tame basin runs the Sedgley–Northfield Ridge rising to 876 ft at Turners Hill and forming the main English watershed as far as Frankley Beeches (841 ft). This ridge separates the two main sections of the Black Country, and despite the existence of some cols which have provided relatively easy passages, it has been a considerable barrier to canals and railways, both of which resort to tunnels.

Much of Birmingham stands on the West Bromwich–Harborne Plateau composed of Keuper and Bunter Sandstones. It is the Keuper Marl that almost entirely underlies the Solihull Plateau whose surface, between 400 ft and 500 ft, has markedly level stretches on which have developed the southern suburbs of Birmingham as well as the

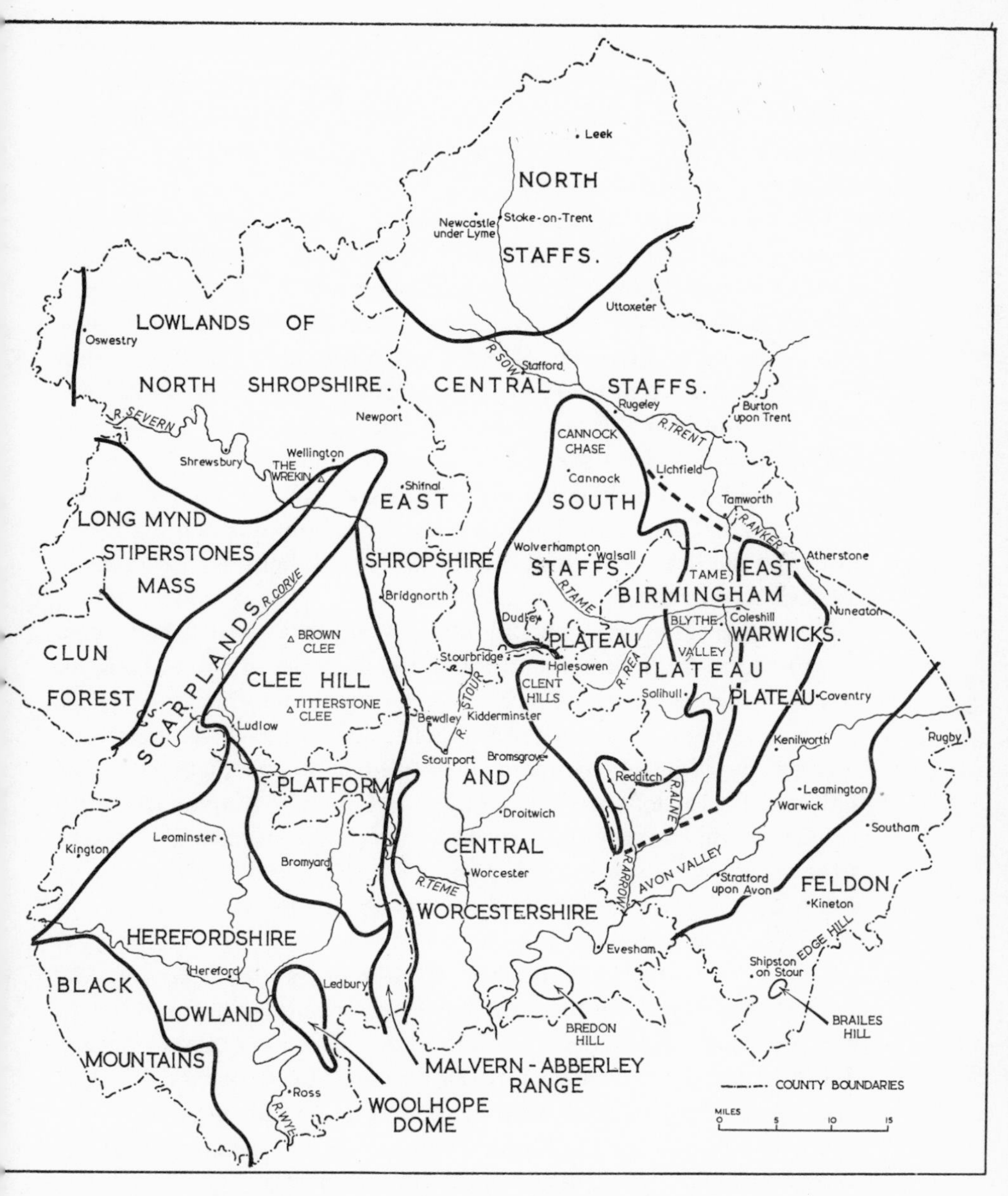

Fig. 35. Major divisions of the West Midlands. The original nucleus of Birmingham adjoined the river Rea between two and three miles from its confluence with the Tame (see pp. 274–5).

mainly residential urban centre of Solihull which is now a county borough.

The East Warwickshire Plateau is much simpler in geology and relief. Much of it is between 400 ft and 500 ft, its highest point being just over 560 ft in Corley Moor. It is composed largely of Middle and Upper Coal Measures, with outcrops of Cambrian shales and quartzites along its north-eastern edge.

These two major divisions of the Birmingham Plateau are almost completely separated, except in the south, by the combined valleys of the Blythe and Lower Tame which follow a south–north direction. Largely because it is cut out of Keuper Marl this valley has considerable width especially north of the neighbourhood of Coleshill which is a converging point for most of the drainage of the whole Plateau.

In the south, at the head of the Blythe valley, there are various cols which, during the glaciation of the Birmingham Plateau, probably served as overflow channels from lakes in the Blythe–Lower Tame basin. The chief col at Kingswood, being only 340 ft high, has provided the best routes for railways and canals between the Plateau and the Avon valley.

The River Trent rises on Biddulph Moor, over 1,000 ft high, and flows in a southerly direction through the Potteries to Stone where its course becomes south-easterly. As far as Stone, the river falls approximately 800 ft and the gradient is about 66 ft per mile, but thenceforward its long course to the Humber shows a further drop of only 300 ft. Its flood-plain is braided and liable to frequent flooding —whence the term 'sulky' Trent—though recent drainage works have improved matters in this respect. Rising above this flood-plain is a low, wide, gravel terrace which has afforded dry settlement sites as well as good agricultural land.

While the gentle gradient of the Trent valley and its relatively low height above sea-level have precluded its tributaries from excessive downward erosion, valleys draining to the Severn and Avon from the Birmingham Plateau have rather different characteristics. In contrast to the broad valleys with gentle gradients shown by the Tame and Rea, streams like the Stour, Arrow and Salwarpe are confined to narrower valleys particularly in their upper portions.

The River Severn itself presents anomalous features particularly in the striking contrast between the upper stretch, above Buildwas, where the river exhibits a fairly open, serpentine valley with a gentle flow, and the middle section to Holt Bridge where the stream is confined to a narrow valley with a straight course and a swift current—especially marked in the classical gorge section between Ironbridge and Bridgnorth. These anomalies are best explained by glacial intervention. Thus the Ironbridge Gorge originated as an overflow channel from ice-dammed lakes formed during successive phases in the retreat of the ice-sheet towards the north-west, and ultimately became the permanent channel for the waters of the upper Severn. As a result, largely of this increased volume of water, the mid-Severn, together with the section below Worcester, have undergone rejuvenation. Apart from other effects this has meant that the gorge from Ironbridge to Bewdley has no appreciable terrace deposits, whereas these are extensive both above and below this stretch and form wide flat expanses of fluviatile sands of considerable agricultural value, sufficiently above flood-level to attract settlement, especially in early times.

The north Shropshire plain between 200 ft and 300 ft is developed largely on Bunter Sandstones and Keuper Marls heavily masked over considerable areas by glacial sands and gravels. It is a land of gentle slopes, broad valleys and sluggish streams, with marshes and meres, the latter particularly in the north and north-east. Eastwards the same kind of relief is continued into the Trent lowlands of mid-Staffordshire underlain principally by Keuper Marls.

Eastern Shropshire forms a lowland based on Triassic sandstones and marls occupying a syncline between the Palaeozoic uplands of south Staffordshire and south Shropshire. This area is more dissected than northern Shropshire and the valleys of the Worf and its tributaries are deeper in contrast to the more open straths of the other lowlands. Southwards the lowland is continued into Worcestershire, being at first fairly narrow, thence broadening out into an extensive plain based mainly on Keuper Marl or Lias stretching west of the Severn to the line of the Malverns, and merging eastwards into the Avon valley. Soil conditions over these areas, while being good on the whole, can vary in quality owing to different combinations of texture, site, drainage and economic circumstances.

The Avon now flows from the neighbourhood of Rugby past Warwick and Stratford, thence through the Vale of Evesham in a series of sweeping curves before joining the Severn. Views regarding its origin have been radically changed in recent years. Far from being a river of some antiquity, it is now believed that up to the onset of the second Welsh glaciation (or Catuvelaunian glaciation correlated with the Riss and Saale of the Continent) there was no Avon tributary of the Severn except possibly a small ancestor near Tewkesbury, while Warwickshire as well as south-east Worcestershire drained to the Trent. A glacial lake (Lake Harrison) then developed covering all that area, being held up by ice to the north-east, north-west (Tame valley) and south-west (Severn valley), but having various outlets across the Jurassic scarp. With the departure of the ice the drainage of the area continued permanently south-westwards to form the Avon as it now is, and the head of the youthful river was cut back with amazing rapidity because the deposits on which it had to work were very soft drifts overlying Keuper Marl. Since then it has become adjusted more easily than the Severn and has formed a wider valley and a wider series of terraces. The most important of these is the second which lies 30–40 ft above the present river-level and extends locally for over a mile in width. Their level surface and light well-drained soils make them peculiarly suitable for cultivation, and they have formed very attractive sites for settlement since prehistoric times.

The Avon basin ranges from 250 to 100 ft in height, but overlooking it to the south is a low gently-undulating plateau 300 to 400 ft high known as the Feldon. This is based on the Lower Lias while on its south-eastern margin it is bounded by a long line of scarps at the junction of the Lower and Middle Lias. These heights mark the general frontier zone with Oxfordshire although the county boundary follows the actual ridge only at and near Edge Hill (about 700 ft high). Fringing this Jurassic escarpment are various detached outliers in Warwickshire, e.g. Brailes Hill with a capping of Upper Lias, while in south Worcestershire, Bredon Hill, with a maximum elevation of 961 ft, has a capping of Oolite. The Lower Lias belt normally produces rather heavy calcareous clays which presumably once bore a dense oak-ash-hazelwood forest. This woodland cover was cleared relatively early so that by the Domesday period (1086) it stood out as an open area in contrast to the wooded Arden country north of the Avon valley on Keuper Marls.

The predominantly upland country of south Shropshire and Herefordshire may be divided broadly into two areas, the north-western and more mountainous area of Silurian and older rocks, and the south-eastern and generally more subdued area of Old Red Sandstone. Exceptions occur in the Malvern–Abberly range (pre-Cambrian and Silurian) of south-eastern Herefordshire which extends northwards into Worcestershire, and also in the Silurian Woolhope dome.

There are thus a number of contrasting units beginning with the Long Mynd–Stiperstones mass in the north-west, a broken upland area, exceeding 1,500 ft. The nucleus is the imposing but bleak plateau of the Pre-Cambrian Long Mynd 1,696 ft high, with its comparatively flat summit. Westwards are the craggy Stiperstones where the quartzites produce many bare rock-surfaces; still farther west lies the Shelve country, a former lead producer. The Long Mynd falls sharply eastwards into the Church Stretton rift to be followed in turn by the remarkable line of hog-backed igneous ridges, the best known being Caer Caradoc (1,506 ft). These are pre-Cambrian rocks, largely heathland, though with some woodland on the lower slopes. North-eastwards this line is continued in the Wrekin (1,385 ft), forming a striking landmark as a lone sentinel above the surrounding plains. South of the Long Mynd–Stiperstones lies Clun Forest, a much dissected plateau of Silurian and Old Red Sandstone rocks having a general level of 1,000 ft, rising to over 1,400 ft in parts.

The landscape changes east of the Caradoc hills to become a series of north-east to south-west scarps and vales based on tilted Silurian rocks, and here Wenlock Edge holds a commanding position with its steep, well-wooded face overlooking Ape Dale to the west. Eastwards it is quickly followed by Hope Dale and then by a much broken line of limestone escarpments rising in parts to over 1,100 ft, to which the general name of View Edge has been given. Corvedale follows on the dip slope of View Edge.

Southwards this type of country is continued beyond the Onny tributary of the Teme in a sinuous series of scarps bearing a number of different names, and it penetrates into north-west Herefordshire, including the area round Kington. The uplands are of little agricultal value; cultivation is limited to the shale lowlands, and especially to the floor of the old Wigmore lake basin of northern Herefordshire.

Another change in the landscape occurs south-east of the Silurian scarp zone in Shropshire where the Old Red Sandstone country is reached with its hills and undulating red-soiled lowlands. First comes the Clee Hill platform surmounted by the twin heights of the

Brown Clee (1,790 ft) and Titterstone Clee (1,749 ft). These hills stand so high because the Old Red Sandstone is here covered by Carboniferous rocks which have been locally preserved by a capping of hard dolerite called 'dhu-stone'. The latter is still extensively quarried for road metal, but although coal from the Middle Coal Measures was formerly obtained in a number of small mines these workings have now practically ceased. Southwards where the additional rocks found in the Clees are absent, the broken upland area of north-eastern Herefordshire focusing on Bromyard is reached where the average height of only 600–800 ft is attained, and many parts are much lower.

Over central Herefordshire stretches a lowland with an elevation of 200–400 ft drained by the Wye and its tributaries, and composed of red marls (Downtonian) which give a gently rolling relief and weather down into deep loams of great fertility. The Hereford lowland forms, in fact, the largest stretch of first-class land in the West Midlands, having a mild and relatively dry climate in addition to its soils, which, although lacking certain plant foods, are very productive under good management. Occasional striking hills of Psammosteid limestone or cornstone (for example, Dinmore and Wormsley Hills) form one of the most attractive features of the Herefordshire landscape rising 200 or 300 ft above the prevailing level of the plain. The whole area is dominated by Hereford city occupying a central position in the lowland at the confluence of the Lugg and the Wye, while the Frome converges on the same area and enters the main river some miles farther downstream. This convergence has brought great spreads of gravels which yielded water and thus permitted closer settlement while they have also provided a belt of lighter soils with a more varied agriculture.

Protruding through the marls, upfolds of Silurian rocks which give rise to a complex scarp and vale relief are well represented not only in the classic Woolhope denuded anticline but also in the foothills west of the Malverns. The latter are formed of resistant gneisses and schists which produce a thin soil cover supporting a scanty vegetation. They form a distinctive north–south ridge rising to a maximum of 1,395 ft and defining part of the boundary between Herefordshire and Worcestershire.

To the south-west of the Herefordshire lowland lie the Black Mountains, only partly within the West Midlands. They form a lofty plateau of almost horizontal sandstones and marls (Upper Old Red Sandstone) rising to nearly 1,600 ft on the Hereford side and reaching 2,624 ft in Brecknock. With its alternation of river valley and barren moorland it is a complex area linked with the Welsh moorlands rather than the English lowlands, and it now forms one of the most beautiful, inaccessible and most thinly peopled corners of the West Midlands.

Evidence of settlement in the West Midlands during prehistoric times is notoriously scanty as compared with many other parts of either Lowland, or Atlantic, Britain. Not only was this area as remote as it could be from the coast, but it had large areas of woods and chilly clayland generally shunned by early man. True, certain parts with lighter soils were heathy and suitable for occupation (for example, the northern Clents), and the terrace deposits of the encircling river valleys were certainly attractive. Thus Baginton, south of Coventry, was a favourite site for settlement; but the extensive woodlands of Arden, Feckenham, Brewood and the Wyre were very considerable obstacles.

In Roman times (*c.* A.D. 43 to 410) important roads crossed the area. The famous Watling Street from the south-east avoided the Plateau as far as it could, but finally had to cross it westwards from Letocetum (Wall) in a col across the South Cannock plateau to reach, about A.D. 48, its goal at Viroconium (Wroxeter). The Ryknield Street traversed the Plateau from south to north, running from the Fosse Way at Bourton on the Water to join the Watling Street at Wall, passing *en route* through Bidford on Avon and Alcester. It crossed the Rea at Lifford and passed near the west-central part of Birmingham, probably along the line of the existing Great Hampton Row before crossing the Tame at Holford, though there is no evidence to suggest any permanent settlement. Joining the Ryknield Street near this point was another road from Worcester and Droitwich (Salinae), probably connected with the salt trade of the latter; but these roads did not materially alter the nature of the area which was one of the most thinly peopled parts of Roman Britain.

Effective settlement began in the Anglo-Saxon period when Anglian groups entered the Avon and its tributaries both via the Trent and Tame and via the Nene and Welland. Moreover, from the south-west came the Hwicce group of West Saxons who spread from the lower Severn and Cotswolds into the Avon valley, so that a cultural overlap occurred there between Angles and West Saxons. Later on, settlement spread to the middle section of the Severn and gradually to other parts of the region.

The original Anglian kingdom of Mercia included most of the territory watered by the Tame and the upper Trent, and its royal capital was Tamworth, at the junction of the Tame and Anker guarding the north-east entrance to the Birmingham Plateau. The area became Christian about the mid-seventh century, and when St Chad became its bishop (669) he made Lichfield the centre of his see. The bounds of Mercia were gradually extended to include the Hwicce settlements as well as some of the border country still farther west, until, under the famous Offa in the later eighth century, the cultural line of demarcation between the English and the Celtic peoples of Wales was defined by the rampart bearing his name.

The division of western Mercia into counties was probably carried out during the tenth century, the focus of each shire being a dominant fortified post. So the Birmingham Plateau was shared between three centres—Warwick with its castle on a Keuper Sandstone bluff guarding the all-important Avon passage-way, Worcester on the terrace deposits of what was then the tidal head of the Severn commanding the route up the Teme as well as up the major valley, and Stafford where the fluvio-glacial gravels secured a dry passage over the marshes of the Sow protecting the route of the upper Trent.

Birmingham had a lowly origin as an Anglian settlement in the far corner of Warwickshire, the actual name recording it as the '*ham* of the people of Beorma or Beormund' who had presumably reached the site via the Trent, the Tame and the Rea. The nucleus of the hamlet was on the south-eastern slope of a Keuper Sandstone ridge forming a conspicuous feature of the local landscape at a point now occupied by the Bull Ring, the original village green, and St Martin's Church. It possessed many advantages—a good local water-supply, building stone, natural drainage, good soils. Moreover, the existence of a fault between the Keuper Sandstones and Marls near the base of the slope provided springs which filled the moat around the manor house and, later on, helped to increase the supply of water for use in the cloth-making and leather trades which were important in the fifteenth and sixteenth centuries. A long street, known as Digbeth, led down from this site to an important east–west fording place over the Rea where Deritend Bridge was eventually built, while on the other side of the river on a terrace lay the small hamlet of Deritend. Although quite small, the Rea occupies a wide valley and it has had a notorious reputation for flooding until quite recent times when much of its water has been confined to brick or concrete channels. This crossing increased in importance since roads gradually converged on the western side from Walsall, Wolverhampton, Dudley and Halesowen and diverged on the other side to Alcester, Stratford, Warwick and Coventry, a road pattern which is still discernible.

In the Domesday record (1086) the manor of Birmingham was still insignificant compared with adjoining manors, for example Aston, Erdington, Edgbaston, Northfield and King's Norton, which have all been absorbed within the existing county borough. It was the possession of the river-crossing that undoubtedly gave medieval Birmingham a definite advantage over its immediate rivals; this enabled it to make good use during the twelfth and thirteenth centuries of market rights secured from the Crown, so that it became the acknowledged centre for the exchange of local products, and this in turn stimulated the rise of local industries. Such commercial activity is reflected in the greater status held by Birmingham in later taxation returns such as the 1327 Lay Subsidy Roll in which its assessment was much higher than that of any other local centre; it had by then outstripped them all.

Lying north-west of Birmingham is the Black Country, whose development has had marked repercussions on the growth of the city itself. To begin with, that area was isolated and backward, and settlements were few; at the time of Domesday, it had a lower density of population than any other part of the Birmingham Plateau. Dudley, its castle perched on the Castle Hill controlling routes between the Tame and Stour valleys, was the key centre, though it formed then, as now, a detached portion of Worcestershire within Staffordshire. Halesowen, also in Worcestershire, occupied a favoured spot in the amphitheatre-like valley of the upper Stour, later to be the site of an abbey. Wolverhampton near the western edge of the Plateau on a hill of Upper Coal Measure sandstone, began as an Anglo-Saxon settlement, and from the ninth century grew around its famous collegiate church of St Peter. What was to become the Black Country was still a picturesque wooded or heath region whose mineral resources were not appreciated.

The really important centres were outside the Plateau, and the most densely peopled territories at the time of Domesday were the Feldon and Edge Hill fringe of south-east Warwickshire together with the Evesham vale which continued to be 'so plentifull of corne' according to Leland (1538) as to form the granary of Worcestershire. Among individual towns Coventry was supreme, although just why it should have become so is not easy to explain. Apart from the fact that it held a key position in the upper Avon it had little to commend it on purely physical grounds; nor was it on a Roman road, although it was not far from the Watling Street. No doubt the re-establishment as a Benedictine priory in 1043 of an earlier religious house by Leofric, Earl of Mercia, who gave it a very generous endowment, proved a powerful stimulus. This was further helped by the deliberate policy of granting privileges of tenure to those inhabiting the earl's part of the town as well as giving various concessions to merchants coming to reside there. Many people migrated to Coventry which became, by the end of the fourteenth century, the fourth largest city in the country with a population of 7,000 and ranking after London, York and Bristol. It contained a great variety of craftsmen in addition to many wool and cloth merchants.

In the early eighteenth century a very different picture is obtained. Birmingham and surrounding districts including Dudley, Wolverhampton, Walsall and Halesowen had become important centres of ironworking; nails were a staple product, while Birmingham had begun to make articles of brass, in the casting of which the possession of good local supplies of Upper Bunter sands was very useful.

Even before Abraham Darby's epoch-making discovery of how to smelt iron ore with coke at Coalbrookdale in 1709, the West Midlands possessed many essentials for ironworking. Charcoal was fundamental in the blast-furnaces for the production of pig iron and also in most of the forges for producing bar iron. For reheating the bar iron before it was rolled and then cut in the slitting mills, coal was used exclusively as the fuel. Similarly coal was used by the smiths in the production of finished articles, for example, nails, locks and bolts, chains and various agricultural implements. The coal consumed in

Birmingham and the adjoining parts came primarily from the Thick Coal (or Ten Yard or Thirty-Foot seam as it was alternatively called) which outcropped regularly over wide areas both in the northern section of the field, including Dudley, Tipton, Bilston and Wednesbury, and also in the southern sector at, for example, Netherton, Gornal and Stourbridge. This was the thickest and richest seam in the whole country, and the coal was very cheap to extract. Iron ore also existed in the Coal Measures, the Gubbin ironstone, immediately below the Thick Coal, being the chief source. Limestone was available from the Silurian inliers of Dudley Castle Hill, the Wrens Nest and Sedgeley. The fact remains, however, that in the early eighteenth century, there was a pronounced shortage of pig iron in the Birmingham Plateau since the Coal Measure ironstones could only produce the poorer coldshot quality pig and not the tough pig needed for the manufacture of the best merchant bar iron. The latter came via the Severn waterway mainly from the Forest of Dean which had a surplus of the tough metal suitable both for producing the best wrought ironwork and for blending with the poorer qualities. Some thousands of tons were despatched annually up the Severn to Bewdley, the great iron-marketing centre, for the many forges on the Stour and its tributary, the Smestow Brook. The Stour basin (including Lye, Cradley, Halesowen and Stourbridge) was the chief centre of the heavy iron industry in the West Midlands, and the main source of supply for the many nailshops and other smithies on the outskirts of the coalfield. Not far behind in importance was the Coalbrookdale area which had many of the advantages possessed by the Stour valley, while it also produced the non-sulphurous 'sweet-clod' coal which enabled Darby to achieve success.

Birmingham had become a town which was alive and active; no longer was it an unimportant centre in the heart of an isolated plateau. True it still suffered from some defects of location, but it had consolidated its position as the leading centre of trade and industry *within* the Plateau and was beginning to widen the area of its services. 'Bremicham, full of inhabitants and resounding with hammers and anvils, for the most of them are smiths' (Camden) made the nails, bits, horseshoes and agricultural implements needed by an agricultural community, and 'knives and all manner of cutting tools' (Leland) as well. Something like 15,000 swords were made for the Parliamentary forces during the Civil War at a mill on the Rea in

Digbeth. Meanwhile as the earlier textile and leather trades declined, 'new' trades became more evident, including the manufacture of guns, buttons, toys and brass articles. These, requiring the application of a high degree of skill to a limited amount of raw material, reflected Birmingham's position away from the coalfield. It became usual, for example, for the semi-finished gun-locks and parts manufactured in Bilston, Wednesbury and Darlaston to be brought into Birmingham for the final finishing and assembling processes carried out in the small workshops of independent gunmasters.

The population, which was probably not more than 1,500 in the mid-sixteenth century, and 5,000 in 1650, increased to about 15,000 by 1700. How far this increase may be directly attributed, as often asserted, to the freedom from religious restrictions enjoyed in Birmingham after the Restoration is open to question; probably the influx was rather the result of the town's industrial and commercial prosperity. It is certainly true that the absence of craft and trade guilds meant that no restrictive influences existed to discourage the introduction of new trades or limit the freedom of entry into any particular branch of industry or commerce. It is also true that, largely in consequence of its earlier insignificance, Birmingham was then governed, as it continued to be until 1838, as a small country town, and it possessed no charter.

Birmingham and South Staffordshire were thus in a favourable position to benefit fully by the successive developments of the eighteenth century—the successful smelting of iron ore with coke, Watt's improvements on Newcomen's steam-engine, the introduction of canal transport as well as of better roads. Thenceforth intensive concentration of all phases of the iron industry within the coalfield became the rule, and South Staffordshire—the Dudley section of the coalfield south of the Bentley Faults as contrasted with the northern section of Cannock Chase—could leap ahead, as here existed the three fundamentals, cheap supplies of Thick Coal, ironstone and limestone. Thus grew up the true Black Country of the nineteenth century as described by Mackinder—'a great workshop both above ground and below; at night it is lurid with flames of the iron-furnaces: by day it appears one vast loosely-knit town of humble homes amid cinder heaps and fields stripped of vegetation by smoke and fumes'. Figures bear out vividly enough the extent of this material progress, for by 1806 there were 42 furnaces in the Black

Country, and by 1858 a maximum of 182 had been built, of which 147 were in blast.

Fundamentally related to these developments were the essential improvements in transport, both within the industrial area and as connecting links with the outside world. Canals dominated the situation until 1840 when railways assumed the leading role. The Birmingham canal system is a network of waterways chiefly within the mining area, but forming a complicated chain between Birmingham and Wolverhampton through Smethwick and Tipton with extensions north-eastwards through Walsall towards the Trent and Mersey Canal, and southwards to Halesowen and the Stour valley. The essentially adverse physical conditions underlying the system have now become increasingly evident, although, in their heyday the canals were wonderfully prosperous. The traffic is still relatively large compared with some other inland navigations, but it is essentially local, travelling an average distance of only about eight miles, and consists of bulky raw materials, of which coal going to various works on the canal banks forms one-half.

The railways of the Plateau are divided between the Midland and the Western systems which interlock in the Black Country along the axial line between Birmingham and Wolverhampton. The main trunk line from south-west to north-east, leading via the open Tame valley, gives the area a very important connexion with the industrial region of the East Midlands and beyond.

In the new national scheme for motorways Birmingham plays a vital role as the hub of the system, and it is significant that the first link to be completed, the M1, was opened in November 1959 between London and Birmingham.

At the present day the whole region is dominated by the West Midland Conurbation—Birmingham and its two large suburban areas of Solihull and Sutton Coldfield, together with the Black Country, the whole covering 270 square miles with a total population of 2·34 m. people. Between 1931 and 1951 this conurbation had a rate of population increase (15·7%) much greater than that of any other such area and now it ranks amongst the most prosperous industrial units in Britain. The Black Country containing about 1 m. people is essentially based on the Productive Coal Measures, south of the Bentley Faults. The title is now in fact a misnomer for gone are the supplies of Thick Coal and iron ore from the visible coalfield, and the only collieries working this coal are the deep ones in the concealed measures to the west at Baggeridge, and to the east at Sandwell Park and Hamstead. Cannock Chase is today the chief centre of coal production, with 4·85 m. tons in 1958 as against just over ½ m. tons south of the Bentley Faults. Very little pig iron is now produced within the Black Country, except at Bilston to which all the raw materials have to be imported—the iron ore from the Jurassic belt, the coke from South Wales, and the limestone from Derbyshire. Steel is brought for engineering and other works from Lanarkshire, the north-eastcoast and South Wales. The Black Country still possesses large supplies of Coal Measure clays, particularly fire-clays, the products of which are exported very widely; but there can be no doubt that its chief assets are the accumulated momentum of tradition and skill, the ordinary capital invested in the area, and its central position relative to the rest of the country. Indeed it is far from being the moribund district it is often pictured to be; it has, on the contrary, continued to prosper since the eighties of last century. Some of the old staples remain such as the making of locks and keys at Willenhall and the adjacent towns of Wolverhampton, Wednesfield and Walsall; nuts, bolts, screws and nails at Darlaston and Smethwick; chains and anchors at Old Hill and Cradley; springs at West Bromwich and Wednesbury; and glass at Stourbridge. In addition, new industries have entered such as electrical engineering at Wolverhampton, while the heavier branches of non-electrical engineering have gravitated to the centre and south of the Black Country, where constructional engineering, the making of cranes and pumps, for example, is well represented. Other industries now existing are concerned with food and drink, clothing, chemicals and rubber. Rapid changes are also taking place in the physical landscape so that it is no longer true to say, as it was even in 1945, that one-eighth of the area is 'derelict' land. Bull-dozers levelling tip-heaps, mechanical shovels filling in clay-pits, and other modern methods of reclamation are altering all that, and much land once waste is now being built upon, or at least made presentable.

Birmingham has had a phenomenal development since the eighteenth century and it now occupies an area six times as large, and houses a population sixteen times as great as in 1801; in 1961 its area was 51,147 acres, and its population 1·105 m. This population, too, is much more mixed than it was, comprising elements from all parts of the British Isles as well as up to 70,000 coloured people—*c.* 92% being West Indians, Pakistanis and Sikhs. As a manufacturing town it has gone through various phases during which have occurred both the consolidation and adaptation to meet modern needs of the older staples including guns, jewellery and the metal trades, as well as the introduction of new ones, particularly cocoa- and chocolate-making, the production of electrical goods, cycles and, above all, motor vehicles. The modern concentration within central Birmingham is most impressive; here there are some 138,000 workers representing about one-quarter of the total industrial manpower of the conurbation working in roughly a thousand factories and giving a density in some parts as high as 40,000 per square mile. The industries of the centre are particularly the older ones in which fairly small units are still very numerous, especially the jewellery trade which has its own 'quarter'. Larger industrial establishments also exist but are more often found some few miles from the centre along the canals or railways, and also in the Rea and Tame valleys at such places as Selly Oak, Bournville, Small Heath, Tyseley and Witton, where are situated non-ferrous metal factories as well as the newer electrical and motor works, the largest of the latter being seven or eight miles away at Longbridge on the south-western side. So great has the motor vehicle trade now become that at least one-fifth of the population employed in manufacturing industry within the conurbation is directly engaged in it, and it has therefore been responsible for a remarkable integration of industrial effort.

In administration the changes in Birmingham have been no less dramatic; after securing its municipal charter in 1838, Birmingham has successively been recognized as a county borough (1887), as a city (1889), and as the centre of a bishopric (1904), while its area has increased enormously since the creation of Greater Birmingham in 1911. Birmingham is now the second city in the kingdom in population, and in area it is three times the size of Glasgow and twice the size of Manchester or Liverpool. It is faced with many problems in the field of economic and social planning, the most crucial being concerned with the rebuilding and reorganization of its central areas which contain much unsatisfactory housing, and, being congested with industry, are too small to cater adequately for the shopping and other needs of the vast urban community which now exists. The city is, indeed, undergoing a dramatic transformation. Following the recommendations of the Barlow Report of 1939, official policy is against any further extension of the urban area into the surrounding green belt despite the fact that practically all the land available for building within the city is already taken up so that overspill problems are becoming more and more acute. Already population and some industries have been moved into towns in the adjoining counties as well as to areas outside the region, for example, to South Wales and Merseyside. The problems are still far from being solved, and while some advocate the building of one or more new towns beyond the green belt in order to accommodate the surplus above a 'planned' maximum of 1,060,000, others would prefer to see some extension of the city's boundaries. Dawley, in eastern Shropshire, and Redditch, south of Birmingham, are now being developed as 'new' towns. Meanwhile Birmingham's responsibilities to its region have grown steadily in range and complexity, since, apart from its economic functions, it has to cater for the many other needs, cultural, medical and intellectual, of the West Midlands.

Outside the conurbation there is plenty of variety to produce a well-balanced region. Among urban units Coventry stands out, having undergone a dramatic industrial expansion in recent years, and in 1961 it had a population of 291,000. This ancient city, although overtaken by Birmingham, has continued to grow by adapting itself to changing circumstances; it was, in fact, the pioneer in the development of the cycle and motor industries which, with artificial silk, machine tools and general engineering, have displaced its earlier trades. Moreover, the major disaster which the city suffered during the Second World War created a remarkable opportunity for planning which has brought it fame. Expansion has also occurred at Bedworth and Nuneaton so as to form a minor conurbation with Coventry, the combined population of the three places being over 390,000. Coventry now extends southwards to within two miles of Kenilworth, while, on its westward side, it is only a very deliberate restrictive policy that preserves, so far, a green belt between it and Birmingham.

Of the various county towns, while all show some modern growth, Worcester's progress has been very striking; its population (65,865) has doubled within the past century. Once primarily a cathedral, administrative and market town, it has well-established glove and porcelain industries to which have been added in recent times various metal and engineering works. Its accessible position has been a great asset, as witness its metal-box industry which uses tinplates made in South Wales, and its leather industry which uses materials coming from India and Africa through Liverpool and Avonmouth. Warwick and Leamington have a combined population of 59,268, and while they are important as residential centres for both Birmingham and Coventry, they are also producers of motor-car and aeroplane components. Similarly Stafford has become an important manufacturer of electrical goods. Further west Shrewsbury, centred on its famous river loop, has combined many functions. Since Anglo-Saxon times it has played a leading role in the life of the area, being fortress, route centre, market, and later to some extent at least an industrial town based on its small coalfield. Today its road and railway network gives some measure of its pre-eminence and its independence, and it had made possible the recent outburst of industrial activity on the north side of the town linked with industrial Lancashire and Yorkshire. Hereford in the centre of its county has its cathedral, assize court, market and network of communications, while the making of cider and jam are old-established industries. Despite its relative remoteness, several new industries have gathered around converted war-time factories so that now Hereford also makes furniture of various kinds as well as brassware for builders, domestic electrical equipment and petrol-feed systems.

Elsewhere a population map indicates low densities generally associated with agricultural areas, and conditions in these are very varied, ranging from market gardening to arable farming and dairying, and from sheep and cattle rearing to feeding and fattening. Surrounding Birmingham and the Black Country conditions are often favourable for market gardening owing to the existence both of a huge market and of lighter soils from Bunter or Keuper Sandstones, as in the Bromsgrove and Lichfield areas. In the former, income is mainly derived from the growing of brassicas, runner-beans and peas, while in the latter potatoes are also very important, and these, with roots and green crops, occupy anything from a quarter to a half the total area of some farms in this district. Two intensively cultivated areas of particular significance comprise the Lower Keuper belt of north Worcestershire, with such centres as Ombersley and Hartlebury, and the Avon valley, above all the Vale of Evesham. The first grows sugar-beet, for which a factory exists in Kidderminster, as well as potatoes and the usual market-garden crops, notably peas. The Evesham area is particularly famous: it was initially favoured with good soils that gave early crops, and a situation near big markets. Essentially its industry rests on highly skilled arduous hand cultivation together with accurate timing of operations and careful choice of varieties. While its small-holdings are characteristic, it also has some quite large farms. Nearly every type of vegetable is grown, asparagus probably being the most notable; the Pershore plum is also important.

Arable farming is well-developed in east Shropshire with the adjoining parts of Worcestershire and Staffordshire in a belt from Market Drayton to Stourport. The light free-working loams are derived from Triassic sandstones, chiefly Bunter, with some glacial material; the rainfall is relatively light (26–30 in.) particularly in the harvesting period of August and September. Farms are generally larger than in most Midland districts, and the chief crops raised are barley and wheat as well as roots, now primarily sugar-beets, which go to the factories at Wellington and Kidderminster, while potatoes are increasing in significance. An essential feature of the agricultural system is the fattening of sheep, mainly Clun and Kerry with Shropshire and Suffolk crosses, which were formerly folded on turnips but are now largely grass fed. Beef-raising is also important, although dairying is increasing.

The most important dairying areas extend in a great crescent around the conurbation, being associated with the heavy Keuper Marls and to some extent boulder clays, in central Staffordshire, large parts of Warwickshire and Worcestershire, with a very important extension in northern Shropshire. Much of the land in these districts was in permanent grass, but during the Second World War much was ploughed to produce oats and mixed corn, and green crops, especially kale. There has been some reversion since then, but the present practice seems to indicate that something like 40% of the total area of the farmland will be treated as under rotation, and the other 60% will remain in grass. As to the livestock, while Shorthorns may still form the type herd, the increase of Friesians and Ayrshires is very marked. In the Lias area of south-east Warwickshire—the old Feldon—the heavy four-to-five-horse land presented acute problems of cultivation after the decline of corn prices in the 1880's, and it had deteriorated into rather poor grassland. But it has been proved that with deep ploughing by modern mechanical tackle it can produce heavy crops of wheat while it also produces good leys so that many fields are now under a rotation of three years in cereals and about three in grass. Thus the area has increased its beef cattle.

In the hill country of south Shropshire and the lowland of Herefordshire the emphasis is on sheep and cattle rearing. Most of the sheep are Clun and Kerry breeds which go to other parts of the West Midlands or farther afield for crossing with Shropshires or various Down breeds; the cattle are, of course, the well-known white-faced Herefords, famous as beef-producers in most parts of the world. The Clun region of south-western Shropshire has demonstrated in a remarkable manner what can be achieved with modern mechanical methods at heights of up to 1,000 ft providing the soil is reasonably good. In fact much of the area has deep fertile soils derived from the Old Red Sandstone, but before the Second World War the hills were largely derelict and covered with bracken. As a result of ploughing with tractors it has been shown that they are capable of much production, so that large areas have been reclaimed and now possess big stocks of cattle in summer.

Herefordshire is famous for many things besides cattle. Nature has been very bountiful, and it is a land of rich red soils, green pastures, hop gardens and cider orchards. On the other hand, it is a rural and isolated area with, at least until 1940, a small and declining population; indeed the whole county contains only 130,910 people of whom nearly a third live in Hereford itself. The agricultural output has thus been curtailed; but, despite this, there are about 4,000 acres of hops, mainly on the eastern side of the county, and if to these are added the 3,000 acres just over the border in Worcestershire, the total represents over one-third of all the hops grown in the country.

The West Midlands forms then a well-balanced region with a diversity of landscapes ranging from the densely peopled industrial and sometimes dismal areas, through the more pleasant and varied farming lands with a much lighter population, to the more remote and rather empty, but scenically attractive stretches, particularly in the west. To these assets are added its central position in southern Britain, the momentum acquired from its past development, and the possession of a large city. It is thus clear why the area now forms one of the most favoured and most stable regions in Great Britain.

Selected Bibliography

W. W. Bishop. 'The Pleistocene Geology and Geomorphology of Three Gaps in the Midland Jurassic Escarpment.' *Philosophical Trans. of the Roy. Soc. of London,* Series B, **241**, 1958, pp. 255–306.

H. C. Darby and I. B. Terrett (ed.). *The Domesday Geography of Midland England.* Cambridge, 1954.

F. H. Edmunds and K. P. Oakley. *The Central England District.* British Regional Geology. London (H.M.S.O.), 2nd ed., 1947.

B. L. C. Johnson. 'The Distribution of Factory Population in the West Midlands Conurbations.' *Trans. Inst. Br. Geographers,* **25**, 1958, pp. 209–24.

H. A. Moisley. 'The Industrial and Urban Development of the North Staffordshire Conurbation.' *Trans. Inst. Br. Geographers,* **17**, 1951, pp. 151–65.

A. H. Morgan. 'Regional Consciousness in the North Staffordshire Potteries.' *Geography,* **27**, 1942, pp. 95–102.

R. W. Pocock and T. H. Whitehead. *The Welsh Borderland.* British Regional Geology. London (H.M.S.O.), 2nd. ed., 1948.

Publications of West Midland Group on Post-War Reconstruction and Planning. *English County: A Planning Survey of Herefordshire,* 1946; *Land Classification in the West Midland Region,* 1947; *Conurbation: A Survey of Birmingham and the Black Country,* 1948.

F. W. Shotton. 'The Pleistocene deposits of the area between Coventry, Rugby and Leamington and their bearing on the topographic development of the Midlands.' *Philosophical Trans. of the Roy. Soc. of London,* Series B, **237**, 1953, pp. 209–60.

M. B. Stedman. 'The Townscape of Birmingham in 1956.' *Trans. Inst. Br. Geographers,* **25**, 1958, pp. 225–38.

M. J. Wise (ed.). *Birmingham and its Regional Setting. A Scientific Survey.* Published for the British Association for Advancement of Science, 1950.

The West Midlands, a regional study. Department of Economic Affairs, London, H.M.S.O., 1965.

CHAPTER 16

THE EAST MIDLANDS

K. C. Edwards

The East Midlands refers to an area, entirely indefinite as to boundaries, in which the term has an accepted currency among the people concerned. This is the area which extends outwards in all directions from the trio of large towns, Nottingham, Leicester, and Derby. The triangle formed by these towns represents the core, so that the approximate limits of the East Midlands lie at varying distances from it. These limits include Chesterfield and Retford in the north, Lincoln, Grantham, Stamford and Wellingborough on the east, Northampton in the south and Hinckley, Burton-on-Trent, Uttoxeter, Ashbourne and Bakewell on the west. To claim that the inhabitants of the East Midlands have developed a regional consciousness would be an exaggeration, yet it is not unfair to suggest that human activities, economic and social, operating within this territory impart to it a measure of coherence.

A brief glance at the landform map (fig. 36), shows that in the northern half of the area the outcrops of all the rocks, ranging from Carboniferous strata to the Lower Jurassic, have a fairly regular outline, succeeding one another from west to east. South of a line drawn through Nottingham and Derby the Triassic rocks, particularly the Keuper Marl division, spread out towards the south-west and occupy the greater part of the surface. The outcrops of the Jurassic rocks similarly swing to the south-west and become irregular in outline. Moreover, within the broad tract of the Keuper Marl to the north-west of Leicester is a small area of Pre-Cambrian rocks, the oldest in the East Midlands, which give rise to the Charnwood Forest hills, and immediately to the west is an outcrop of Coal Measures indicating the presence of the Leicestershire and South Derbyshire coalfield.

The main features of the relief broadly reflect the pattern of the outcrops. In the north-west the Carboniferous rocks of the Derbyshire Peak District form the southern end of the Pennines. The centre of this area is a dome-like massif of Carboniferous Limestone giving

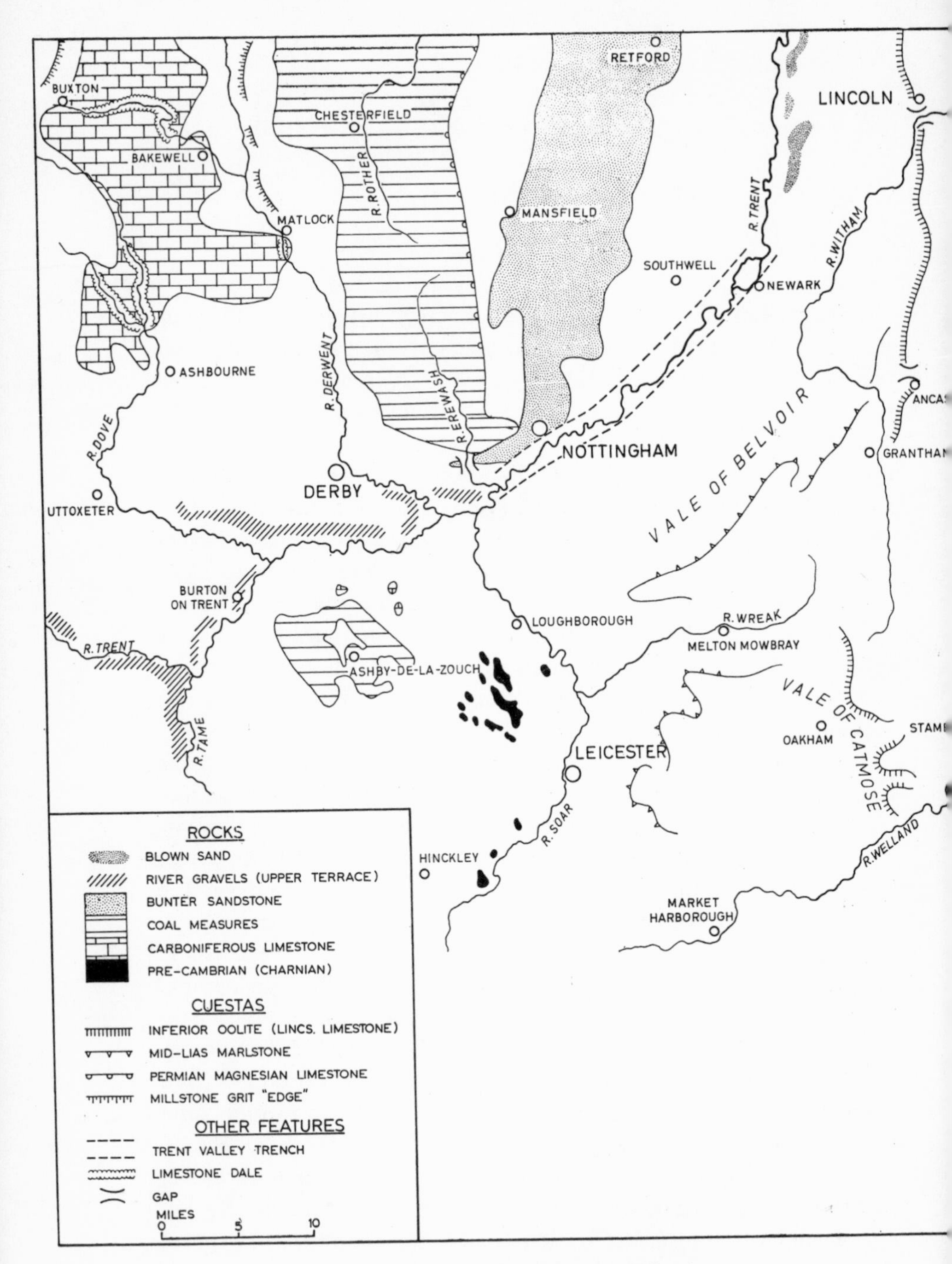

Fig. 36. Landforms of the East Midlands.

an undulating plateau surface with a general elevation of over 1,000 ft. Towards its margins, however, this surface is deeply notched by steep-sided valleys known as the Derbyshire Dales, for example Dovedale, Millers' Dale, and Monsal Dale. Surrounding the limestone, except on the south, are the Millstone Grit moorlands with their inward-facing grit scarps called 'edges', such as Froggatt Edge and Stanage Edge on the east, Axe Edge on the west. These grit moors, often covered with peat and heather, form even higher ground, especially to the north where the flattish summit of Kinderscout exceeds 2,000 ft. Between the limestone and the gritstone is a narrow belt of shales forming lower ground. Where this is followed by a river, such as the Derwent above Matlock, it gives rise to a broad valley.

The scenery of the Peak makes a strong appeal to those wishing to escape from the crowded industrial districts which surround it—Manchester, Sheffield, the West Riding, Nottingham and Derby. The area offers scope for many and varied recreations: it is good ground for rambling, camping, rock-climbing and ski-ing; there are naturalists' haunts; caves and underground passages in the limestone, especially around Castleton; historic houses like Chatsworth and Haddon Hall and prehistoric monuments like the stone circle of Arbor Low. Few upland areas combine outstanding scenery with such a wealth of interest, and in 1949 it became the first of Britain's national parks.

An important feature of the gritstone moors are the numerous reservoirs supplying water to the thickly populated lowlands. Heavy precipitation, a low rate of evaporation, relatively impermeable rocks affording abundant run-off, and valleys suitable for dam construction provide favourable conditions for water storage. The largest of the schemes, consisting of a series of reservoirs on the upper Derwent, is one in which Sheffield and the three chief towns of the East Midlands, Nottingham, Leicester and Derby, jointly participate.

Flanking the Millstone Grit on the east are the Coal Measures which form the continuation of the Yorkshire coalfield into Derbyshire and Nottinghamshire. The Coal Measures, composed of clays and shales containing layers of resistant sandstone, present rather broken country ranging from 200 to 600 ft, the northern part of which is drained by the Rother to the Don, and the southern part by the Erewash, a tributary of the Trent. The valleys of the Erewash and Rother, separated by a low watershed, afford an important route through the heart of the coalfield, one that is used by the former Midland Railway connecting Leicester and Nottingham with Sheffield.

Eastwards from the coalfield are the successive outcrops of Permian (Magnesian) Limestone, Bunter Sandstone, Keuper Sandstone and Marl, and Lias Limestone and Clay. As a group these rocks lie unconformably upon the Coal Measures beneath and are tilted gently to the east or south-east. They form a series of low cuestas with westward-facing scarps which vary greatly in prominence. Their dip-slope surfaces are sufficiently broad however to exhibit differences in relief, soil and land use, differences which are heightened by the almost complete absence of drift deposits except in the extreme east. The entire area lying between the Coal Measures and the Lincoln Cliff is a broad lowland of comparatively gentle relief, across which in the form of a wide arc runs the Trent valley. Within the East Midlands, especially from Nottingham to Newark, this valley takes the form of a shallow steep-sided trench, within which the flood-plain is at least a mile in width. This feature was carved by a mighty predecessor of the Trent in glacial times, the present river being an obvious misfit. The lowland as a whole is made up of different belts of country, each distinctive in character.

The narrow belt of Magnesian Limestone, yielding a medium light well-drained soil, is largely under arable farming. In places it is breached by consequent streams which have cut small gorges and in one of these, at Creswell Crags, are the caves noted for their late palaeolithic remains. At Bolsover the limestone escarpment, crowned by a castle, presents a bold feature overlooking the Chesterfield section of the Rother valley and the Derbyshire moors beyond.

The Bunter Sandstone country is an undulating tract of cultivated land interspersed with large stretches of woods, heath and bracken which are the remnants of the once extensive Sherwood Forest. Owing to the porosity of the rock there is little surface drainage, while the light sandy soil is infertile unless counter-measures are adopted. It is a thinly settled area containing the several large estates known as the Dukeries. Modern coal mining, however, has given it a new significance.

The adjoining Keuper Marl country which extends eastwards to beyond the Trent presents a very different landscape. The relief is more marked, partly owing to dissection by numerous streams and partly to the occurrence in the marl of thin resistant sandstone bands called 'skerries'. The soil is of a stiff clayey nature and is suited to grain crops and the surface is more fully cultivated than the Bunter Sandstone. Accordingly it supports a far greater number of villages and farms. Certain districts such as Tuxford and Southwell are noted for their orchards, chiefly of apple, plum and damson, although in general, mixed farming with arable and grassland in roughly equal proportions determines the pattern of land use. Broadly similar conditions prevail in the belt of Lias Clay which falls largely within Lincolnshire where it is known as the Western Clay Lowland. Stretching eastwards from the Trent a little below Newark is a considerable accumulation of gravel and blown sand. This area forms a distinctive unit largely devoted to the cultivation of carrots, making very profitable use of the light soil around the villages of Thorney, North Clifton and North and South Collingham.

Southwards and westwards from Nottingham, the broadening expanse of Keuper Marl extends over south Derbyshire and much of west Leicestershire, giving lowland scenery typical of the so-called Midland Plain. Characteristic, too, are the wide gravel-filled floodplains of the larger rivers, Trent, Soar, Derwent and Dove, which cross it in various directions. Of these the Trent valley is especially important in providing a natural routeway which gives access from the East Midlands to Birmingham and the Black Country. Apart from Charnwood Forest and the adjoining coalfield which form higher ground, small isolated hills representing inliers of older rocks break the surface here and there. Some are of Carboniferous Limestone such as Ticknall and Breedon Hill, others like Enderby and Croft are of igneous origin, and most of them have been extensively quarried. The agricultural land is largely under grass, though differences in farming exist between Leicestershire and south Derbyshire. In the latter area grassland is devoted more exclusively to dairying. More than three-quarters of all the livestock are dairy cattle, and milk production is the farmers' main concern. Improvements both in milk yields and stock-carrying capacity have been made in recent years by greater attention given to fodder cultivation and a marked increase in the use of temporary pasture. Milk is handled in bulk by railside depots at Ashbourne, Uttoxeter and Egginton, all in the Dove valley. In west Leicestershire where, partly owing to exposures of Keuper Sandstone and partly to a considerable though intermittent cover of glacial drift, soil conditions are more diversified, there is more arable land. Yet livestock, including dairy cattle and pigs, remain the chief source of income. Small market centres such as Ashby-de-la-Zouch, Market Bosworth and Lutterworth are now only of local importance.

A few miles north of Ashby, around the little town of Melbourne, is a small district of intensive market gardening. Vegetables, greens, rhubarb and small fruit, especially strawberries, are raised on holdings ranging from 10 to 50 acres for the markets of Derby, Loughborough and even more distant centres. Market gardening is long established here, and there was a period in the middle of last century when many of the growers specialized in the output of quickset hedges for fencing the railway tracks which were then being laid all over the country.

Charnwood Forest, a small hilly area lying to the west of the Soar has many distinctive features. It is mainly composed of rocks of Pre-Cambrian age (slates, grits, quartzites and volcanic ash) in which intrusive materials also occur. In broad plan these rocks form a series of four narrow uplands separating three longitudinal depressions. All these trend north-west to south-east and follow the direction of the ancient Charnian folds. After prolonged erosion, the whole area during Triassic times was buried beneath deposits of Keuper Marl, but later uplift has resulted in only a partial removal of the Marl thus revealing a fossil landscape of pre-Triassic date. While the Marls form the surrounding lowland, except in the north-west where Charnwood is flanked by the Coal Measures, they also cover the floors of the depressions and smaller valleys and fill the hollows between the higher hills. Thus the Marls in the interior lie at a greater elevation than those of the lowland. Only in one valley, that of the Black Brook which drains the northern part of the innermost depression, is their removal complete.

A striking contrast is therefore presented between the rugged and rather barren hills with their sharp, projecting rocks and the gentle outlines of the valleys rendered green by their crops and pasture. Most of the uplands reach to over 700 ft but in the west Bardon Hill, the highest summit, rises to 912 ft. Rocky areas interspersed with farmland provide the keynote to the scenery of Charnwood, but there are other distinguishing elements. The impermeable nature of the surface, resulting in a considerable run-off, has enabled reservoirs to be built in the lower valleys to supply Leicester and Loughborough with water. Old quarries abound, their scars now largely hidden by vegetation. Some of these yielded the Swithland purple slates which can be seen in the roofs of local villages while others provided granite, or more strictly syenite, for making road setts and chippings. Quarrying for the latter purpose is still active at Mountsorrel, an eastern outpost of Charnwood. In early times the seclusion of the interior attracted monastic foundations of which several ruins remain. Indeed, Charnwood is still valued not only for its agriculture, but also for its unspoilt and open country much appreciated by the people of the neighbouring industrial centres, especially of Leicester.

In the south-eastern portion of the East Midlands the Lias Clay lowland is almost everywhere bordered on the east by rising ground, reaching generally to over 400 ft, formed by the Middle Lias Marlstone. West of Grantham the latter gives rise to a bold escarpment overlooking the Vale of Belvoir. In fact this feature continues southwestwards across Leicestershire as the dominant element in the Jurassic scarplands of central England. In places it rises to a height of 600 ft, but is broken in the neighbourhood of Melton Mowbray by the valley of the Wreak, and near Market Harborough by those of the Soar and the Welland, which are separated by a low watershed. Between these passages, which are used by the Midland railway routes from London to Nottingham and Leicester respectively, lies a block of upland country part of which is in Leicestershire and the remainder in Rutland. This block, together with the similar areas of higher ground formed by the Marlstone to the north and south, as well as the Lias Clay lowland to the west, all carry a considerable spread of boulder clay which has however been removed from most of the valleys.

A great deal of all this country is devoted to pasture, referred to as either 'lowland' or 'upland' grassland, and is famed for the fattening of livestock. These pastures have been remarkably persistent in fertility, and it is claimed that even without treatment they will support one bullock or twenty sheep to the acre for years. Towns such as Melton Mowbray and Market Harborough, together with Oakham in the Vale of Catmose, are active market centres, though the first two also have some industry. They are also famous hunting centres associated with the names of Belvoir, Cottesmore, Quorn and Pytchley. In the Vale of Belvoir, besides the grazing of cattle and sheep, the making of Stilton cheese is a speciality in some villages such as Long Clawson and Nether Broughton, but much is now factory-made at Melton Mowbray.

In the Marlstone area, especially between Grantham and Melton Mowbray, ironstone is worked considerably, chiefly by open-cast methods after stripping the overburden. The ore is sent to the blast-furnaces at Holwell near Melton Mowbray or to other smelting centres in the East Midlands.

Industry in the East Midlands is mainly concentrated upon the Nottinghamshire and Derbyshire coalfield, the smaller Leicestershire and south Derbyshire field, and in the three large towns of Nottingham, Leicester and Derby, together with the smaller centres with which they are closely related, for example, Mansfield (53,222), Loughborough (38,621) and Long Eaton (30,464). There are also a few outlying centres of manufacturing such as Newark (24,610) and Hinckley (41,573), and, in the Derwent valley, Belper (15,563) and Matlock (18,486).

The broad pattern of industry is relatively simple. In addition to coal mining and iron production, the latter being almost entirely confined to the major coalfield, engineering of many different kinds is widely distributed and is perhaps the most characteristic activity. Almost as widespread in occurrence is the manufacture of hosiery and knitted wear, which despite its heavy concentration in Leicester and Nottingham, is carried on from Mansfield in the north to Hinckley in the south. The East Midlands is by far the most important area in the country for these products. Other textile industries include lace-making in the Nottingham district and the production of rayon in Derby. Apart from these industries, there is a great diversity of general manufacturing with many instances of specialized production in particular places.

The East Midland section of the great Yorkshire, Derbyshire and Nottinghamshire coalfield occupies much of east Derbyshire and the adjoining area of Nottinghamshire. No physical distinction can be made between this part and the Yorkshire portion of the field, though some differences occur between the two in the use and disposal of the coal produced.

In length the Nottinghamshire and Derbyshire field extends for about 35 miles from the Yorkshire border to its southern extremity near Nottingham, while its breadth varies from 12 to 20 miles. Its western edge is limited sharply by the outcrop of Millstone Grit, but on the eastern side the coal-bearing strata pass beneath an unconformable series of Permian and Triassic rocks. Under these rocks is the so-called 'concealed' coalfield in which the coal-bearing strata dip more gently than on the exposed field and are accessible at a workable depth for many miles eastwards. The Coal Measures are subdivided into Upper, Middle and Lower groups. The Upper Measures have been removed by denudation, except in parts of the concealed field where, however, they contain no coal. The Middle Measures, consisting of 2,000–3,000 ft of sandstones and shales, contain numerous workable seams and occupy the greater part of the exposed field. Owing to the pronounced dip of the rocks, many of these seams appear at the surface. Beneath the Middle Measures and outcropping along the western edge of the coalfield are the Lower Measures which contain, with one exception, only a few rather thin seams. Outstanding among the coals of the Middle Measures is the Top Hard Seam which is the equivalent of the famous Barnsley Bed farther north in Yorkshire. With a thickness varying from 5 to 6 ft, the Top Hard is a composite seam yielding good house coal, high grade steam coal, and in some places gas-making coal. Its thickness and quality are maintained over an area exceeding 200 square miles, and many collieries have been sunk with the prime object of working this seam. Other valuable seams are the Deep Hard and Deep Soft, both steam-raising coals and the High Hazles, a famous house coal.

Geological conditions afford important advantages over many other coalfields in Britain. The seams are comparatively little disturbed by faulting and they are not severely affected by minor folds. Although they are not very thick, they persist for great distances without marked deterioration in quality. Also the many types of coal available enable a diversity of markets to be supplied. The field as a whole, however, is poor in coking coal; this is largely restricted to a few seams in the area lying between Mansfield and Chesterfield. On the older exposed field some of the seams are approaching exhaustion, but the concealed portion provides scope for mining over an indefinite period, the actual reserves and limits of the field having yet to be precisely determined.

In addition to the favourable physical conditions, the productivity of the coalfield has been enhanced by a general absence of labour difficulties owing to a tradition of good relations between miners and employers established long before the coal industry was nationalized. Again, a high degree of colliery mechanization has promoted a high output per man. Production at about 45 m. tons annually represents one-fifth of the national output and this amount is raised by only one-seventh of the country's miners.

Marked contrasts in landscape occur between the exposed and concealed sections of the coalfield. The former was naturally developed first because the seams appeared at the surface or only a little below, and it was not until 1859 that the earliest colliery was sunk through the Permian strata at Shireoaks near Worksop, heralding the advance on to the concealed field. In east Derbyshire the valleys of the Rother and its tributary the Doe Lea Brook, as well as the Erewash along the Nottinghamshire border, greatly influenced the distribution of both collieries and railways. The Erewash in particular, having cut its course along the crest of an anticline in the Middle Coal Measures, made valuable seams accessible on either side of the valley. Today the whole area is covered with collieries, works, railways and the straggling settlements of the mining community. To the already hilly surface have been added mountainous spoil heaps. At intervals from north to south, generally on rising ground, are former market towns which have been transformed into industrial centres: Chesterfield, Alfreton, Ripley, Heanor and Ilkeston.

Coal, however, was not the only economic mineral in this area, for bands of good grade ore found in the Coal Measures gave rise to iron-smelting at a number of places in the larger valleys: Stanton, Ilkeston, Clay Cross and Staveley. Limestone as a flux for the blast-furnaces was readily obtainable from central Derbyshire. Supplies of pig iron thus encouraged the growth of forge and foundry work and many branches of engineering. The production of foundry iron rather than steel has, in fact, remained a characteristic feature of East Midland heavy industry. Though the local ore has long been exhausted and many blast-furnaces dismantled, large units remain active at Stanton, Staveley and Sheepbridge, using Northamptonshire ironstone. At Stanton, Staveley and Clay Cross the production of seamless iron pipes is a speciality, though the last-named works has ceased its smelting operations.

For the existing ironworks a supply of coke is vital and at least part of their requirements are obtained from the Chesterfield area. In this district coke-ovens are a feature of some of the collieries as they are also at the Stanton and Staveley works. The large coke-producing plant near Chesterfield, known as the Avenue Works, completed in 1956, should also be noted. Coal from several collieries is brought there for conversion into coke and various by-products; the coke, however, is not for metallurgical use but for the ordinary industrial and domestic market. The output is over 2,000 tons daily.

At the southern end of the coalfield industrial development has been greatly influenced by the proximity of Nottingham. Textile working in particular occurs near the city and relies largely on the availability of female labour in the mining districts. Hosiery and knitted goods are made at Ilkeston, Heanor and Langley Mill as well as in Mansfield, Sutton-in-Ashfield and Kirby-in-Ashfield. The spinning of both cotton and wool is carried on at Pleasley Vale just north of Mansfield, while clothing factories using the newer fibres are now to be found at Alfreton. Engineering is similarly widespread, sometimes dating from pioneer enterprises such as the famous Butterley works at Ripley (1790) and the Clay Cross works (1842) in which George Stephenson was a partner. Among smaller industries it should be noted that Coal Measures clay is used for making pottery at Denby and Langley Mill.

To the highly industrialized older portion of the coalfield the concealed field of central Nottinghamshire presents a striking contrast. While a number of collieries and their related settlements were established on the fringe of the Permo-Trias cover during the period 1860–1900, the sinking of deep mines farther east took place only after the First World War. In any case they were few in number and until 1939 were all located on the Bunter Sandstone in the area of Sherwood Forest. The new collieries planned after the Second World War extended mining operations still farther east and the latest, at Cotgrave, is the first to be sited across the Trent. The new colliery at Bevercotes near Retford, opened in 1965, is the world's first fully-automated deep coal mine. These mines, however, have had relatively little effect upon the countryside apart from railway links and the necessary housing. In some cases the existence of a mine is betrayed only by the appearance of the spoil heap above the forest trees. The collieries are large, relatively far apart, electrically operated and therefore smoke-free. Most significant of all is the fact that they have attracted no additional industries. On the other hand the extension of mining has given increased importance to Mansfield, the nearest town. Not only has it become the chief shopping centre for the colliery communities, but its own industrial pursuits have expanded considerably. These facts are reflected in a rapid growth of population from 21,000 in 1901 to 53,000 in 1961.

Meanwhile other changes have taken place in the Sherwood Forest area. Stretches of heath and bracken have given way to coniferous plantations, though a number of the ancient woods of oak and birch remain. The great estates have in part been broken up and the ducal mansions put to other uses. In the present age of social welfare, rest homes, hospitals and sanatoria have taken their place, enjoying the comparatively pure air and open views. Here and there, in response to Nottingham's growing demands, new pumping-stations raise large quantities of water from the highly porous Bunter Sandstone. The agricultural land, characterized by poor thin sandy soils and large arable fields, has profited from expenditure on fertilizers and from the cultivation of sugar-beet. Although sheep are still essential on many farms for manuring the land, increasing attention is being devoted to cattle and milk production. Thus the Forest with its particular assembly of characteristic features, farms and woodlands, public health institutions, pumping stations and coal mines, presents a landscape which reflects in a striking manner many different trends in present-day life.

Among the secondary mineral resources worked in the East Midlands are limestone, fluorspar, gypsum and gravel. In Derbyshire the Carboniferous Limestone is extensively quarried and crushed, largely for use in the heavy chemicals industry. The chief workings lie around Buxton, at Wirksworth, Matlock and Millers' Dale, while in the Hope Valley at the junction of the Limestone and Edale Shales a large cement factory is favourably placed to serve northern rather than Midland markets. Another mineral found in the Derbyshire limestone is fluorspar which is obtained from the spoil heaps of old lead-mines and from at least one mine worked solely for the purpose at Eyam. It is sent to Sheffield for use as a flux in the making of special steels.

Gypsum occurs in the upper layers of the Keuper Marl east of the Trent in Nottinghamshire which is one of the main sources of supply in the country. It is mined at several places south-east of Nottingham and also at Newark and is used for the manufacture of plaster and plaster of Paris. As a reflexion of the post-war demands of the building industry, a large plaster-board mill, the only inland factory of its kind, has been erected at East Leake beside the railway between Nottingham and Loughborough. Gypsum is also worked at Chellaston near Derby, where in the form of alabaster it was used in the Middle Ages and after, for statuary and ornamental work for which the 'kervers' (carvers) of Nottingham and Derby became famous throughout Europe.

The gravels of the Trent, like those of the Thames and Severn, are extensively worked to provide concrete aggregate and road-making material. Extraction is mostly from river terraces as at Hoveringham, Attenborough and Sawley in the Nottingham area and from the terraces of the Dove at Hilton near Derby.

The production of oil in the East Midlands, though very small, should not be overlooked. For the past twenty years crude petroleum has been raised from wells distributed in groups at different localities in east Nottinghamshire and adjacent areas. The oil occurs in small dome structures in the Carboniferous rocks beneath the Trias and Mesozoic strata which form the surface. The depth at which the oil occurs is usually about 3,000 ft. As the wells in one group decline in yield, other centres, located by geophysical methods, are brought into production. The chief centres are around Eakring, the oldest, Kelham Hills west of Newark, Egmanton near Tuxford, and Plungar in the Vale of Belvoir. The most recent discovery, in which oil was struck at a depth of over 4,000 ft, lies to the east of Gainsborough. The entire output is sent to the oil-shale area of Scotland for refining. Although the amount is insignificant, these diminutive oil-fields are of value as a training ground for geologists and technicians prior to working abroad.

The Leicestershire and South Derbyshire coalfield, though small, is highly productive. It covers a length of about fifteen miles and a maximum width of eight miles. An anticlinal axis of Charnian trend running north-west to south-east through the old market town of Ashby-de-la-Zouch divides the field into two coal-bearing basins, leaving Ashby, itself on the barren measures, between the two. The Coal Measures of both the eastern and western basins contain considerable proved reserves at no great depth beneath the Trias cover. In fact, in the eastern basin nearly all the collieries, including large mechanized drift-workings, are located on the 'concealed' area, the seams, such as the Main Coal, being thick and close to the surface. On the other hand the large extension of the western basin, known as the Coton Basin, has only recently been proved. The total output from the field is over 7 m. tons annually, nearly twice the amount of pre-war years.

In addition to its coal the western basin contains valuable fire-clays and pipe-clays. Extraction of these clays forms another basic industry and the manufacture of pottery, earthenware, glazed pipes and refractory products comes second only to mining. Some 25% of the country's output of salt-glazed pipes comes from this district. Most of the works are situated around Swadlincote (20,000), Gresley and Moira. The concentration of this activity has promoted a dismal yet distinctive landscape. Clay-pits, some of them enormous, spoil heaps, kilns, and works, emitting smoke and fumes which inhibit vegetation growth, and derelict buildings and waste areas all contribute to the general disfigurement of the surface. These depressing conditions are further accentuated by the severe effects of mining subsidence. By contrast the eastern basin, lacking good quality clays, is concerned only with brick-making for which the Triassic marl is used. The town of Coalville (26,159), which serves the mining area is also closely related to Leicester, being a centre for engineering, including textile machinery, as well as hosiery.

The importance of the two coalfields is matched by that of the three large towns of Nottingham, Leicester and Derby (fig. 37).

Nottingham occupies a very striking geographical position both as regards its general situation and the local conditions of its site. It not only stands at one of the major crossing points of the Trent, but it is also at the point of contact between the great coalfield and industrial area on the fringe of the Pennines and the extensive agricultural low-lands which lie to the east and south. As a place of exchange between contrasted economies Nottingham derives great importance. Moreover, although the Trent ceases to be tidal a little below Newark, Nottingham through the centuries has been the effective head of navigation. Despite its position, however, the most direct route from London to the north, either by road or rail, crosses the Trent not at Nottingham but at Newark.

Although traces of Roman settlement have been found in the district, no evidence of any township at Nottingham has been brought to light and the earliest community of which there is record, dates

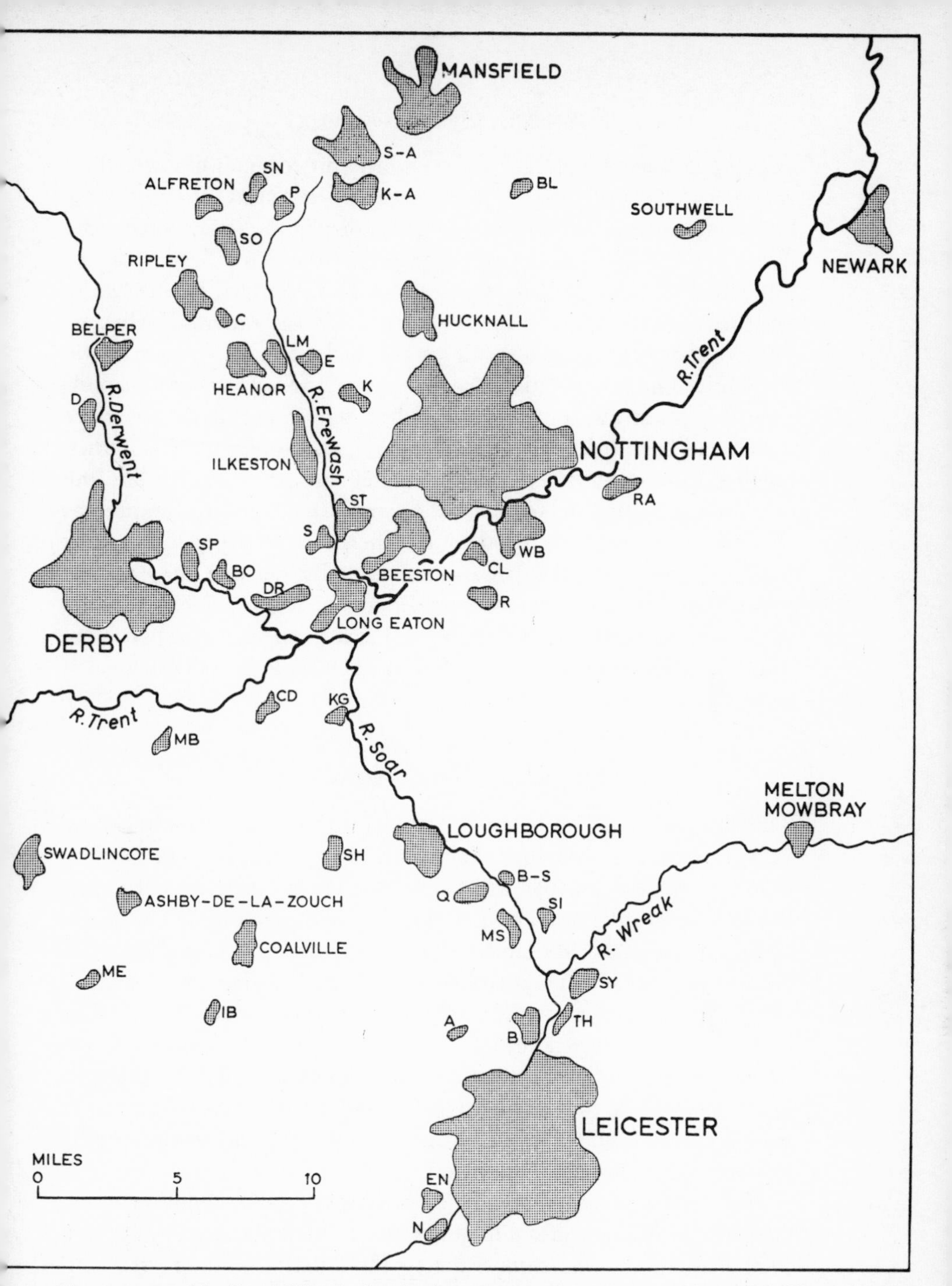

[Fig]. 37. The core area of the East Midlands, showing large towns, urban centres and industrial villages. Smaller [urb]an centres and industrial villages with a population of 2,500 or more are shown. A few places adjacent to the [lar]ge towns, such as Clifton, West Bridgford and Radcliffe-on-Trent in the case of Nottingham are almost entirely residential.

[A.] Anstey. B. Birstall. BL. Blidworth. BO. Borrowash. B–S. Barrow-on-Soar. C. Codnor. CD. Castle Doning[ton]. CL. Clifton. D. Duffield. DR. Draycott and Breaston. E. Eastwood. EN. Enderby. IB. Ibstock. K. [Ki]mberley. K–A. Kirkby-in-Ashfield. KG. Kegworth. LM. Langley Mill. MB. Melbourne. ME. Measham. [M]S. Mountsorrel. N. Narborough. P. Pinxton. Q. Quorn. R. Ruddington. RA. Radcliffe-on-Trent. S. [Sa]ndiacre. S–A. Sutton-in-Ashfield. SH. Shepshed. SI. Sileby. SN. South Normanton. SO. Somercotes. SP. Spondon. ST. Stapleford. SY. Syston. TH. Thurmaston. WB. West Bridgford.

from Anglo-Saxon times. The site embraced two small hills less than a quarter of a mile apart at the southern extremity of the Bunter Sandstone outcrop, which provided firm, dry ground overlooking the Trent flood-plain. On the more easterly of the two hills an Anglian settlement was established and this was later occupied by the Danes and became one of the Five Boroughs. On the other hill with its massive sandstone crag affording natural defence, a Norman castle was built. The latter with its considerable garrison occupied a position of great strategic value controlling movement across the river which had been bridged as early as A.D. 920. Nottingham as a town resulted from the fusion of the two settlements, the one civil and the other military, its dual origin being symbolized down to the present day in the two maces carried by the sheriff. Inevitably the shallow depression between the two hills eventually became the market-place, the largest of any town in England. Medieval Nottingham was noted for its diversity of trades and crafts, especially ironworking, using charcoal from Sherwood Forest. The fame of its craftsmen is recalled by the old-time jingle:

The little smith of Nottingham
Who doth the work that no man can.

The invention of the stocking-frame in 1589 by William Lee of Calverton, a neighbouring village, laid the foundation of the hosiery industry which developed into one of the leading activities of today, not only in Nottingham and Leicester but also in many of the smaller towns of the East Midlands. Later on, with the introduction of Strutt's device for making ribbed hose and with the coming of Arkwright and Hargreaves to Nottingham, the hosiery and cotton industries became firmly established. An adaptation of the stocking-frame led to the production of machine-made lace for which Nottingham became world-famous. This industry, no longer as important as it was, remains strongly localized in the city itself and in the nearby towns of Beeston and Long Eaton.

With coal immediately at hand, the Industrial Revolution transformed Nottingham into a major centre of manufacturing, and with the industrial development of the coalfield itself, relations with Nottingham became increasingly close, giving it further importance. Towards the end of the nineteenth century outstanding enterprise on the part of a few local men resulted in the growth of three large firms making cycles, tobacco and cigarettes, and chemicals and pharmaceutical products respectively, further broadening the economic basis. Industrial expansion caused the spread of the town, first across the flood-plain as far as the Trent; then in other directions, linking up with the outlying centres of Arnold, Carlton, Beeston, Long Eaton and Hucknall and with the mining districts of the Erewash valley. Today Greater Nottingham as a continuous built-up area is by far the largest urban complex in the East Midlands, with a population of not less than 500,000. The population of the city itself was about 10,000 in 1750, 29,000 in 1801, 240,000 in 1901 and is now over 310,000.

The original site of Leicester was on a gravel terrace affording dry ground on the right bank of the Soar. This supported a camp of the Coritani tribe which was afterwards occupied and developed by the Romans. Here the Fosse Way, striking across the country from the south-west to Lincoln, made the last of its several crossings of the Soar and also fixed the terminal point for the Via Devana coming from the south-east. The importance of Ratae Coritanorum to the Romans is shown by the considerable remains which have been exposed to view near the Central Station. Long after the Roman period the town was re-established by the Danes as another of the Five Boroughs. The town subsequently grew as a market centre and as a focus of the wool trade which in one form or another has continued in importance to the present day. Largely to serve the existing hosiery industry a little wool spinning is still carried on.

For a long period the town was confined to the area east of the river and Stukeley's map of 1722, when the population was only 7,000, shows that it had grown very little beyond the limits of the Roman enclosure. From the end of the eighteenth century expansion was rapid, first eastwards on to higher ground, and later on across the Soar to the west. It is interesting to recall that to the north and south of the present commercial centre are large public open spaces of ancient origin. On the north, Abbey Park forms part of the land once attached to the abbey where Cardinal Wolsey died, while on the south, Victoria Park is a survival of the Cowhay, a gift of Simon de Montfort, Leicester's 'shining patriot' of the thirteenth century.

The growth of modern manufacturing was encouraged by the opening of the Leicester canal (1794) connecting the town with the Soar Navigation through Loughborough to the Trent, and particularly by the famous Leicester and Swannington Railway (1832) which enabled coal to be brought to the town from the area west of Charnwood. From its status as a small country market-town in 1800 with a population of under 20,000, Leicester, like Nottingham, developed into a major industrial centre. Its population reached 100,000 soon after 1870 and today it is approaching 300,000. In addition to the leading industries of hosiery, footwear and engineering, there is an exceptionally wide range of general manufacturing. Few cities of its size have such a broadly based economy and partly because of this Leicester has the distinction of being one of the most consistently prosperous cities in Britain.

If Leicester is essentially a city of the lowlands, in fact of the Midland Plain, Derby is a town at the junction of highland and lowland. Its position on the Derwent lies only a mile or two below the point where the river leaves the hill country of the Pennines and enters the Trent lowland. This situation has made Derby a natural focus of communications. Routes from east to west, from Lincoln and Nottingham to Chester or Merseyside, in passing round the southern end of the Pennines inevitably lead through the town, while from the south the Derwent valley provides easily graded routes through the hills to Manchester or Sheffield. The first organized settlement, about A.D. 80, was that of a Roman military station (Little Chester) which lay on the great highway called Ryknield Street leading from Bath to Yorkshire. At this point the road crossed the Derwent and eventually from Little Chester roads were built in other directions, notably to Chester (Deva), to the Roman bathing resort at Buxton and along the Derwent tc the limestone area of the Peak where lead occurred. The dual function of Little Chester was to exercise control over the native tribes of the Peak and to protect the valuable lead workings.

With the withdrawal of the Romans the settlement was abandoned and after a long interval the succeeding Danish borough was located on an adjacent site on the right bank of the Derwent. Although Derby was an important market and trading centre during the Middle Ages, its modern industries were founded in the eighteenth century. Its interest in textiles, as in Nottingham and Leicester, resulted from the introduction of the stocking-frame but there was also John Lombe's silk mill erected in 1719, the first real factory organized on a basis of mass output. In 1775 Strutt and Arkwright built their cotton and calico mill. The making of porcelain, the renowned Crown

Derby ware, began about 1750 while the growth of ironworking laid the foundations of later engineering.

The historic importance of Derby as a route centre was further emphasized by the coming of the railways, especially by the Midland Railway system which originated in the area and made Derby its headquarters as well as its centre for locomotive and carriage building. Within twenty years after the lines to Birmingham (1838) and Nottingham (1839) had been built, the Midland Company linked together the three large towns of the East Midlands and gave them connexions with London and other major centres of the country. Since several of the lines served to open up the great coalfield, a powerful stimulus was given to industrial growth. In Derby itself large engineering concerns grew up adjoining the railways, especially to the south of the town on the gravel terrace bordering the Derwent flood-plain. Of these the largest is the Midland Railway works and the most famous is Rolls Royce (motors and aero engines), but others making foundry products, castings, tubes and various kinds of machinery are also important. To the east, actually on the flood-plain, are the large rayon works of British Celanese, established in 1920 and since greatly extended. Derby's population, which was 12,000 at the beginning of last century and 30,000 at the dawn of the railway era, is now 132,000, but 175,000 in the total built-up area.

Taking a comparative view of the functions of the three large centres, both similarities and differences may be noted. All three are county towns with important administrative, financial, market and retail functions; all are major centres of transport; and they are all centres with higher educational and cultural institutions and have large-scale provision for entertainment. As in all large cities these so-called service functions employ more people than are found in any other occupational group. In Nottingham, the largest of the three, which also has a more populous hinterland than the others, the proportion of the total occupied population employed in this way is 49%, compared with 43% in Leicester and 42% in Derby.

Cities are as much a product of human enterprise and organization as they are of their geographical position and their access to resources. Thus the three leading towns of the East Midlands, while exhibiting a broadly similar industrial structure, show significant differences in detail. Derby, with its three outstanding concerns, Rolls Royce, Midland Railway (British Railways) and British Celanese (now

Courtaulds), is rather less diversified industrially than Nottingham and Leicester, although the other forms of engineering and textile production are important. Leicester's engineering is more varied and includes a considerable output of textile machinery, much greater than in the case of Nottingham, as well as machine tools and radio and electrical equipment. Textile manufacturing in Leicester is dominated by hosiery to a greater degree than in Nottingham, while the footwear industry, one of Leicester's major pursuits, is virtually confined to that city and small places in the surrounding district. Nottingham's activities, however, are even more varied. Engineering includes a very wide range of production, with only one firm of outstanding size, Raleigh Industries, largely concerned in making cycles. In textiles considerably fewer workers are employed than in Leicester, and hosiery, though highly important, is less predominant for there is also the lace industry. A distinctive feature, moreover, is the much larger number employed in the textile finishing trades than in Leicester. The large concerns producing tobacco and cigarettes, and chemicals and pharmaceuticals, together with the telephone works at Beeston, add further variety to manufacturing in Nottingham.

Notwithstanding these differences, the three towns share two economic advantages. In the first place they derive benefit from their diversified industries and in the second, although some of their products are widely exported, the bulk of the output consists of consumer goods destined for the home market which is less subject to violent trade fluctuations.

The East Midlands is then an area of great variety and growing importance. The greater part of the surface is put to agricultural use in one form or another and there is as much diversity of farming as there is of manufacturing. The industrial districts, except for the Nottinghamshire and Derbyshire coalfield, are rather restricted in size and occupy a relatively small proportion of the total area. Their distribution is largely a reflexion of the important part played by the old Midland Railway in bringing them into existence during the nineteenth century. The Midland main lines provided an axis, running in general from south-east to north-west, on which lay the chief towns, Kettering, Leicester, Loughborough, Nottingham, Derby, Chesterfield, all became industrialized. Along this axis, which provided for the movement of coal in one direction and iron ore in the other, centres of heavy industry at Holwell, Stanton, Clay Cross and Staveley were added. Today, however, a new feature is to be seen in the cultural landscape. Even more conspicuous than large factories are the huge generating stations now being erected at intervals on the banks of the Trent from Burton to beyond Gainsborough. Not only are they something new upon the landscape, but they indicate a new phase in economic development. Fuelled from the nearby coalfield, consuming many millions of tons annually, and using the river water for cooling purposes, they form a new industrial axis, one of power production, broadly aligned from south-west to north-east. The enormous output of electricity to be supplied to the national super-grid, far in excess of local demands, represents a large-scale contribution to the power requirements of the country as a whole. The East Midlands, already the greatest producer of coal, is about to become the greatest producer of electricity.

Selected Bibliography

P. W. Bryan (ed.). *A Scientific Study of Leicester and District.* Published for the British Association for the Advancement of Science, London, 1933.

J. D. Chambers. *Modern Nottingham in the Making.* Nottingham, 1945.

K. C. Edwards. 'The East Midlands: Some General Considerations.' *East Midland Geographer*, **1**, 1954, pp. 3–12.

K. C. Edwards. 'The East Midlands' in *Studies in Regional Planning* (ed. G. H. J. Daysh). London, 1945.

K. C. Edwards with H. H. Swinnerton and R. H. Hall. *The Peak District.* London, 1962.

W. Edwards. *The Concealed Coalfield of Yorkshire and Nottinghamshire.* (Geological Survey.) 3rd edition. London (H.M.S.O.), 1951.

W. G. Hoskins. *Leicestershire.* The Making of the English Landscape. London, 1957.

R. B. Jones. *The Pattern of Farming in the East Midlands.* Nottingham, 1954.

C. E. Marshall (ed.). *Guide to the Geology of the East Midlands.* Nottingham, 1948.

W. A. Richardson. *Citizen's Derby.* London, 1949.

H. H. Swinnerton (ed.). *A Scientific Survey of Nottingham and District.* Published for the British Association for the Advancement of Science. London, 1937.

R. G. Waddington. *Leicester: the Making of a Modern City.* Leicester, 1939.

CHAPTER 17

LINCOLNSHIRE

K. C. Edwards

Lincolnshire is by tradition a great agricultural region and, having a high proportion of its surface devoted to arable land besides supporting large numbers of high-quality livestock, it still ranks as one of the most productive farming areas in Britain. In addition several other industries, each somewhat localized, now help to diversify both its landscape and its economy.

Lying between the Wash and the Humber, with some 25 miles of its coastline from Donna Nook to Gibraltar Point confronting the North Sea, Lincolnshire forms a somewhat blunted peninsula. Its peninsular character was more strongly emphasized in early times by the presence of extensive tidal marshes spreading inland from the Wash and similar marshy tracts at the head of the Humber and along the lower Trent. Furthermore, even in Roman times the coastline between the Wash and Spurn Point stood farther out to sea than at present. Under such conditions access to the area by dry land from the rest of England was largely restricted to routes entering across the Kesteven Plateau in the south-west. The relative isolation which resulted was progressively broken down during historic times by the drainage of the marshes, the reclamation of the Fens and by the bridging of the Trent and Welland which formed natural boundaries to the west and south respectively. To the north, however, the absence of the long-discussed Humber Bridge still prevents rapid movement between Lincolnshire and east Yorkshire, but apart from this, modern communications, first the railways and more recently improved road transport, have virtually brought an end to the comparative remoteness. Also forming part of Lincolnshire is the Isle of Axholme lying to the west of the lower Trent. This is an area of artificially drained alluvium and peat, in the midst of which small outcrops of Keuper Marl provide flood-free ground which supports the villages of Belton, Epworth and Haxey, and the little town of Crowle.

Lincolnshire is a truly lowland area. Three-quarters of its surface is under 100 ft in altitude and much of this is only a little above sea-level, while the rest is all below 600 ft. Thus from the central tower of Lincoln Cathedral on a clear day a splendid impression of the landscape can be obtained, with the belfry tower of St Botolph's Church at Boston, familiarly known as Boston Stump, easily discernible to the south-east nearly 30 miles away. This is not to say, however, that the surface is devoid of relief, for the Wolds as well as parts of the Kesteven Plateau are areas of comparatively elevated and well-dissected country.

The build of Lincolnshire is based on a simple plan which reflects the nature and arrangement of the surface rocks (fig. 38). The latter

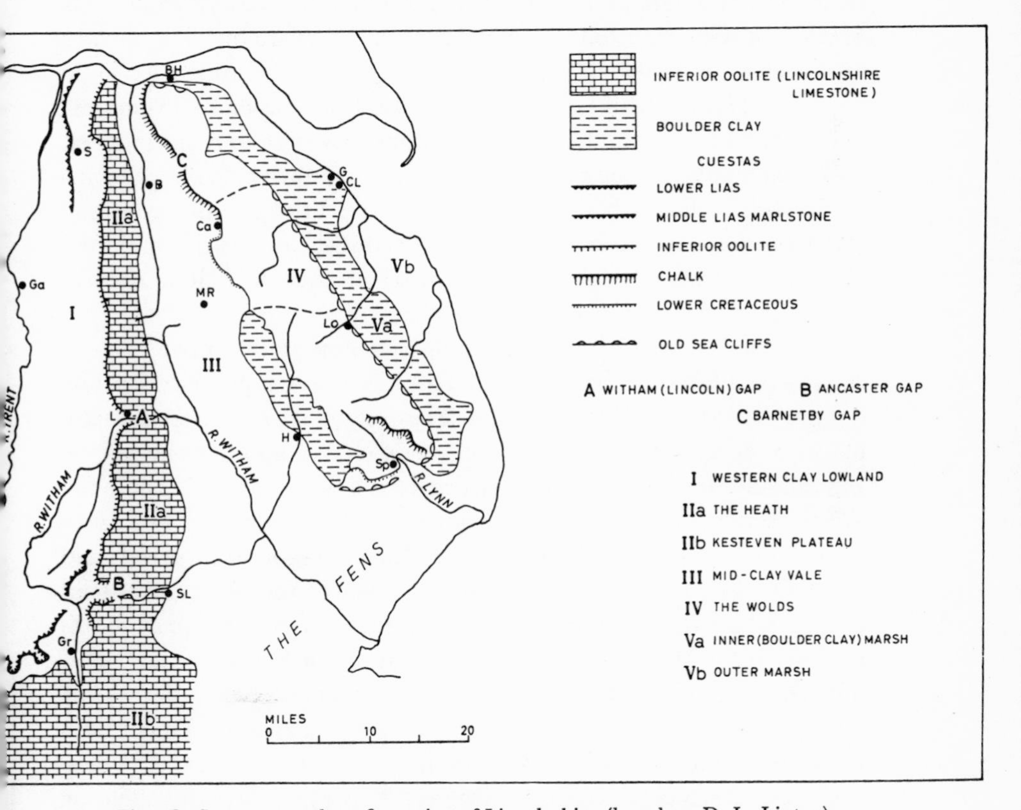

Fig. 38. Structure and configuration of Lincolnshire (based on D. L. Linton).
rigg. BH. Barton on Humber. Ca. Caistor. Cl. Cleethorpes. G. Grimsby. Ga. Gainsborough. Gr. Grantham.
. Horncastle. L. Lincoln. Lo. Louth. MR. Market Rasen. S. Scunthorpe. Sl. Sleaford. Sp. Spilsby.

consist of Mesozoic strata which dip gently eastwards, their outcrops thus forming a succession of belts running approximately from north to south. The more resistant rocks, that is the Oolite Limestone, which form the Lincoln Heath and the Cretaceous rocks, largely chalk, which give rise to the Wolds, provide two belts of higher ground each with a west-facing escarpment. Alternating with these are outcrops of weaker strata, chiefly clays, which form three belts of low-lying and almost featureless country. Thus to the west of the Heath is a lowland of gentle relief composed of Keuper Marl and Lias Clay covered in places by boulder clay. This may be termed the Western Clay Lowland. Towards its northern and southern extremities, however, more prominent features are produced by layers of limestone containing workable iron ore, the Lower Lias Limestone with the Frodingham Ironstone in the north and the Middle Lias Marlstone with ironstone in the south near Grantham. The former produces a cuesta which runs southwards from the Humber for some ten miles, giving abrupt cliffs at Burton-on-Stather where it is undercut by the Trent. Similarly the marlstone gives rise to a pronounced escarpment overlooking the Vale of Belvoir and provides a commanding site for Belvoir Castle. While the Frodingham ironstone forms the basis of the steel industry at Scunthorpe, the ore obtained from the marlstone ironstone is smelted at the Holwell ironworks near Melton Mowbray in Leicestershire.

Between the Heath and the Wolds a succession of Upper Jurassic clays—the Oxford, Ampthill and Kimmeridge Clays—forms a low flat belt called the Mid-Clay Vale. These clays are obscured in many places, however, by an irregular cover of chalky boulder clay, while west of the Witham is a well-defined belt of fenland. Beyond the Wolds is the low coastal plain known as the Marsh, the surface of which consists mainly of boulder clay along the foot of the Wolds and of marine and estuarine clays and silts towards the coast. It should be noted that while the Heath runs more or less from north to south, the axis of the Wolds is almost north-west to south-east and in consequence the Mid-Clay Vale broadens from about two miles at the Humber to some fifteen miles where it approaches the Fens. Both the Heath and the Wolds also broaden southwards and in so doing each loses its typical cuesta character. These two 'upland' belts in fact demand a little more attention, as do the Marsh and the coast.

The Heath north of Lincoln is narrow, seldom exceeding three miles across. This is partly because of the relatively pronounced dip of the Lincolnshire Limestone. Farther south, however, as the dip diminishes, the outcrop steadily widens until around Grantham, where the rock is thicker and almost horizontal, the Heath passes into the Kesteven Plateau, most of which has a general elevation of well over 400 ft. The Heath escarpment, known as the Cliff, or Edge, persists as a prominent feature for nearly fifty miles, with only two significant breaks, the Witham Gap at Lincoln and the more elevated Honington or Ancaster windgap. Its crest, however, except in the extreme south, never quite reaches 300 ft.

The Heath forms a truly distinctive belt of country. It is so called because, prior to the great agricultural improvements of the nineteenth century, it was an area of heath, fern and gorse, with patches of poor grass on which a few sheep could be grazed. The limestone surface, being highly permeable, is lacking in streams and is without marked relief. Virtually without woodland and devoid of settlements, except for individual farms, it is nevertheless almost wholly under cultivation, principally wheat, barley and roots, which in turn give support to large numbers of sheep. The farms are large, mostly over 300 acres. Large fields, too, are characteristic, often bounded by dry-walling derived from the local stone. Villages occur only on the margins of the Heath. On the west they are sited at regular intervals on the scarp slope, taking advantage of the spring-line at the junction of the limestone with the underlying clays. North of the Lincoln Gap the spring-line is close to the foot of the slope, whereas to the south it is nearer the crest, so the villages are placed accordingly. On the east another line of villages occurs in relation to springs or streams issuing from the dip slope near its junction with the Oxford Clay. Along the dry open surface of the Heath itself runs the ancient Ermine Street which over much of its length is maintained as a modern highway.

North of Kirton-in-Lindsey, where the Heath is lower, an accumulation of wind-blown sand, extending from the Trent valley, gives rise to poor sandy warrens. Some of these have been redeemed by afforestation but the coniferous plantations add a strangely unfamiliar element to the landscape.

The Kesteven Plateau differs markedly from the Heath. Its western edge is extremely broken, the Lincolnshire Limestone presenting no regular escarpment comparable with the Cliff. The Plateau itself is higher and its surface furrowed by relatively deep valleys. A considerable mantle of boulder clay provides heavier and moister land and, although grain and roots are the chief crops, beans are also important. There is as much grassland as arable and woods are numerous, while fields are mostly defined by hedgerows. Water is more plentiful and thus the villages are uniformly distributed.

The Wolds present a landscape of great interest in which three distinct sub-divisions can be recognized, the Northern, Central and Southern Wolds respectively. Of these only the first, which extends from the Humber to just north of Caistor, has a typical escarpment rising in places to 300 ft, and a dip-slope surface inclined to the east, containing dry valleys (fig. 38). At Barnetby the Northern Wolds are cut by a dry gap utilized by the railways leading from Lincoln, Gainsborough and Scunthorpe to Grimsby and to New Holland, the ferry terminus for Hull. The Central Wolds, from the neighbourhood of Caistor to the valley of the upper Bain, are wider and considerably higher than the northern portion. The scarp, though much dissected by combes, is a bold feature reaching in places to over 500 ft, its height being due not so much to the Chalk alone as to the presence of Lower Cretaceous strata beneath it. Occurring in these strata is the Claxby Ironstone which is worked from a mine south of Caistor and sent to Scunthorpe. Beyond the scarp the Chalk forms a plateau some ten miles wide cut by comparatively deep valleys some containing streams. The general slope, however, is north-eastwards and not strictly in accordance with the dip.

The Southern Wolds, the largest of the three divisions, is different again. Here the western slope is heavily masked by chalky boulder clay which extends from the floor of the Mid-Clay Vale, and is therefore less steep than would otherwise be the case. Moreover only the eastern portion of the Southern Wolds is composed of Chalk, the western and southern parts being formed by Lower Cretaceous sandstone (Spilsby Sandstone) and clays. The Spilsby Sandstone locally includes the Roach Ironstone which may eventually prove of economic importance. Between the high ground formed by the sandstone and the slightly more elevated outcrop of the Chalk is a longitudinal depression floored by the weaker clays. This feature, mainly the work of the River Lynn (or Steeping), widens into a broad vale opening on to the Fens. Along its eastern side the Chalk forms an interior escarpment which reveals the famous Red Chalk underlying the white.

The highest ground in both Central and Southern Wolds, ranging from 450 to 500 ft, forms a more or less continuous erosion surface regardless of the rocks of which it is composed and should therefore be considered as a single element of the relief. As a rather bleak stretch of poor land it is followed by the winding course of an ancient ridgeway known as the Bluestone Heath Road.

The Wolds as a whole have at least one characteristic in common. Their eastern edge, even though it is largely masked by boulder clay, generally drops abruptly to the coastal plain and is a line of old sea-cliffs which were cut before the deposition of the boulder clay and the other sediments now forming the Marsh. The foundation of the Marsh itself therefore is an extensive wave-cut platform of chalk.

The Northern Wolds, being lower and less hilly than the rest, have deeper and more productive soils and for this reason have long been termed the 'good' Wolds. Some 70% of the surface is under arable cropping, including high-quality malting barley, potatoes and sugar-beet as well as market gardening. In contrast to this, the remainder of the Wolds, with their thin chalk soils and steep slopes, are classed as 'poor' Wolds, supporting a sheep and barley economy from which crop yields are on the whole rather inferior. The grassland, too, is of indifferent quality.

The surface of the coastal plain or Marsh consists of boulder clay along the Wolds margin, but farther east, as far as the coast, this is covered by layers of salt marsh clay and marine silt. The total thickness of these glacial and post-glacial deposits varies from 70 ft to over 100 ft. In the neighbourhood of Grimsby and Cleethorpes, however, the boulder clay itself extends to the Humber shore, thus separating the main belt of marine silt from the smaller area bordering the upper Humber. Farther south another prolongation of the boulder clay reaches the village of Anderby only a mile or so behind the coast. Thus an obvious distinction arises between the Inner Marsh on boulder clay and the Outer Marsh on marine silt and clay, the latter being slightly lower (often less than 10 ft above present sea-level), and even flatter than the former. Very broadly the junction of the

Inner and Outer Marsh can be traced south of Grimsby by a line of villages which include Humberston, Fulstow, North and South Cockerington, Gayton-le-Marsh, Maltby-le-Marsh, Mumby and Burgh-le-Marsh. In both sections of the Marsh, patches of other superficial deposits diversify the surface including glacial gravels as in the Louth and Alford districts and a tract of sandy material around North Somercotes in the Outer Marsh.

Southward beyond Wainfleet the Marsh merges into the Fens which it superficially resembles. The Marsh differs, however, from the Fens in that large areas of peat, old river gravels and the alluvium of large streams, which characterize the latter, are absent.

The Lincolnshire coast consists of a low, flat and fairly regular shoreline backed in places by sand dunes which seldom exceed 40 ft in height. The position and character of the shoreline have varied greatly in post-glacial times, and even within the historical period substantial changes have occurred. Over broad stretches of time such changes have been due to alterations in sea-level while over restricted periods they have resulted from the varying effects of erosion and accretion. Protection of the marshland behind the coast has always been necessary since most of it lies below the level of spring tides. An early defence was the clay embankment known as the Commissioners' Bank, or so-called Roman Bank, which ran from Donna Nook to Gibraltar Point. It can still be followed between Sutton and Skegness and for a few miles north of the latter place it carries the coast road. The dunes have accumulated on the seaward side of the Bank and their migration inland has been prevented by the growth of sea-buckthorn and marram grass.

Since the thirteenth century the coast has suffered severe erosion. It is estimated that in the neighbourhood of Mablethorpe and Sutton the shore has retreated up to half a mile in the past 400 years, at least five medieval churches having been lost to the sea in that time. Even at Cleethorpes, just within the Humber, some land has vanished, and in modern times a signal station has thrice been moved back beyond the reach of the sea. While erosion was formerly active all along the coast as far south as Skegness, in recent years accretion has predominated northwards of Mablethorpe as well as southwards from Skegness. The stretch of shore between these places remains liable to serious inroads as is shown by recurrent damage to the sea defences. Here the beaches are somewhat lower and, after heavy storms, interesting relics may be exposed at low tide, indicating that the coastline once stood farther to the east. Such are the remains of Roman salt-workings and pottery sites at Ingoldmells and the submerged forest of Neolithic date at Sutton. These examples are evidence of several oscillations of sea-level in post-glacial times. At present along this section of the coast the beach provides insufficient sand for the dunes to afford adequate protection, consequently sea-walls are also necessary. It was here that the violent storm surge in 1953 caused the greatest damage and flooding on the entire coast, but the reconstructed works are much stronger, while new groynes appear to have stabilized the beach.

Southward of Skegness accretion goes on and the spit at Gibraltar Point affords shelter in which salt marsh is actively forming owing to the presence of sand and silt and the spread of the grass *Spartina townsendii*. Around the Point the shore has advanced eastwards by over 1,000 yards in the past 150 years. Accretion to the north of Mablethorpe has resulted in the silting of a few little havens such as Saltfleet and Tetney, the latter providing the outlet for the disused Louth Navigation.

It is significant that the stretch of shore which offers no natural opportunities for port development, and along which erosion is active, is the section in which, apart from Cleethorpes, the Lincolnshire coastal resorts are concentrated. At Skegness (12,843), Mablethorpe and Sutton-on-Sea (5,389), as well as smaller places like Chapel St Leonards and Ingoldmells the sandy beaches and bracing air attract large numbers of holiday-makers and coach excursions, above all from the Midlands and Yorkshire. As with Cleethorpes the growth of these resorts began with the coming of the railway but has been greatly accelerated in the days of motor transport. Cleethorpes (32,705) occupies a totally different position, being sited on a low cliff of boulder clay which is subject to erosion by tidal currents in the Humber. The cliff is now almost lost to view owing to the building of coast defence-works as well as a promenade.

The outstanding importance of agriculture in Lincolnshire is broadly the result of four more or less inter-related factors: favourable conditions of climate and soil; the control and maintenance of drainage over all the flat, low-lying areas; the progressive development of farming systems since the advent of modern husbandry in the late eighteenth century; and the stimulus provided by the present-day commercialization of agriculture which is reflected in the use of the most up-to-date techniques and the consequent increase in yields per acre.

As in other parts of eastern England the relatively low rainfall favours crop production, especially cereals and root crops. Well over half the county has a mean annual precipitation of under 25 in. and in only a few places in the higher Wolds does it slightly exceed 30 in. Temperature conditions are similar to those in other eastern counties, giving a period completely free from air frost between early May and the beginning of October. Sunshine is also liberal in amount, especially in June, about 6·75 hours per day, a figure distinctly higher than the mean for the country as a whole.

While a great variety of soils can be found, three main types may be distinguished and these together cover by far the greater part of Lincolnshire. First, there are the deep, moist, medium-heavy clay soils such as those of the Western Clay Lowland and those derived from the different kinds of boulder clay found on the Kesteven Plateau, in the Mid-Clay Vale and along the Wolds-Marsh border. Secondly, there are the shallower, drier and more friable soils of the Heath and Wolds, derived from limestone and chalk respectively. Thirdly, there are the rich, artificially drained soils formed from fen silts, tidal silts and alluvium, characteristic of the Marsh, the fens of the Mid-Clay Vale, the lower Trent and Isle of Axholme, including the warplands. The last-named are comparatively small flat areas where the old practice of enriching the land with deposits of silt was undertaken, by admitting tidal water to the fields by artificial channels and afterwards draining it off. The silt is locally known as 'warp' and the channels 'warping-drains'.

For the maintenance of present-day farming a great deal of Lincolnshire, owing to its flatness, is entirely dependent upon various systems of land drainage. This work, which includes the provision of a network of channels (called drains or dykes), pumping-stations and sluices, as well as the regulation of streams such as the Ancholme and the lower Witham, has been evolved over a long period of time. The winding course of the Car Dyke which skirts the western edge of the Witham Fen is thought to be a Roman drainage channel, possibly used for navigation as well. The drainage of the Isle of Axholme by the Dutch engineer Cornelius Vermuyden in the seventeenth century, though not altogether successful at first, is the best known among the earlier projects of modern times. The Ancholme Levels at the northern end of the Mid-Clay Vale are the result of intermittent yet progressive efforts at land drainage over the past 700 years, though their final conversion from grazing lands liable to flood to areas of permanent cultivation is hardly yet completed. The regulated channel known as the New River Ancholme which forms the basis of the present-day drainage dates from 1637. The general effect of all these undertakings, including the reclamation and drainage of large areas in the Marsh, has been to add enormously to the cultivable area of the county and to transform the local economy. Little more than a century ago many low-lying parts were still inaccessible and sparsely peopled, livelihood often being derived from fishing and wild-fowling. Nowadays such areas not only support a much greater population but are intensively worked and produce a great diversity of crops which bring highly remunerative returns. Another consequence of artificial drainage is the regularity of pattern imparted by the system of channels and ditches to fields, fences, roads and even settlement forms. Innumerable bridges and frequent right-angle turns on many roads are further details which give a distinctive appearance to the scene.

Compared with the flat low-lying parts, the two 'upland' belts of Lincolnshire exhibit entirely different conditions as regards surface drainage. The Heath and the Northern Wolds are naturally well drained owing to the permeability of the underlying rock, so much so that a shortage of water may occur in dry seasons. On the Kesteven Plateau and on the rest of the Wolds, however, less permeable rocks, including boulder clay, promote a fairly satisfactory run-off in the form of streams, although in many areas the underdrainage of fields is necessary.

'Rich grazing-lands are the glory of Lincolnshire,' wrote Arthur Young in 1793, and equally to be admired at that time were the many thousands of fine cattle and sheep while in addition the Wolds were famous for horse breeding. Today Lincolnshire is predominantly an arable county and the transformation, which took place mainly in the nineteenth century, was largely due to land improvement involving large capital expenditure by landowners and to the development of modern systems of farming. Even before this, in common with the rest of eastern England, the introduction of turnips and clover into the crop rotation had resulted in higher yields of grain and stimulated a substantial increase in cultivation. This was a leading factor in the agricultural revolution as was the advent of artificial fertilizer later on. In Lincolnshire the changes led not only to a vastly greater output of wheat and barley, especially on the Heath and Wolds, but also the increased provision of winter fodder resulting from the four-course rotation gave a new impetus to livestock production within a general system of mixed farming. Interest in stock breeding enabled farmers to develop new strains of animal, two of which, the Lincoln Longwool sheep and the Lincoln Red Shorthorn cattle, have become famous in other pastoral countries of the world such as Australia, New Zealand, Argentina and the U.S.S.R., and they still provide a valuable export.

More recently the system of mixed farming has changed again, in emphasis rather than in principle. Thus more attention is now devoted to wheat than to barley, except in parts of Kesteven and the Wolds where good malting barley is raised. Sugar-beet as a cash crop has largely superseded the turnip, and potatoes, too, have increased in importance. Livestock, on the other hand, especially since the sharp decline in the numbers of sheep, are no longer of outstanding importance. Both beef and dairy cattle as well as sheep are, however, still numerous in the Wolds and Marsh, and Louth is the busiest livestock market in the county. These differences of emphasis have chiefly been brought about in response to changing economic conditions, but there are many reminders of the earlier phase and some traditional features survive. Thus in the Wolds large farms often from 300 to 800 acres with their substantial buildings, including huge barns and granaries and a capacious fold-yard, testify to the continuance of some form of mixed farming adapted to present-day needs. Again, many Wold farmers rent pastures on the Marsh or the Ancholme Carrs for summer grazing, a reminder of the time when, as in Arthur Young's day, such farmers often owned their own Marsh land.

The productive capacity of the land continues to be enhanced by greater use of fertilizers, especially of phosphates and nitrates, by deeper ploughing on the poorer soils of the Heath and Wolds, by the selection of heavier yielding seeds for which Sleaford and Spalding are important centres of production, and by further mechanization.

Lincolnshire engineering firms, in close touch with farmers, supply many ingenious devices which save labour without loss of efficiency. Moreover, by devoting greater attention to cash crops such as sugar-beet, potatoes and vegetables, cultivation has become more intensive. This applies not only to the areas of small holdings such as the Isle of Axholme and the lower Trent but also to other parts, for example, the 'good' Wolds where market-garden crops, especially cabbage, broccoli and carrots, are now grown, and the tract of light soils around North Somercotes, south-east of Grimsby, where market gardening has also developed. No better instance of commercialized agriculture can be noted than the 8,000 acres under potatoes in the parish of Nocton, stretching from the Heath to the Witham fens, worked by Smith's Potato Crisps and situated conveniently near their Lincoln factory.

The extent to which land in Lincolnshire, excluding the Fens, is devoted to farming is shown by the Ministry of Agriculture statistics for 1957, which fairly represent the position in recent years. These figures indicate that out of a total area of over 1·4 m. acres, some 1·2 m. acres, or nearly five-sixths of the entire surface, are under crops and permanent pasture. Of the farmland, moreover, some 928,000 acres, or nearly three-quarters of the total, are arable, with less than 300,000 acres under grass. Even more striking is the fact that on the cultivated land two grain crops, wheat and barley, and two root crops, potatoes and sugar-beet, together occupy 70% of the acreage. Oats have only a subordinate place. On the other hand, while the area under permanent grass is relatively small, it should be remembered that considerable amounts of fodder crops such as roots, peas, beans and kale are grown.

The predominance of farmland in Lincolnshire is further emphasized by the small amount of woodland. Wooded areas occur as a rule only where soil or slope conditions are unfavourable to the plough. Thus deciduous or mixed woods are a feature of the steeper portions of the Cliff and the Wolds scarp, and on the heavy boulder clay country in South Kesteven. In districts of poor light soils coniferous plantations have been established such as the Laughton Forest on the blown sands of the northern Heath mentioned earlier, and also on sandy tracts around Woodhall Spa and Market Rasen. Even including the more recent cases of afforestation, woodlands occupy barely 3% of the total area.

In most cases soil conditions are a major factor affecting the distribution of crops. Thus although wheat and barley are the outstanding grains, the latter normally exceeding the former in total acreage, the two crops are largely complementary in occurrence. Barley prefers the lighter soils of the Wolds and the Heath, while wheat is chiefly found on the heavier land of the Mid-Clay Vale, the Isle of Axholme and the Northern Wolds. Wheat, however, is grown more widely than barley, and on the Heath north of Lincoln it occupies the greater acreage. Much of the barley is fed to stock but considerable quantities are grown for malting.

Sugar-beet shows a fairly wide distribution but because of the difficulty in lifting, the heavy soils of the boulder clay and the damper parts of the Western Clay Lowland are avoided. Little is grown on either the Wolds, except for the northern portion, or the Marsh, partly because these areas are less accessible to the factories. Potatoes on the other hand are concentrated on the deep rich soils of the Isle of Axholme, the Ancholme Carrs and the lower Witham Fens, and in the last-named district 'earlies' as well as 'main crop' are produced.

Cattle are fairly evenly distributed, but they are noticeably fewer in number on the Heath and 'poor' Wolds where sheep are more dominant, reflecting the survival of the old sheep-barley-turnips economy. Lincolnshire is not important for milk production, but dairy cattle are kept in relatively small numbers in the Western Clay Lowland and along the Wolds-Marsh border.

Industrial activity developed relatively late in Lincolnshire and was mainly stimulated by the growth of railways in the fifties and sixties of last century when several lines built across the Trent enabled coal, foundry iron and other materials to be conveyed into the county from the Midlands and the North. Thus agricultural engineering, already established at Lincoln, expanded rapidly in that city and in other centres such as Grantham and Gainsborough and later on at Stamford. Similarly local industries involving the use of farm products, such as flour-milling, malting and brewing, grew in importance. This was in fact the period of the final breaking down of Lincolnshire's isolation. These two forms of activity, engineering and the treatment of agricultural products, thus provided a dual basis from which many of the present industries are derived, though almost all of these have become more specialized in character.

Apart from ironstone Lincolnshire is not rich in economic minerals. Extractive industry is mainly confined to stone-quarrying as at Ancaster where the Lincolnshire Limestone provides good freestone; the making of cement at Kirton-in-Lindsey and South Ferriby, using the limestone and chalk respectively, with adjacent clays; and brick- and tile-making at Barton-on-Humber. It is of interest to note that some years ago deep borings at Spital, north of Lincoln, and at Stixwould near Woodhall Spa revealed the presence of coal. The former discovery implies a continuation of the Nottinghamshire and south Yorkshire coalfield into Lincolnshire, and the latter, with seams at a depth of near 4,000 ft, suggests an entirely separate occurrence. Whether either source will ever be worked is problematic.

At present large-scale industrial activity is mainly confined to three areas: the Scunthorpe district in the extreme north, Grimsby and the Humber shore, and Lincoln.

The growth of iron and steel production at Scunthorpe has been particularly rapid since 1900, and in that area traditional agricultural interests have had increasingly to compete with industry for both land and labour. The basis for this development is to be found in four inter-related factors: the accessibility of the Frodingham Ironstone, the availability of coal and coke from south Yorkshire, the growth of the Humber ports, and the provision of rail and road transport linking the district with Sheffield and various steel-consuming centres (fig. 39). Transport in particular is the essential key to modern industrialization and in this case the promotion of a new axis of movement, connecting the orefield with Sheffield and the south Yorkshire coalfield to the west and the ports of Grimsby and Immingham to the east, has been vital. For this route the line of the former Manchester, Sheffield and Lincolnshire railway crossing the Trent at Keadby Bridge, within a mile or so of the orefield, became of major importance. The bridge itself was rebuilt in 1916, the new structure being made to carry both railway and road, the latter becoming the trunk road (A18) linking Grimsby and Scunthorpe with Doncaster. Here, too, the Stainforth Canal, still important for coal and coke traffic, reaches the Trent. The nodality thus given to Keadby has resulted in some local industrial development based also on navigation connexions with Hull, but its proximity to Scunthorpe has prevented the growth of a separate town.

The ironfield with ore beds varying from 18 to 30 ft in thickness,

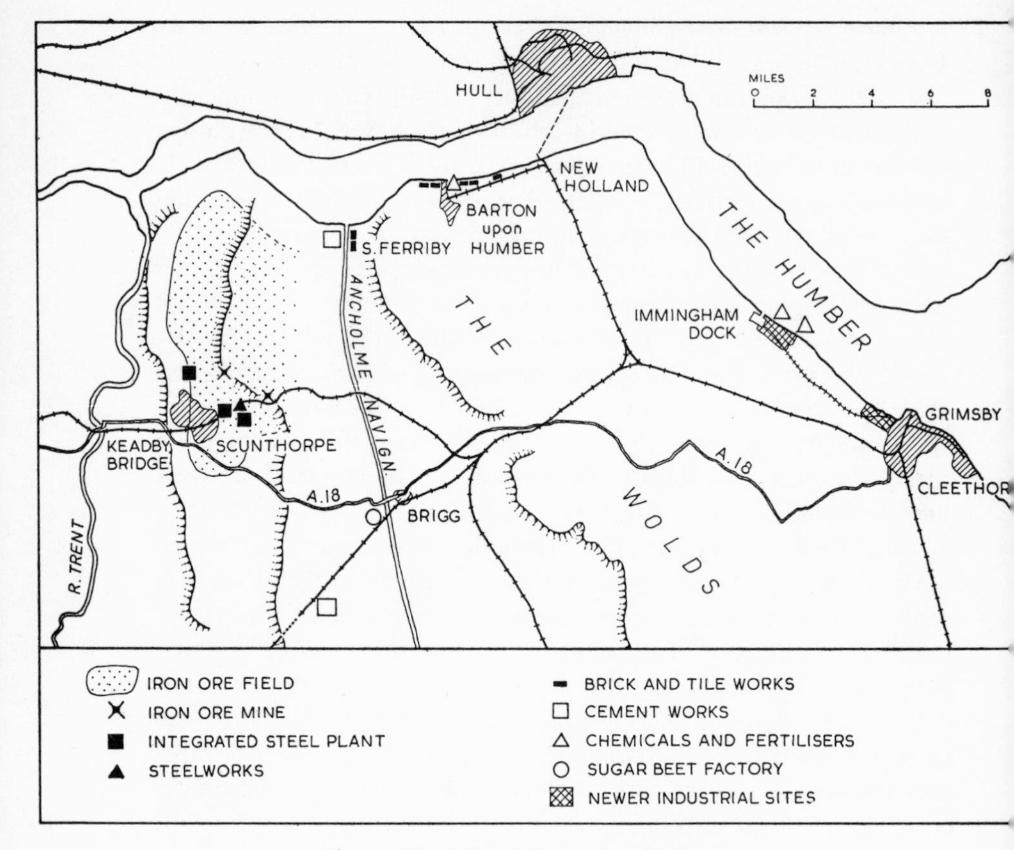

Fig. 39. The industrial area of north Lincolnshire.

extends for about six miles along the outcrop of the Frodingham Ironstone. The iron content is poor, averaging only 20% and because of the high proportion of lime, which admittedly reduces the amount of flux required in smelting, siliceous ore must also be used. The presence of manganese and the low phosphorus content, however, render the ore suitable for making basic steel. Much of the readily accessible ore, won by opencast working, has already been extracted, so that from now on exploitation has to contend with increasing difficulties. Since the beds dip eastwards the overburden becomes thicker, while the ironstone itself becomes more powdery, a condition which hampers pig-iron production; it also contains more sulphur which is deleterious to steel making. Although technical measures can overcome these difficulties, their application inevitably affects production costs. In the meantime underground mining has become necessary and two mines, at Santon and Dragonby, have been in operation for some time. Much of the future output will be derived from these sources. At present the total output is some 4·5 m. tons per annum, but as the Scunthorpe works use an admixture of both local and Northamptonshire siliceous ores, some of the former is sent to other smelting centres. Smaller quantities of high-grade ore, imported chiefly from Sweden, are also used.

At Scunthorpe steel production, using the basic Open-Hearth method is carried on by large-scale continuous processing in three integrated works, giving an annual output of over 2 m. tons or about 10% of the country's total. Another works produces steel castings, for some of which special quality steels are used. Recent improvements in plant equipment includes the erection in 1954 of a blast-furnace with a capacity of 6,000 tons a week, said to be the largest in Europe.

Scunthorpe itself was created by the iron industry, for when the local ore was first exploited in the early sixties, it was merely a hamlet lying close to the village of Frodingham. Even the start was on a modest scale for, until the furnaces of the Trent Iron Co. appeared in 1864, the ore was sent to Elsecar near Barnsley on the Yorkshire coalfield to be smelted. Other firms soon followed however and by 1890, when steel was first produced, Scunthorpe and Frodingham had grown into a single community with a population of just over 3,000. Subsequent growth was remarkably rapid. In 1900 the population was 4,000, by 1931 it was 34,000, by 1951 it reached 54,000 and it is now 67,257. In 1936 the town of Scunthorpe was incorporated as a municipal borough. In many respects its rise is comparable with that of Middlesbrough except for the fact that unlike Middlesbrough it has no direct access to the sea.

Scunthorpe has every appearance of being a modern town having few buildings earlier than the late-Victorian period. St John's Church, built in 1891, stands close to one of the great steelworks and affords a contrast to St Lawrence's, the thirteenth-century church of Frodingham, a short distance away. The town is almost wholly dependent on the steel industry which absorbs over 48% of the employed population. Most of the other activities, such as the making of fertilizers and chemicals from basic slag, tar distillation as a by-product of the coke ovens, the preparation of road-making materials, and several forms of engineering, are equally dependent on the basic industry for their materials. More recently the introduction of clothing and footwear manufacturing, giving employment to women, has contributed a little towards diversifying the economy.

Grimsby, despite the overwhelming importance of the fish trade, has attracted considerable industrial development which is increasingly making use of unoccupied land along the Humber shore as far as Immingham. Grimsby still has a claim to be regarded as the leading fishing port of Britain and perhaps of the world, for although a greater weight of fish per annum is landed at Hull, Grimsby's catch, of which a large proportion is prime fish, is of higher value. Grimsby is strictly a modern port, for the ancient harbour at the mouth of the Freshney, prominent in the Middle Ages, stagnated for centuries as the result of silting. New docks attracting increasing numbers of fishing vessels were built soon after the Manchester, Sheffield and Lincolnshire railway provided connexion with inland industrial areas in 1848, and, following the advent of steam trawlers about 1890, in which Grimsby led the way, an enormous growth in the fish trade took place.

The docks, of which the latest is the No. 3 Fish Dock opened in 1934, are built on the tidal flats beyond the shore and are entered by locks. While the port is mainly concerned with demersal fish, especially cod, haddock and plaice, there is a considerable herring trade which is handled at a tidal basin at the harbour entrance. This allows the drifters and luggers to deliver their catch at any time of day or night, whereas trawlers must await the tide. When landed, fish is sold by auction, and the quay, known as the Pontoon, is really a combination of wharf, market and packing-shed, with the railway alongside, continuing round the docks for over a mile.

Grimsby's general commerce, by no means negligible, consists of imports of timber and pulp, dairy products and iron ore from Scandinavian and Baltic countries and the export of coal and manufactured goods. The fish trade naturally encouraged the growth of ancillary industries: salting and curing establishments, box-making, ice manufacture in the world's largest ice factory, ship repairing and marine engineering. Fish processing and packing, and the production of fish-meal and fertilizer are other related pursuits. All these activities have helped to promote an industrial community which relies heavily upon the harvest of the sea. Paper-making, jam-making and brewing are also important, while in recent years the manufacturing of clothing has been established.

During the past hundred years the growth of Grimsby has been paralleled on a smaller scale by that of Cleethorpes, the adjacent resort and residential town. The two form a continuous built-up area with a combined population of 129,470, which is by far the largest urban unit in Lincolnshire.

Immingham Dock, about six miles above Grimsby, is situated at a point where the deep-water channel of the Humber swings close to the Lincolnshire shore. It was opened in 1912 and was primarily designed for the export of coal and the import of timber, including pit props, as well as iron ore for use at Scunthorpe. Post-war developments both here and along the shore towards Grimsby have resulted in Immingham becoming part of the industrial complex of which the great fishing port is the focus. The new industries which include the production of fertilizers, especially superphosphate, and chemicals such as titanium oxide, owe their location largely to the facilities for importing raw materials and to the availability of large quantities of water required for processing derived from the chalk substratum.

Lincoln, the third largest industrial centre, is also the historic cathedral city and the leading administrative, cultural, retail and market centre (fig. 40). Lincoln's long history is marked by four periods of exceptional prominence. In the Roman era, as *Lindum coloniae*, it was one of the principal towns in the country, first as a military centre and later as a civil township; under the Danes it was one of the five fortified boroughs of the East Midlands, having important mercantile connexions with the Scandinavian lands; in medieval times it flourished as a major centre of the wool trade; then, after centuries of comparative decline, the railway age gave stimulus to industrial development. Today it is chiefly by such activity that its population of 77,065 is supported.

While the old city occupies an elevated strategic site on the north side of the Witham Gap and on the sharp slope leading down to the river, the modern industrial zone is concentrated on the floor of the gap, closely following the railways and the canalized Witham in a general east–west direction. Flour-milling, seed-crushing, the making of fertilizer, and the manufacture of farm equipment reflects

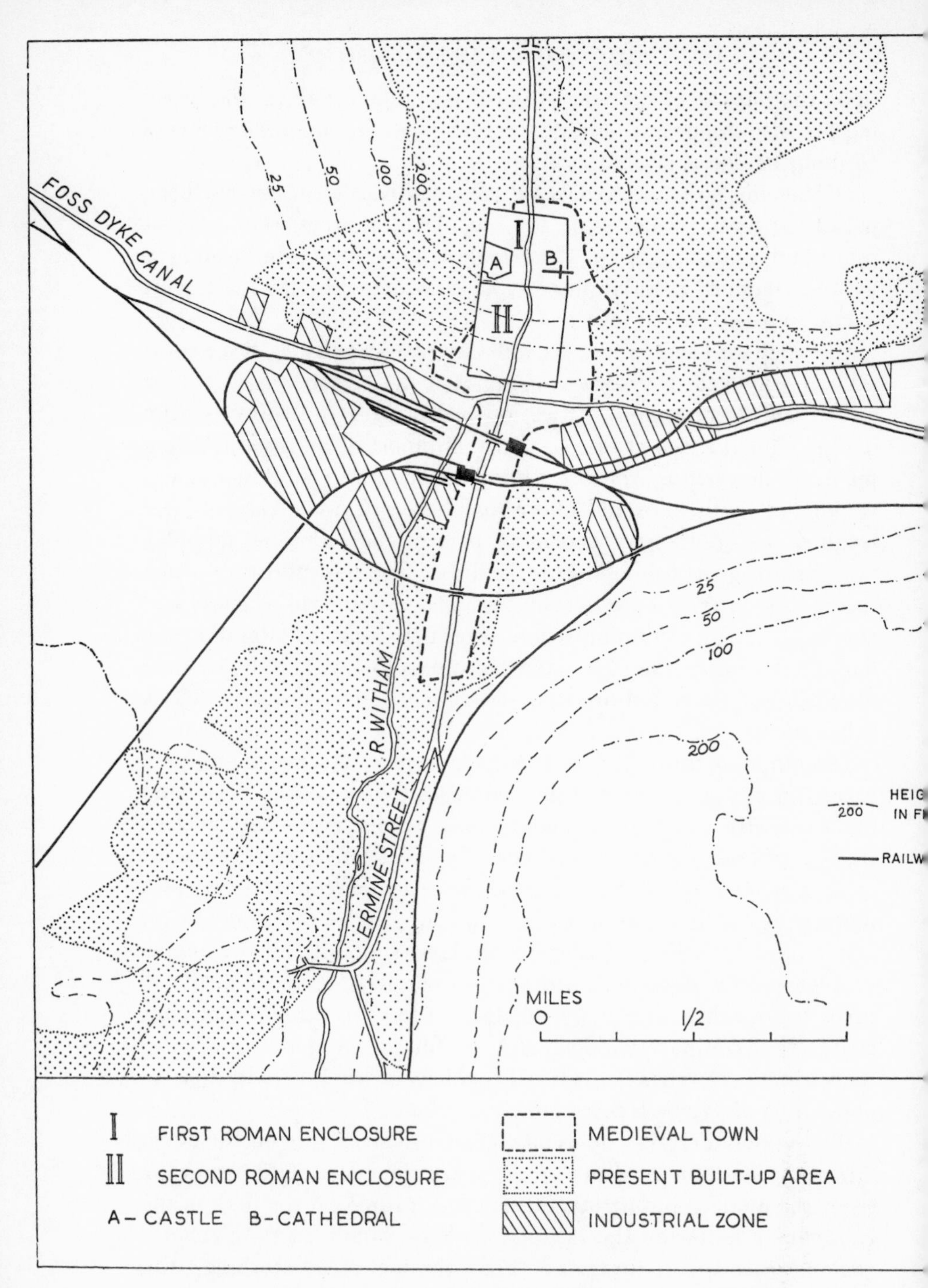

Fig. 40. The growth of Lincoln (After F. T. Baker and J. W. F. Hill).

the traditional connexions with agriculture but engineering in many specialized forms is the outstanding pursuit, providing occupation for over one-third of the employed population. Some of the engineering products, such as road-rollers, stone-crushing plant, and mechanical excavators, denote Lincoln's persistent interest in all kinds of work on the land. More important, however, is the output of oil engines, gas turbines, electric pumps, vacuum brakes and many foundry goods.

Grantham (25,030), Gainsborough (17,276), and Stamford (11,743) are smaller towns. As an old market town Grantham has profited from its position on the main railway route from London to the North and as a junction for several branch lines. While continuing to be the market for much of the Kesteven Plateau, it is also an industrial town with several engineering works, road-rollers and reversible ploughs being typical products. The making of suède leather and footwear is a more recent enterprise. Gainsborough, which earlier depended on its market and the river trade, though some revival of the latter should be noted, has long been known for its large engineering works producing tractors and agricultural implements. Another form of engineering, the making of packing and wrapping machinery, has now somewhat reduced the town's dependence upon agricultural engineering. Stamford, another ancient market town, missed the opportunity, perhaps happily, in the fifties of last century, of being on the Great Northern main line which was instead constructed through Peterborough. Eventually, however, a branch line reached the town and this enabled yet another agricultural engineering works to be established, and it remains the only significant industrial unit in Stamford today.

Lincolnshire affords many examples of the newer forms of processing the products of agriculture. Often these, like the older pursuits of malting and flour-milling, are most conveniently located at the larger and more active market towns, for in such places labour, power and transport facilities are most readily available. Thus, at Brigg, fruit and vegetable canning and the making of confectionery and preserves are found, together with one of the three Lincolnshire sugar-beet factories; the others are at Bardney just east of Lincoln, and at Spalding. Similar activities are found at Louth (11,490), together with the preparation of quick-frozen foods, while the processing of potatoes at Lincoln, Boston and Stamford may also be noted.

Since the majority of people live in rural areas Lincolnshire has few large towns. Apart from Lincoln, Grimsby (including Cleethorpes) and Scunthorpe, the only ones with a population of more than 20,000 are Grantham and Boston. On the other hand, the interests of the agricultural community have always been best served by local market towns of which there are many, though they are all small. A number of these, including several which were never served by a railway, for example, Caistor, Burgh-le-Marsh and Market Stainton, have long ceased to hold regular markets and have virtually stagnated. Others, often with small-scale industries, continue to flourish.

The distribution of market towns, past and present, reflects the value of sites at or near the junction of contrasting types of country. Thus Barton-on-Humber, Louth and Alford lie along the Wolds-Marsh border while Caistor, Market Rasen and Horncastle are their counterparts where the Wolds descend to the Mid-Clay Vale. Likewise Spilsby stands at the southern end of the Wolds within reach of the Fens. Sleaford lies near the western edge of the Mid-Clay Vale, while Bourne and Stamford are close to the western limit of the Fens. The position of Brigg, the leading market centre in the north, is doubly advantageous because of the proximity of the Heath and the Wolds on either side, though its modern growth has been inhibited by the development of Scunthorpe only six miles away. Quite different from all these towns is Woodhall Spa, a small resort situated in the Mid-Clay Vale on a sandy tract now covered with heath and pinewoods. It developed on the site of a spring containing bromine and iodine, discovered in the nineteenth century and found to be beneficial to sufferers from rheumatism and nervous complaints.

With the expansion of industrial activity in Lincolnshire many people have forsaken the land in favour of the large urban centres. This movement, however, has not seriously affected agricultural production, for on the whole the reduction in labour, except for an inevitable seasonal shortage, has been offset by greater efficiency on the farms which, in turn, has resulted in increased yields. Moreover, except for a few places, the rural landscape remains typical throughout the county.

Selected Bibliography

J. Bygott. *Lincolnshire.* London, 1952.

K. C. Edwards. *Lincoln, A Geographical Excursion* (Geographical Association), 1953.

J. W. F. Hill. *Medieval Lincoln.* Cambridge, 1948.

J. W. F. Hill. *Tudor and Stuart Lincoln.* Cambridge, 1956.

D. L. Linton. *The Landforms of Lincolnshire,* and K. C. Edwards, *Changing Geographical Patterns in Lincolnshire* (Lincolnshire Studies, Geographical Association), 1953.

H. H. Swinnerton and P. E. Kent. *The Geology of Lincolnshire.* Lincoln, 1949.

CHAPTER 18

FROM PENNINE HIGH PEAK TO THE HUMBER

ALICE GARNETT

More than forty years ago the late Professor C. B. Fawcett defined a region which he designated Peakdon. A compact area, extending about twenty-five miles from west to east, and approximately thirty miles from north to south, this region traversed the geological grain of the Pennine flanks from the Ouse to the south Pennine summits in the High and Low Peak country (fig. 41). It thus was designed to include virtually the whole of the basin of the River Don and of its tributaries the Dove, Dearne and Rother, and with these a considerable segment of the upper Derwent basin and of the south Pennine highlands as far as the summits of Bleaklow, Kinderscout and Axe Edge. The southern and physically less obviously defined boundary trended from west to east south of the Wye valley, approximately to Clay Cross.

The existence in effect of a region so defined with Sheffield as a natural regional capital may well be debated. Rather it might be held that it includes a mosaic of fragments of other units each with diverse or distinctive personality. If there are any grounds for regarding the area as a composite whole these can only have applied since the industrial revolution when the mark of the now dominating metallurgical industries became widely impressed—most of all during the last hundred years when improved means of communications broke down the local and regional isolation of earlier times.

In the present context part of the southern and eastern fringes of Fawcett's region are considered in other chapters. The River Wye will be taken as defining approximately the southern boundary within the highlands, and the eastern boundary follows the dip-slopes somewhat west of the line drawn by Fawcett. But in most other respects the regions are identical.

Viewed first in relation to a wider setting within England, the fundamental factor of regional isolation becomes, in an historic

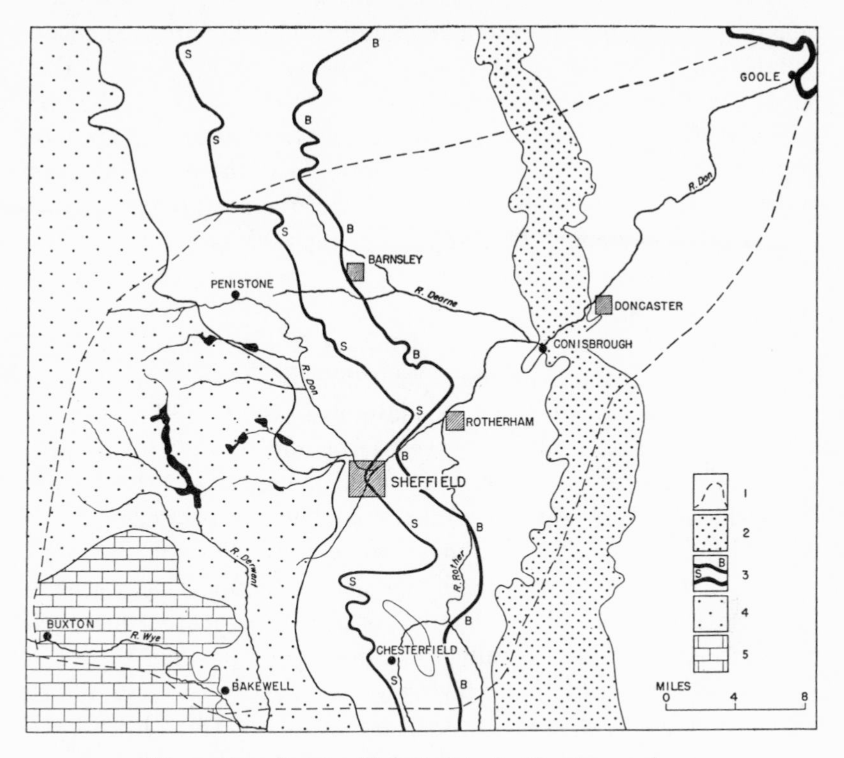

Fig. 41. From Pennine High Peak to the Humber: geology

1. Approximate boundary of region; 2. Magnesian Limestone; 3. Coal measures with (B) Barnsley and Top Hard Seam and (S) Silkstone and Black Shale Seam; 4. Millstone Grit; 5. Carboniferous Limestone. Reservoirs solid Black.

sense, very obvious. The Don valley—no more than a Pennine cul-de-sac from many points of view—is no counterpart of the neighbouring Trent valley to the east, whilst its main tributary systems (the west to east directed Dove-Dearne valleys) are functionally no equal of the Aire-Calder valleys of the Ouse system to the north. It is not surprising that for a long period the south-west corner of Yorkshire centring on the elbow turn of the Don valley came to be identified as a local *pays* known as 'Hallamshire'; i.e. one of those interesting early 'shires within a shire' that were recognized where forest, hill and moor brought isolation and made lands difficult to administer from the central authority of those days.

Much of the lofty High Peak frame is uninhabited today, and this in some measure is equally true of the one-time fenland of the Doncaster–Ouse confluence. Despite these areas, however, the population of the region is well over 1 m., mainly included within Sheffield, Britain's sixth largest city, set amidst a growing and now considerable-sized conurbation. From another point of view the Don valley between Sheffield and Rotherham must now rank as one of the most impressive economic units of Europe in terms of the vast investment of capital per square mile that now characterizes this highly specialized industrial zone. Geographically, then, Peakdon today is a region of marked economic contrasts, many of which are closely related to a physique of some complexity.

From extensive areas of marsh not 25 ft above sea-level in the lowest reaches of the Don basin (reflecting some effects of the aftermath of glacial and periglacial conditions in the Ouse basin generally), the land rises gently both westwards and southwards across a section of the drier and lightly dissected low cuesta of the Keuper Sandstones and Magnesian Limestone. The scarped edge of the latter, in places 500–600 ft above sea-level, overlooks farther westwards the more accidented landscape and increasingly bold relief of the grits and shales of the South Yorkshire coalfield, whose well-dissected surfaces rise steadily to altitudes of from 600 to 800 ft. Westwards again, the Pennine highlands rise sharply from a significant and abrupt break of slope at about 800 ft to 1,000 ft. This break is cut sometimes across Coal Measures but more often across the Millstone Grit series, and leads to broad uplands and plateaux often exceeding altitudes of 1,200 ft and rising to 1,500 ft and 2,000 ft in the west. These surfaces are deeply trenched by the long arcuate course of the Derwent river system and its important southern tributary, the Wye.

Much of the land presents *par excellence* a landscape of cuesta relief, both large and small in scale. Despite this apparent simplicity, structure and morphology are together indicative of an involved physical history, producing at times complex physical patterns and local features difficult to explain (fig. 41). The basic structural unit is the low anticlinal Pennine axis trending south-south-east from the High Peak, and seamed irregularly in the south with igneous beds. This gives rise to a characteristic sequence of eastward-dipping rocks from Carboniferous limestone to Coal Measure shales and sandstones, with, overall, a simple north to south geological grain. A complex secondary pattern of minor flexures, however, developed later in Carboniferous times at right-angles to those of the main orogeny and this gave rise to subsidiary cross folds curving from west to east, with, in addition, the Don faults marking an important and much-faulted sharp monoclinal flexure, approximately between what are now the sites of Conisbrough and Sheffield. All of these structures were partially planed off before the Permo-Triassic top cover was deposited against the older land-mass and, as a result of this, and of later erosion, individual outcrops of both Lower and Upper Carboniferous rocks often follow a markedly sinuous course from north to south, successively weaving a pattern of westward- and eastward-advancing embayments. The pattern of these embayments, certainly within the Coal Measure Series, has had important local economic consequences and often largely accounts for the contrasting landscapes that are quite sharply juxtaposed within the coalfield, as between industry with mining and undisturbed areas of farmland.

Most writers infer the existence of a second very gently inclined and uncomformable top cover of Upper Cretaceous rocks dipping eastwards on which, finally, the contemporary drainage pattern was initiated. This Chalk cover has been entirely stripped from the region now under discussion, but remnants of a one-time simple consequent drainage pattern on such a surface still seem evident here and there. The danger of over-facile generalization regarding hydrographic origins in the region are very evident, however, in the Don system itself. The sharp elbow turn of the valley at Sheffield is indicative at once of a composite origin. Below the bend the valley is closely guided by the faulted monoclinal flexure already noted as an important structure. Above the bend the valley might at first be regarded in a general sense as subsequent, but in detail it is found to be incised into and across whole series of alternating shale and grit bands sometimes in significant incised meanders (as near Wortley) that are quite out of adjustment with local structure. Yet the upper Don has long been recognized as an aggressor, having captured the head-waters of the adjacent consequent Dearne. Aggressor or no, the Don itself seems also to have lost drainage to the upper Derwent, farther west. The consequent and beheaded Dearne may initially have set the position where the Don between Conisbrough and Doncaster cuts a characteristic gorge through the Magnesian Limestone, but in this siting, there is only partially selective adjustment to local structures. The long south to north vale of the Rother–Doe Lea system follows a subsequent direction roughly parallel to the Magnesian Limestone scarp, but in detail, again the Rother shows discordances in many parts of its course as it cuts to and fro across the grain of the Coal Measures exposed in bevelled fold structures. The extent to which these valleys were affected by the direct or indirect effects of glaciation is still a matter for conjecture. The Dearne–lower Don basin presumably held large proglacial lakes but how far these and associated spillways affected the middle Don basin is still uncertain.

The range and diversity of the relief from the High Peak to the Humber, lead to important local differences of climate which, together with the differences of soils and vegetation, give rise at times to sharp changes in land use. These changes, in association with the effects of other industrial, social and economic factors produce a cultural landscape of much diversity, whose differences will be considered with respect to the several sub-regions that form: the Gritstone and the Limestone highlands; the exposed coalfield; the concealed coalfield and Humberhead Levels.

Much of the Pennine highland closely approaches or exceeds the limits of habitability in Great Britain. Kinderscout and Bleaklow in the west exceed 2,000 ft, and extensive areas elsewhere exceed 1,200 ft to 1,500 ft. Of great natural beauty, much of this region now is included within a national park; it falls into two quite distinctive areas roughly separated by the course of the vale of Hope and the Derwent valley. The northern and eastern section comprises lofty Millstone Grit plateaux, with associated gritstone scarps and steep shale slopes; the southern section is one of more rolling and slightly less elevated country largely composed of Carboniferous limestone. Both regions show polycyclic features of relief developed within an extensive high upland surface generally presumed to be of Tertiary subaerial origin.

The gritstone plateaux of the north are often edged by very impressive sheer rock cliffs whose boulder and rock-strewn bases flank very flat watersheds above. At lower altitudes, where drainage is generally free, both valley sides and divides may be covered with bilberry or heather moorland, but at higher altitudes (though in places even as low as 1,300 ft) interfluves are commonly mantled with sodden desolate cotton-grass bog or deep peat, some of which is much eroded and gullied. The high altitudes and exposure emphasize the bleakness and inhospitable character of these uplands, where hill fog, cloudiness and low air-temperatures combine with heavy snowfall, high mean annual rainfalls (50–65 in.), and high wind velocities to make them, at least in winter, as formidable a climatic divide as they are inevitably so orographically. This inhospitality is reflected in the remarkable rapidity of the change from high population density at the conurbation fringes to quite uninhabited moorland a short distance above within only a few miles' range.

All but devoid of settlement though these gritstone moorlands now are, their role has changed through human history. Considerable evidence exists to show that there was selective Mesolithic occupation restricted to parts of the moors near the vale of Hope and the moorlands approaching the environs of Sheffield. This continued with waning regional significance as part of a wider highland settlement pattern through Neolithic, Bronze Age and Iron Age periods. Today the only impressive evidence of these phases is that of Iron Age hill forts, for example, at Mam Tor (1,670 ft), crowning isolated hill-tops, where now only derelict and blackened stone walls or occasional roving sheep indicate their present limited use. In the more sheltered valleys and hollows there are, however, dispersed hill farms up to altitudes well above 1,000 ft supporting stock rearing and sheep farming as the general husbandry. Built of local sandstones in the solid traditional style the buildings blend into the hill sides. Some may represent the perpetuation of sites used only by summer pastoralists centuries ago, whilst at two farmsteads Neolithic finds suggest the use of the sites since prehistoric times. But hill farming today is precarious and shows evidence of decline and decay.

The Millstone Grit valleys now conserve one very important national asset in the abundant water resources guaranteed by the heavy rainfall seasonally well distributed on ground favourable to surface run off. Now stored in vast reservoirs in the Derwent and other valleys, this water supply is largely exported for use in distant industrial cities. The Derwent valley reservoirs have a capacity of some 10,400 m. gallons and from them there is gravitational feeding to the consuming centres of Derby, Leicester, Nottingham and Sheffield. A further 6,700 m. gallons is fed to Sheffield from more local reservoirs in smaller valleys all within thirteen miles of the city centre.

Quite different from the sombre and at times desolate gritstone moors are the rolling hills and more vivid green pastures, the white stone walls, and compact villages of the Carboniferous Limestone plateau south and west of the Hope–Derwent basin. The limestone dome dips steeply under the shales and grits along its eastern and northern flanks; elsewhere it presents an undulating upland surface with summit ridges at altitudes of from 1,300 ft to 1,500 ft, separating broad basins at about 900 ft to 1,000 ft. Into this somewhat softly moulded surface deep gorges have been incised, but only in the case of the Wye valley has such a feature very conspicuous length and continuous surface drainage.

To the naturally drier surfaces characteristic of any limestone region there is here added the factor of a somewhat drier highland climate, for the mean annual rainfall on the limestone is not more than 40 in. to 50 in. at the highest altitudes in contrast to the wetter gritstone heights to the north. The main water-table lies several hundred feet below the surface except locally where it is upheld by beds of igneous rock. These dark impervious rocks, locally known as 'toadstones', have been of great importance not only in relation to groundwater circulation, but also in the associated veins of minerals and ore bodies to which they gave rise. The working of rich veins of lead has been a major industry and in some localities certainly dates back to Roman times and reached a heyday in the eighteenth and early nineteenth centuries. In a new guise exploitation survives in the twentieth century in the renewed working of some of the veins for the one-time gangue minerals such as fluorite, barite and calcite.

Few regions in Britain at comparable altitudes can show in their landscape so many elements indicative of changing modes of using the land since prehistoric times. Though finds associated with Neolithic, Beaker and Bronze Age farmers and herders of the first and second millenia B.C. cannot match the abundance of evidence on the similar limestone plateau south of the Wye gorge, there are indications of an expanding farming population in the last millenium before Christ. Long after that time Roman occupation—civil and military—left roadways and trackways notably between the two important Roman centres of Buxton and Brough, and many sections of these Roman routes are still followed by modern roads. Evidence of early settlement in succeeding centuries is seen in the survival of classic types of settlement patterns in varying forms of 'green' or 'street' villages, interesting in their diversity and number at so high an altitude (900 ft to 1,000 ft), and in their siting at spring and well sites maintained either by the junction of limestone and shale grits or by the occurrence of perched water tables where toadstones occur within the limestone rocks. The village patterns can still be identified though the initial site factors no longer have significance.

The population was for long supported by the dual economy of lead mining and subsistence farming, supplemented by immense flocks of sheep roving the extensive limestone pastures; lead and wool were very important sources of wealth in medieval times. Despite the altitudes, around the villages open fields were cultivated of which there is still much evidence in the survival of narrow strip-shaped fields following the archaic reversed 'S' pattern fossilized into the landscape by enclosure. These are clearly differentiated from the square-shaped fields of late enclosure and give interesting evidence of the considerable extent of medieval ploughing in this highland region and of the altitude at which it took place. Large sections of the limestone plateau are now treeless, but this characteristic is essentially man-induced. On the precipitous slopes of the narrow incised dales, ample evidence is preserved of the ash-hazel woods and comparatively dense tree growth that is natural to the limestone, whilst their ready growth when encouraged as windbreaks round farmsteads, or as the regrowth of coppice and woodland following the run of derelict lead workings, both tell the same story. In so isolated a highland region the heavy demands for timber for lead smelting, the effects of stock grazing, and domestic needs, together must have brought devastation to woodlands at an early date.

The nineteenth and twentieth centuries have seen new changes in the cultural landscape. Occasional large and isolated farms remote from the ancient villages now characterize parts of the plateau and, despite the continuing difficulties imposed by inadequate water supplies and poor soils, considerable herds of cattle rather than sheep now predominate. An expansion of ploughing and the improvement of pastures increased the stock-carrying capacity of the region, and the Second World War further intensified the new agricultural trend. In addition, unusually varied and widely scattered small rural industries give a unique character to this region. Some of these are long established, for example, the occasional cotton mills in the recesses of the Wye and Derwent valleys, or the shoe factories at Middleton and Eyam. Others, however, are more recent developments as in the cement works near Castleton, the ubiquitous quarries and lime-workings, a plastics factory, and specialized steelworks.

The only two small towns that fall within that part of the Peak region here considered are the market centres of Buxton and Bakewell, at the western and eastern margins of the limestone dome and at the upper and lower sections of the Wye valley respectively. Buxton (1,000 ft), one-time spa and a tourist centre, with a population of 19,236, is the highest town in Great Britain. Bakewell (3,603) bids fair to develop and extend its composite role of market town and dormitory suburb.

Access from the interior Pennine plateaux into the arcuate course of the Derwent valley is still somewhat limited, and towering gritstone scarps have for long effectively isolated the valley no less than the plateau core from the Coal Measure lowlands beyond. These high gritstone edges were pierced only in 1893 by the second longest railway tunnel in England (Totley: 3 miles 950 yd), and the age of road transport brought other improvements into the Peak District, mainly during the twentieth century. Only then could the highland region in any sense come effectively within the Sheffield sphere of influence since when the commuting of labour and development of dormitory suburbs—especially in the Derwent valley—have become increasingly common. Significantly, despite the apparent importance of this valley as a linking routeway threading the highland region, there has never been a railway traversing the valley continuously from north to south.

The composite basin of the Middle Don, Dearne, and Rother rivers, extending from the Pennines to the Magnesian limestone scarp, across the South Yorkshire and North Derbyshire coalfield, comprises one of the most important industrial regions of Great Britain. The truncated folds and faults give to the generalized geological grain of the Coal Measures that here trend from north-north-west to south-south-east many important if local deviations of direction. Across these structures fragments of a well-recognized upper erosion surface grade from 700 ft to 550 ft eastwards, with an extensive lower surface at 450 ft. to 300 ft. Prior to the Industrial Revolution this must have been a densely wooded countryside of great natural beauty, as it still is in some localities, but much of its surface today shows devastation from mining, tip heaps, subsidence, and industry, and the cumulative effects on vegetation and buildings alike of long-continued air pollution.

There are important differences within the main divisions of the Coal Measures that have locally affected land use and scenery. The lower Coal Measures, in the west, in some respects differ very little from the Millstone Grit series and their massive sandstones repeat, if at lower altitudes, the cuesta, tabular and hog-back features of the highland scenery. Economically they include very little coal; more important are the beds of gannister, averaging 3 ft in thickness, which with the fire-clays have had very great value for the manufacture of refractory materials.

The Middle Coal Measures, on the other hand, are composed of softer sandstones and shales so that the relief grades eastwards into more subdued scenery. These measures also include the important coal seams; of their total thickness of some 800 ft or more, north of Sheffield, 60 ft are in seams of coal more than 2 ft thick. Some of these have been of great economic importance—notably the Silkstone (a famous gas and coking household coal), the Parkgate (a 3-ft bed of semi-anthracite coal), and the Barnsley or Top Hard (a seam of remarkable thickness locally, averaging 6 ft, and, near Barnsley, as much as 11 ft). On the Barnsley seam more than any other, the high productivity of the field has depended. Apart from their vast coal resources, the Middle Measures were of great early importance for the ironstone nodules that are included adjacent to some seams, and which were worked from bell-pits, some of which, with adjacent spoil heaps overgrown with vegetation, can still be identified in the landscape. The Upper Coal Measures like the Lower include little productive coal, only one of the few seams having a thickness of as much as 4 ft. Though comprising similar elements, the scenery becomes more and more subdued to the point of monotony as the eastern limits of the exposed field are reached.

In response to both the local physical factors and the wider regional relationships, the area has experienced a unique historical and economic development. For early prehistoric man it seems to have held none of the attractions that then favoured the Pennine Highlands, and the first evidence of the spread of cultivators or herders into the middle Don valley is associated with Iron Age hill forts—as, for instance, on Wincobank hill (between Rotherham and Sheffield), overlooking the then forested and marshy plain of the River Don, now occupied by the heart of the heavy steel industry. Later, Roman routeways bifurcated near the site of Sheffield, and there were civil and military bases near Rotherham. But the earliest effective colonization of the region seems to have been associated with Anglian settlement prior to the eighth century A.D. Place-names indicate some clearing of forest at that time. The choice of the sites of the early settlements again shows a high degree of selectivity, in this case in the very frequent use of outcrops of drier Coal Measure sandstones which, in addition, were often also etched out as isolated knolls and low hills. But in a wider regional sense the coalfield zone was still a backwater right off the main stream of human movement in Britain.

None the less, despite isolation, through medieval times there was a growing occupation of the land associated with the dual development of subsistence agriculture and numerous cottage industries, including local iron smelting and forging, long before coal working was of any importance. The earliest record of iron smelting and forging relates to Kimberworth, near Rotherham, in 1167, whilst Subsidy Rolls in 1297 record a peasant craftsman in Sheffield using local iron and keeping one cow, and show that by 1378 local ironmongery and cutlery manufactures were widely characteristic over the area from Sheffield as far as Barnsley. For such industries the forest resources in wood and charcoal had great value, as later, too, had the abundant water power from streams turning wheels in the steeply graded valleys, whether working bellows or grindstones—the latter again in plentiful supply from the local gritstones.

Shallow medieval coal working in south Yorkshire goes back at least to the thirteenth century (fig. 42a), and during the succeeding centuries coal production expanded widely, following the outcrops in a proliferation of small local collieries. The great modern industrial development of the region in association with heavy industries came, however, only with the spread of deeper mining in the eighteenth and nineteenth centuries, particularly between Barnsley and Sheffield, with Barnsley very fortunately located between rich seams of Silkstone and Barnsley coals (fig. 42b). But these changes depended not only on new technological skills but also on the development of more efficient means of communication by turnpike roads, canals and railways, which again stresses the significance of the

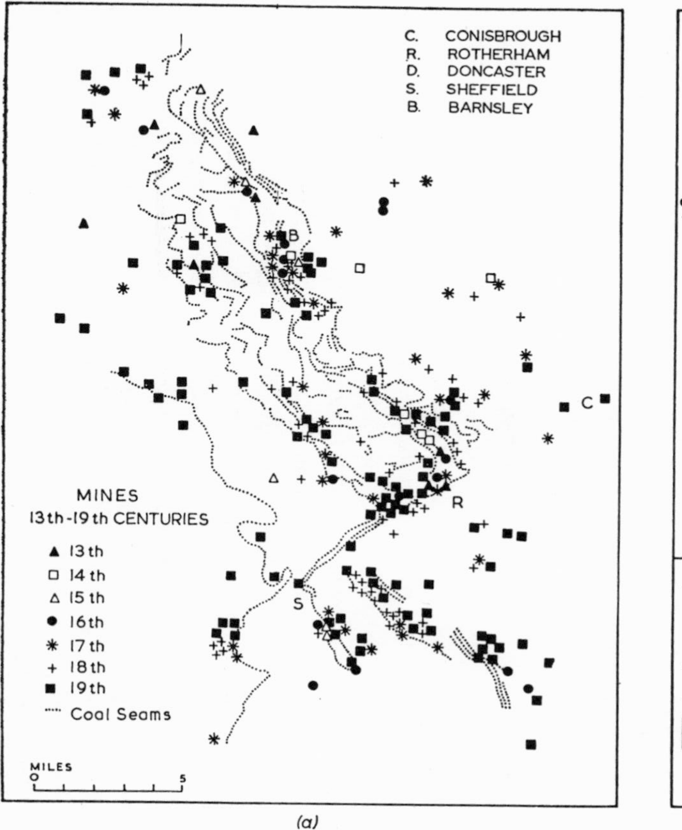

(a)

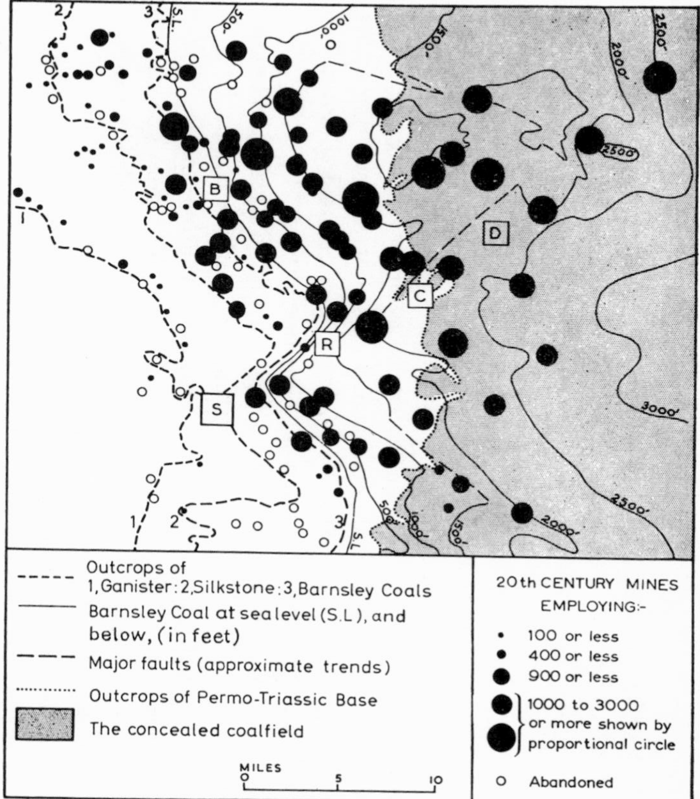

(b)

Fig. 42. Distribution of mines in the south Yorkshire Coalfield.
(*a*) The exposed field from the thirteenth to the nineteenth century inclusive; (*b*) the exposed and concealed field in the mid-twentieth century.

Y

initial physical isolation of this coalfield. Trains of pack-horses were the sole means of transport over steep tracks, often inexorably bad, until well into the eighteenth century; a direct canal link with the sea from Hull to Sheffield was completed only in 1819, and Sheffield, first served only by a branch railway from Rotherham in 1838, had no direct link with the south via Bradway tunnel and Chesterfield until 1870. Once the general difficulties of communication with the outer world had been met, the exploitation was assured of a field not only of uncommonly rich coal-seams, but also with structures ensuring such gentle easterly dips over most of the region that mining could be comparatively easily developed.

Within this setting, the growth of Sheffield into a great steel manufacturing centre, with a population of nearly ½ m., is in some respects a natural expression of industrial evolution, but from many points of view it is quite anomalous. Like the region generally in which it is placed, Sheffield was early associated with ironworking, closely connected with the local ore resources, but there were many other ironworking centres in Britain that might equally have grown to industrial eminence. Even four centuries ago, Sheffield was importing high-grade ores from Sweden, Russia and Spain, since the local ores were unsuited to steel manufacture and such imports had here to be brought to a site as far inland as any in Britain, and this before roads worthy of the name existed. Physically, too, when later urban growth came apace, the ruggedness of the relief provided a site of exceptional character and difficulty. Clearly a local reservoir of skilled craftsmen, with their social tradition and inheritance, must have done much to localize the steel industry in its earliest phases.

The precise site of the early village township was not without some physical advantage in relation to the needs of feudal times. At the elbow-turn of the Don, where the then marshy flood-plain is only ½ mile wide, the river follows a gap cut between two prominent spurs in the Coal Measure grits at Spittal and Park Hills respectively (fig. 43). Towards this gap there converges, somewhat radially, the deep and steeply graded valleys from the south and south-west between whose courses bold spurs rise in long inclines from only 150 ft at the Don flood-plain to more than 1,000 ft within a distance of a few miles, and the flanks of these spurs are at times quite precipitous. One of the spurs (Hallam ridge) projects forward conspicuously from the

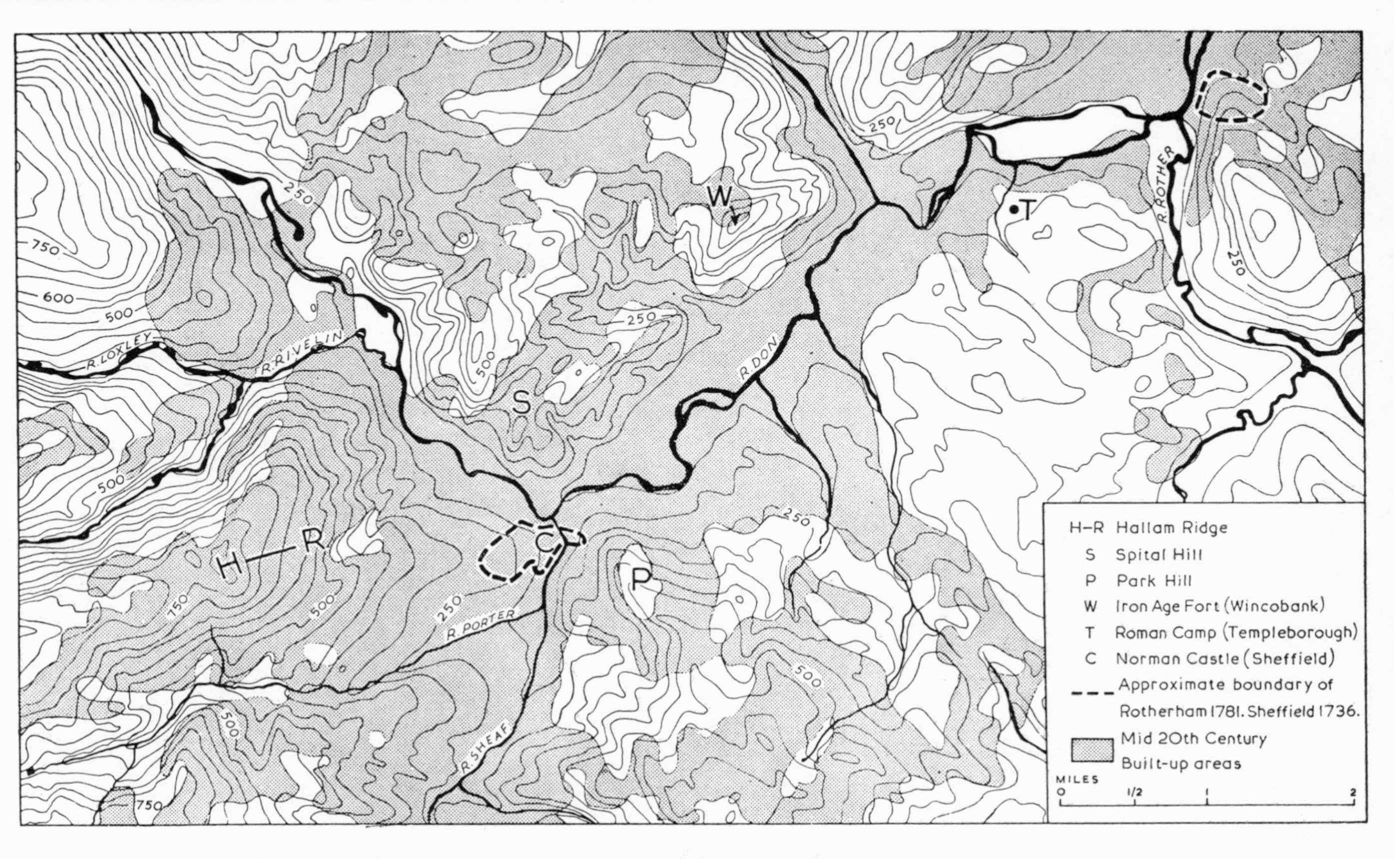

Fig 43. Urban growth since the eighteenth century at Rotherham and Sheffield in relation to the relief. Reproduced from the Ordnance Survey Map with the sanction of the Controller of H.M. Stationery Office. Crown Copyright Reserved.

south-west, tapering right into the Don gap. Here it terminates in a low bluff, the northern end of which is trimmed where the meanders of the Don undercut its base, and the eastward-facing flank of which makes a steep descent to the confluence of the tributary Sheaf. Both rivers, Don and Sheaf, could be forded or bridged, as they soon were, but routes generally in such country kept to the valley sides or followed the spurs, as the Roman route from High Peak clearly showed in its course along the Hallam ridge as far as the site of Sheffield, and thence, probably, along the southern flank of the Don valley to Rotherham.

A Saxon manor is thought to have existed on the easily defended Hallam spur-end, but the first certain evidence of the importance attached to the site was the building of a Norman castle, in the twelfth century, using the low, dry, and steeply edged bluff with its defensive double water frontage. A market (still so named) lay behind the castle up-slope, together with a church near by (now the cathedral site), and in due course a High Street was alined up the spur slope with tofts and crofts at intervals along it. The truncated sequence of sharply dipping sandstones and shales from which the spur-end is sculptured gave an abundance of shallow well and spring points adequate for a small settlement if not later for a growing industrial town. Close by, on the gullied northern spur flank, was one of the Town Fields. To this day, road patterns and names and even building lines in the heart of the city in many cases still reflect some of these medieval embryo urban and rural origins.

By the end of the sixteenth century the little village-town was exporting cutlery for which trade Sheffield had by then acquired a high reputation. Foreign ores for steel were imported to Hull and thence by boat to the then river port of Bawtry. Though the steel industry had to depend on imported ores, for the growing industry there were some important supplementary local advantages. Fuel from the woodland forests was abundant for the early charcoal furnaces, even if in the late seventeenth century it is recorded that every ton of pig iron required 26–27 cwt of charcoal for its manufacture. The furnaces at first needed hill-top sites well exposed to high winds and such 'bole hills', still so named, occur within the city. When the use of water-power from steeply graded streams later became more general, for working bellows, hammers and grindstones, the Hallam ridge was well placed. Chains of hammer-ponds and tilt-wheels appeared in the deeply sunk valleys into which the industry gravitated on either hand. A few of these ponds and wheels survived even into the twentieth century. The transition from a dependence on local water-power to that of coal found Sheffield again well sited, for the highly productive Middle Coal Measures (including the Silkstone seam) could be worked near the very heart of the town, and, not far distant, there were important supplies of gannister in the Lower Coal Measures, essential for furnace linings, and dolomite in the Magnesian Limestone, needed later by heavy industry.

The development of Sheffield's main industry was associated with a remarkable growth of population matched in its momentum by few other cities. From a small village still of only 3,000 inhabitants as late as the early seventeenth century, the numbers grew to 12,000 by the mid-eighteenth and 46,000 in the early nineteenth century. This rate of growth was continued, numbers triplicating to 135,000 by 1851, and again to 409,000 at the opening of the twentieth century. The population, now more static in numbers, exceeded 500,000 for some decades. The city is remarkable in terms of its size for the exceptional degree of specialization which its occupations show. Light trades traditional to the city, such as cutlery, tools, and tableware, continue in some importance, but the heavy industries making special steels, steel alloys and the complex processing and machining of these steels in engineering workshops account for two-fifths of the employed population. These have supplied a very high proportion of the United Kingdom's output in alloy steel manufacture of rods and bars in coil, hot and cold strip, tyres, wheels and axles, and tool and magnet and high-speed steels. The bulk of the more highly alloyed steels is made now in electric arc furnaces as a highly specialized production to exact specifications for orders designed to meet special and peculiar technical and industrial needs. Alloys of this kind may be extremely expensive and are 'precious' metals that may cost more than fifty times the price of ordinary mild steel. Thus today, Sheffield's specialized basic industries have come to depend more than ever before, on the one hand, on an array of imported ores, alloys and metals (ranging from everyday sources of pig iron and scrap to rare ferro-alloys); on the other hand, on the inherited traditional local craftsmen's metallurgical skills, now backed by a wealth of accumulated knowledge derived from local technological and scientific research in these distinctive industrial fields.

The urban expansion in area still in progress, includes already more than sixty square miles of hills and valleys, and has led to interesting features of urban morphology and metamorphosis within the site. By virtue of its situation Sheffield has never been the regional capital of a wide tributary area as has been true of many other British towns of far smaller size and importance. As a result the administrative and service centre and area is surprisingly small and coincides roughly with the early town site at the end of the Hallam spur. To this day despite its size, Sheffield is little more than 'a great workshop in the nook of the Pennines' (Rudmose Brown) and reflects this in its urban structure.

Light industries within the town, especially cutlery, have always clustered on or around the lower Hallam ridge fringing the early town site. Suburban and residential areas have expanded intermittently during periods of prosperity in the nineteenth and twentieth centuries, with crowded brick terrace dwellings in the valleys and brick and stone houses on the ridges. Beyond are modern villa estates reaching out towards the windswept moors at altitudes approaching 1,000 ft, with, at the same time, well-planned modern factories set out in some of the tributary valleys; but near the heart of the town there are still some industrial vistas recalling the squalid conditions associated with the paleotechnic era of urban growth. These will soon vanish in present urban development.

The last major functional unit to develop was that associated with heavy industry, and this, requiring extensive flat land for workshops, together with immediate access to road, canal and railways, grew naturally along the broad flat Don flood-plain ultimately to coalesce with Rotherham. Thus in broad terms, if not in all details, the historical uncontrolled growth of the city has brought into being a generalized functional pattern that is distinctive and not altogether out of harmony with the physical requirements of the site as viewed today. These characteristics, however, present some new problems in other respects. No other town of its size in Britain includes so great a range of relief in a setting that in winter has a marked degree of climatic severity. Snow in the higher suburbs may lie long on the ground. At other times anticyclonic weather in such sharpened relief can produce deep and marked inversions of temperature in the valleys and within the stagnant stable air that there collects serious air pollution from both industrial and domestic sources can result.

Begrimed buildings and deteriorating pastures have afforded evidence of these insidious conditions in and around the city.

Rotherham (85,346), now a part of the overall area of the conurbation, merges by ribbon growth into Sheffield on the one hand, and extends down the Don valley towards Mexborough on the other. From some points of view it may seem surprising that it did not outstrip Sheffield at an early date, for the district had early importance. It was near here, on one of the few fragments of river terraces in the Don valley (providing a dry site at Templeborough), and not at Sheffield, that a Roman camp was placed. Later, it was nearer to Rotherham than Sheffield that the earliest ironworks were recorded. At Rotherham, as at Sheffield, an early market settlement grew up, similarly sited on the well-drained terminal slopes of a spur, but in this case overlooking the confluence of the Rother with the Don (fig. 43). Unlike Sheffield, however, no feudal castle was built at Rotherham, and though gaining an early reputation for cutlery and smithies, it lacked the outstanding local resources in water-power on which Sheffield's industries came largely to depend. It lagged behind also, later during the great industrial surges, despite its better position in relation to the productive coalfield and communications with other towns; Rotherham lay, in fact, on the main post-road that by-passed Sheffield by many miles, and also where canal and railway linkages with the outer world were in due time more easily effected. Today one-third of its employed population is engaged in metalwork, mainly associated with heavier branches of iron and steel manufactures and engineering and iron brass foundries.

In relation to the economic changes in the middle Don basin the remaining regions of the exposed coalfield present both similarities and differences. Thus, in the Rother basin there was an early association with the working of local ores preserved in the fold structures, though much of the vale long remained predominantly an agricultural backwater with Chesterfield little more than a market town for such. Particularly where eroded folds bring Lower not Middle Coal Measures swinging far to the east, the rural characters still remain. New phases of mining and metalworking and new-found accessibility brought the development here not of steelworks but of blast-furnaces and large ironworks at many points (for example, Renishaw, Clay Cross, Sheepbridge and Staveley) all with good access to coking coals, though, later, Chesterfield developed large-scale engineering works. In some cases, as at Staveley and Clay Cross, coal mining and the metallurgical industries have developed collaterally in close association.

The upper Don basin, on the other hand, followed another industrial trend with the maintenance of rural characteristics assured by the dominance of Lower Coal Measure series over the basin. Early settlements were largely up-slope, keeping out of the narrowly incised Don valley. Thus Wortley, an early iron-forging centre using local ore and timber, was on a divide. Penistone, a remote hill town of some size for its location, was renowned not for iron but for its textiles, until in 1864 Cammell Laird started steel manufacture here as an implantation from Sheffield, and thereby obliterated the older industry. Stocksbridge represents only a modern expansion of the Sheffield industry into the rural highland setting. Elsewhere, as at Deepcar and Oughtibridge, the important manufactures associated with this area have been concerned with the exploitation of the refractory materials (for crucibles, silica bricks, furnace linings, etc.) made possible by the former local wealth of the Lower Coal Measures in gannister and fire-clay beds, essential to the modern metallurgical industry. These refractory industries remain, though they now also depend on imported materials, and have spread up to the head of the Don valley on to the divide between the Don and Holme drainage at 1,050 ft.

The Dearne basin presents a third contrast in the much greater emphasis on coal mining and coal production that there applies owing to the local abundance of very valuable and easily worked seams. Shallow mining was already extensive in the Barnsley region in the seventeenth century, and the early large-scale development of mining towards the end of the eighteenth century with the sinking of many new collieries between 1811 and 1840, gave a great lead to Dearne valley coal production, helped by the opening of canals both to the Aire and Calder and, by the Dearne and Dove, to the Don. Its continued significance is reflected today in the vast nets of mineral railway lines and the almost continuous ribbon of colliery settlements and associated industrial growth from Barnsley to Mexborough forming virtually a Dearne valley conurbation. Within this, until quite recently, Barnsley has been the administrative hub of the south Yorkshire coal-mining industry—a role that it is now losing to the growing centre of Doncaster on the concealed field, twenty miles to the east. Lying upslope well above the flood-plain of the Dearne, Barnsley is an old, indeed an historic market centre, rooted in the local region with a market charter that dates from 1240. Before the domination of coal mining it was associated with a variety of quite different industries including linen, paper-making, ironfounding, and glass-blowing, towards new versions of which, together with other new industries, it will undoubtedly again turn, thus modifying its present dominant industrial function but not losing its focal and commercial prestige.

The ultimate headstream of the upper Dearne lies west of the Middle Coal Measures and above the main Calder–Dearne water-parting and presents an interesting change to an entirely new economic region though technically still within the same physical unit. At Denby Dale, Clayton West and Skelmanthorpe textile and not metallurgical industries link this area with the West Riding woollen towns, and the economic boundary is quite sharply defined in this respect. Some might regard this occurrence as a natural 'overspill' expansion from the woollen textiles of the Colne basin. More probably, as the evidence at Penistone has already hinted, it may represent the now shrunken limits of an industry that was once of much wider extent and which only recently elsewhere in the region has been subordinated to the metallurgical and mining industries. These latter have progressively advanced and expanded into the remoter limits of the physically defined Don–Dearne basin, spreading up to, but in this case not surmounting, the upper Don–Dearne water-parting.

Though vast areas of the exposed coalfield admittedly present a picture of differing degrees of industrial devastation, in all these regions farming still makes up a substantial area of the Coal Measures landscape, especially where outcrops of the Lower Measures locally swing eastwards into the industrial zone. In such cases arable farming generally occurs in association with mixed husbandry for dairying, and farms near towns are often very small with holdings of no more than 20 acres. In the south-west, notably round Sheffield, arable farming is often carried to unexpectedly high altitudes (for example, even for wheat above 900 ft) and yields may be very high. This occasional intensive farming in part reflects growing urban needs and is helped by the relatively dry early autumn conditions on the south-east Pennine borders. Rainfall in September is generally light. Claims on the use of farmland within the coalfield have often conflicted in recent years with growing industrial and housing needs; as when, in addition to serious effects following industrial contamination of both soils and vegetation by air-pollution, or the waterlogging of surfaces through mining subsidence, there is added the temporary restriction of farming in favour of the working of extensive areas of open-cast coal. More than 28 m. tons has been gained by this means since about 1940.

East of the exposed coalfield the physical and cultural landscape changes sharply. Below Conisbrough for four miles the river occupies a sinuous shallow gorge cut across the Magnesian Limestone cuesta that here forms a distinctive belt of somewhat subdued relief not more than 300 ft to 400 ft in altitude and occasionally broken by narrow gorges. The variable soil thicknesses often include fertile loams and clays and with a mean annual rainfall not exceeding 23 in. there is extensive arable and mixed farming. Villages often reflect in their field and house patterns a medieval past, their traditional pale yellowish grey limestone cottages and pan-tiled roofs contrasting sharply with the begrimed gritstone and brick buildings farther west. Conisbrough, clustering round the old Norman castle built at a natural defensive point on a limestone promontory near the western end of the gap, no longer has the dominating importance that it held in medieval times.

Beyond, eastwards, the basin opens into the low Triassic plain of the Humberhead levels which formed part of the proglacial Lake Humber. Glacial water-laid drifts mantle the surface together with post-glacial alluvium, to form a very featureless and somewhat waterlogged plain mostly less than 25 ft above sea-level. Close to Thorne at Hatfield Moors and Thorne Waste are remains of ancient peat bogs, now drained, whose alluvial floors have been related to late Neolithic transgressions. Interesting evidence from pollen analysis suggests cultivation here as early as the beginning of the Bronze Age which supports other evidence of selective prehistoric settlement on tiny patches of dry land in this region. But through much of prehistoric and historic times, lake and fenland conditions extended over most of the region, except where very low islands of Keuper and Bunter Sandstones just break through the drift surfaces as at Thorne, and these, together with outwash delta gravels made low but invaluable dry surfaces that throughout history have controlled the distribution of settlement. The Roman road from London was deflected to follow these dry stepping-stones round the Humber Fens and the site of the Roman station at Doncaster (Danum) depended on the existence at that point of the lowest firm ground on the Don where a crossing could be effected. In such conditions the Fenland was used only by fishers and wild-fowlers, and later it was reserved as a chase.

Not until the early seventeenth century was the landscape materially altered when the plans of the Dutch engineer Vermuyden to drain the Fens had been effected. Until then the Don, Torne and Idle rivers meandered over the surface between the Trent, Ouse and Aire. The Don was taken by the new 'Dutch river' to the Ouse confluence at Goole and the Idle and Torne were directed by new courses and drains into the Trent north of Axholme. With pumping and the extensive practice of warping from the tidal silts of the Ouse, the land was largely transformed into profitable farmland with some dispersed settlement.

But underlying both Magnesian Limestone and fenland in the Doncaster region the concealed coalfield extends, still highly productive at depths of more than 1,500 ft in the eastern basin (fig. 42b). The sinking of pits through the Permo-Triassic top cover to these and greater depths has mainly been effected during the first two decades of the twentieth century and already by 1925 reached as far as Thorne. The field is particularly rich, the Barnsley seam, still averaging a thickness of 8 ft, providing at least 60% of the output. The sinking, working and maintenance of pits here is, however, very costly. The problem of water seepage is often serious, whether from the water-bearing Permo-Triassic rocks, or from worked-out seams that, westwards, lie at more shallow depths, or from Fen marsh and quicksands. Furthermore at a depth of 3,000 ft the working of the Barnsley seam can become very dangerous from pressure fracture and spontaneous underground burning. Pits must therefore be planned to give a very high output. Gone, therefore, is the proliferation of small collieries characteristic of the old field, and of haphazard building of miners' rows. Instead a rich agricultural landscape remains undisturbed except where now only occasional very large spoil heaps and pitheads bring a new element into an older order, or modern new villages and small mining-towns bear the mark of orderly planning with rapid growth; for example, Adwick-le-Street (18,212) and Bentley (22,952).

With increasing distance eastwards the location of settlements is now closely related to sites more than 25 ft above Ordnance Datum; mining has reintroduced in new forms the long-standing problems of Fen drainage since subsidence can be expected to occur through a depth equal to two-thirds of the thickness of the coal won from the ground. On this basis much of the area is liable to a general subsidence of from 15 ft to 23 ft and the total extraction of coal could lead to the permanent resubmergence of the Fens. Reductions in the gradient of the Don already make the river tidal up to within two miles of Doncaster. Though mining underneath rivers and drains is now restricted, it continues under the plains on either hand, and this causes subsidence which involves the reversal of surface drains. Further permanent pumping in advance of mining is necessary in the control of a drainage system and land surface that becomes increasingly artificial. But in terms of the ultimate history of settlement and industry in the Humber Levels this will be a transient short phase, for present estimates suggest that the coalfield may be substantially exhausted within the next century.

The development of the concealed coalfield has been the major factor in the very rapid twentieth-century growth of Doncaster into a town now larger than Rotherham. Despite its Roman origins and medieval growth at an obviously important focus for communications for a wide region, it remained for long a market centre serving no more than local agricultural needs. The advent of the railway era led to rapid urban expansion around the ancient nucleus in an elongated strip from north-west to south-east. This follows the firm dry sandstones and gravel outcrops that thread the carr lands in a very low watershed, not 50 ft above sea-level between Don and Trent. The development of the concealed coalfield accentuated this growth particularly in recent decades; Doncaster increased in population by 20% between 1931 and 1951, and has now a population of 86,402. It has attracted and developed a wide range of industries, including textiles, railway works, light and heavy industries, plastics, glass and foods, some of them well adapted to the reservoir of female labour that a rapidly expanding mining area ensures. These industries blend with the town's older role as the local market and with its growing status as a regional centre and administrative hub of the coalfield, usurping in this respect the status for so long held by Barnsley to which Doncaster is now linked by almost continuous industrial ribbon growth.

Few regions of comparable size in Great Britain can portray better than this the changing values that the environment may present in relation to man's changing technical skills and social and traditional needs, and, from Pennines to Ouse, the diverse physical landscapes reflect very clearly a changing but dominating pattern of the works of his hand.

Selected Bibliography

R. N. Rudmose Brown. 'Sheffield, its Rise and Growth.' *Geography*, **21**, 1936, pp. 175–84.

Mary Walton. *Sheffield, its Story and its Achievements*. Sheffield, 1948.

C. B. Fawcett. *The Provinces of England: A Study of some Geographical Aspects of Devolution*. London, 1919; revised edition, 1960.

D. L. Linton (ed.). *Sheffield and its Region: A Scientific and Historical Survey*. Published for the British Association for the Advancement of Science, 1956.

G. D. B. Gray. 'The South Yorkshire Coalfield.' *Geography*, **32**, 1947, pp. 113–51.

E. Charlesworth. 'A Local Example of the Factors Influencing Industrial Location.' *Geographical Journal*, **91**, 1938, pp. 340–51.

W. H. Wilcockson. 'Some Variations in the Coal Measures of Yorkshire.' *Yorks. Geol. Soc.*, **27**, 1947, pp. 58–81.

A. C. Dalton. 'Glacial Evidence of the Sheffield Area.' *North-Western Naturalist* (New Series), **24**, 1953, pp. 38–54.

J. B. Sissons. 'The Erosion Surfaces of Drainage Systems of South-West Yorkshire.' *Yorks. Geol. Soc.*, **29**, 1954, pp. 305–42.

K. C. Edwards and F. A. Wells. *A Survey of the Chesterfield Region*. Chesterfield Regional Planning Committee, 1949.

Report of the Sheffield Town Planning Committee. *Sheffield Replanned*. 1945.

CHAPTER 19

WEST YORKSHIRE

F. J. Fowler

West Yorkshire is a fairly compact industrial area in the middle of the West Riding of Yorkshire. It is rather more extensive than the conurbation to which the title 'West Yorkshire' is customarily attached. Its boundaries are arbitrary, but within the triangular-shaped area with a base along the Pennines from Skipton to Holmfirth and with its sides tapering eastwards to meet near Knottingley, the cultural landscape serves to distinguish it from the peripheral areas (fig. 44). That landscape is essentially industrial, an intermixture of urban and rural, substantial manufacturing towns and textile and mining villages, mills, factories, collieries and houses. Nevertheless, outside the more heavily built-up areas, both within and without the conurbation, much land is still agricultural and some parts are still well wooded. The total population of about 1·8 m. is found in communities ranging from groups of a few hundred people to the regional capital, Leeds, with more than half a million inhabitants; communities which combine pride in their social and economic self-sufficiency with an attraction towards, and some degree of dependence upon, some larger town.

The greater part of West Yorkshire lies within the valleys of the Rivers Calder, Colne and middle Aire, but it extends northwards into the valley of the middle Wharfe and southwards into that of the Dearne, a tributary of the Don. In successive cycles of erosion, these valleys have been cut into a surface, which loses height towards the Vale of York, leaving tiered slabs of country with pronounced westward- or northward-facing scarps. In the extreme north-west around Skipton the Lower Carboniferous Limestone series are exposed, but over much of the area the valleys are cut into the Millstone Grit and Coal Measure series (fig. 45). The Millstone Grit series makes up the country west of Halifax and north of Leeds and Bradford. Composed of conglomerates, grits, sandstones and shales, the diversified character of the scenery is due in large part to the sculpturing of these rocks

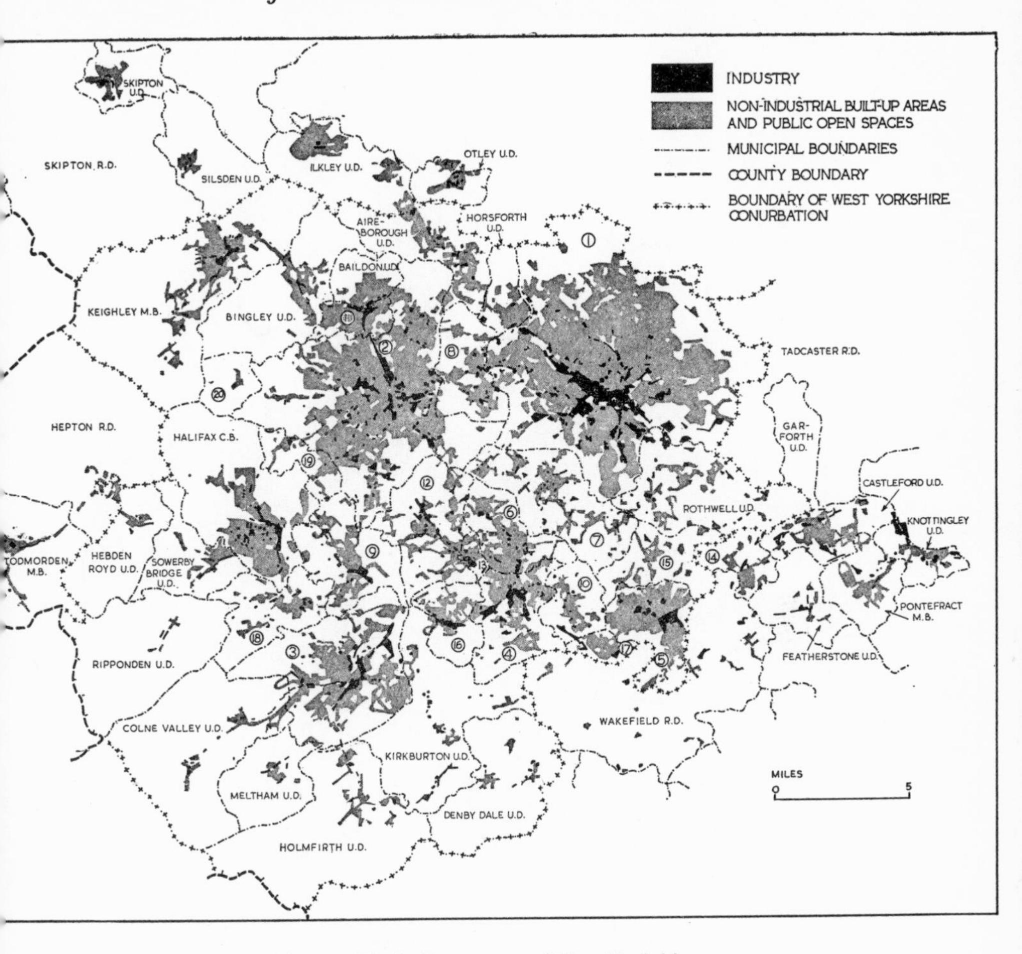

Fig. 44. The built-up areas of West Yorkshire.
Leeds C.B. 2. Bradford C.B. 3. Huddersfield C.B. 4. Dewsbury C.B. 5. Wakefield C.B. 6. Batley M.B. Morely M.B. 8. Pudsey M.B. 9. Brighouse M.B. 10. Ossett M.B. 11. Shipley U.D. 12. Spenborough U.D. Heckmondwike U.D. 14. Normanton U.D. 15. Stanley U.D. 16. Mirfield U.D. 17. Horbury U.D. 18. Elland U.D. 19. Queensbury and Shelf U.D. 20. Denholme U.D.

and their different degrees of resistance to weathering agents. Across the gritstone country lie the broad valleys of the Aire and Wharfe, and the gorge-like valley of the Calder, and above the steep or precipitous valley slopes are bleak, open, often heather-clad moorlands, with an average elevation of 1,300 ft to 1,500 ft. For recreation many nearby town-dwellers are attracted to the unspoilt localities, for example, to Rombalds (Ilkley) Moor, to Haworth Moor (the

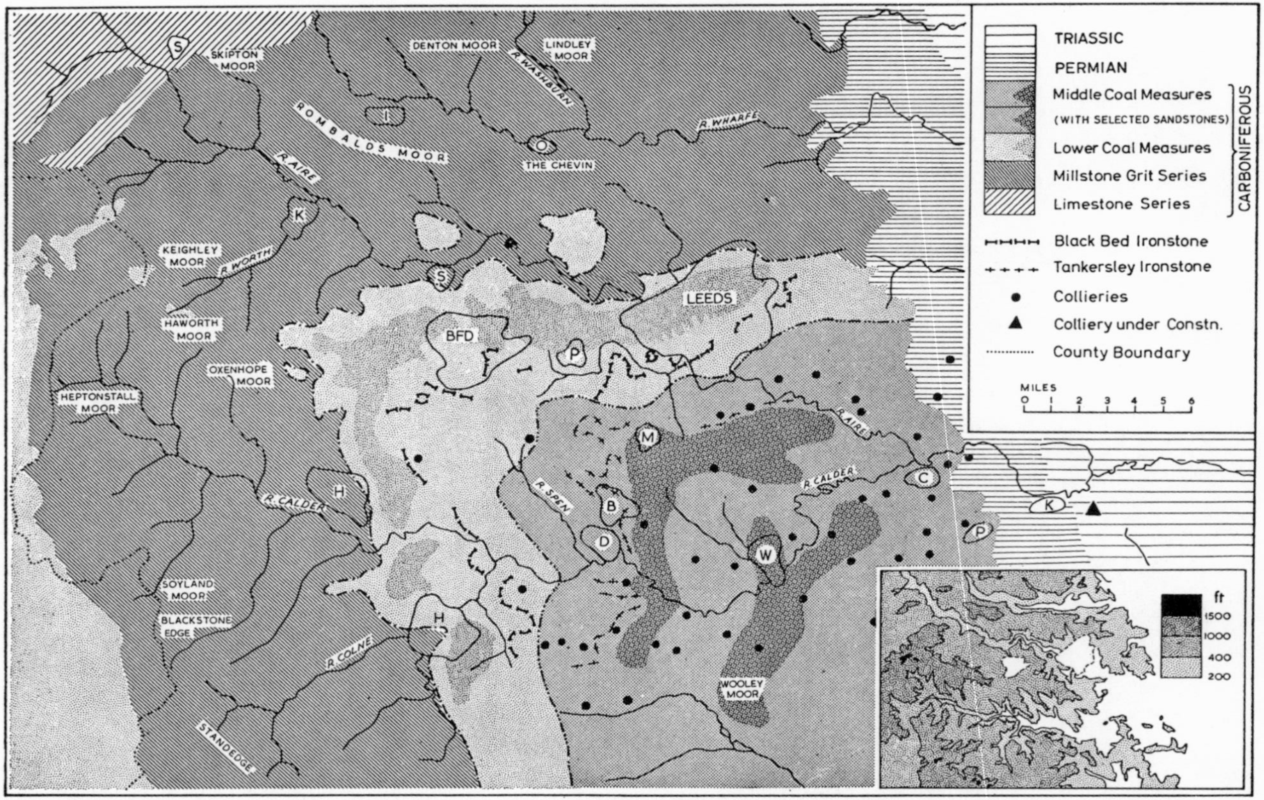

Fig. 45. Physical features of West Yorkshire.

Brontë country), and to Hardcastle Crags (near Hebden Bridge).

The northern edge of the Coal Measures lies between Leeds and Bradford, with two outliers at Baildon and Horsforth. West of Leeds there is the normal junction of the Millstone Grit and Coal Measures, but to the east the coalfield is cut off by the northern boundary fault. Since the conditions which obtained in Yorkshire during Millstone Grit times were continued into Coal Measure times, there is no abrupt change in the character of the rocks; gradually the grade of the material of the sandstones and shales of the Coal Measures becomes much finer than in the Millstone Grit series; shales become increasingly important and coal seams become more numerous, of greater thickness and more persistent over wide areas. The Elland Flagstone of the Lower Coal Measures is the most extensive of all the sandstones in the Coal Measures: it is very well developed around Halifax and Elland. On account of its thickness, up to 200 ft, and its resistant nature, it forms the summit of the higher points around Leeds, Bradford, Halifax and Huddersfield. It yields valuable building stone and is easily split into flags which were formerly widely used for pavements and roofing. In the Middle Coal Measures, sandstones decrease in importance and none is comparable in thickness and extent to the Elland Flagstone, but there are several beds which, whilst not continuous over the whole area, give rise to marked features in the landscape. For example, the Birstall Rock stands up between the valleys of the River Spen and Batley Beck; the Thornhill Rock outcrops on high ground between Morley and Thornhill; south of Wakefield, the Woolley Edge Rock forms the prominent feature of that name; farther east, the Ackworth Rock outcrops around Ackworth Moor Top. All these rocks have been extensively quarried.

Within the Coal Measures area the grand wild country, typical of the upper reaches of the Aire and Calder valleys, gradually gives place to different scenery. The moorland hills with their flat tops change to uplands of a more genial character and less elevation with fields and woodlands stretching to the summits and no longer confined to the valley bottoms. The main valleys widen out and the hills take on a more rounded appearance, and both the Aire and Calder begin to meander conspicuously across the flat bottoms of their valleys. Eastwards of its confluence with the Calder, the River Aire cuts through the Magnesian Limestone between Castleford and

Ferrybridge. The belt of limestone is some four to six miles wide and lies on the peneplaned surface of the Coal Measures: it presents to the west an escarpment which is fairly well marked in places but masked in others by a thick covering of drift. The rich arable lands on the limestone and on the Triassic and drift deposits in the Vale of York are in striking contrast to those on the relatively infertile soils of the Coal Measures. This outcrop might well define the eastward boundary of West Yorkshire but for the fact that coal mining has extended beyond it and given rise to industries at Ferrybridge and Knottingley.

On the basis of the industrial character of West Yorkshire it is possible to distinguish two major zones, a textile zone and a coal-mining zone. The boundary between the two is by no means clear cut, but it is roughly along an arcuate line extending from east Leeds, through Wakefield from whence it curves south-westwards towards Holmfirth. The textile zone to the west has a very distinct character which springs from its long-standing associations and specialization in producing woollen and worsted goods. This distinctiveness is apparent in the cultural landscape, and in all the towns within the zone more than 10% of the population are directly employed in producing textile fabrics or in making clothing. The region holds an unchallenged position at the head of this industry, accounting in 1951 for 19% of Britain's textile and clothing workers and 83% of those engaged in the woollen and worsted trades.

The textile industry is a group of related trades and the proportion of the total textile force employed in each is: woollen and worsted 52%, clothing 26%, cotton 6%, synthetic fibres 4%, carpets 3%, others 9%. Throughout a long period of development, some specialization in various districts has emerged, but no clear dividing line can be drawn between the woollen and worsted branches for the two are inextricably mixed. In general, the worsted centres are to be found north-west of a line running between Stanningley and Cleckheaton, and the woollen centres to the south-east. Both branches are fairly well represented in the Halifax–Huddersfield district. The heavy woollen district around Dewsbury, Batley, Ossett and Morley, concentrates on 'mungo' and 'shoddy'—tailors' clippings and rags, which are torn up to obtain fibres for re-use, often with the addition of new wool fibres, in making woollen yarns. Other specializations are carpets at Halifax and Heckmondwike, combing and dyeing at Bradford, and the clothing trades in Leeds. The most important cotton centres are Todmorden, Ripponden and Hebden Bridge, an overflow from the Lancashire area, but cotton weaving is carried on in Huddersfield, Halifax and Skipton. Brighouse is the main centre for silk spinning, but several towns manufacture fabrics of silk mixed with other fibres.

To the south and east of the textile zone, coal mining supplants the textile industry as the dominant occupation of the population of numerous villages and small towns. Here is the most productive part of the 'West Yorkshire' Coalfield, i.e. the administrative division of the northern extension of the Yorkshire-Nottinghamshire-Derbyshire Coalfield. In the prevailing landscape there are no marked features to distinguish this zone from its extension into South Yorkshire, but its economic and social affinities are with the textile zone rather than with the major coal-mining centres such as Barnsley or Doncaster, or with its surrounding agricultural districts. This coal-mining zone in West Yorkshire lies within the spheres of influence of Leeds and Wakefield.

An examination of the distribution of built-up areas within West Yorkshire reveals very little difference between the textile and coal-mining zones except that the former includes all the larger towns. Some degree of continuity of built-up areas is discernible in the valley of the Aire between Leeds and Keighley, and along the valleys of the Calder and Colne between Wakefield and Halifax and Huddersfield. On the western periphery there is a considerable area of open land, and within the ring of large towns, Leeds, Bradford, Halifax, Huddersfield, Dewsbury and Wakefield, the built-up areas merge one into the other only along the main arteries of communication. The textile zone is almost co-extensive with the 'West Yorkshire' conurbation. In the coal-mining zone, there is a cluster of towns and industrial villages—Castleford, Pontefract, Knottingley, Normanton—and open land is no more conspicuous than that within the relatively empty central core of the textile zone. Thus the exclusion of this coal-mining zone from the conurbation can be justified only on the basis of the form of its economic development, and in defining the boundaries of the conurbation for the purpose of the 1951 Census, great importance was attached to uniformity of economic structure within an area and less to the degree of continuity of the built-up areas. Based upon the latter criterion alone, there is little justification for the restriction of the West Yorkshire conurbation to the textile zone: it would have included the cluster of towns mentioned above. If physical, social and economic considerations are taken into account the two zones must be included within the West Yorkshire region. Moreover, coal mining formerly extended much farther west, and coal was one of the major factors influencing the development and location of the textile industries.

The part played by coal in the growth of the textile industry cannot be overstressed. Within the Coal Measures there are more than thirty seams of coal and most of the productive thicker seams are found in the Middle Coal Measures: in contrast the Lower Coal Measures contain few seams of comparable value. The Halifax Hard and Soft Beds near the base, and the Better Bed, Black Bed and Beeston seams near the top of the Lower Coal Measures, outcrop frequently along many valley sides. The Beeston seam is the most important and persistent of these seams and it outcrops in south Leeds, westwards towards Bradford and then southwards through Mirfield where it is known as the Whinmoor seam. Of the many productive seams within the Middle Coal Measures, the best known is the Warren House seam which outcrops a few miles north and west of Wakefield and southwards towards Barnsley where it is known as the Barnsley seam. Both the Beeston and Warren House seams are extensively mined today east of Wakefield, but the prosperity of West Yorkshire was built in the nineteenth century upon coal mined to the west of Leeds and Wakefield, and in the earliest period of the use of steam-power the thin coal seams of the Lower Coal Measures came to play a most important role in the wool textile industry.

In its domestic phase the making of cloth was a widespread occupation, with spinning and weaving especially important in many hillside or upland hamlets, and fulling and dyeing in premises sited in the valley bottoms. Cloth-making was relatively more important in the gritstone country within the basins of the Calder and the Colne, but the Coal Measure country to the west of Wakefield had a share in the industry. If any regional differentiation can be made it is that the gritstone country had little but wool and cloth-making as a basis for the livelihood of its inhabitants. The resources of the Coal Measure district were more varied and environmental conditions generally were more favourable for the growth of market towns with their varied trades: Leeds, Wakefield, Gomersal, Almondbury, Huddersfield and Halifax, all possessed cloth and wool markets.

The expansion of the industry, and with it a transformation in the settlement pattern, came about with the introduction of power-driven machinery. From about 1775 scribbling mills, and from about 1830 spinning mills, housing new carding machines and spinning mules respectively, began to be established near streams to form the nucleus of many of the present valley-site textile villages and towns. The application of power to weaving and to wool combing followed shortly afterwards, and from about the middle of the nineteenth century the fate of many upland weaving hamlets and villages rested upon the decision to use, or not to use, the power loom. By this time, steam had become more important than water power and therefore coal was of critical importance and the problem of supply was one which could not always be solved. In the Millstone Grit country local poor thin coal seams were mined, but they were soon exhausted and the salvation of many high-lying weaving villages depended upon the facilities for the transport of coal from elsewhere. Thus their fate depended on their proximity to canals which were confined to the main valleys, or to railways: many branch lines in the Keighley–Halifax–Huddersfield district were constructed after 1850 to meet the demands of mill-owners for coal. Along the western fringe of the Lower Coal Measures coal supply was much easier, and there was marked industrial development as manufacturers selected new sites for mills on or very near to coal pits. Development of power-combing was most conspicuous in the Bradford area and many worsted mills were established in the Aire valley between Leeds and Keighley. Farther eastwards on the Middle Coal Measures the supply of coal was even easier, and many new coal pits met the needs of manufacturers eager to use steam-power, and the heavy woollen district witnessed remarkable growth. It can be said that wherever the textile industry was already established on the coalfield it reaped the benefits of its advantageous position and its expansion was relatively more rapid than in the gritstone area, though it must be emphasized that another factor, a piped water supply which became available almost everywhere, aided its growth.

As might be expected the demand for coal was heavy and as the thinner coal-seams of the Lower Coal Measures were soon exhausted, the thicker and more numerous seams of the Middle Coal Measures eventually came to supply all the power needs of the textile industry. In the part of the coalfield within the textile zone, peak production was achieved in the late nineteenth century. Today very little coal is produced in the zone and only the numerous burnt-out waste heaps bear witness to the former importance of coal mining. Much of the present production, and most of the reserves of coal, lie eastward and the maintenance of the quantity of West Yorkshire coal depends upon the continued development of mining in the Pontefract–Micklefield district. Reorganization of the industry based upon 'colliery units' has taken place in post-war years, but only one new shaft-sinking is in progress: the new colliery is at Kellingley about two miles east of Knottingley. Nearly all the coal produced in the coal-mining zone moves by rail, road and canal into the textile zone, and even beyond into Lancashire. Though most of the coal within the textile zone has been exhausted, there still remains abundant resources of fire-clay, an alkali-free refractory clay which usually forms the 'seat earth' of several coal seams, mainly in the Lower Coal Measures. The extraction of fire-clay, by underground mining or from open-cast coal-workings, is localized along narrow outcrops from Leeds to Elland, and in the Denby Dale district, and the clay is used for the local manufacture of salt-glazed goods, drainage and sanitary ware, and refractory goods.

It is difficult to visualize the rapid industrial development of West Yorkshire throughout the nineteenth century without an iron industry which was based upon local coal and ironstones. Within the Coal Measures thin bands of nodular ironstones occur fairly frequently, but only the more persistent bands were exploited commercially. The most important of these is the Black Bed ironstone which overlies the Black Bed coal so closely that coal and ironstone could be worked from the same gallery. About 70 ft to 100 ft below the Black Bed coal is the Better Bed coal, characterized by a remarkably low sulphur content which made it most suitable for iron-working and helped to make Yorkshire wrought-iron one of the best in the country. The outcrop of this Black Bed ironstone near the northern and western limits of the coalfield is shown on figs. 45 and 46. The other important source of iron was the Tankersley ironstone, a shelly bed some 8 to 12 in. thick, rich in calcium carbonate and therefore self-fluxing.

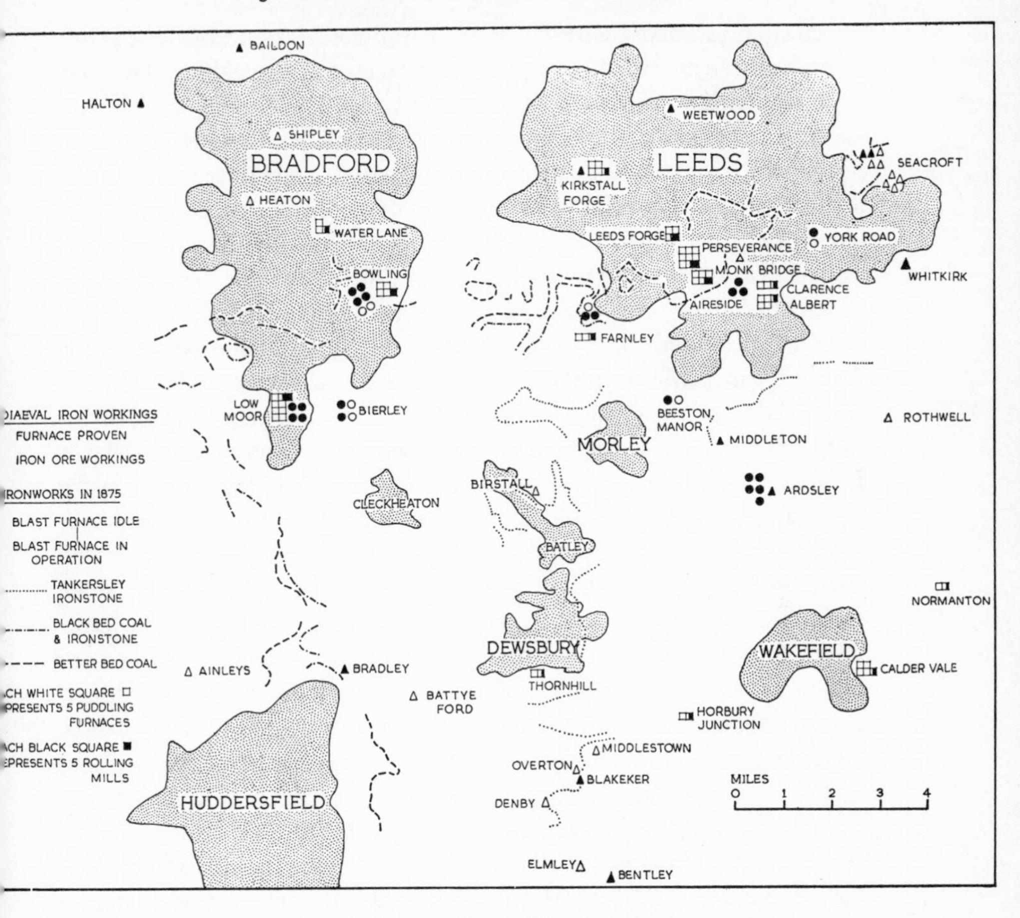

Fig. 46. Iron working in West Yorkshire.

It is a matter of conjecture when the iron industry began in West Yorkshire, but by the fifteenth century all the above-mentioned outcrops of ironstone appear to have been worked. The Tankersley ironstone appears to have been the more important and gave rise to smelting at centres such as Seacroft, Meanwood, Kirkstall (all within the present boundaries of Leeds), East Ardsley, Colne Bridge, Flockton and Emley. More than two-thirds of the recorded workings were monastic, but all these early centres declined in importance with the growing shortage of local charcoal and the industry did not revive until the end of the eighteenth century when the use of coke as fuel spread to the area. The new methods in iron manufacture put a premium on ironstones near which a supply of coal suitable for coking was available and so there was a reassessment of site values. The first coke ironworks were started at Seacroft (Leeds) in 1776, and in the succeeding decade others began at Birkenshaw, Bowling and Low Moor in the Bradford district, and at Shelf near Halifax. The Bradford district was specially favoured in this context: coking coal and ironstone outcropped in close proximity, and as the coal was being mined extensively it was an easy operation to take out the ironstone. Fire-clay for refractories was at hand, and the nearby Millstone Grit country provided heat-resisting sandstones and grits suitable for furnace sands. Carboniferous limestone for a flux could easily be imported from the Craven District after the completion in 1775 of the Shipley to Skipton section of the Leeds–Liverpool Canal with a branch to Bradford. It is not surprising, therefore, that the industry made rapid progress in this district. In the Leeds district the Seacroft works were short-lived and no further effort was made to use the Black Bed ironstone until the middle of the nineteenth century. Since the demand for iron was great, it is not easy to explain the later development. One possible reason is that in the Leeds district very little coal was taken from the Black and Better Bed seams: the much thicker Beeston and Middleton seams supplied all the local needs. Eventually new shafts were sunk to the ironstone and smelting began at Farnley, Beeston and Hunslet. Five new furnaces were opened at East Ardsley in 1868 in an effort to increase pig-iron production to meet the growing needs of wrought-iron manufacture which had spread outside the Leeds–Bradford–Halifax districts to places such as Thornhill, Horbury, Wakefield and Normanton. Nevertheless, West Yorkshire had a considerable deficiency in pig iron and increasing quantities began to be imported from outside. Iron smelting reached its zenith about 1875 when it began to decline in face of growing competition from Lincolnshire and Northamptonshire pig iron at a time when the more accessible local ironstones were becoming exhausted: the making of wrought iron also declined as the metal was largely superseded by steel. Steel was made in several centres but never attained real importance, and by 1932 the decline of West Yorkshire as an iron and steel area was virtually complete. The last blast-furnace had been dismantled and only two wrought iron and one steelworks remained.

The iron industry and the presence of coal inevitably gave rise to engineering and associated iron founding which spread beyond the nineteenth-century iron centres into the whole of West Yorkshire. In turn the metal trades benefited the textile trade. In the Middle Ages, Colne Bridge became important in supplying bar iron to the makers of cloth shears and of iron wire who had made their appearance in Halifax and Brighouse: the iron wire was used in making cards for the hand carding and combing of wool. Centuries later came the demand for textile machinery, and its manufacture began in Leeds, Keighley and Bradford, and spread to most of the textile towns. In turn, this gave rise to miscellaneous engineering industries in Leeds, Wakefield, Huddersfield and Halifax. Significantly this growth helped to diversify the economy in parts of the coalfield and left the textile industry virtually unchallenged in the gritstone country.

Any assessment of the factors which influenced the rise and development of the textile industry in West Yorkshire would be incomplete without reference to the supply of water, and fortunately suitable water, in sufficient quantity, has always been widely available. The distribution of rainfall is closely related to the configuration of the land, and the higher moorlands in the west, coming under the direct influence of the prevailing winds from the west or south-west, have an annual rainfall often exceeding 50 in. (fig. 47). Lower down the valleys, in the lee of the highland, the rainfall decreases steadily to about 25 in. in the east of the region. The gathering grounds of the main tributaries of the Rivers Calder and Colne and the northward-flowing tributaries of the River Aire lie within the Millstone Grit country, which, lacking minerals and with a covering of sour soil and poor vegetation, is of little direct value to man except as a grazing ground. However, in one respect, this gritstone country has been a valuable asset in that it provided soft lime-free water that has influenced the location and development of the textile industries. The heavy and constant rainfall, the nature of the shales and hard impervious gritstones that give abundant surface run-off, and the relief, all combine to make excellent gathering grounds of streams that serve the textile zone. Furthermore the existence of a thick accumulation of spongy water-holding peat which covers most of the moorland catchment areas results in a regular water yield and helps to keep surface evaporation very low. Thus there are countless streams and rivulets which rarely dry up. In providing lime-free water, the Coal Measures present similar characteristics to the gritstones

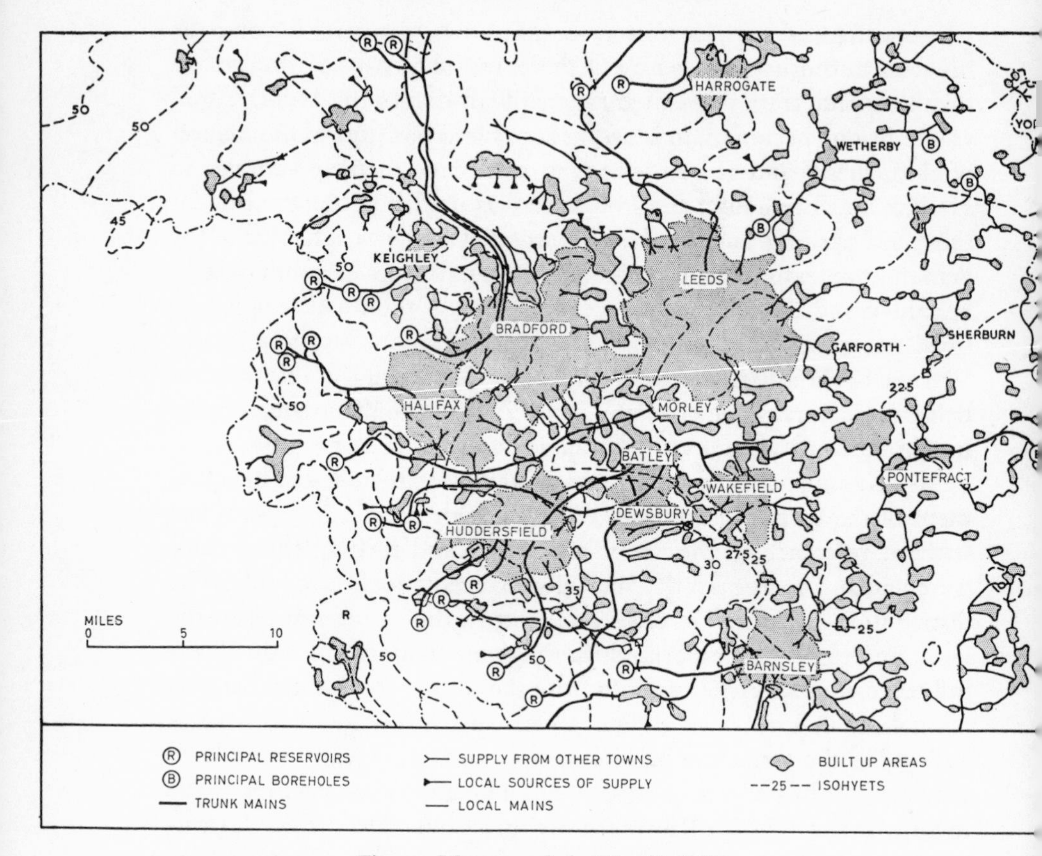

Fig. 47. Water supply in West Yorkshire.

except that the water-holding peat is absent and the rainfall is lower.

In the early days of making woollen cloth, water was essential for many processes such as washing and scouring of raw wool, the fulling or felting of woollen fabrics, and the dyeing of yarn or cloth. Domestic workers in their hillside hamlets or villages had water literally on their doorsteps, and the larger streams satisfied the requirements of the fulling mill and dyehouse. Later, when machines were first introduced, many streams were easily harnessed to provide the necessary water-power and this enhanced the value of the Pennine sites as compared with those parts of the country with less varied relief. Then with the development of large-scale water storage and a modern system of supply, it is not surprising that the majority of West Yorkshire towns looked to the gritstone area in the west to find their water-supplies. That system developed without any co-ordinated plan by the local authorities concerned and the resulting pattern of reservoirs and pipe-lines is very complex over the textile zone. Only Ossett, of all the 'woollen' towns, supplies most of its area from a local borehole. Bradford has found it necessary to supplement its western catchments from gathering grounds within the basins of the Wharfe and Nidd: Leeds takes its water from the Washburn tributary of the Wharfe, and from the Ure valley in the North Riding of Yorkshire. Within the coal-mining zone no town or village draws water directly from gritstone catchment areas: some villages are supplied by the town of Wakefield, but a large area around Pontefract is supplied from boreholes into the Triassic sandstone formations. The Pennine reservoirs, in addition to serving the great needs of industry and population within the greater part of West Yorkshire, also supply much of the water used in the canal network. Many of the canals have lost their importance as a means of transport but, with the rivers, they are a source of considerable supply of water to many industries and power stations on their banks. Many textile firms can now use canal and river water for raising steam only because the increasing pollution of such water has rendered it unsuitable for most textile purposes.

It will be obvious from the preceding paragraphs that the economic foundations of West Yorkshire were laid in the nineteenth century when population increased rapidly and building development sprawled haphazardly over the countryside. Industries forged ahead while social needs were neglected, so that the area now presents many problems to those engaged in planning its future. Fortunately there are no insurmountable difficulties to be faced in planning the layout of the countryside. First, little provision has to be made for expanding population; secondly, while there has been a tendency for nearby towns to merge with each other, the process has not reached the stage that is apparent, for example, in the West Midlands conurbation; thirdly, except in Bradford and Leeds, congestion of industry and residential buildings is localized. The central core of the conurbation is ringed by its six largest towns and although ribbon development has taken place along the main roads more than half the total area is made up of extensive tracts of semi-rural country, for example, between Queensbury and Mirfield, between Spenborough and Morley, and between Morley–Batley and Wakefield–Rothwell (fig. 44). Thus there is little need to open up the urban sprawl, but the preservation of all the open areas which serve to separate the larger towns is essential if the identity and character of the urban areas are to be retained. Unfortunately many parts of this core area are more or less derelict as a result of active and abandoned mineral workings and associated railway lines and industrial sites, and since these relics are often intermingled with housing their reclamation will not be easy. A beginning has been made to hide or remove some of the spoil heaps, and some abandoned industrial sites have been put to new use, as for example, at Tingley where the largest gas- and coke-producing plant in the area now occupies the site of an abandoned colliery. In the coal-mining zone, the removal of spoil heaps and the reclamation of areas flooded owing to subsidence are problems that await solution.

The West Yorkshire conurbation is contained within a 'green belt' which stretches, to the north, along the Wharfe valley from Ilkley to Wetherby and includes Rombalds Moor and Otley Chevin: to the east it follows the Magnesian Limestone roughly along the line of the Great North Road: to the south it embraces a belt of rural countryside between Wakefield and Barnsley: to the west the relief has confined building to ribbon-like strips along the valleys and the high unspoilt moorland areas form a link between the Peak District and the Yorkshire Dales. Recent encroachments into this encircling 'green belt' have not been marked except around Leeds and Bradford where some most attractive localities have encouraged residential building. Northwards of Bradford the built-up area has spread to Keighley in one direction, and in another towards Baildon and Guiseley and into the valley of the Wharfe. North-east of Leeds, building has also spread on to the southern slopes of the valley of the Wharfe: the City of Leeds boundary has recently been extended at the expense of Tadcaster Rural District to provide space for a relatively insignificant overspill envisaged during the next decade. The overspill from all the other towns is expected to be very small.

The provision of better communications within West Yorkshire will continue to attract attention. The road network is adequate in spite of diverse physical conditions but the condition of many roads is such that they are inadequate to carry the ever-increasing dense traffic. The conurbation is the origin and the destination of much traffic, and its geographical position results in much through traffic also, especially in an east–west direction. In recent years there has been a considerable increase in the westward movement of coal from the collieries to the textile towns and this has added to the general congestion at bottlenecks. The roads into industrial Lancashire must find a way through the Pennines and the main road to Manchester, via Huddersfield and Standedge, is very much exposed and hence traffic delays owing to snow, ice and fog, are not infrequent. An alternative route is required. The shortest route eastward from Leeds to Hull is impeded by a wooden toll-bridge across the Ouse at Selby: a modern bridge must be planned. Easy and rapid access to the south of England now awaits the construction of the 'Yorkshire Motorway', while improvements to the main outlets to the coastal resorts, via Skipton, York or Selby, are necessary to cater for heavy seasonal traffic. In contrast, the complex railway system is more than sufficient for present needs. Many routes are duplicated as a result of uncoordinated construction by competing railway companies, and with the unification of control and declining rail traffic several lines have become redundant and now, with numerous disused canals, scar the landscape. Little use is made of any of the canals west of Leeds and Wakefield—the former connexions with Lancashire via the Calder and Colne valleys are severed—but eastwards canals are much used to carry coal to Goole and to large power-stations at Ferrybridge, Wakefield and Leeds, and petroleum, cement and other bulky commodities to depots in Leeds.

The economic prosperity of West Yorkshire was built up in the nineteenth century and plans for the future must ensure its survival as a major industrial area. Unfortunately a detailed analysis of its industrial character reveals its continuing dependence upon old and generally static industries such as textiles, clothing, coal mining and the metal trades. Apart from these, manufacturing industry is relatively poorly developed: it is the older branches which predominate to the exclusion of newer branches of industry. Older branches of mechanical engineering and the making of textile machinery predominate in the engineering trades, whilst its chemical industry is represented principally by coal carbonization and the manufacture of dye-stuffs. The area has not shared in rapid expansion of modern manufacturing industry such as plastics, electronic equipment and goods connected with atomic energy, motor-cars or aircraft, and it accounts for only a small proportion of the national production of such commodities as electrical equipment, rubber, furniture or household goods, artificial fibres and light chemicals.

After the First World War, West Yorkshire felt the effect of economic depression, but in the thirties it did not suffer unduly from the exceptionally severe unemployment experienced in some parts of the country. On the other hand, it did not enjoy the prosperity of the Midlands and south-eastern England. It stood in fact in an intermediate position between prosperity and depression and that feature has persisted until recent times. Since the Second World War the demand for its coal, metal products, wool 'tops' and yarns, woollen and worsted goods has remained fairly steady and the result has been full employment except in one or two localities. Nevertheless, the area has failed to keep pace with the national rate of increase in employment and it falls behind the country as a whole in the field of recent industrial development. It could well be described as a 'museum of industry' and the view has been expressed that it is in danger of becoming an industrial backwater or even a depressed area of the future. This economic situation is reflected in the relatively insignificant population changes in recent years. Between 1931 and 1951 the total population increased by little more than 2%; since 1881 the rate of increase has been smaller than that of the country as a whole.

The national 'Distribution of Industry' policy must take part of the blame for this lack of development of new manufacturing industries since 1946. Not only have new industries been excluded, but several established industrial firms wishing to expand their activities have had to look elsewhere for industrial sites; several textile firms, for example, have factories in North-East England. This failure of the area to secure a proportional share of new industries, in particular those which are expanding most rapidly today, has meant that almost nothing has been done to alleviate the problem of excessive specialization in the industrial structure. In this connexion some distinction must be drawn between the larger cities and the rest of the area. The former with the greatest concentrations of population and offering the widest range of employment have been more fortunate in securing the major share of expanding service industries. In contrast, in many small textile or mining towns or villages, there is a very strong dependence upon one industry, in some cases upon one mill or one colliery. However, in the textile zone, the various branches of the industry offer a wide range of employment, and, if coupled with various engineering and sometimes chemical industries, the dependence upon one basic industry is not too serious. A striking feature of all the textile towns is the high female employment rate, and in the early post-war years the demand outgrew the supply of women who were willing and able to work in the mills. This factor alone partly explains why several firms moved out of West Yorkshire, and it is the reason for the daily influx of women into the textile zone: by motor-coach they travel from districts where coal mining offers the only type of employment. Some redistribution of industries within West Yorkshire would help to reduce specialization within the coal-mining zone, but only the introduction of new industries can alleviate the problem of specialization in the textile zone and at the same time stimulate a more vigorous industrial expansion in the whole area. Unfortunately, coal mining with its inevitable problems of subsidence and unsightly landscape is not conducive to the attraction of other industries: the textile zone is better in this respect though it may not be apparent to industrialists unfamiliar with the area. Local authorities, however, have now taken the initial steps of selecting sites for industrial development. Within the West Riding County Council area it is hoped to establish centres for new industries, other than clothing or textiles, at selected nodal points in the mining area, and to attract new industries to places such as Skipton, Silsden and Todmorden, and also to the heavy woollen district where local pockets of unemployment have appeared. The aspirations and needs of the different localities are many, varied and sometimes conflicting. The success of planning the countryside and reshaping the industrial character of West Yorkshire is far from assured and solutions might be easier if there were hopes of a reorganization of local government towards an administrative framework more compatible with the human geography of the area.

Selected Bibliography

County Council of the West Riding of Yorkshire. *County Development Plan, Report of the Survey*, 1951, *and First Review*, 1960.

W. B. Crump. 'The Wool Textile Industry of the Pennines in its Physical Setting.' *The Journal of the Textile Institute*, **26**, 1935.

T. W. Freeman. *The Conurbations of Great Britain*, Chapter 6, 'West Yorkshire'. Manchester, 1959.

H. Heaton. *The Yorkshire Woollen and Worsted Industries*. Oxford, 1920.

Wilfred Smith. *An Economic Geography of Great Britain*, Chapter IX, 'Wool Textile Manufacture'. London, 1949.

H. C. Versey. *Geology and Scenery of the Countryside round Leeds and Bradford*. London, 1948.

CHAPTER 20

RURAL YORKSHIRE

G. DE BOER

The somewhat tenuous geographical unity of this area rests mainly on two grounds. First its several strongly and contrastingly characterized regions, the western moors and dales of the Pennines, the eastern uplands and the plains encircled by them, are articulated both by the drainage and by the communications to the medial lowland, the Vale of York. The drainage is gathered thither by river-capture or glacial diversion; the communications likewise converge there, channelled along the dales from the west by high moorland ridges, swung inland from the east by the Humber on the south and by the curve of the coast westwards from Whitby to Teesmouth on the north.

Secondly, its north–south extent corresponds with the gap between the coalfields of Durham and south Yorkshire. A predominantly rural landscape therefore prevails allowing of an intimacy between the human and physical aspects of landscape that focuses the more sharply the distinctiveness of each region. The directness of geographical relationships has not been blurred over large areas by the momentum of manufacturing industry or by urban sprawl. Farms, villages and market-towns are the typical settlements. York, at the focus, is the only large town that really belongs to the area. The other two large urban tracts, Teesside[1] and Hull, both situated on the edge of the area, are linked in large measure to the industrial activities of adjacent areas; they are the seaward terminals of the west–east bundles of communications that, stretching from coalfield to estuary across the open ends of the Vale of York, provide a regional boundary where a clearly definitive relief feature is most lacking (fig. 53).

The extensions of these routes farther westwards into the trans-Pennine corridors of Stainmore and the Aire Gap embrace that section of high Pennine plateau named, because of its structural stability, the Askrigg Block (fig. 48). Separated by the Craven and Dent

[1] This is so clearly linked to Durham that it is described in chapter 23 rather than here.

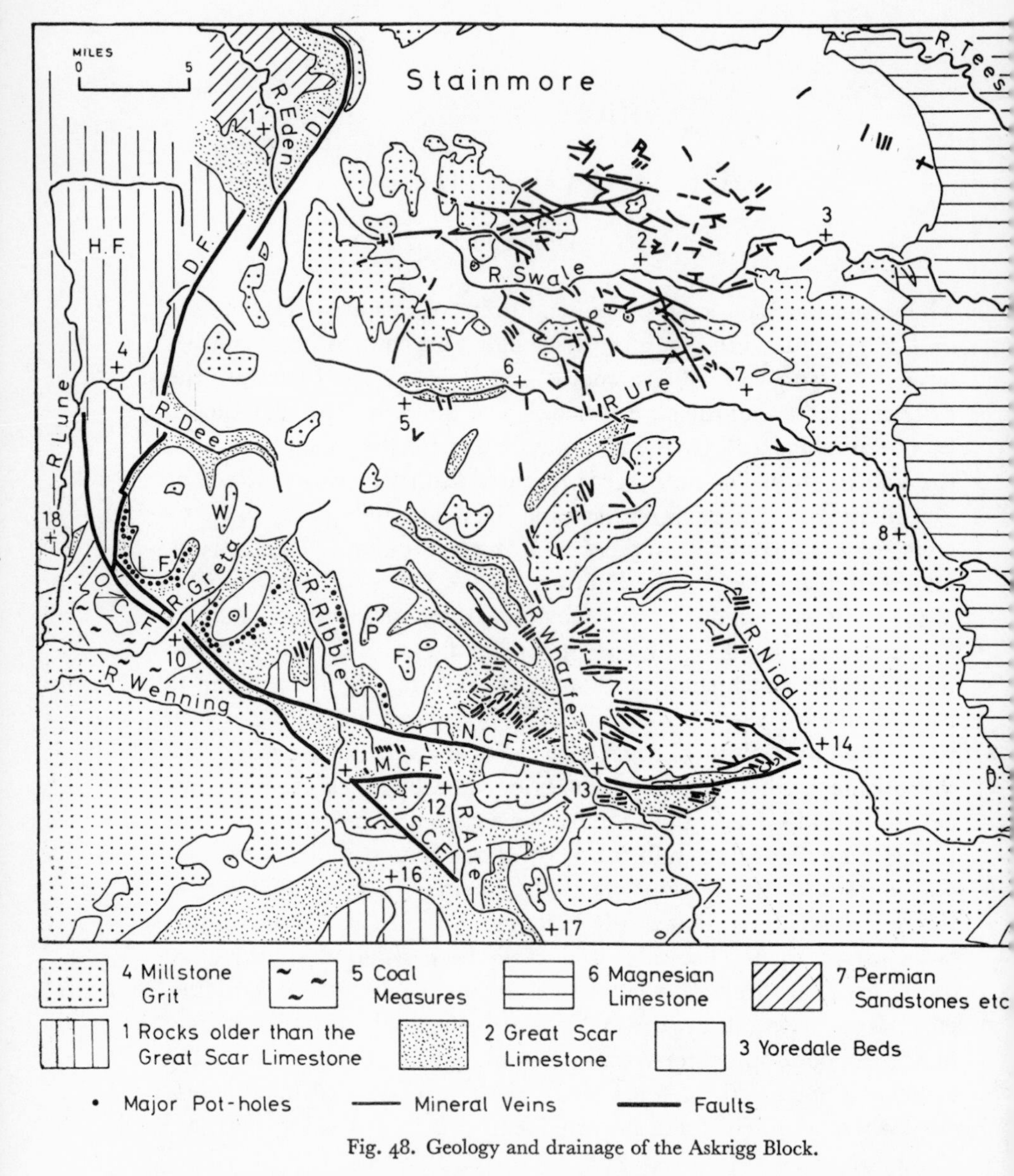

Fig. 48. Geology and drainage of the Askrigg Block.

(In the legend, the geological formations are numbered from the oldest to the youngest.)

1. Kirkby Stephen. 2. Reeth. 3. Richmond. 4. Sedbergh. 5. Hawes. 6. Askrigg. 7. Leyburn. 8. Masham· 9. Ripon. 10. Ingleton. 11. Settle. 12. Malham. 13. Grassington. 14. Pateley Bridge. 15. Harrogate. 16. Hellifield. 17. Skipton. 18. Kirkby Lonsdale. H.F. Howgill Fells. L.F. Leck Fell. W. Whernside. I. Ingleborough. P. Penyghent. F. Fountains Fell. D.L. Dent Line. D.F. Dent Fault. S.C.F. South Craven Fault. M.C.F. Middle Craven Fault. N.C.F. North Craven Fault. O.C.F. Outer Craven Fault.

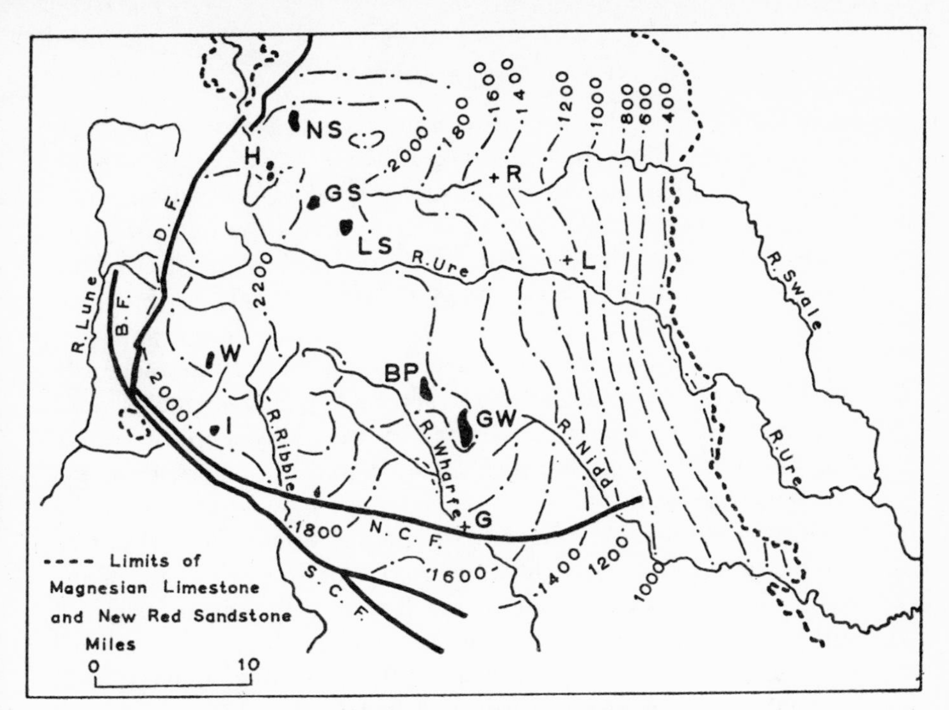

Fig. 49. Generalized contours of the Askrigg Block.

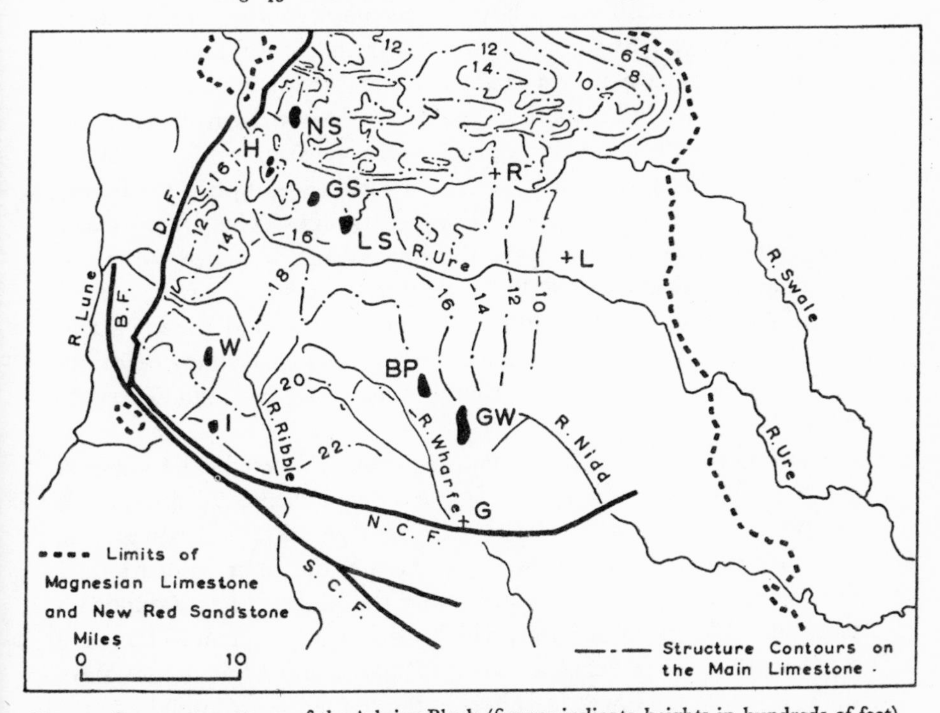

Fig. 50. Structure contours of the Askrigg Block (figures indicate heights in hundreds of feet).

G. Grassington. L. Leyburn. R. Reeth. N.S. Nine Standards Rigg. H. High Seat. G.S. Great Shunner Fell. L.S. Lovely Seat. B.P. Buckden Pike. W. Whernside. I. Ingleborough. G.W. Great Whernside. D.F. Dent Fault. B.F. Barbon Fault. N.C.F. North Craven Fault. S.C.R. South Craven Fault.

faults from the much more structurally complicated areas to south and west, the Block tilts slightly eastwards. It owes its tectonic stability to a foundation massif of tightly folded Pre-Cambrian, Ordovician and Silurian slates and grits, across the upturned and eroded edges of which are laid evenly the six hundred feet of the Great Scar Limestone surmounted by the Yoredale Series which repeat up to eleven times the upward sequence limestone, shale, sandstone, with sometimes coal, and produce the very characteristic persistence of limestone scar and structural bench along the terraced hillsides. The Yoredales are covered by the Millstone Grit; where this passes unconformably under the Magnesian Limestone is the regional boundary to the east.

Generalized contours suggest two main divisions: first an eastern ramp where ridge crests climbing westwards from the base of the Magnesian Limestone at about 400 ft may represent an exhumed sub-Permian surface, second a western plateau where flat-topped fell and ridge summits rising to over 2,000 ft are possibly remnants of an uplifted and dissected late Tertiary peneplain (fig. 49). Morphological development appears to have taken place in three main stages. First, the higher fells were roughed out from the uplifted peneplain by the carving of a platform at about 1,300 ft that shoulders the dales scooped in it in the second stage. Near the southern edge of the Block the rivers tumble through narrow gorges sunk into the floors of the dales during the third stage. Glaciation has applied the finishing touches. Several dales are U-shaped troughs. Their floors carry processions of drumlins and successions of breached terminal moraines and drained lake flats. Boulder clay spreads in a patchy mantle over the fell-sides and partly masks the solid rocks. Many waterfalls or 'forces' have resulted from glacial diversion steepening stream-courses across the alternately weak and resistant Yoredale rocks.

The rocks are folded into a faulted anticlinal belt in the north, a gentle half dome in the south (fig. 50). The near radial drainage is partly related to the generalized relief, partly to these structures. The latter are not directly reflected in the relief, but their truncation by the two erosion surfaces, the ramp and the plateau, has produced a particular pattern of outcrops which leads to a significant change of scenery as the dip of the rocks down the flanks of the half dome causes the features of the lower members of the sequence, prominent in the south, to be replaced to the north and east by those of the higher beds. These changes can be traced outwards from Ingleborough which forms a kind of centre-piece to the pattern; it is the only major hill completely surrounded by a shelf of Great Scar Limestone, here largely coincident with the 1,300 ft platform. Round it Leck Fell, Whernside, Penyghent and Fountains Fell stand linked into a half-circle by the edge of the nearby continuous cover of Yoredales beyond which the Great Scar Limestone is more and more completely concealed. Only within the half-circle has erosion exposed the foundation rocks in the valley floors so that the full thickness of the Great Scar Limestone, cut right through, forms the flanks of the dales. Its strongly marked jointing is emphasized in the vertical lines of these 500 ft high precipitously terraced light grey walls; its equally distinct bedding in the horizontal pavements of the surface of the 1,300 ft platform above. From this as from a plinth rises the central mound of Ingleborough, its slopes of Yoredale rocks characteristically terraced by the lesser scars of the repeating limestones. Only within the half-circle are the celebrated major karst features where considerable-sized streams, on running off the Yoredales on to the limestone, plunge into vertical shafts up to 360 ft deep to emerge from cave passages reaching back into the base of the great scars just above the foundation rocks. Outside the half-circle, the Great Scar Limestone is carried progressively lower by the dip, and outcrops are confined more and more within the dales. It forms the floors and lower part of the valley walls of Dentdale, Littondale, and Wharfedale. Here the bigger side streams have usually slashed steep-walled gorges right across the more restricted limestone shelf; the smaller are swallowed by narrower clefts. The main streams sometimes sink into their limestone beds and flow underground. Still farther from the centre, in Wensleydale where the Great Scar Limestone only appears in the floor and in Swaledale where it hardly appears at all, all the hillsides below the Millstone Grit summits are shaped into Yoredale terraces. The thicker limestones, particularly the Main Limestone near the top of the Yoredales, reproduce on a smaller scale the swallow holes and gorges of the Great Scar Limestone. The dip northwards causes the Grit caps of the fells to be thicker and wider in the north-west; the dip eastwards and the descent of the Grit base southwards together explain the continuous spread of gritstone country in the south-east, which joins up with the crag-fringed gritstone moorlands of Rylstone Fell, Barden Moor and the Forest of Knaresborough, south of the North Craven Fault. Nidderdale lies wholly within the gritstone country. Near the head, the dale cuts right through the Grit and into a thick Yoredale limestone; here the Nidd flows underground. These gritstone moors are the only part of the area used for water catchment on a considerable scale.

The mantle of vegetation colours the regional differences of structure and relief. The acid soils of the sandstones, grits and shales carry heather moors, the higher wetter western areas cotton-grass mosses over peat beds. Birch scrub clings to the steep sides of the gills. These sombrely tinted plants contrast sharply with the green turf of the limestone areas with which they alternate in strips over the Yoredale terraces, wherever the differences are not masked by drift. Ash and hazel wood spreads in narrow bands along the lower scars and clothes the steeper limestone slopes and gills. Draining and clearing of the lake flats and flood-plains of the valley floors have produced the richest grassland of the upper dales.

This environment has always closely conditioned human activity and settlement. Stock farming and lead mining are the persistent themes. The conditioning becomes closer and the landscape changes accordingly as height increases up the ramp from east to west. Arable fields, bounded by hedgerows as well as walls, farms and villages linked by a continuous net of side roads spread over the divides between the dales about as far west as a line joining Richmond, Masham, and Otley—markets serving particularly this eastern fringe. Animal rearing is the main activity; easier conditions and ampler winter feed from fodder crops give scope for dairying and fattening. Pastures enclosed by dry stone walls are continuous across the ridges for a little farther west. Beyond, where tongues of moorland on the continuously high ridges of the western plateau separate the pasture lands of each dale is found the most distinctive landscape.

The dale floors at between 500 ft and 800 ft, are beyond the practicable limits of crop-growing; here farmers, with only meadow and grazing, are mainly concerned with the breeding and rearing of sheep particularly the hardy Swaledale breed for lowland fattening, and of young dairy cows as replacements for the herds of the urban milk suppliers. Milk production has become very important since the advent of motor transport; cheese is now made in a factory near Hawes instead of domestically. Most farms have meadow on the floor and lower slopes of the dale, enclosed pastures on the higher slopes, and moorland higher still. The meadow grows the only locally produced winter feed, and hay production governs the annual rhythm of stock movement between the higher and lower pasturage. Because of elevation it is usually April before the meadow grass begins to grow. The meadows have first to feed until mid-May the ewes brought here for lambing. The sheep then go up on to the moor, and the grass is left to grow for mowing. The late start, together with the dryness of April, May and June, delay haymaking until the wet period of July–August; the need therefore to get the hay in quickly explains the highly characteristic barns ('laithes') dotted about the floors and lower slopes of the dales, usually at least one to every three or four meadows. The hay is stored and fed to the cattle that winter here, and go up to the higher pastures in mid-May. As soon as the meadows recover from mowing, the milk cows graze the aftermath, returning to the barns in mid-November; the heifers in calf graze what is left and go under cover at the beginning of December. Most of the sheep spend all the year on the moor or in the high pastures; hay is carried up to them in February and March to supplement the grazing then at its poorest. The pressure on winter feed is eased by the annual reduction of the animal population by the autumn sales of livestock and by the removal of some sheep for wintering outside the area.

Along the bigger dales on the east and south from Swaledale round to Ribblesdale, small villages on bluffs and terraces above the floodplains, and on alluvial fans near tributary confluences, punctuate the valley floors every three or four miles. In the western dales farms are scattered singly over the lower slopes from Chapel-le-Dale round to the Rawthey valley.

A ring of small market towns—Barnard Castle, Reeth, Leyburn, Pateley Bridge, Grassington, Settle, Ingleton, Kirkby Lonsdale, Sedbergh, Kirkby Stephen—stands just outside the plateau on the west and south-west near the mouths of the dales, on the eastern side as far up the dale as permits of reasonable lateral routes across the ridges of the ramp. Hawes is the only market town within the ring, and stands where the radial streams of south and west converge headward to focus routes upon Wensleydale, the only through valley. The one moderately large town, Harrogate (56,332), lies in the more structurally complicated area of the Forest of Knaresborough. south of the Craven Fault, and therefore off the Askrigg Block. Spa and resort, dormitory and conference town, it is located where a great variety of mineral springs issue, distinct and unmingled, from sandstones interleaved between shales of Yoredale age nosing up through the gritstone in the core of a sharp anticline.

There is little industrial activity. There is quarrying of limestone and of the foundation rocks for roadstone in a number of areas particularly near railways. There are some scattered mills; cotton at Linton, and paper at Settle, wool at Sedbergh, Otley, Burley-in-Wharfedale and Addingham, hemp at Pateley Bridge and Summerbridge. The most important industry, lead mining and smelting, after nearly two millennia of activity is virtually extinct save for some reworking of old dumps for fluorspar or barytes. Production rose to a peak in the 1850's and 1860's when 4,000 workers were employed and then declined rapidly. The population of the former mining areas is now a third or a quarter of that of a century ago. There were three main centres of mining: Greenhow Hill near Pateley Bridge, the tableland behind Grassington, pimpled with mine dumps and pricked by scores of shafts, and Swaledale. Here, particularly at the places where the main mineral veins cross the tributary valleys Gunnerside Gill, Hard Level Gill and Arkengarthdale, the hillsides, perforated by long levels, are veritable 'badlands' of hush gutters—steep, rocky ravines produced by the repeated releasing of torrents of water along the outcrop of the lead vein. The industry, tied by geological occurrence to difficult locations, and flourishing before the advent of mechanical transport, made use of locally available water-power and building materials. What remains today—stone-arched levels, reservoirs, water-courses, and ruined stone-built smelt-mills, peat sheds, and dressing floors—is, therefore, as intimately related to the landscape as the features of the farming economy.

The beauty of this landscape of fell and dale, beck and force, stone buildings and stone-walled fields, has been recognized in the creation of the Yorkshire Dales National Park, and the accommodation of visitors makes an increasingly important contribution to local incomes.

The Jurassic upland of the north-east (fig. 51) is in some respects a half-scale version of the Pennine area; its landscapes are as strongly characterized and respond as closely to the varied character of the pile of sedimentary rocks out of which the area is fashioned; vegeta-

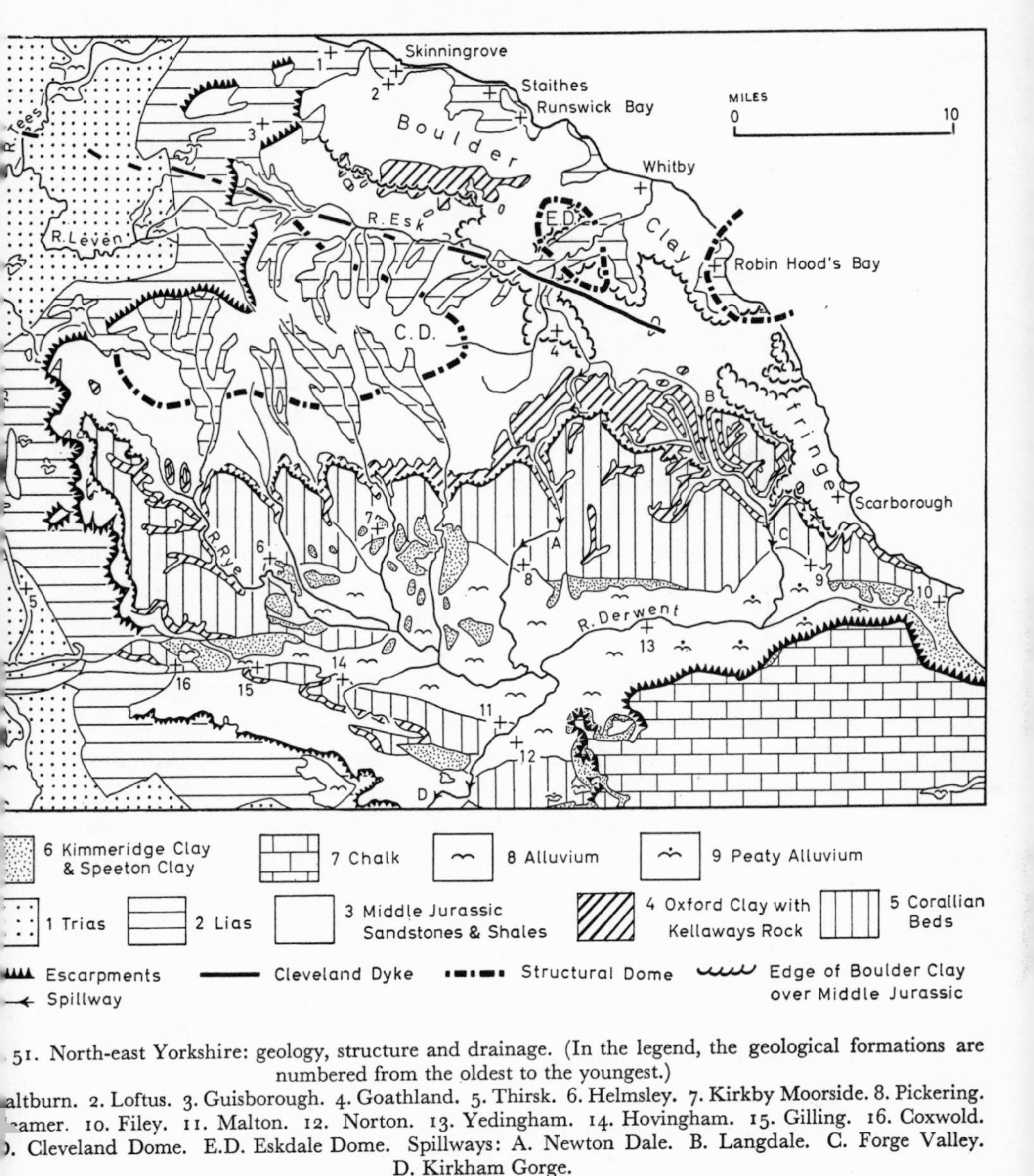

51. North-east Yorkshire: geology, structure and drainage. (In the legend, the geological formations are numbered from the oldest to the youngest.)
altburn. 2. Loftus. 3. Guisborough. 4. Goathland. 5. Thirsk. 6. Helmsley. 7. Kirkby Moorside. 8. Pickering. -amer. 10. Filey. 11. Malton. 12. Norton. 13. Yedingham. 14. Hovingham. 15. Gilling. 16. Coxwold.). Cleveland Dome. E.D. Eskdale Dome. Spillways: A. Newton Dale. B. Langdale. C. Forge Valley. D. Kirkham Gorge.

tion colours the contrasts between them, sometimes in the absence as well as in the presence of related relief features; indeed, the regularity with which moorlands and conifers have become associated with sandy rocks, and grassland, deciduous trees, or cultivation with clays and limestones is part of the distinctiveness of the area.

The major part of the upland, an eastward-tilted scarp-edged plateau, consists of a thick sequence of Middle Jurassic sandstones, grits, and shales with thin coal seams, covering Lower Jurassic Liassic shales, the whole folded into a series of domes and basins. Massive sandstones near the base of the Middle Jurassic give a hard, often craggy, edge that throughout the moorlands forms the characteristic steep, sometimes rocky, rim of the bounding scarps and interior dales whence the name Cleveland (=cliffland) derives. Above the edge, the sandy Middle Jurassic rocks are modelled into the sombre-hued, peat- and heather-clad flatly convex ridges well described by the older name for the district, Blackamore. Below the edge, the concave shale flanks of the dales carry bracken on their steeper upper slopes with patches of oak, birch, and mountain ash woodland at the narrow dale heads, enclosed and cultivated fields on their flatter lower slopes.

Though subdued by erosion, the structural domes are significant relief features. The only ground over 1,400 ft is at the centre of the biggest, the Cleveland Dome. Erosion has bitten away its north-west corner almost to the centre, producing the boldest relief of the whole district, the embayment rimmed by the great escarpments to which the name Cleveland Hills particularly and most appropriately applies (fig. 53). Radial consequent streams have sliced open the moorland flanks of the dome into a slashed-sleeve pattern of Liassic inliers, each widely gaping lozenge-shaped cut showing a green lining of farmland set in a steep-edged frame of Middle Jurassic moorland. Farther east, where the Esk which consistently ignores structure bisects the Eskdale Dome, a wide strip of woodland and fields parts the high moorland lips of the gash. The sea has eaten into the soft Liassic centre of the Robin Hood's Bay Dome forming a green boulder-clay-lined theatre of lowland steeply rimmed by the Middle Jurassic moorlands behind.

The boulder clay is a legacy of the girdle of ice fronts round the upland that produced the systems of glacial lakes and spillways, the classic examples of their type. The huge looped trench of Newtondale drained Lake Eskdale into Lake Pickering; Langdale and Forge Valley linked formerly independent streams into the present upper Derwent. The boulder clay sheet covers the northern and eastern sides of the upland rounding the relief and filling pre-glacial valleys. Existing streams have either grooved narrow, heavily wooded ravines in the filling, often two streams to one former valley in twin parallel ravines separated by a narrow medial ridge of boulder clay, or have cut new courses, vertically walled rock gorges often with waterfalls. A landscape enriched by enclosed fields, farms and villages spreads over the better soils of the boulder clay to the edges of the spillways and is there sharply juxtaposed with the bare moorland of the unglaciated core. Vivid examples of this are the five-mile-wide carpet of pastures and ploughlands, lying where the ice once lay over the northern flank of the upland to the Whitby–Guisborough road, and the green tongue of farmland where the ice lobe poked into the Goathland recess.

The moorland rocks pass southwards under the Oxford Clay; a narrow, intermittent strip of farmland sometimes only two fields wide traces the outcrop along the scarp that separates off a ragged fringe of Corallian rocks in the south. Place-names accurately delineate the physical features. The streams flowing south off the Cleveland Dome cross the fringe in narrow thickly wooded gorges, cutting it up into a range of steep-edged, flat-topped plateaux aptly named by John Phillips the Tabular Hills; they slope gently southwards to the Vale of Pickering. The flared, trumpet mouths of the gorges trim the scarp into northward-jutting salients or 'nabs' (=beaks). The name Hambleton Hills (=scarred hills) faithfully portrays the landslips and cliffs of the western escarpment. In the whole upland, limestones have a significant part in the landscape in this fringe only. The lowest beds of the Corallian are sandy; moorland covers their outcrop along the scarp crest. Where the tableland surfaces pass southwards on to the limestone above, enclosures, cultivation and settlements at once appear. The streams threading the gorges sink underground where they encounter these limestones.

All this variety and distinctiveness of physical feature inland characterizes also the coastline along which cliffs nearly 700 ft high, promontories, wide bays in boulder-clay-filled pre-glacial hollows, and narrow post-glacial inlets follow each other in picturesque succession.

Farming and forestry, mining and quarrying, fishing, each have made their distinctive contribution to the landscape. Of recent years the traditional pattern of sheep (Blackface and Blackface-Leicester crosses) and store-cattle rearing has been modified by the increased importance of dairying. Nowhere is there a complete dependence on grass comparable to that of the Pennine dales; in the less rigorous climatic conditions, a substantial proportion of enclosed land everywhere is devoted to fodder crops. Dry stone wall field boundaries do not predominate; they are found mainly near the edge of the moorland, hedges being general elsewhere. Each sub-division of the upland has its own arrangement of fields, stone-built farmsteads and cottages, which here, where suitable clays and shales for tile-making occur, have red pantiled rather than stone-slabbed roofs as in the Pennines. The farmlands, roads and nucleated villages avoid the narrow gorges and ravines of the boulder clay fringe of north and east and of the Corallian fringe of the south; they are spread over tableland and divide. Only through-roads remain on the bare moorland ridges and plateaux of the central area. Farmlands and settlements descend into the wide dale floors and are followed thither by tracks which twist steeply down the Middle Jurassic rim. Damp stream-side sites in the axis of the dale are avoided. Villages on terraces and benches fringe Eskdale; single farmsteads standing on the flatter slopes beneath the Middle Jurassic rim are strung out in elongated loops round each of the 'slashed-sleeve' dales; in Danby Dale and Bilsdale the isolated parish church stands centrally within the loop.

Large areas of sandy Tabular Hills plateau, especially east of Pickering, and smaller belts of moorland, scarp-face, and dale-side have now been afforested.

The upland has yielded a varied assortment of minerals, including coal, shale and clays for brick- and tile-making, whinstone from the Cleveland Dyke and building stones. Limestone quarrying is active near Pickering and Thornton-le-Dale. Jet, alum, and ironstone have been the most distinctive products. Jet occurs sporadically in the Jet Rock of the Upper Lias. Much, including the small quantity used today, has been gathered from the shore, but in the boom period of the 1860's and 1870's it was mined from scores of short adits driven side by side along the outcrop; they perforate the cliff face and trace a dotted line of dumps round the scarp face of the north-western embayment. Carving and polishing were centred on Whitby.

The manufacture of alum from the Alum Shales occurring in the Upper Lias above the Jet Rock, an important industry for nearly three centuries, was quickly killed by the development in 1860 of a cheaper process using Coal Measure shales. The alum works were situated on the edge of cliff, gorge, or escarpment, where the vast quantities of spent shale and overburden, here least thick, could most easily be dumped. The alum workings at Boulby, Kettleness, Sandsend and Saltwick Nab have completely remodelled the cliffs and headlands there.

The modern phase of ironstone mining, with only one significant exception, has concentrated on the four seams of the Ironstone Series of the Middle Lias, in particular upon the uppermost, the Main Seam. It is thickest (11 ft) and richest (about 30% iron) where it crops out in the north face of Eston Nab and Upleatham Hill. At the southern limit of working, roughly a line joining the River Leven to Staithes, there are about 4½ ft of ironstone of 25% iron content.

Large-scale mining awaited railway building. The Whitby–Pickering railway was opened in 1836; mining of the lower seams from 1837 made Grosmont and Glaisdale for a while small industrial centres. A long mineral railway was built in 1860 over the Cleveland scarp to Rosedale to the exceptional and rich (43% iron) magnetitic deposits in the very top of the Lias, now worked out. The main development began with the discovery in 1850 of the Main Seam on Eston Nab and continued hand in hand with the construction of ironworks on Teesside and of a close net of railways in Cleveland. An annual production of 5 m. tons was maintained from 1872 to 1915, much of it from quarries and adit-mines near the escarpment. A structural basin centred on North Skelton carries the Main Seam below sea-level; the mines here had shafts about 700 ft deep, but production finally ceased in 1964. The steelworks on a coastal plateau at Skinningrove is a vigorous outlier of the Teesmouth industrial area.

These industries of the railway age have brought to the mining areas, particularly in the triangle Staithes–Guisborough–Saltburn, a landscape reminiscent of the Durham coalfield—winding-gear and spoil heaps, mining villages of terraced brick-built cottages, and allotments and pigeon-lofts.

In recent years borings near Whitby and Robin Hood's Bay have discovered natural gas in the Magnesian Limestone underneath the Eskdale Dome, and potash and other salts higher in the Permian. The gas has been piped to Whitby since 1960 and now serves Pickering and Malton as well; a pilot plant is to be installed to test the practicability of exploiting the soluble salts.

Guisborough (12,079) in the little vale between the northern escarpment of Cleveland and the outlier of Eston Moor is the only inland market town within the area. The fishing villages and towns of Staithes, Runswick, Whitby, Robin Hood's Bay, and Scarborough, perched insecurely but picturesquely on steep slopes of landslip or boulder clay overlooking the bay or post-glacial inlet into the north-west angle of which they huddle for shelter from northerly and north-easterly gales, have a powerful attraction for visitors. Whitby (11,662), its hinterland greatly restricted by the relief, began as a small fishing settlement and harbour of refuge offering in its deep post-glacial inlet all-round protection from gales. Shipping coal to coastal alum works stimulated trade and led to shipbuilding, later to whaling. Today fishing and boat-building are still active, but it is primarily a holiday resort. Scarborough (42,587), from a similar small beginning on the south side of its castle-crowned promontory, of less importance as a port, developed earlier into a resort and, more accessible, is now much larger.

The grandeur of the coast and its picturesque fishing villages, and the open sweeps of moor inland setting off the verdure of ravine and dale amply justify designation as a National Park.

A narrow tail of Jurassic rocks, dubbed by William Marshall in 1788 the Howardian Hills, wraps round the basin-like western end of the Vale of Pickering and forms a narrow isthmus of plateau linking the Hambleton Hills and Yorkshire Wolds. Here, also, landscape responds to rock type. On the northern edge, the Corallian limestones carry a mile-wide band of arable land and pale grey villages between Hovingham and Malton; south of this a belt of woodland clothes the steep banks where the Calcareous Grit and Oxford Clay crop out; south of this again, parks and mansions strung out between Coxwold and Birdsall mark the sandier Middle Jurassic rocks, poorer agriculturally but having attractively varied relief, the result of intricate faulting. The isthmus is severed by the Coxwold–Gilling fault-trough in the north, in the south by the Kirkham spillway, the outlet of Lake Pickering still used by the Derwent.

Flat-floored, hill-girt except at the eastern end, the Vale of Pickering looks exactly what it is, a lake bed. Morainic mounds near each extremity mark the positions of the ice-dams that impounded Lake Pickering in this strike vale hollowed pre-glacially out of the Kimmeridge Clay cropping out between the Corallian rocks and the

Chalk. In Marshall's 'mudland quarter of the west', i.e. west of the Malton–Pickering road, hillocks of Kimmeridge Clay weathering to heavy soils rise through the sheet of silt, and allow in the centre of the Vale island-site villages which are linked by a fairly close road network. The easily flooded remainder, Marshall's 'marshland quarter of the east', is threaded by the Derwent, flatly graded between the Forge Valley and Kirkham spillways. Near Yedingham, a river-crossing and the only mid-vale village of the 'marshland', it divides into the 'Marishes' to the west, level silts bearing a scatter of single farms, and the 'Carrs' to the east, peaty and liable to flood. Here, farms and roads are most widely spaced. Mixed farming is the general rule throughout the Vale, with some cash root-cropping and an emphasis on dairying near the coast. A peripheral loop of villages interspersed with market centres, Malton, Helmsley, Kirkby Moorside, Pickering and Seamer, edges the Vale floor. Filey, a fishing settlement turned resort, shelters from northerly gales behind Filey Brigg, a rib of Lower Calcareous Grit. Bungalows and caravans, and a large holiday camp line the boulder clay cliffs of Filey Bay farther south as far as the 400 ft cliffs of the north side of Flamborough Head.

Thence, the scarped outer face of the great chalk crescent of the Yorkshire Wolds smoothly fronts the Vale of Pickering to the north; on the north-west it is fretted into deep embayments where it crosses the belt of faulting of the Howardian Hills; it presents westwards to the Vale of York a scalloped alternation of rounded bastions and wide obsequent valley mouths above the narrow Jurassic outcrop which, reduced by upwarping and erosion before the deposition of the Chalk to the merest fringe north of Market Weighton, widens southwards into a narrow terrace (fig. 52). The inner side slopes gently towards the Hull valley–Holderness lowland and below 200 ft is covered by a thin, almost featureless skin of boulder clay. The crescent, which reaches 808 ft at its highest point, is dissected by winding, steep-sided, flat-bottomed dry valleys; its wide northern arm is seamed longitudinally by the wide mature groove of the Great Wold valley, which is flexed into a double bend by the same fault that recesses the escarpment.

The distribution of villages reflects the unattractiveness in the pre-enclosure period of the thin dry soils of the high Wolds above the edge of the boulder clay for only three stand here. A single line,

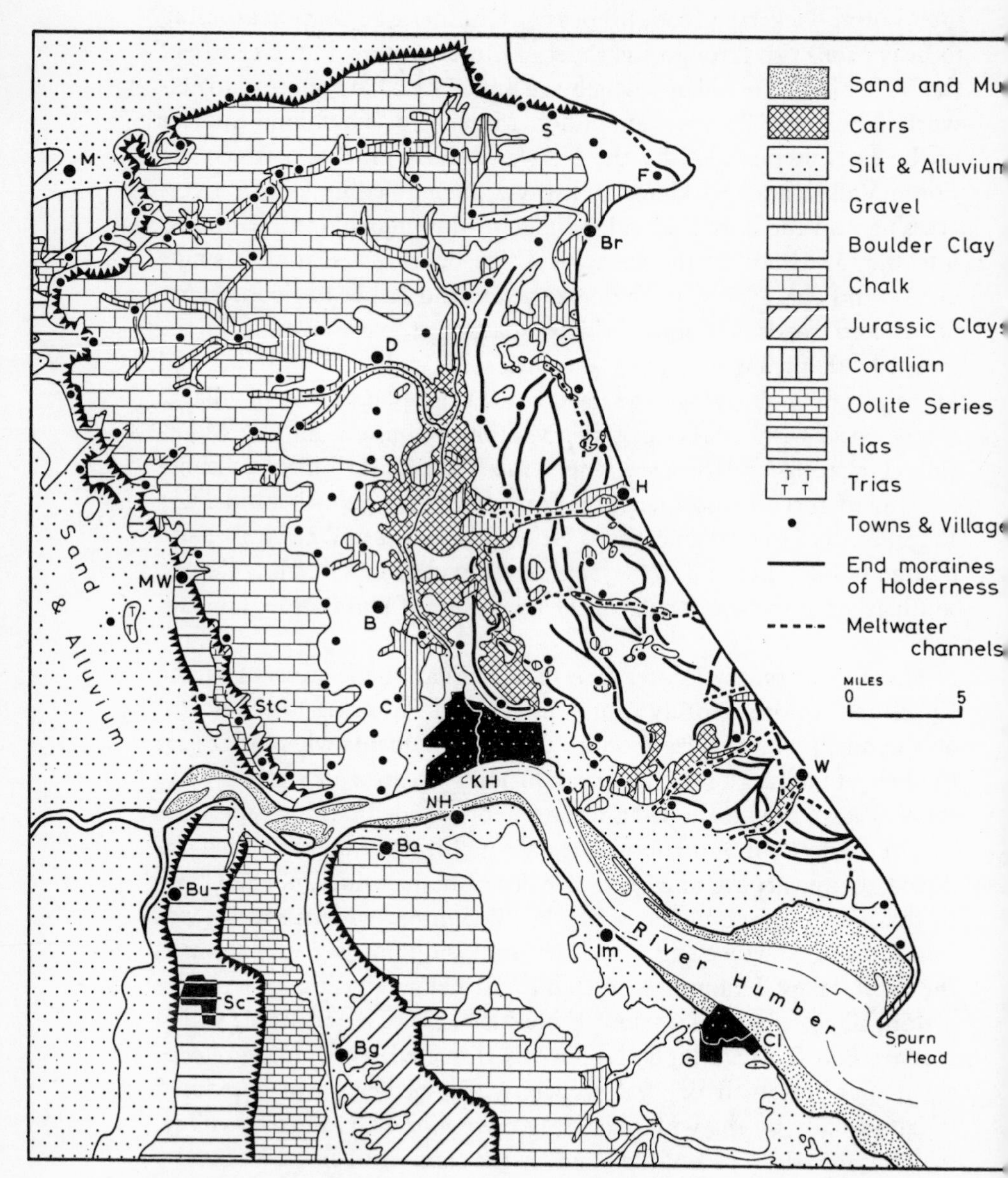

Fig. 52. Geology and settlement of south-east Yorkshire and north Lincolnshire.
B. Beverley. Ba. Barton on Humber. Bg. Brigg. Br. Bridlington. Bu. Burton upon Stather. C. Cottingh **Cl.** Cleethorpes. D. Driffield. F. Flamborough. G. Grimsby. H. Hornsea (m. Hornsea Mere). Im. Immingh **Dock.** KH. Kingston upon Hull. M. Malton. MW. Market Weighton. NH. New Holland. S. Spee Sc. Scunthorpe. StC. South Cave. W. Withernsea.

standing in valley mouths and embayments on the west and north-west, follows the scarp foot. There is a double row along the inside of the crescent, one row near the upper margin of the boulder clay, the other where the slope flattens into the Hull valley. Further lines are dotted along the Great Wold valley and along the morainic fringe of Flamborough Head; Flamborough, a fishing village, is centrally placed near the tip, equidistant from the little bays used as landing-places. Enclosure, tree planting, and husbandry, successful by about 1850, transformed the Wolds. Instead of open fields, rabbit-warrens and sheep-walks, there are now large hedgerows, arable fields, and wide, straight roads dotted with substantial farmsteads set in hollow squares of screen plantation. A four- or five-course rotation, involving folding sheep on to turnips, still prevails, though since the war the turnip acreage has been reduced and temporary grass increased. Barley is the chief cereal of the higher Wolds; wheat is more important lower down.

The River Hull, which rises from springs concentrated by convergent dips into the angle of the Wolds near Driffield, and the Humber below the great bend, originated together as a late-Glacial stream of meltwater flowing southwards along ice fronts and moraines, curving through Holderness. As this ice lobe contracted eastwards it uncovered sheets, hills and ridges of gravel and boulder clay that form islands in the Hull valley or are festooned in complexes of terminal moraines in Holderness. Radial sub-glacial streams fanning out westwards cut through these ridges steep-sided channels which still carry most of the drainage inland to the Hull valley. Marshy flat-topped fills of peat occupy these hollows in which lay formerly many small lakes; only Hornsea Mere survives.

Villages usually stand rather prominently on low eminences of morainic drift; the church and older cottages are often built of erratic boulders. Only after much draining and embanking were isolated farms built on lower ground. Beyond the string of villages marking the southern edge of the boulder clay stretches the flat expanse of Sunk Island, formerly Humber shoal and saltmarsh, but now consisting of mainly hedgeless arable fields studded with isolated farms. Reclamation, bit by bit, began about 1670 and is likely to continue into the future.

The agriculture of Holderness has long been characterized by high proportions of arable land under wheat, by the importance of peas and beans, and by the practice of bare fallowing to clean land so heavy that the growth of root crops is difficult. Wheat occupies up to one-third of the arable, cereals together nearly two-thirds. Peas for drying are replacing beans, and fallowing has nearly disappeared. Animal populations are high; pig-breeding and fattening, helped by ample supplies of fodder barley, is now the major source of income from stock. In the warp and siltlands of Humberside permanent grass and livestock are much less important, indeed some farmers, dispensing with animals, rely on cash crops of wheat, barley, peas and, since the war, potatoes. Wheat, the main cash crop of the Hull valley, shares the arable there with fodder crops, seeds, oats, beans, and turnips. These fodder crops, together with a high proportion of permanent grass in the poorer drainage conditions of the valley, are associated with a dense population of pigs, poultry and cattle, especially dairy cattle near Beverley and Hull.

Market gardening is important between Beverley and Hull, in naturally light loams near Woodmansey, and on the heavier clays near Cottingham, the latter artificially lightened before 1920 by a long period of application of ashes and nightsoil from Hull. Following the arrival in 1932–36 of immigrant Dutch growers, it has become the largest horticultural area in Britain to use the Dutch large-paned, unheated glass-house and frame. It markets its produce not only locally, but also in the industrial towns of the West Riding, north-east England, and Clydeside, for which its position—it is the most northern market-gardening area that serves distant markets—is advantageous. It benefits also from having adjacent the wholesale marketing facilities of Hull, which were originally developed to deal with imported produce. The main crops are lettuce, tomatoes, cucumbers, melons, and cauliflowers.

There are no fishing villages along the wasting boulder clay cliffs of Holderness which sweep in a great curve from Flamborough Head to the long dune-capped spit of Spurn Head. The sea, advancing about 5 ft a year, has annihilated many agricultural villages and crept near to others. Rail construction about a century ago turned Hornsea and Withernsea into small resorts which now have to maintain costly sea-walls. The motor-car has added shacks, bungalows and caravans to the other coastal settlements. Three market towns have grown up on the route along the foot of the dip slope of the Wolds, from the Humber to the sea. Bridlington (26,007), fishing town and resort as well as market resembles, in its sheltered position and evolution, Scarborough and Filey. Driffield and Beverley are located at places on this route easily reached from both the Wolds and Holderness. Driffield (6,890) stands where ridgeways across the Wolds converge on a point where the marshy Hull valley can be skirted. Near to Beverley (16,024) the main route from Holderness over the Wolds to York is guided across the Hull valley by a large island of boulder clay that provides a stepping-stone. In the Middle Ages Beverley was a resort of pilgrims whose gifts built the exquisite minster, a market, a small port, and a cloth-making town. Today it is the county town of the East Riding and a market town; it builds trawlers and manufactures vehicle accessories.

Its huge basin of about 9,000 square miles makes the Humber both a focus of waterways and also wide enough to be a major obstacle to north–south movement. Only the Hull–New Holland ferry crosses it. The sites of the Roman and early medieval ferries and of the proposed bridge are where its broad stream narrows into a 1¼-mile-wide gap cut along a line of faulting and folding through the Jurassic and Cretaceous escarpments. The great bend below Hull where the late-Glacial ice-front swung the Humber out of its consequent course along the dip, brings close to the north side a deep channel which farther downstream is guided to the south side off Immingham Dock by a slighter reverse curve.

The varying fortunes of the Humber ports are an example of an almost Darwinian testing of the fitness of possible sites for commercial activity. Hull, Kingston-upon-Hull (303,268), alone of the older ports, has remained continuously active. The commercial advantages of the mouth of the River Hull, a commodious and sheltered haven approached by the deep channel, outweighed the handicap of low banks flooded by the tide even occasionally today. A port Wyke (= creek) upon Hull, first mentioned about 1160 rapidly grew to national importance and by 1280 had some claim to be third port of the kingdom. Acquired by Edward I from the monks of Meaux Abbey in 1293, improved and renamed Kingston-upon-Hull, it remained literally 'upon Hull' until 1809 when the first dock opening directly on to the Humber came into use. Further dock construction has re-aligned the port along seven miles of the Humber where the channel hugs the shore. Hull, still the third port of the kingdom, handles well over half of the commerce of the Humber ports. The main imports are petroleum, grain, food, fruit and vegetables, timber, oilseeds and nuts, wool, iron and steel scrap; the main exports, iron and steel goods, machinery, textiles, vehicles and vehicle parts. The export of coal and coke has declined abruptly, from 2,210,000 tons in 1957 to 313,000 tons in 1959; it recovered to 470,000 tons in 1960, to 1,931,000 tons in 1964.

Three-quarters of the fish landed in England and Wales enters the Humber; about a half of this comes to Hull. The Hull fleet consists almost entirely of large trawlers which fish distant waters.

Industries processing imported raw materials edge the dock estates and line the River Hull—flour-mills, seed-crushing-mills producing vegetable oils and feeding-stuffs, factories producing margarine, soap, paint, cocoa butter, glue, and processing fish. Lining the Humber gap, there is shipbuilding at Hessle, metal-refining and cement-making at Melton (situated between Wolds chalk and Humber mud), aircraft-building at Brough.

Almost all the battlefields of Yorkshire lie in the bottle-necked strike-vale traversed by the Ouse, a routeway and a meeting-place; the centre of Yorkshire and the half-way stage between London and Edinburgh (fig. 53). The vale owes its origin no doubt to river erosion and capture along the weakly resistant Triassic sands and marls, its present relief and drainage to its glacial and post-glacial filling. An ice-tongue thrusting down the vale diverted the rivers entering from the west and made them cut steep-sided gorges in the Magnesian Limestone. As it withdrew it left behind a hummocky mantle of boulder clay, particularly evident in the northern bottle-neck, and a succession of terminal moraines curving across the vale at Escrick, at York, and near Boroughbridge, Easingwold, and Thirsk. They rise about 30 ft above the not quite flat sheets of sands spread out by meltwater from the dwindling glacier, the almost completely flat sheets of clays deposited in the lakes that lapped its edge; these deposits occupy the central vale north of York and spread out widely south of the city. Fingers of estuarine warp and alluvium poke up along the lower courses of Ouse, Wharfe and Derwent.

There is much variety of agriculture, landscape and settlement. On the faintly defined, boulder clay-spread dip slope of Magnesian Limestone on the west there is much parkland and mixed farming noted for malting barley and potatoes. There is also mixed farming

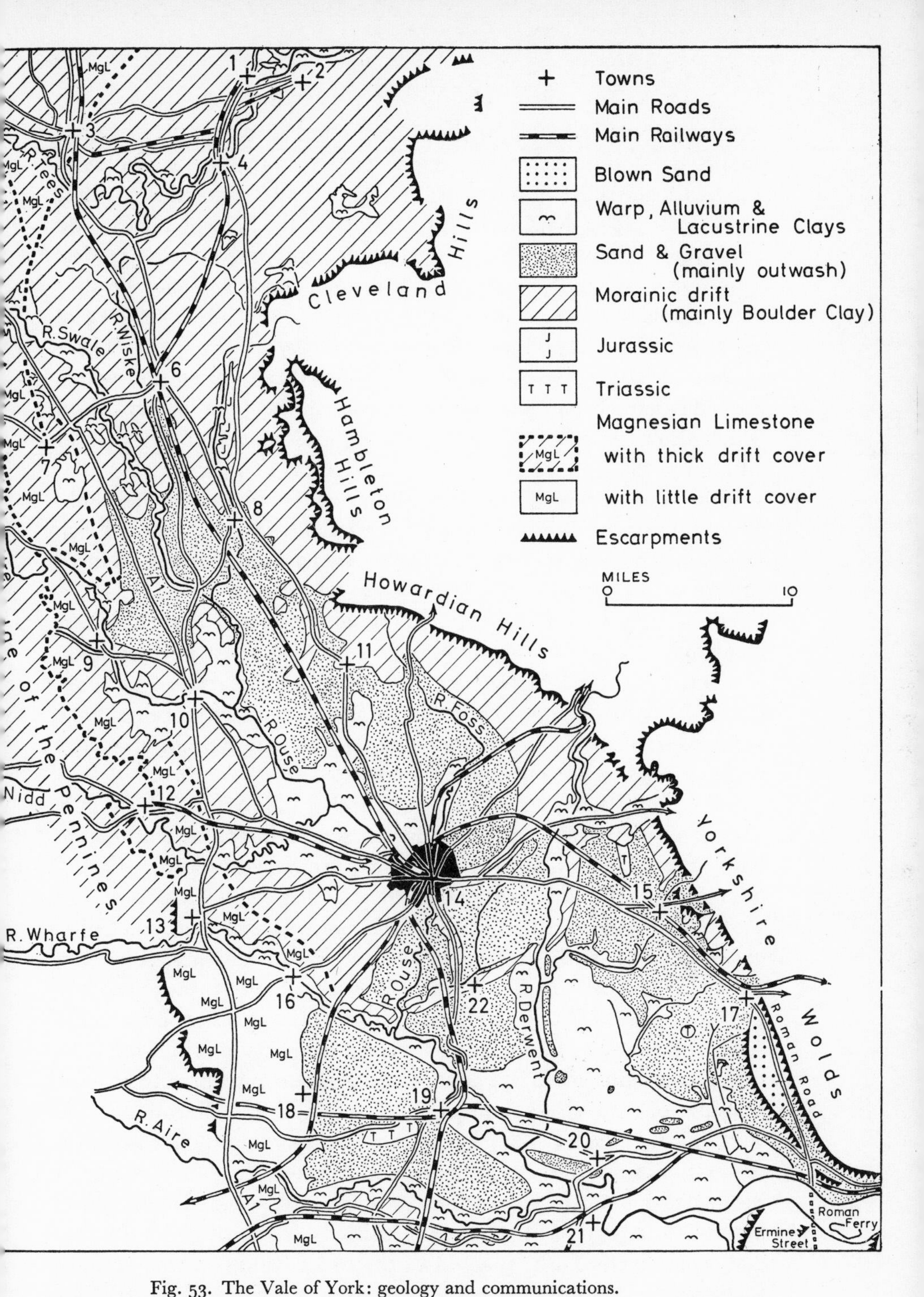

Fig. 53. The Vale of York: geology and communications.
1. Stockton. 2. Middlesbrough. 3. Darlington. 4. Yarm. 5. Catterick. 6. Northallerton. 7. Bedale. 8. Thirsk. 9. Ripon. 10. Boroughbridge. 11. Easingwold. 12. Knaresbrough. 13. Wetherby. 14. York. 15. Pocklington. 16. Tadcaster. 17. Market Weighton. 18. Sherburn in Elmet. 19. Selby. 20. Howden. 21. Goole. 22. Escrick.

on the quite strongly undulating morainic drift of the vale. Cattle are pastured on the heaviest and most poorly drained areas of alluvial and lacustrine soils; with better drainage, cereals, particularly winter wheat, may be grown. Remnants survive of the ancient heaths and birchwoods of the areas of outwash sands. Improved during the last 150 years by marling, i.e. spreading Keuper Marl, Lias clay or alluvium, these sandlands have become since about 1920 an area of intensive cash-cropping of potatoes, sugar-beet and carrots. There are similar patches of woodland, heath and root cultivation where the sands have been blown on to the edge of the Jurassic terrace on which also there is between South Cave (fig. 51) and the Humber an offshoot of the Cottingham horticultural area. The warplands, areas of tidal alluvium deposited naturally or artificially along the banks of the Ouse and Humber especially near Howden, are areas of rich cash-cropping, mainly potatoes, sugar-beet, peas and cereals. Horticulture is important in the Selby–Cawood district.

The villages on the Magnesian Limestone and on the Jurassic terrace are built of stone; those on the floor of the vale are built of brick, and stand on swells of boulder clay or on gentler rises of outwash sands. They are strung like beads along the loops of moraine, edge the terraces of the Derwent, and, along the Wharfe and Ouse, occupy riverside sites on the outsides of bends.

Communications are drawn out in the Vale of Mowbray, the northern bottle-neck dominated by the escarpment of the Hambleton Hills, into a sheaf of parallel strands connecting market towns along the foot of each flanking slope, Catterick, Bedale and Ripon on the west, Northallerton and Thirsk on the east. At the south end of the wider Vale of York proper, the Humberhead marshes filling the centre pushed the Lincoln–York road east on to the Jurassic terrace, the Great North Road (A1) west on to the Magnesian Limestone. Also on these drier margins, where the streams have firmer banks of rock or moraine, there is a cordon of market towns standing at river crossings or near valley mouths, points of entry into the vale: on the west, Boroughbridge, Knaresborough, Wetherby and Tadcaster; on the east, Easingwold, Pocklington and Market Weighton. Right in the centre of the vale, where formerly tide-water reached through the marshes to a loop of moraine that assisted the crossing of both vale and river, stands York (104,468) itself, at the focus of a web of main roads converging from the market towns on the rim of the vale. This nodality is the abiding basis of its civil, military, ecclesiastical and commercial importance. In Roman and medieval times it was the great northern capital, the military stronghold against the Scottish threat, the centre of the civil administration of Yorkshire and of the ecclesiastical administration of the Northern Province, a market, port and wool town. A long period of decline from 1500 until the railway era helps to explain the survival of so much medieval building, especially the gates and walls, through the period when they were regarded as tiresome relics of barbarity until they became the valued antiques and tourist attractions of today. The railway reasserted the nodality of York by making it the centre of a second, an iron web, the main strand of which, unlike the roads, followed the axis of the vale and put the city for the first time on the main route between London and Edinburgh. York has remained a market and shopping centre. Its wealth of architecture, its splendid museums, the success of its arts festivals, and the foundation of a university in 1963 combine into a strong claim for the city to be the cultural capital of Yorkshire.

Selected Bibliography

S. E. J. Best. *East Yorkshire: a study in Agricultural Geography*. London, 1930.

British Association. *York: A Survey*. York, 1959.

G. H. J. Daysh (ed.). *A Survey of Whitby*. Eton, 1958.

F. Elgee. *The Moorlands of North-Eastern Yorkshire*. London, 1912.

C. Fox-Strangways. 'The Jurassic Rocks of Britain, Vol. I. Yorkshire'. *Mem. Geol. Surv.* London, 1892.

R. G. S. Hudson and others. 'The Geology and Scenery of the North-West of Yorkshire.' *Proc. Geol. Assoc.*, **44**, 1933, pp. 228–55.

P. F. Kendall and H. E. Wroot. *The Geology of Yorkshire*. Privately printed in Vienna, 1924.

C. A. M. King. *The Yorkshire Dales*. *Scarborough*. British Landscapes through Maps. Geographical Association, 1960, 1965.

W. H. Long and G. M. Davies. *Farm Life in a Yorkshire Dale*. Clapham, Yorkshire, 1948.

A. Raistrick. 'Notes on Lead-Mining and Smelting in West Yorkshire.' *Trans. Newcomen Soc.*, 7, 1926–27, pp. 81–96.

A. Raistrick and J. L. Illingworth. *The Face of North-West Yorkshire*. Clapham, Yorkshire, 1949.

CHAPTER 21

LANCASTRIA

GEOFFREY NORTH

This area forms a fairly compact physical unit on the western shores of the Irish Sea from the estuary of the Lune to the Dee. It is drained principally by two river-systems: the Ribble and the Mersey with its two main tributaries, the Irwell and the Weaver. The area is, however, by no means entirely lowland in character. Except in the south where it merges into the plain of central Cheshire, the region is flanked by uplands, on the south-west by the Welsh hills and on the east by the high moorlands of the Pennines. The latter, with its two western extensions, Bowland and Rossendale, form as essential an element of the Lancastrian scene as the lowlands. Together, lowland and upland impart an overall unity to the area, for in its economic development the resources of both have been complementary. There is perhaps no other area of comparable size in Great Britain which exhibits such a wealth and variety of industrial and commercial activity together with an almost equally diverse agriculture. Over 6 m. people live here in settlements that range from the isolated moorland farmstead or Victorian industrial hamlet to the breezy seaside resort or sprawling conurbation. And yet some sense of regional unity prevails: to what extent this is a reflexion of the physical geography of the region is a problem that invites detailed comment. But no such consideration would be complete unless it is viewed against the background of industry, the character and development of which has so profoundly affected the life of the whole region during the past two hundred years. This, the main theme of discussion, is the reason why there are so many apparently conflicting elements in the present landscape: why, for example, towns with decaying economies and declining population are to be found within a few miles of thriving cities wrestling with problems of overspill; or why the same processes of industrial growth have rendered land derelict in one locality and have stimulated land reclamation projects for intensive agricultural use in another.

The broad contrast that exists between upland and lowland in this region is primarily a reflexion of its solid geology (fig. 54). The Pennines are composed of Carboniferous rocks, including gritstones, mudstones, shales and coals of the Lower and Middle Coal Measures, while the north-east part of the Welsh Hills is made up of Silurian rocks, chiefly slates, mudstones and sandstones that give way in the east to the Coal Measures of the Carboniferous series. The lowland, much younger in origin, is floored by Permo–Triassic sediments. These beds yield sandstones, marls, pebblebeds and salt, but few of these occur as surface outcrops for they are obscured by a thick mantle of drift left by Quaternary ice from the Irish Sea, the Lake District and southern Scotland. In the Pennines ice and melt-waters made spillways and overdeepened many existing cols and valleys, for example the Cliviger Gorge. Post-Glacial deposits have added to the variety in local relief, as in the extensive stretches of deltaic sands within the Manchester embayment, and the sands and peaty mosslands of south-west Lancashire and north Cheshire.

The effect of such a physical make-up on the economic evolution of the region has been considerable, but it has varied from one period to another. Most of the Pennine area remained relatively remote and inaccessible almost until the time of factory industry. The main range, Rossendale and Bowland, was difficult to cross: in fact, no canal ever penetrated Rossendale, and Bowland Forest remains remote to this day. The road and railway network has always been more dense on the plain, though even here the mosslands and dunes were uninhabited until well into the nineteenth century. Nevertheless, the lowlands lay across major north–south routes which, from medieval times, had stimulated the growth of town life at important river crossing-points as at Chester, Preston and Lancaster. The medieval site of Manchester, on the Irwell at the heart of the embayment, proved well chosen once certain economic forces became active.

In economic resource, the region yields, or has yielded, much. The gritstones of the Rossendale anticline, massive and accessible in prominent outcrops along the steep valley-sides of the upper Irwell, provided excellent building-stone which, at the height of its exploitation towards the end of the nineteenth century, supplied districts far

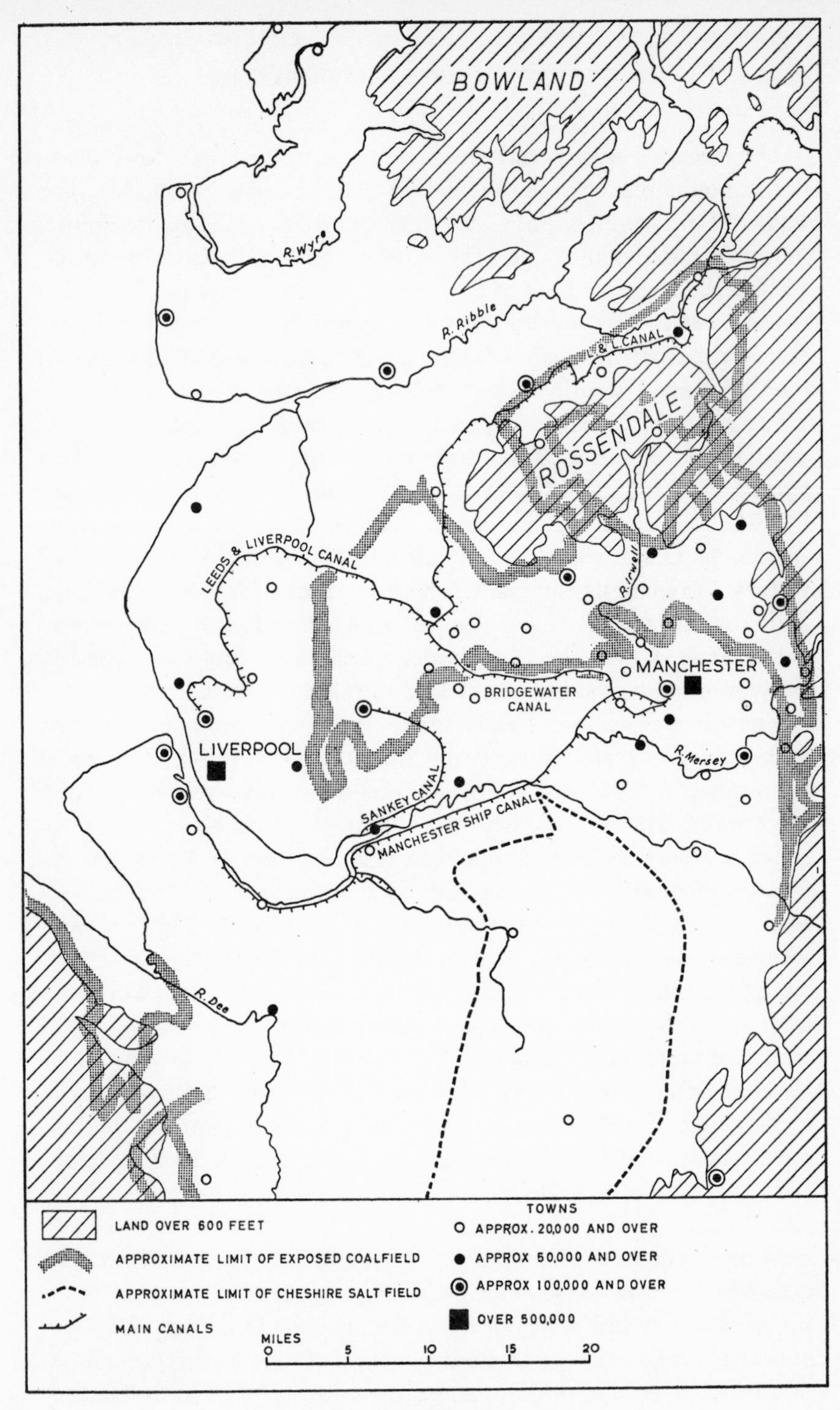

Fig. 54. Lancastria.

beyond Lancashire and Cheshire. Increased cost of working brought about the decline of quarrying, though the hills of the central part of the anticline are still severely scarred, in strong contrast with the Forest of Bowland. In the lowlands, the sands have been widely used in brick industries, as in north Manchester, and the Shirdley Hill sands in the glass industry at St Helens. In mid-Cheshire, rock salt, formed under desert conditions in the central depressions of the Trias sea, provided the raw material for the salt and chemical industries of such towns as Northwich and Winsford.

But the main industrial resources have been associated with the provision of power-supplies which have been exploited in a variety of ways during the past two hundred years. Without doubt, the availability of water-power was the most important physical factor influencing the early location and development of the local textile industry. Though the oft-quoted influence of humidity and water-softness cannot be entirely discounted, their importance was merely incidental to that of water-power. East Lancashire is no more humid than many parts of the coast, though water draining off the grits is admittedly softer in its natural state than that obtained from underground sources on the plain. By the middle of the nineteenth century, however, every accessible and sizeable stream in east and south-east Lancashire had suffered pollution by industrial effluents to such a degree that only their upper reaches still offered naturally soft water. This, coupled with the fact that the water was free and its use unrestricted, accounted for the survival and concentration, not of textile-making as a whole in such relatively inaccessible parts, but of dyeing, bleaching and finishing in particular where the need for soft water was greatest. But for many years at the beginning of the nineteenth century the search for sites where water could be easily and abundantly harnessed to the new power-driven machinery dominated industrial location. An examination of the first edition of the Ordnance Survey for Lancashire (1844–50) shows this emphatically. Mills were on sites as high as 1,000 ft along narrow stream valleys at breaks of slope, either at the outcrop of more resistant bands of gritstone, or at knick points. Some mills survived long after the introduction of steam-power and if coal could be locally mined or brought up the valley by rail, they provided the nucleus of many industrial hamlets that exist today. Most, however, declined and though they are now, in their derelict state, the preserve of the industrial archaeologist, without them much subsequent industrial development would never have taken place.

Nearly all new textile mills built during the 1820's used steam-driven machinery: this demanded a site either near the pithead or one where coal could be brought cheaply. In this the textile industry was by no means unique. From that time onwards, coal mining provided the foundation upon which the whole industrial structure of the region was built. Accessible seams, especially where the drift cover was thin, were the first to be exploited and contributed greatly to the location and rapid growth of towns such as Wigan, Bolton, Bury and Oldham. Not all industrial towns concentrated on textiles; mining and engineering became important industries in St Helens, Prescot, Wigan and Leigh. Not all industry was centred on the coalfield proper; canals attracted some, and although railways largely superseded canals as the main carriers of industrial raw materials, three canal systems in particular influenced the pattern of industrial development long after the coming of railways. The Sankey Canal, completed in 1755, linked up St Helens with the mouth of the Weaver, which had been canalized in 1732, and established a link between the salt of Cheshire and the coal of south Lancashire. This stimulated the development of industry, particularly chemicals, at either end of the system as well as at the point where it crossed the Mersey at Widnes and Runcorn. The Bridgewater Canal, constructed in 1761 to carry coal from Worsley into Manchester, had an equally lasting influence and made possible industrial development along an east–west axis in the heart of Manchester. Finally, the Leeds and Liverpool Canal which reached Wigan in 1771 and was extended to Yorkshire by 1816, stimulated in east Lancashire in particular the growth of textile manufacturing along its banks.

In 1830, however, a new and more vital phase in the economic development of this region began when Stephenson's railway was completed between Liverpool and Manchester. It was a remarkable achievement in many ways: not only was this line the first ever of its kind but it had been constructed in the space of four years, in the face of many physical difficulties. Chat Moss, to the west of Manchester, had long seemed a real barrier to any improvement in communications; pessimism among engineers and city fathers alike was strong. And yet, by much draining and embanking, it was crossed. A contrasting, though equally serious problem, was that which involved the excavation of long cuttings through the New Red Sandstone as the railway entered Liverpool.

After this time the railway network spread rapidly across south Lancashire and by the middle of the century the initial processes of industrial evolution were complete.

Above all, this industrial region was well integrated. There was a clear interdependence of coalfield and factory, but the rail and canal had permitted dispersion. There was also integration within industries, of which the textile industry afforded the finest example. By the end of the nineteenth century, carding, spinning and doubling dominated the textile industries in a broad belt of towns that extended from Wigan to Stockport, while to the north of the Rossendale anticline weaving was the main textile activity in towns from Preston to Colne. Neither zone concentrated exclusively on one main process, though there was a marked separation between the two which only overlapped in Rossendale, where combined premises prevailed. Liverpool handled the raw cotton, Manchester the finished yarn and cloth. No other cotton-manufacturing region in the world has ever displayed such a remarkable separation of its component industrial processes.

Links were also created between different manufacturing industries which became closely interdependent. The textile industry was always the main stimulus: it encouraged, for example, the development of textile engineering in Manchester, Oldham, Stalybridge and Wigan, and the expansion of the chemical industry in the middle Mersey and Cheshire which supplied products for the finishing trades.

This high degree of integration, in so many forms, did much to foster the commercial life of the region, centring it on two towns: Liverpool and Manchester.

Liverpool had not assumed real importance as a major port until the middle of the eighteenth century. It had grown up on a low plateau-like upland of Triassic sandstone but remained relatively isolated from the rest of Lancashire until the lowlying, badly drained mossland that virtually encircled it to the north and west had been crossed by road and canal. With its deep, tidally scoured channel, the port was able to take full advantage of the rapidly expanding West Indian and African trade, in complete contrast to Chester which, by the end of the Middle Ages, had declined as a port with the silting-up of the Dee. But it was the rapid and varied industrialization of its hinterland that permanently established Liverpool as one of the greatest ports in Britain.

Industrially this hinterland extended from the textile mills of east Lancashire to the salt-fields of Cheshire. A supply-point for important raw materials, notably cotton, was necessary. More important still was the fact that many industries, especially textiles, concentrated on production for export through the port. In 1840 cotton goods accounted by value for one-third of Britain's exports and in 1913 they still represented one-quarter. Similarly, Manchester grew up as the main commercial centre for finished goods, at the hub of a communications system in the centre of the embayment which enabled it to expand outwards towards the industrial towns around it. Itself an industrial centre of no mean significance, it had attracted the finishing and mercantile branches of the textile trade, and also clothing, textile engineering, and an increasingly wide range of lighter industries and distributive trades. Small wonder, then, that the city grew too rapidly to permit planning. As in Liverpool, the merchants left the city centre to live on its outskirts and thereby set in motion a movement, particularly into the Cheshire countryside, that has increased in momentum ever since. They left behind a city with its main shops, warehouses and offices at its heart, ringed first by factories and then by a grid-iron zone of brick terraced artisan houses. This grim and depressing inner area is now beset with many social problems.

By the end of the nineteenth century, therefore, much of the industrial landscape as we know it today had come into existence. In 1894, however, the completion of the Manchester Ship Canal, 35 miles in length from Eastham Locks on the Mersey estuary to Salford Docks, marked the beginning of a new and vigorous phase of industrial growth that is now perhaps the most dominant feature of modern large-scale industry in the region. At no point is the canal less than 28 ft in depth and ocean-going vessels of up to 10,000 tons can penetrate the heart of the Manchester conurbation. Its primary purpose—to carry raw cotton to Manchester—has long since been overshadowed by the demands for raw materials made by other industries that have been established along its banks. This last example of canal building in this region provides a canal-side location that is likely to be of abiding significance. Iron and steel-making, originally established at Warrington at the tidal limit of the Mersey river, was extended to Irlam on the canal where the construction of coal and ore wharves has made the plant the biggest in Lancashire. Food processing, chemical industries and oil refining have since grown up along the canal and there are today few stretches where the powerful influence of the canal in modern industrial growth cannot be seen (fig. 55). In many respects, the canal saved south-east Lancashire: it brought a degree of industrial diversity—as in the huge industrial estate of Trafford Park, established in 1896—that continued to attract employment at a time when other industries in Lancashire were in decline.

But there were weaknesses. Too many towns were dominated by one industry, particularly by cotton. In east Lancashire, towns such as Nelson, Colne, Padiham, Burnley, Blackburn, Accrington and Haslingden all had, before the economic depression of the 1930's, at least two-thirds of their working population employed in textiles. Only in Rossendale was there industrial diversity within the cotton area to a degree that one associates with present-day industrial life. Here footwear manufacture had succeeded a declining woollen industry in the 1870's and, at a time when cotton manufacture was locally in a state of temporary depression, it had taken over many former cotton mills. At first by concentrating on cheap felt slippers and then by extending its range to include women's cheaper footwear and children's shoes, the industry established a position strong enough to withstand the effects of the loss of overseas markets upon which it had originally depended. Through the department store it captured the home market in cheaper-quality footwear during the period between the two world wars. Long before Lancashire became familiar with cotton mills used for other purposes this was seen in Rawtenstall where employment in footwear was as great as in cotton. But in the spinning towns to the south, such as Bolton, Bury and Oldham where other industries, notably engineering, had also taken root, cotton remained the chief employer of labour. Consequently the industrial landscape of these towns reflected their staple industry —either weaving sheds or tall spinning mills clustered along canal and railway, interspersed with tightly packed rows of terraced houses. Farther south other formerly dominant industries have left their mark. In south Lancashire, as around Wigan, St Helens, Prescot and Leigh, the progressively southward movement of coal-mining has left

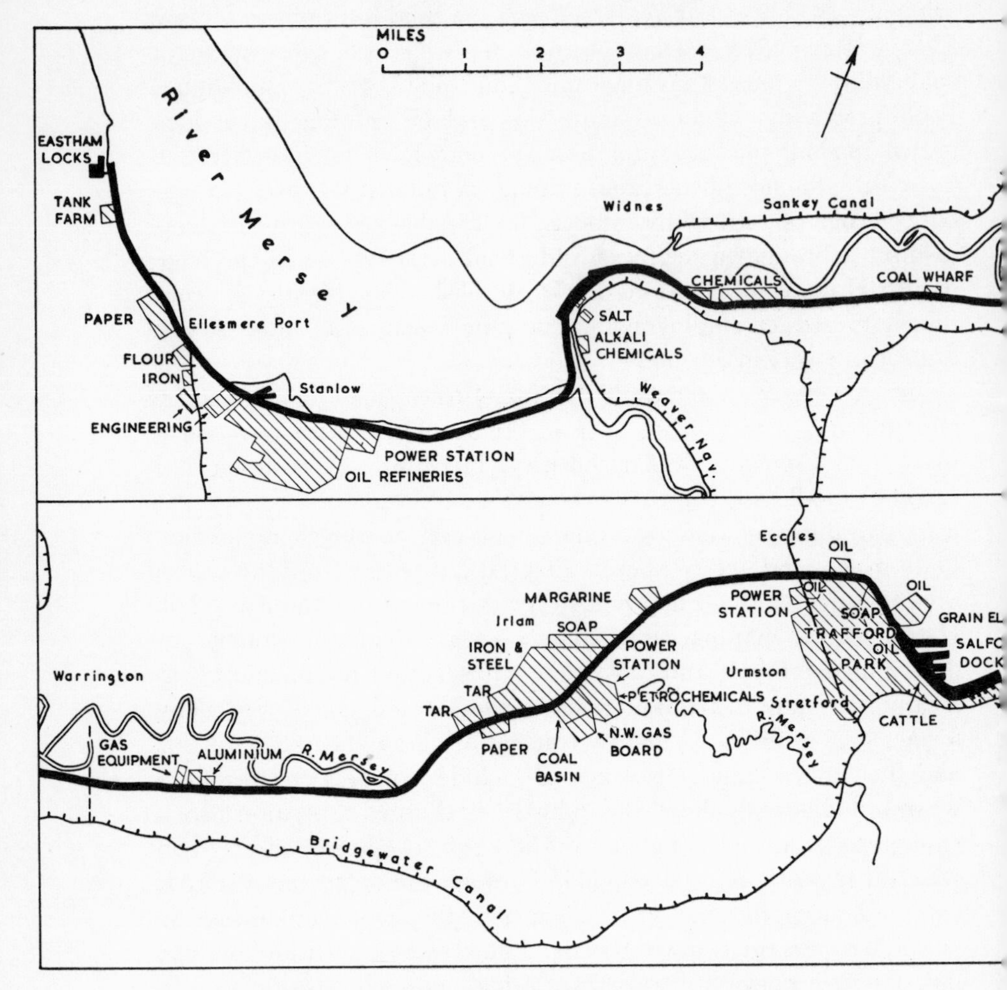

Fig. 55. Manchester Ship Canal and canal-side industries.

much land derelict. Other industries now decayed, for example the chemical industry of St Helens and the iron industry of Wigan, have made their ruinous impression on the landscape and there has been much dumping of waste from glass-making plants. Spoil heaps, subsidence and the growth of flashes have turned this part of Lancashire into an area of almost unparalleled visual horror. Again, in the middle Weaver basin, unrestricted exploitation of the salt-field brought even more serious problems through subsidence.

It is important to stress that many of the various economic forces that produced this wave of industrialization during the nineteenth century have ceased to exist. This is precisely why so much of the Lancastrian scene is Victorian in appearance. Even Manchester and Liverpool, where industrial activities are at their most diverse, succeed in creating an atmosphere that at once reflects a distinctive economy that expanded rapidly at one particular period of time. Liverpool, with a broad industrial belt ranged behind the docks from which it draws its raw materials, has preserved the atmosphere of a port. Manchester displays its specialized functions as the commercial centre of South-East Lancashire housed in a riot of blackened Gothic buildings. On the other hand, this does not necessarily mean that the economy is now in a fossil state: its visual aspect belies much that internally is still changing rapidly.

Modifications are taking place as a result of many factors, not the least of which is the changing importance of local resources. Coal mining, in particular, has suffered severe contraction largely as a result of the exhaustion of workable seams and the need to rationalize the industry at a time when competition from other fuels is increasing. Except in the Manchester area, there is now very little mining in the eastern parts of the Lancashire coalfield; pits around Oldham, for example, have long since ceased to provide power for the industry that chose the coalfield as its site. Even in the western half of the field, there has been an increasing concentration of pits on the extreme southern edge, to the south of Wigan and St Helens, and Leigh has increased in relative importance. Before 1913 the total production of the Lancashire coalfield was 26 m. tons; in 1959 it had fallen to less than 12 m. tons and faces still further contraction, though rationalization has brought an increase in mining at certain places, notably at Parkside (in the St Helens area) and Agecroft (in the Manchester area). Thus, Lancashire which had never been able to meet its own coal needs once the full industrialization of its countryside was under way, now relies on outside supplies, which come chiefly by rail from the Yorkshire coalfield, to meet half its total coal requirements. In the Flint and Denbighshire field there has been similar contraction and concentration. There is now very little coal mining in Flintshire; to the south in Denbighshire pits have moved to the concealed part of the field along its western and north-western edges. With these changes other industries, particularly along Deeside, have declined and the integrated iron- and steel-plant at Shotton relies on outside supplies of raw materials, particularly via the Mersey where Wallasey has extensive wharves from which ore is despatched by rail to Shotton.

But the most far-reaching changes have taken place in the textile towns. The loss of the export trade in the face of overseas competition either by manufacturers in former market areas, such as India, or by more effective competitors, notably Japan, in remaining markets, has brought inevitable decline. In 1913 Lancashire exported 7 m. yards of cotton cloth; in 1959 it produced barely 2 m. yards. Contraction of the industry and an improvement in the quality of products to guarantee sales in the home market are two of the many steps that have been taken (with limited success) to preserve industrial life in towns that once depended almost exclusively on textile-making. But everywhere the symptoms of continued industrial decline are apparent. First, the absolute decline of the cotton industry has reduced the population dependent on cotton by two-thirds. A decline in numbers employed in cotton continued right through the 1950's when 87,000 left the industry, and under the Cotton Industry Act of 1959 even more drastic, though rational, contraction is taking place. The Act called for a reduction of 49% of the spindles and 40% of the looms, most of which were actually idle when the Act came into force. Even so, this involves the closing of over 200 weaving sheds, principally in Burnley, Nelson, Colne, Blackburn, Accrington and Rossendale, and of about 80 spinning mills, over half of which are in Bolton and Oldham. Geographically this may well produce interesting results: the central parts of the old weaving-area, notably Blackburn, Burnley and Accrington, have suffered greater contraction than towns farther east so that weaving is no longer a dominant industry. In contrast to this, certain towns in south-east Lancashire, such as Ashton and Hyde, still retain many looms and so their relative importance as weaving centres has increased. Secondly, diversification has come, either in a natural or enforced way, to all textile towns decreasing the relative importance of cotton. Blackburn is already a town where more are employed in light engineering than in cotton and each year the total of new industries that have taken over former cotton mills increases. Little of this diversification has taken place under the Development Area Act of 1945, for the weaving area, most dependent on cotton, did not receive assistance until 1953. Thirdly, the diversification has never kept pace with the decline of the cotton industry; this is best seen in the census returns that show the overall decline in population between 1931 and 1951—in many weaving towns, for example, it was over 10%. The scale of this contraction demands an even greater rate of industrial diversification than has hitherto taken place, if many towns are to preserve their effective status. Most new industrialization has taken place as a result of the opportunities offered by cheap premises and redundant labour, though most textile towns suffer from congestion and out-of-date property and, economically, are not as attractive as they would at first seem. Since so few new factories have been built, these towns still *look* like cotton towns, grim and depressing, though they are not without a strong sense of local pride rooted in a more prosperous past.

Industrial changes in the region have not everywhere involved contraction. In South-East Lancashire, centred on Manchester, the whole range and character of industry is such as almost to defy depression on anything like the scale that so sorely afflicted other parts of the region in the early 1930's. Manchester now boasts, within a great industrial belt that runs east–west through the heart of the conurbation from Irlam to Oldham, a population of about 2½ m. and an immense range of industrial activities. Many of these, such as the finishing and clothing trades, are still derivatives of the textile industry. They are housed not only in the industrial premises of Victorian times but also in the former dwellings of its merchants, as many examples in Salford and Cheetham show with their peeling stucco. Though these houses may well indicate a very thriving aspect of the economy, where there is an acute shortage of industrial premises, they have still further contributed to the aesthetic horrors of the broad zone that encloses the city centre. In complete contrast to this is Trafford Park which has expanded so rapidly since the opening of the Manchester Ship Canal, whose total volume of traffic approached 18 m. tons in 1958. There are now some 200 factories on the estate, employing nearly 60,000 workers in a variety of industries that include processing of foods and raw materials and, more especially, heavy engineering which accounts for 40% of the total employed labour. Into this estate workers move daily from all parts of the conurbation and this has greatly added to the area's traffic problems. The Ship Canal has continued to attract industrial development along its banks, particularly industries based on imported oil. In 1954 the existing docks at Stanlow were extended and new ones added at the entrance to the canal: tankers of 30,000 tons can unload there and this has greatly stimulated oil-refining at Stanlow and Ince, and petrochemicals at Carrington Moss. Wood pulp and paper industries, established alongside the canal have also expanded during post-war years, particularly to the north of Ellesmere Port, near Warburton and at Trafford Park. The middle Mersey chemical industry, especially at Runcorn, still makes much use of the canal, and the advantages of central situation between the Lancashire coalfield and the salt of Cheshire are greater than ever. The production of heavy chemicals is highly localized in the Widnes–Warrington area but an increasingly wide range of chemicals is offered to the industries of surrounding districts such as soap, glass, textiles, linoleum, paint, leather, rubber and the metal trades. Unlike the textile industry, the chemical industry, an old-established and geographically concentrated industry, has been blessed, through technological improvements, with a rise in demand for its products. Here is the second largest chemical industry in Great Britain, and increase in employment since the war shows it to be one of the fastest-growing industries in the north-west. Even so, over-production and out-of-date methods of working have caused some contraction on the salt-field itself. The demand for and export of coarse salt has declined and there has been a general contraction of salt-working to Middlewich and Sandbach, the traditional fine salt areas. Northwich, once an important coarse salt producer, has turned over to heavy chemicals, whereas Winsford still awaits further expansion of its developing chemical industries before its population ceases to be dependent for much employment on other places, particularly Northwich and Middlewich.

Merseyside, with a population of about 1½ m., seems to combine the two features of contraction and diversification which characterize so much of Lancastrian industrial life, but peculiar to itself are the activities related to its port. Industries based on the handling and break of bulk of imported raw materials dominate the industrial zone behind the dock area. Shipbuilding and marine engineering, on the Wallasey and Birkenhead side of the Mersey, have been long established. But Merseyside industry is too dependent on Liverpool as a port. Periodic recessions in world trade and a fall in freight rates have both caused ships to be laid up; and increased overseas competition, particularly from Japan and Germany, has seriously added to the problems that face the shipbuilding and engineering industries. It is not surprising, therefore, that at the beginning of 1960, a period when the country as a whole enjoyed a high level of employment, 4·2% of workers on Merseyside were unemployed: this was twice the national average. Industrial diversification is more than a temporary expedient: the pre-war experiments on trading-estates at Fazakerley and Speke have met with success and have been extended to include other new centres at Kirkby, to the north-east of Liverpool, and at Bromborough on the Cheshire side of the Mersey. Ellesmere Port, also peripheral to the conurbation, is being considered, along with Kirkby and Speke, as a likely site for motor-car manufacturing to give employment to 21,000 workers from the centre of Merseyside. All these developments are bringing changes in the distribution of industry in Merseyside as the relative and absolute importance of portside industries in the central area declines.

It is precisely this need to re-employ workers that is directing Liverpool's overspill policy. Overspill is a pressing problem in both Liverpool and Manchester. The flight of the middle classes from the city centres has been accompanied during the last thirty years by an even greater movement of the working classes. As the economic core of Liverpool declines and new centres of industry grow up by the Manchester Ship Canal and along major roads—such as the East Lancashire Road—there is perhaps some case for it. On the other hand, the urbanization of so attractive an agricultural landscape as Wirral gives cause for real concern. In the Manchester area, where the total population of the conurbation was estimated at the 1961 census to be over 2,400,000, unsatisfactory housing conditions more than economic depression have compelled the city-dweller to sprawl farther and farther into the country-side, particularly into Cheshire. Villages such as Cheadle Hulme, Wilmslow and Bramhall have changed into dormitory towns. This rapid outward spread has been made possible by the network of communications that radiate from the city: roads and suburban electric railways take thousands of commuters into the city daily, and traffic conditions worsen. Added to this there have been positive experiments by the city authorities to re-house much of its population, for example, in Wythenshawe where nearly 90,000 now live. Though many improvements in the general standard of housing have undoubtedly taken place—especially in Salford where congestion, lack of land and bad housing were all at their worst—the general expansion outwards is not merely to be regretted but strongly to be deplored. Much valuable agricultural land is lost, the city is surrounded by characterless dormitory suburbs and inter-county planning difficulties are aggravated as Cheshire's pride is hurt by what it can only recognize as piracy. The towns of the northern ring are not equipped either economically or socially to face great influxes of city people who remain outside the old community: in Worsley, for example, over 60% of its new inhabitants travel daily into Trafford Park and central Manchester. This emphasizes the real nature of the present problem, that the best opportunities for employment are still offered by Manchester's industrial collar, which exists around a ruined area that could well be the scene of more vigorous planning measures. There is a strong case for the reclamation of derelict land in south Lancashire for re-housing schemes; there are schemes in an incipient stage but they are likely to prove very costly.

The present distribution and character of industrial estates (fig. 56) summarizes in many ways what is happening to urban and industrial life in the region. Trafford Park shows clearly the abiding advantages of canal-side location for industry that has vigorously responded to modern technological progress. In Burnley, east Manchester, Wrexham and Bolton, industrial estates represent the need to diversify industry where the staple industry has passed into decline. Others around Liverpool are the outcome of both depression and overspill, while Wythenshawe illustrates the positive measures taken by Manchester to solve overcrowding in the city centre. With the exception of Trafford Park and others along the canal, such as Partington, the industrial estates owe their precise location to road services that have permitted, in recent years, an increasingly flexible distribution of industry, particularly of a lighter type, here as in the rest of Britain. Not all can be counted dazzling successes of the stature of Kirkby: at the other end of the scale there is one that boasts only a single pickle factory.

Industrial development and continued change have had their

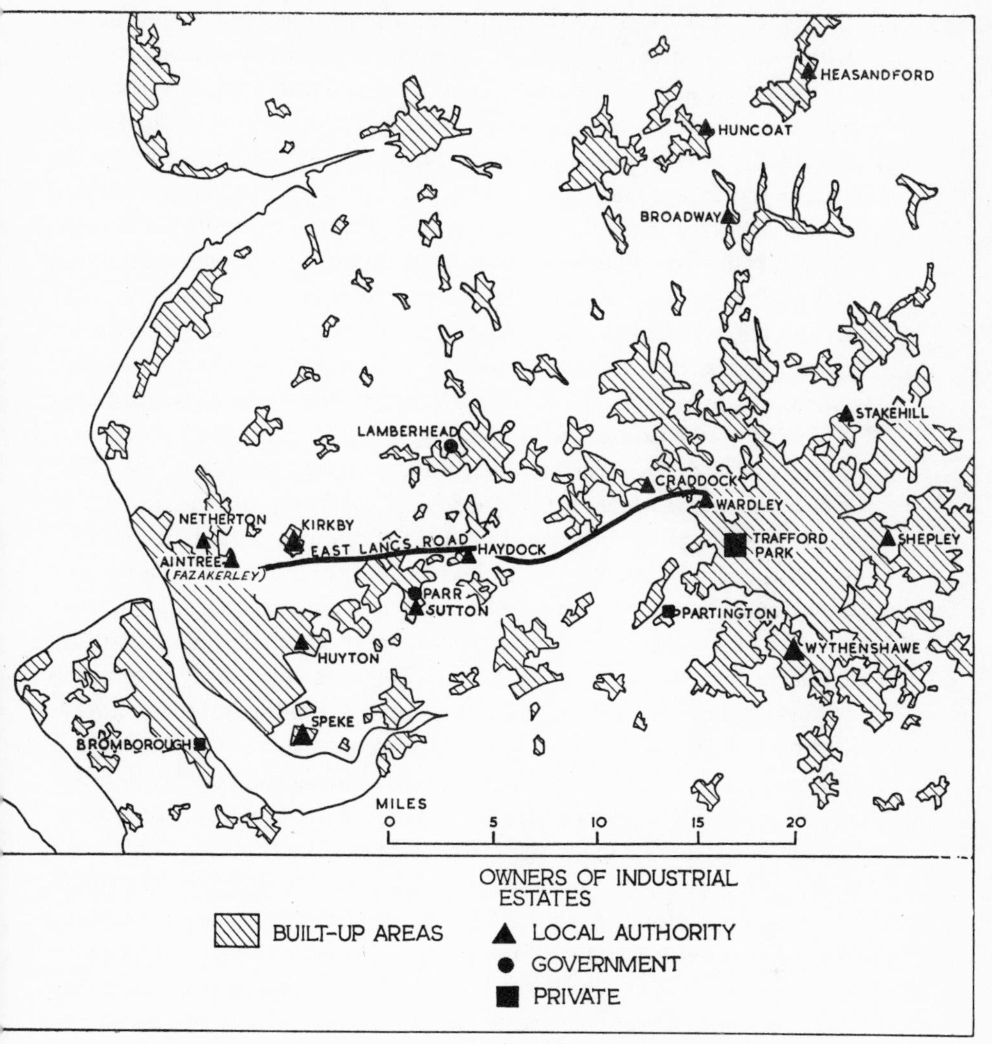

Fig. 56. Industrial estates of South-East Lancashire and Merseyside.

effects on the Lancastrian scene outside the industrial areas as well as within. Not the least significant of these has been the growth of holiday resorts along the Lancashire coast, now the playground of industrial workers from far beyond the Mersey and the Pennines. From Fleetwood to Lytham there is an almost unbroken line of building, at the centre of which is Blackpool (152,113). A thriving and characterful resort, it has grown to dominate the economic life of the Fylde coast. Good road and rail links have made it easy of access from the industrial hinterland. With an excellent beach, washed by tides that would cut away the boulder-clay cliffs were it not for the promenade wall, Blackpool has always been able to offer seaside amenities superior either to Morecambe, where tides daily expose the dreary expanse of mudflats in the bay, or Southport, where coastal deposition has made that resort virtually lose sight of the sea. St Annes, immediately to the south of Blackpool, is indeed a creature of the Industrial Revolution. Here enterprising Rossendale footwear manufacturers bought land and built on it to create a residential town, with some of the amenities of a resort, for retired people and east Lancashire commuters. Southport (81,976), too, provides a dormitory for executives who travel daily by excellent rail services to Liverpool and Manchester. Fleetwood (27,760), still a port, sends its fish by fast rail services into Lancashire. Lytham, on the other hand, once a small port with small-scale shipbuilding, lost its trade with the silting-up of the Ribble and with the rise of Preston as a port; it now enjoys a quiet, respectable existence as a residential resort for retired east Lancashire industrialists and maintains its Victorian calm in the face of Blackpool's sprawling vulgarity. But, as in many other parts of Lancashire, the effects of rapid industrial growth have not been entirely beneficial to the Fylde coast resorts. They too have fallen victim to over-specialization on trades associated with holiday-making, and during the winter months there is severe unemployment, equal in January 1960, for example, to that of Merseyside, which qualifies them for assistance under the Local Employment Bill.

Even in predominantly rural areas, industrialization has made its mark. The wide range of agricultural activity is in many ways a reflexion of the variety in local physical conditions, such as the contrast between sheep and stock cattle raising in the Pennines, and arable farming on the plain of south-west Lancashire. On the other hand, the effect of many natural physical differences has been obscured by changes that have come with the growth of nearby urban markets. Demand for food from these areas has done much towards the reclamation of land formerly considered physically unattractive. In other areas it has permitted the survival of agriculture where for every other reason it might well have declined. The boulder clays and loams of the plain are not everywhere naturally suitable for agriculture: in certain areas where local drainage was bad there had developed since glacial times peaty mosslands. With a relatively high rainfall, well distributed throughout the year, these acid peats rose to a level higher than the surrounding land, created their own water-table and so aggravated the difficulties of local drainage. Such mosslands were particularly extensive east of the Wyre estuary, for example Pilling Moss; in a broad belt to the east of Southport, where Martin Mere was the most extensive; to the north-west of St Helens, as in Reeds Moss and Holiday Moss; and along the Mersey where, to the west of Manchester, Chat Moss, Barton Moss and Carrington Moss were the largest. Reclamation, involving the burning of the vegetation, the draining and cutting of the peat, and the use of fertilizers, did not take place on any substantial scale until the nineteenth century, that is, not until demand from urban centres made possible profitable returns from improved agriculture. Though the process is not as yet complete, the transformation of the landscape has been most remarkable. The Martin Mere area is the scene of most intensive agriculture not dissimilar to the best of Fenland farming.

South-west Lancashire as a whole is now a region of highly intensive arable farming, producing roots, cereals and vegetables in rotation. During the 1950's there was a conspicuous increase in pea growing, a response to the increased demand for canned and frozen peas. The Fylde area, too, is one where agriculture has been transformed by reclamation and land improvement to supply the special needs of industrial areas. There is much arable farming plus a marked specialization in egg production. The Ribble valley, with its heavier soils, has many farms rearing cattle. Others as in small tributary valleys such as the Whitewell are prosperous, raising sheep, cattle, pigs and poultry, and are highly mechanized on account of the difficulty of obtaining farm labour. Intensive market gardening is found to the south of Manchester, as around Timperley: this is yet another example of the effect of nearby urban markets which, in this case, has caused a gradual change-over from extensive arable farming in what is a naturally productive area. Farther south, even in rural Cheshire proper, industrial growth has made its mark. Though the heavier loams favour intensive dairy farming (in contrast to the lighter loams of south-west Lancashire where arable farming has persisted), the importance of good road and railway services that link the area to the urban markets, both to the north and to the south, cannot be discounted.

Pennine farming, on the other hand, is neither intensive nor highly productive. Acid upland soils, remoteness and a high incidence of rainy and overcast days have always made farming very difficult, restricting it at first to stock raising and hay production. The very uncertainty of farming helped to foster, as a supplementary occupation, the domestic textile industry, the forerunner of the modern factory industry in these valleys, and may well have encouraged the existence of small farms that, as economic units, lead a very precarious existence at the present time. On the other hand, their extinction has been prevented by exactly the same factors that have stimulated more progressive agriculture on the plain. Small though the industrial towns are in these valleys, they nevertheless guarantee a demand for milk that has enabled the Pennine farmer to survive. Dairy farming, supported by imported hay and lime fertilizers, is often carried on to altitudes of 1,000 ft. Many farmers are still producer-retailers of milk, though in post-war years the marketing has been taken over in many areas by larger dairies. Stock raising for the Ribble valley and Cheshire still continues, but the farmer has to rely on government assistance as well as on supplementary income earned often by members of his family in the industrial towns which he supplies.

Thus, even in Pennine farming, there is that strong link between upland and lowland, town and country, industry and farm, so characteristic of the region as a whole. It is, moreover, a link that has long existed in spite of the changes that, particularly in industrial development, have created so many problems that beset the region in so many different ways. It is to be hoped that these problems will never be solved purely in favour of the sprawling city.

Selected Bibliography

T. W. Freeman. *The Conurbations of Great Britain*. Chapter 4. Merseyside, Chapter 5. The Manchester Conurbation. Manchester, 1959.

L. P. Green. *Provincial Metropolis*. London, 1959.

Lancashire and Merseyside Industrial Development Association: *Industrial Reports*, Manchester, 1947–56.

R. Millward. *Lancashire*. The Making of the English Landscape. London, 1955.

Wilfred Smith (ed.). *A Scientific Study of Merseyside*. Published for the British Association for the Advancement of Science, Liverpool, 1953.

CHAPTER 22

CUMBRIA

F. J. Monkhouse

There can be little doubt concerning the regional identity of the corner of north-western England enclosed between Morecambe Bay and the Solway Firth. In the west the blunt headland of St Bees projects its red rocks boldly into the Irish Sea; in the east the fault-line scarp of the northern Pennines rises steeply from the Vale of Eden. Only in the south-east is the boundary at all vague, where the Shap and Howgill Fells form an upland link between the Lake District hills and the Pennines.

Parts of three counties (fig. 57), whose boundaries meet at the Three Shire Stone on Wrynose Pass, lie within the area. Most of Cumberland and Westmorland are included, though not their Pennine fringes; Cumberland indeed by a quirk of history extends on to the Pennine plateau in the parish of Alston, which to emphasize its anomalous position belongs to the diocese of Durham. Lancashire is represented by the Furness Fells and the lowlands bordering the northern indentations of Morecambe Bay. To this part of north-western England can be applied the name Cumbria, after the British kingdom of the *Cumbri*, which considerably antedates the county of Cumberland itself, a Norman creation first appearing under its modern name in 1177.

The heart of Cumbria is the Lakeland Dome, in the words of John Ruskin, a 'large piece of precious chasing and embossed work'. It forms a compact area of some 800 square miles of mountains and radiating valleys, smooth fells and bold crags, tiny mountain tarns and long deep lakes. This Dome is composed of a mass of Lower Palaeozoic rocks (fig. 58). The oldest, known collectively as the Skiddaw Slates, consist of a series of darkish slates interspersed with coarse grits, of Upper Cambrian and Lower Ordovician age. Most of the northern hills, in particular Skiddaw, Saddleback and the Grasmoor group, are formed of these rocks, and an outlying mass in the south-west rises in the rounded hump of Black Combe. Though

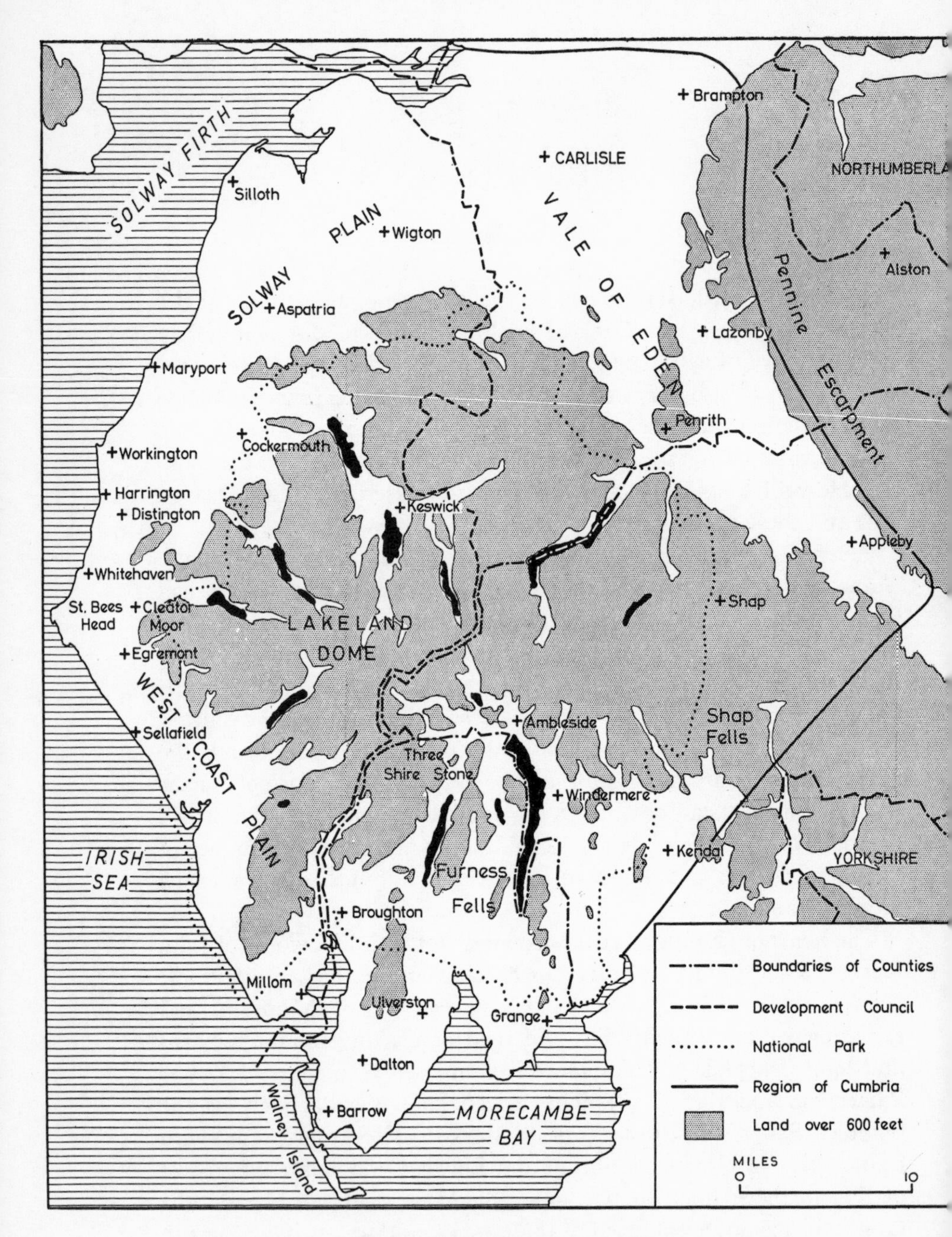

Fig. 57. Physical features of Cumbria.

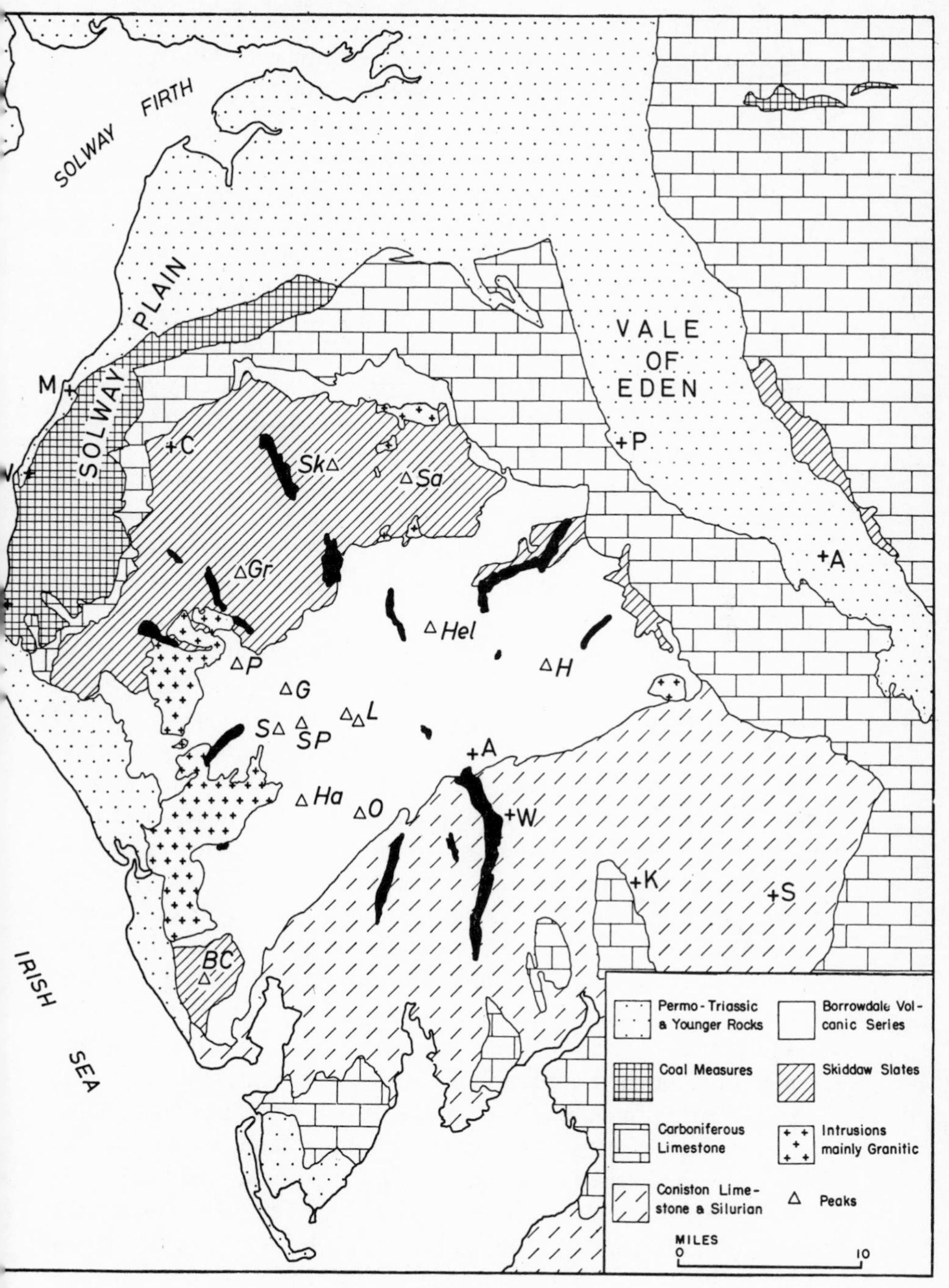

Fig. 58. Geology of Cumbria.

slopes are in many places steep, the hills are generally smooth in profile, for the homogeneous rocks weather more or less uniformly.

A vast thickness of lavas, fine-grained ashes and coarse agglomerates was ejected during a long period of volcanic activity in Ordovician times; these are known collectively as the Borrowdale Volcanic Series. They offer a most varied resistance to denudation, with the result that the mountains in the centre of the Dome (Scafell, Scafell Pike, Great Gable and Pillar) have strikingly rugged outlines. To the south of the Borrowdale Volcanics, younger rocks form the more subdued country around Windermere and Coniston, where few eminences exceed 1,000 ft. There is first a narrow outcrop of Coniston Limestone (usually included in the Ordovician system) along the line of a prominent tear-fault, succeeded by an extensive series of slates, shales and grits of Silurian age. In addition, intrusive masses of granite and other complex crystalline rocks have been revealed by denudation; these include Carrock Fell to the north of Skiddaw, Shap Fell in the east, and others in the Ennerdale and Eskdale districts.

These rocks of central Cumbria have been profoundly affected by several orogenic phases, the first of which, the Caledonian, occurred towards the close of the Silurian. The rocks were folded into a broad ridge, with a trend more or less from east-north-east to west-south-west approximately through the position of Skiddaw. This uplifted mass was subsequently reduced to an undulating peneplain. In the north, along the main line of uplift where denudation was concentrated, both the Silurian and Borrowdale Volcanic rocks were removed, revealing the Skiddaw Slates. In the centre only the Silurian rocks disappeared, so exposing the Borrowdale Volcanics, while in the lower land to the south the Silurian rocks were preserved. This in fact is the present broad pattern of the rocks, in spite of later vicissitudes.

On to this worn-down surface Carboniferous rocks were deposited, remarkable for their variability both in thickness and lithology. The Carboniferous Limestone accumulated to a great thickness in the flanking depressions in north-eastern Cumberland, in the Vale of Eden, and in Furness; but it thinned out on the actual margins of the Dome, which probably survived as an island protruding from the 'Carboniferous sea'. The Millstone Grit is but poorly developed and can be recognized only near Whitehaven. The Upper Carboniferous includes a thousand feet of productive Coal Measures, together with the 'Barren Red' or Whitehaven Sandstone series.

Towards the end of the Carboniferous, the Hercynian orogeny uplifted the rocks once again more or less along the previous Caledonian fold-trends. Prolonged denudation removed the Carboniferous rocks except from the outer margins. Once again a worn-down surface was formed and on this Permo–Triassic rocks were deposited, in places upon the Carboniferous Limestone, elsewhere on the older rocks; they survive as the Penrith Sandstones in the down-warped trough of the Vale of Eden, the St Bees Shales and the massive dark red St Bees Sandstone. Jurassic and perhaps even Cretaceous rocks were also probably widely deposited, though now these are represented only by a low-lying plateau of Lower Liassic shales and limestones to the west of Carlisle.

In mid-Tertiary times the whole region was once more uplifted into an elongated dome with an east–west axis through Scafell Pike. At the same time, the long, continuous Pennine faults, probably initiated during the Hercynian orogeny, were resurrected and so emphasized the trough of the Eden valley. Another prominent fault trends north-eastward from near Maryport, with its down-throw to the north; though hardly appreciable on the surface, this defines the boundary between the Palaeozoic rocks (particularly the Coal Measures) and the basin to the north infilled with New Red and some Liassic rocks.

A prolonged period of denudation followed as the rivers flowing outwards from the elongated Dome cut down their valleys, and the newer rocks, which had determined the initial direction of the drainage, were stripped away from the higher parts. The Carboniferous and younger rocks survive as a discontinuous rim around the Dome and more extensively in the Solway Basin in the north, while a long re-entrant of Permo–Triassic sandstones extends southward along the Vale of Eden; numerous minor though distinctive cuestas have developed in these younger rocks. The river systems are indeed a superimposed legacy of the vanished cover. The Derwent flows to the west; the Ehen, Esk and Duddon to the south-west; the Crake, Leven and Kent to the south; the Eamont and Lowther to the north-east via the Eden; and the Petterill and Caldew to the north, also ultimately to the Eden. It has been stressed, however, that the Dome was elongated along a west–east axis, and as a result this drainage pattern is not truly radial.

The effects of the Quaternary glaciation are to be seen dramatically in the Lakeland landscape. The glacier-straightened and deepened valleys contain sixteen major lakes, mostly long and narrow, radiating from the drainage axis of the Dome. They owe their existence in part to morainic damming, in part to glacial overdeepening; Wastwater, for example, 258 ft deep, has its surface at 200 ft O.D. The high-lying combes with their steep back-walls, craggy arêtes and tiny tarns from which streams fall abruptly in cascades to the floors of the main valleys, the truncated spurs along the valley-sides, and the numerous roches moutonées are other familiar glacial contributions to the mountain landscape. The effects of glaciation were not merely erosive, even in the uplands, for much boulder clay was deposited, sometimes in level sheets (as on the floors of Langdale and Borrowdale), elsewhere in a chaos of hillocks (as in Ennerdale), or as terminal moraine near most valley-mouths. The bordering lowlands were much more thickly plastered with drift, laid down both in undulating sheets and as swarms of drumlins. The last can be seen particularly clearly between Wigton and Carlisle, and in the Vale of Eden between Penrith and Appleby.

Since the Quaternary snowfields and glaciers finally disappeared, the weather and the rivers have continued their slow but inexorable work of earth-sculpture. Slopes of frost-shattered scree stream away beneath each rock buttress, the torrents carry down their burdens of debris, and the lacustrine deltas they build are slowly filling the lakes; some indeed have already disappeared, as in Kentmere, others have marshy 'bottoms' at their heads which appreciably reduce their former lengths. Bassenthwaite and Derwentwater are separated by an alluvial flat deposited by the River Greta and the Newlands Beck.

On the slopes of the Dome the rocks are at or near the surface; the soils are thin and immature, siliceous and base-deficient, and there is much bare rock and scree. Many of the flatter summit-ridges are covered with horizontal sheets of crumbly, gravelly material, the product of rock disintegration *in situ*, while others are covered with angular blocks of rock of all sizes. Stony clays swathe most of the valley floors and lower slopes, patches of alluvial soil occur at the lake-heads, and occasionally lacustrine clays have been deposited in now vanished proglacial lakes. Even on some valley floors torrents have deposited sheets of gravel and in places boulders, as at Wasdale Head, and elsewhere the alluvial flats are marshy and reed covered.

Though records are limited, it seems that the western foothills receive a mean annual precipitation of 40 in. to 80 in., and most of the central Dome has over 100 in.; indeed, on either side of Stye Head the annual total averages 150 in., and in 1928 no less than 250 in. were recorded. Some of the precipitation may fall in the form of snow, but this is extremely variable from year to year and from place to place.

The combined result of relief, soils and rainfall is a vegetation cover over much of the Dome which at its best can be broadly described as 'moorland and fell-grazing', characterized by stretches of *Calluna vulgaris* (inaccurately called 'heather-fell'), with bilberries among the screes, and with poor *Nardus* and *Molinia* grasses. On the flatter summit ridges there are considerable areas of bog, with sphagnum and cotton-grass. Only on the outer limestone hills do tracts of better fescue-pasture appear. The lower slopes are widely covered with bracken, a menace to the hill-farmers, since it ruins pastures, is difficult to check, and is virtually impossible to eradicate.

The agricultural environment is thus limited. A Lakeland farm consists of a few stone buildings, with surrounding valley-lands enclosed in tiny stone-walled fields, and from 500 to more than 2,000 acres of fell-grazing, usually unenclosed but sometimes in huge fields surrounded by dry-stone walls which climb steeply up the mountain sides. In the tiny valley fields are grown some root-crops or patches of oats, but most of the fields are meadow used for early summer lambing and for hay. The fell-grazing can sustain barely one sheep to the acre and only hardy breeds of sheep can be reared: in the west the sturdy little Herdwicks, in the east the Rough Fell breed. After their first winter on the Solway marshes or in such sheltered districts as the Vale of Lorton, the sheep spend the rest of their existence on their own fell-farms, which are usually sold or let with the flocks. Cattle include a few dairy animals for local milk supplies, and an increasing number of dual-purpose Shorthorns and occasionally Galloways, which in summer graze far up the slopes, providing stores for fattening elsewhere.

One alien element in the Lakeland landscape is provided by coniferous plantations, the introduction of which has in fact caused considerable controversy. Indeed, since the formation of the Lake District National Park (fig. 57) further planting has been excluded from the central part of it. The Forestry Commission has developed five main forests: Thornthwaite on the shores of Bassenthwaite and on the hills around Whinlatter Pass, Ennerdale, Hardknott lying between Eskdale and Dunnerdale, Grizedale between Coniston and Windermere, and Greystoke to the west of Penrith. The Manchester Corporation has also planted 10,000 acres on the catchment-area around Thirlmere, a district from which sheep and the public were long excluded, though this policy has recently been ameliorated in favour of the latter.

The Lakeland valleys open outward into the Vale of Eden, the Solway and West Coast Plains, and the coastlands of Morecambe Bay. These marginal lowlands with their gentle slopes, broad river valleys, better soils, and appreciably drier climate are obviously much more favourable for agriculture than is the Dome. Workington on the west coast has a mean annual precipitation of 37 in., while a 'tongue' of rainfall of only 30 in. to 35 in. extends south-eastwards from the Solway Firth to near Penrith. Most of the lowland soils are derived from drift, and considerable areas are covered with rather heavy damp clays, though in parts they are interbedded with sands and gravels. In the Eden valley and in south-western Cumberland tracts of reddish sandy-loams, derived in part from the drift, in part from the underlying New Red rocks, afford some of the best arable soils of the region, though in places, by contrast, the drift-free sandstones bear only a thin dry heath-covered soil. A light clay-cover on Carboniferous Limestone usually affords an excellent workable soil. Further variety in soil-cover is provided by the alluvium along the lower valleys and estuaries of the Eden, the Esk, and the Duddon, and by deposits of marine alluvium along the Solway coast. The last, known generally as warp, ranges from dark-coloured clays to sandy brown loams; in places it is waterlogged or covered with salt-marshes, elsewhere it grows the superb 'Cumberland sea-washed turf' used widely for bowling-greens, and where well drained it affords excellent farmland. Some lowlying areas of both warp and drift, for example Solway Moss and Bowness Moss, are peat-covered; where drained these afford black organically rich soils.

Farming of these lowlands is truly mixed. Except for the higher parts of the Eden valley (which is almost entirely under permanent pasture), arable occupies between 30% and 40% of the total area. Though wheat was once important, the chief crops are now potatoes, oats, and roots grown mainly as stock-feed. A few areas concentrate on horticulture, as to the south-west of Carlisle (including the highly productive holdings of the Land Settlement Association), and in the Brampton district with its light sandy loams. The main emphasis is, however, on cattle-rearing, and a significant development has been the increase in the number of pedigree attested Dairy Shorthorns. Milk production has been stimulated by the opening of a large Milk Marketing Board depot at Aspatria, and of several milk-processing factories. One firm which has manufactured cream, butter and milk-powder for some years in the Carlisle district has recently opened a large-scale processing-dairy at Distington to cope with increased milk-production in West Cumberland. Other stock, such as pigs and poultry, are produced, and there is a large egg-packing depot at Lazonby, to the north of Penrith. Sheep form part of the mixed farming economy; various crosses with Suffolk rams produce fat lambs for the markets at Carlisle, Cockermouth, Lazonby and Penrith. In the Solway plains some flocks of 'lowland sheep' (such as Border Leicesters) are maintained, as well as some Wensleydales in the Furness district and Swaledales in the southern Eden valley.

In all, then, the agricultural economy of Cumbria is one of considerable diversity, largely the result of the fundamental contrast in physical background between the Lakeland Dome and its surrounding lowlands.

The outcrop of the Coal Measures surviving on the north-western flanks of the Dome forms the Cumberland coalfield (fig. 59). This consists of three main districts: a narrow north-eastern portion, a western field extending inland for 5 miles from the coast between Maryport and Whitehaven, and an undersea field which lies offshore for an unknown distance. To both north and south occur deeply down-faulted concealed fields; borings through the post-Carboniferous strata to the south of St Bees have reached coal, and a large 'Solway Basin' to the north of Aspatria lies under at least 3,000 ft of New Red Sandstone. The National Coal Board is concerned mainly with the undersea deposits; 'proved reserves' are estimated at 310 m. tons, and 'probable reserves' within 5 miles of the coast are given as a further 260 m. tons.

Although there was mining by the mid-sixteenth century, and indeed Whitehaven was for long one of the main English coal-ports, this has always been a difficult field to exploit. The inland section is

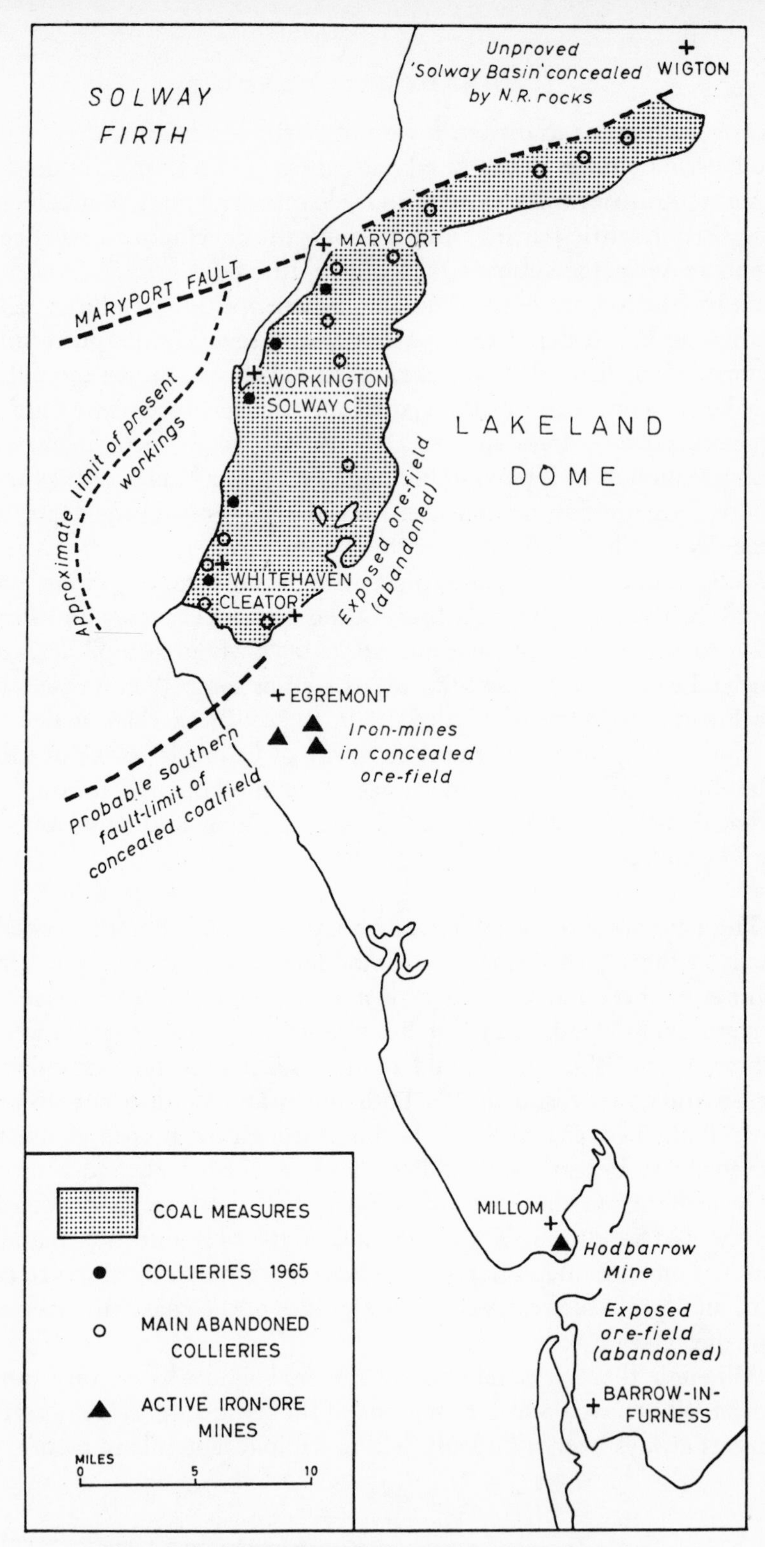

Fig. 59. Coal Measures, collieries and ironworks of Cumbria.
In 1966 all the collieries may close.

heavily disturbed by faulting, so that the Measures are 'locked' into small pockets, each of which has had to be worked by a separate colliery. The seams are thin and discontinuous, and methane is prevalent, resulting in a high accident rate and there have been several major disasters, one of the worst at the Haig Pit near Whitehaven in 1922. The undersea field also presents difficulties, for in places the seams are too near the sea-floor to be worked with safety.

By the beginning of the twentieth century about 2 m. tons of coal were raised each year, and a maximum of 2·3 m. tons was produced in 1923. But by then the best inland coal had been removed; the small collieries closed one by one, so that by 1936 output was down to 1·3 m. tons. Before the war efforts were made to develop further the undersea deposits by sinking the Solway Colliery near Workington, and after the war the Haig Pit was reorganized. But still more collieries have recently been closed, and in 1965 only five were in operation (fig. 59). About 874,000 tons were produced during the fiscal year 1964–65, at a heavy financial loss. Nevertheless, this coalfield has long contributed to the industrial life of West Cumberland, and indeed continues to do so, for a third of its output is consumed in coke-ovens near Workington. The coal is mostly of high-volatile caking quality, and has to be blended with Durham and Welsh coals to produce the necessary quality of metallurgical coke. Some is used for domestic and general industrial purposes, or is shipped to Northern Ireland and Eire from Whitehaven and Workington.

A second contribution to industrial development is the occurrence among the Carboniferous Limestone of high-grade haematite ores, with an iron content of 40% to 62%, a silica content of 5% to 15%, and an extremely low phosphorus content. This haematite was deposited as veins, flats and vertical masses locally known as 'sops' by mineralized solutions, derived from the former overlying Triassic sandstones, percolating along lines of weakness in the limestone. One group of ore-fields is exposed, except for a thin boulder clay cover, within a narrow irregular band from Lamplugh to Egremont; quarries and shallow pits with their spoil-heaps dot the landscape near Rowrah, Arlecdon, Frizington, Cleator and Cleator Moor. Two other fields occur, at Hodbarrow on the shores of the Duddon estuary near Millom and in the neighbourhood of Dalton-in-Furness. These three deposits were vigorously exploited during the nineteenth century to supply the rapidly growing iron industry of West Cumberland and

Furness, and also for export to South Wales. The annual output exceeded 2 m. tons for many years, and more than 3 m. in 1882; probably 150 m. tons have been removed during the last century. But this ore was so readily accessible that it has now been largely exhausted. Soon after the First World War the mines on the west Cumberland field closed, leaving the red desolation of the Cleator district. The Furness output likewise declined to an output of only 116,000 tons in 1938, and now these mines too are entirely worked out. Only Hodbarrow in the exposed fields is still productive.

However, the excellent quality of the ore stimulated exploration for a possible field concealed beneath the New Red rocks to the south of Egremont. Many borings were sunk, mostly with negative results, but one considerable ore-body, now worked at three mines, has been found; the Haile Moor mine, opened in 1941, is one of the most modern in Europe. The output of these three mines, and of Hodbarrow, is however dwindling; in 1937 the output was 857,000 tons, but this had fallen to 281,000 tons in 1964. Moreover, as the best ores have been removed, the iron content is now lower than it was, averaging about 48%, though this is still much better than the Jurassic ironstones of the English Midlands.

Yet in spite of this decline in the output of both coal and ore, these minerals have long afforded the basis of a flourishing metallurgical industry. Moreover, the Carboniferous Limestone outcrops provide furnace-flux, and several large quarries are active; the one at Rowrah, for example, produces annually 125,000 tons for the Workington furnaces, and others are worked near Millom and in Furness. A coastal location has helped industry; the Workington Dock imports 750,000 tons of iron ore each year from North Africa, Sweden, Sierra Leone and Spain.

There are three main iron and steel-making districts: in West Cumberland, South Cumberland, and Furness. During the nineteenth century numerous small companies began operations, and in all forty-five blast-furnaces were active in 1870; a decade later, in fact, Cumberland produced no less than 12% of England's total pig-iron output.

It was near Workington that Henry Bessemer developed his method of steel-making, and the first Bessemer steel plant and rolling-mill were here put into operation in 1877. Later, however, the individal companies along the coast between Maryport and Workington gradually amalgamated to form the Workington Iron and Steel Company, which itself became a branch of the United Steel Companies in 1919. These amalgamations inevitably meant that the small furnaces went out of blast and in due course they were dismantled. They have been replaced by a single integrated steelworks, with three blast-furnaces, two Acid Bessemer converters, and two large rolling-mills, grouped near the coast to the south of Workington, with huge slag-banks extending along the foreshore. Integration also includes an associated company which uses the coke-oven by-products at the nearby Lowca tar-distillation plant and contributes excess coke-oven gas to the West Cumberland grid. A brickworks at Micklem produces refractories. The main product of the Workington steelworks is rails, for which acid steel is particularly suitable, together with sleepers and fish-plates, both for home use and for export. An associated concern, established at Distington at the beginning of the Second World War to make certain special steels, now carries out iron-founding and general engineering, using pig from the main works; indeed, the output from its foundry is the largest of any similar unit in Britain, and it also produces mine-cars, steelworks plant, and paper-making machinery. The Workington Iron and Steel Company, with its associates, is in fact the largest single employer of labour in West Cumberland, with about 5,000 workers.

In South Cumberland, near Millom, two modern blast-furnaces produce annually a ¼ m. tons of haematite pig, and a foundry makes special large-scale iron-castings. These works, on the western shores of the Duddon estuary, are served by the company's own wharf.

On the opposite side of the Duddon is Barrow-in-Furness, with a population of 64,824, yet just over a century ago it was only a village of 300 people. This remarkable growth has been partly the result of its development as a port, protected from the open sea by the long curve of Walney Island. When Gladstone opened the Devonshire Dock in 1867, he referred to Barrow as 'the youngest child of England's enterprise'. The port, now with three main docks, is situated to the south-west of the town, and can handle vessels of up to about 10,000 tons.

Though ore mining and smelting were active in Furness as early as the twelfth century, it was in 1859 that the first blast-furnaces were built on a coastal site to the west of the town, where the present steelworks now stands. The Barrow Iron Works, however, ceased activity in January 1963. The steelworks produces ingots from its open-hearth furnaces, and general engineering is carried out. The main activities comprise shipbuilding and marine engineering, and 11,000 workers are employed by the two associated firms of Vickers-Armstrong (Shipbuilding and Engineering). Though there was much industrial depression in the 1920's and 1930's, and though heavy damage was sustained in the Second World War, the yards today are highly productive and well equipped. There has been specialization in the building of naval vessels, especially carriers and submarines; the first British nuclear-powered submarine was built there. Several passenger ships have been built, including the world's largest all-welded liner (the *Orsova*), and two 47,100-ton oil-tankers have been recently constructed for a Greek firm. Other products are floating-docks, caissons, dredgers, ice-breakers and industrial plant generally. Vickers instal their own marine machinery and produce armaments for warships.

The production of coal, iron and steel is not the only industrial activity in Cumbria, though many others are on a small scale. The attractive green slates from the Borrowdale Volcanics are quarried near Buttermere and Coniston, granites are worked at Shap, Threlkeld and Eskdale, and mines of lead, zinc and copper have been exploited sporadically for centuries, though not in recent years. Some of the towns around the Dome have developed various industries. Kendal, for example, has long been a centre of textile manufacturing, originally based on local wool, and it still produces high-quality woollens, and also footwear. Near by the pure waters of the River Kent supply several paper-mills. In Furness several towns, in addition to Barrow, carry on a variety of industries. Ulverston, for example, though primarily an administrative centre for a large agricultural area, has a modern factory making antibiotic compounds, two firms producing electrical accessories, an old-established tannery, and several light engineering works. Most of the market towns (Penrith, Appleby and Cockermouth) have food-processing industries and agricultural machinery depots.

The border city of Carlisle, the administrative county town, has a population of 71,112. It owes much to its situation near the west coast entry to Scotland, and at the convergence of routes from the south via the Eden valley and Shap, from West Cumberland, and from the east through the Tyne Gap. Its industrial activities are diverse, and include the production of textiles (notably at the Holme Head works which employ a thousand people), food-processing, and the manufacture of boxes and containers. An engineering firm, established for over a century, makes cranes, turn-tables, wagons, railway and dockside machinery.

It is apparent that despite these developments at several towns, the industrial scene in Cumbria has for long been dominated by heavy industry. This dependence is dangerous, especially in times of industrial depression, and moreover the lack of variety not only offers little opportunity to give women work, but causes a general drift of population out of the district. Barrow, in spite of its flourishing shipyards, has been anxious in recent years to diversify its activities. It formed in 1946 a Development Committee, backed by the Lancashire Industrial Development Association, and has successfully established, particularly on an estate east of the town near Roose, paper-making, flour-milling, worsted-spinning, the manufacture of ready-made clothing, knitting-wools, rugs and furnishing fabrics, and the production of laundry machinery. Again, an industrial estate has been developed near Carlisle at Durranhills, with new factories such as one producing fibre-glass crash helmets.

Still more striking have been developments in West Cumberland. After the onset of the world depression in 1930, unemployment became widespread in West Cumberland, involving in some districts as many as 80% of the working population. Under the Special Areas Development and Improvement Act of 1934, West Cumberland became one of the scheduled areas. The Cumberland Development Council (fig. 57), established in 1935, induced the government to create two years later the West Cumberland Industrial Development Company, with the aim of acquiring sites for new industries. The first, the Solway Estate, was established in a 'hard core' area of unemployment to the south of Maryport, and others were developed along the coast near Workington and Whitehaven, at a few centres in the derelict mining and ironworking areas, and in some of the market towns of the county. In addition to government-sponsored factories, several large firms were encouraged and helped to expand their existing installations or to build new ones. The net result is that today more than two-thirds of the working population are engaged in types of employment new to the district during the last fifteen years. All these industries have certain common advantages: spacious sites, adequate labour supplies, local coal, and plentiful water from the rivers or lakes.

The variety of the new industries is impossible to describe in a short space; they include the manufacture of woollen and silk textiles, berets, buttons, furniture, plastics, carpets, canned goods, electrical fitments, hosiery, handbags, boots and shoes, surgical instruments, lawn-mowers, computers, wrapping paper and thermometers. Several metallurgical firms produce non-ferrous alloys and castings, and one factory at Distington has the largest drop-forging hammer in Europe. The firms include many nationally and internationally famous names, and several concerns were established by refugees from central Europe. One enterprise can be mentioned in more detail as an example. In 1940 an immigrant from Austria began to produce fire-lighters at a modest factory near Hensingham; out of those beginnings has grown the huge Marchon establishment on the site of the former Ladysmith Colliery to the south of Whitehaven. Today Marchon manufactures bases for detergents, toilet preparations and cosmetics, sulphuric acid, synthetic alcohol, and cement. One of its main raw materials is anhydrite; though formerly imported, vast reserves were recently discovered in the New Red Sandstone near St Bees Head at Kells, and now 9,000 tons a week are mined virtually on the spot. The company's own ships, *Marchon Trader*, *Marchon Venturer* and *Marchon Enterprise* bring another essential raw material, phosphate rock, from Casablanca in North Africa to Whitehaven harbour; 210,000 tons were imported in 1964.

It is fitting to end this industrial survey with a reference to two establishments operated by the Industrial Group of the United Kingdom Atomic Energy Authority, on the site of a Royal Ordnance Factory on the coast near Sellafield. The twin pencil-slim chimneys of the Windscale atomic reactors, flanked by the massive cooling-towers of neighbouring Calder Hall, the world's first atomic power-station, which put electricity into the National Grid on 17th October 1958, are pointers to a new age.

This coastal fringe of Cumbria is obviously one of Britain's industrial regions. Yet its people are always conscious of 'the bold back-cloth of the fells', living and working as they do between the mountains and the sea. Within a short distance of this thin 'industrial crescent' is the Lake District National Park, of 866 square miles (fig. 57), established in 1951, and administered by the Lake District

Planning Board, whereby it is hoped that the natural heritage of mountain and lake may be preserved unspoilt. One potential threat arises from the fact that the lakes form obvious natural reservoirs; water is extracted from Thirlmere and Haweswater for the Manchester conurbation, while Crummock Water and Ennerdale Water supply the Workington and Whitehaven districts respectively. Manchester is at present (1965) actively exploring possibilities of increasing its supplies from Lakeland, notably from Windermere, but proposals so far have met with widespread opposition, since it is feared that the natural amenities may be thereby impaired. Since the eighteenth century, visitors have sought rest and relaxation in the Lake District, and holiday-makers come in numbers at all times of year. The tourist industry, though not the only industry of Cumbria, as has been shown, is none the less a very important one. Visitors contribute much to the prosperity of such places as Keswick, Ambleside, Grasmere and Patterdale, as well as to that of almost every Lakeland village and farmhouse. Along the coast such resorts as Silloth, Seascale, Walney Island and Grange-over-Sands flourish where the natural amenities have not been destroyed.

While, therefore, Cumbria has an undoubted regional unity, this is nevertheless a unity in diversity, an area of contrasts as remarkable as in any part of Britain.

Selected Bibliography

W. G. Collingwood. *The Lake Counties*. London, 1932.

G. H. J. Daysh. 'Cumberland', in *Studies in Regional Planning* (ed. G. H. Daysh). London, 1949.

G. H. J. Daysh and Evelyn M. Watson. *Cumberland, with Special Reference to the West Cumberland Development Area*. Whitehaven, 1951.

J. E. Marr. *The Geology of the Lake District*. Cambridge, 1916.

Ministry of Fuel and Power. *The Northumberland and Cumberland Coalfields*. Regional Survey Reports. London (H.M.S.O.), 1945.

F. J. Monkhouse. *The Lake District*. British Landscapes through Maps. (Geographical Association) 1960.

F. J. Monkhouse. 'Some Features of the Historical Geography of Elizabethan Mining Enterprise in Lakeland.' *Geography*, **28**, 1943, pp. 107–13.

A. E. Smailes. *North England*. Edinburgh, 1960.

CHAPTER 23

NORTH-EAST ENGLAND

A. A. L. Caesar

North-East England is one of the best-defined regions of Great Britain. On maps of relief, land utilization or density of population it stands out quite clearly, and the boundaries based on these several criteria are almost coincident. But the limits of the region do not coincide with county boundaries. A small part of northern Northumberland, at least the Till valley below Wooler, belongs to the Tweed basin. In the west, the Alston area is in Cumberland, but physically and economically is part of the North-East. Similarly, in the south, all the upper Tees valley must be included and urban Teesside is an integral part of the industrial core of the North-East. The natural boundary here runs across the wide Northallerton Gate and along the northern edge of the Yorkshire moors. The boundary then is physically distinct, especially along the summits of the Cheviot Hills and Cross Fell Edge where it is in high moorland areas, and almost all of it is in sparsely inhabited country in marked contrast to the densely peopled centre. One indication of the region's acceptance as a human unit is given by the frequency of use of the term 'North-East' in the headlines of regional newspapers—a term readily acceptable and intelligible to its inhabitants as denoting a distinct community.

Structurally the main features of the region are the igneous dome of the Cheviot Hills in the north, the Alston Block of the northern Pennines in the west and scarplands developed on rocks from Carboniferous to Jurassic in age to the east and south. The major landforms are closely related to this structure and the lithology of the rocks, but the Quaternary glaciation had a marked effect upon details and caused considerable modifications of the drainage pattern.

The Cheviot area includes a variety of lavas of Devonian age and the red granite of the summits of Cheviot (2,676 ft) and Hedgehope. Broad, rounded hills with gentle slopes are characteristic of the higher parts, some of which are poorly drained and have extensive peat bogs. The flanks have narrow deeply entrenched valleys, some showing exposures of glacial materials. The Alston Block is clearly defined on three sides by fault systems. In the west the great Pennine fault-scarp overlooks the Eden valley; the western part of the northern edge is defined by the Stublick fault system of the Tyne valley as far east as the Hexham area; to the south are the east–west faults associated with the Stainmore depression, which separates it from the Yorkshire Pennines. The highest point is near Cross Fell (2,930 ft) and from it there is a gentle dip of the Carboniferous rocks—limestones, shales, grits and sandstones—to the north, east and south, giving the characteristic plateau skylines. Peat bogs are again common and produce a rather desolate moorland scenery in the higher parts. Landscape details depend largely upon the outcrop of the Great Limestone and the coarser grits. In the eastern part of the Alston Block tabular grit-capped fells form marked features, but farther east hills are lower and more rounded and slopes have a thicker mantle of drift.

The Northumbrian scarplands flank the Cheviot Hills and the lesser structural dome of the Bewcastle Fells. To the south of the Cheviot Hills are the broad, undulating and drift-covered vales drained by the upper Aln and Coquet and the steep scarp of the Fell Sandstone breached by the gorges of these rivers. The outcrop of the Fell Sandstone is most marked in Rothbury Forest, an area of high barren moorland with the angular skylines of tabular sandstone hills. This passes westwards into somewhat similar country in the border district south-west of the Cheviots and includes the upper Rede and North Tyne valleys. It is open, bleak moorland and very sparsely populated. To the east and south of the exposure of the Fell Sandstones scarps are less well marked. Hill ridges are composed of narrow limestone or sandstone outcrops in shales and, as in County Durham, the hills are lower and more rounded, while glacial smoothing is more evident. But in Northumberland the Millstone Grit is much less significant in its influence upon landscape and scenery than it is in County Durham.

With their dip to the south and east the Millstone grits pass under the Coal Measures, consisting largely of coals, shales and sandstones. The western limit of the coalfield is approximately from Amble on the coast of Northumberland to a point to the south-west of Bishop Auckland, but it is irregular and partly defined by faults. The southern limit is structurally more complex; the Butterknowle fault gives a major down-throw of the Carboniferous to the south into an area of complex folding which has not yet been fully elucidated. The exposed coalfield therefore consists of the south-east corner of Northumberland and the central part of County Durham. It includes many thin seams of coal and some twenty workable seams, but few of them can be traced over the whole field. The Northumbrian portion of the field forms in the main a low, gently sloping and drift-covered plateau with minor features formed by outcropping sandstones and limestones. The Durham portion is rather more varied as glacial modifications were greater and the valleys are generally deeper. Steep scarps form marked local features and most of the higher areas are formed by the more resistant sandstones.

In east Durham from the mouth of the Tyne southwards the Coal Measures are overlain unconformably by Permian rocks, giving a concealed coalfield extending south to the area of Hartlepool. From this area the outcrop of the Permian swings south-westwards and passes into Yorkshire. At the base of the Permian are the Yellow Sands, but they have little effect upon surface relief. Above them, again unconformably, are dolomites—the Magnesian Limestone—characterized by their regular bedding. The westward- and north-westward-facing scarp of the Magnesian Limestone is a very well-marked feature of east Durham, rising to between 400 ft and 600 ft, but the low plateau of the gentle eastward dip-slope carries thick drift deposits. Outcrops of the limestone are few except on the coast. At the top of the Permian are the Red Marls, with associated anhydrites and rock-salt, of the south-east corner of County Durham. They merge into the Bunter Sandstones and Keuper Marls of the Triassic which together produce the low ground around the Tees estuary, the only considerable area of estuarine lowland in the North-East. The estuary in turn is overlooked by the high and striking escarpment of Jurassic rocks of the Cleveland Hills with its outliers. North-facing to the south of the estuary, it is etched by the work of scarp streams and curves away to the south to form the eastern side of the Northallerton Gate. Southward lies the high plateau of the Jurassic moorlands with its deep valleys, the low, gently undulating and drift-covered country of the Northallerton Gate and the Vale of York, and the limestone dales of the Yorkshire Pennines. Each produces a characteristic landscape and each is quite distinct from the landscapes of the North-East.

One further feature of the solid geology is important in details of surface relief—sills and dykes. Much the most important is the Whin Sill, the rock of which is a quartz-dolerite of Carboniferous age. It outcrops in north Northumberland and is very well shown in the Farne Islands and on the coast at Bamburgh and Dunstanburgh. In mid-Northumberland the outcrop swings to the south-west, crosses the North Tyne, and then runs nearly due west to the north of Haltwhistle. In this stretch the military values of its craggy north-facing cliffs were appreciated by the Romans in the siting of Hadrian's Wall. In the Alston Block the outcrop swings to the south and where the Tees cuts down to it lies High Force.

In the coastal area of County Durham south of Seaham Harbour there are traces of Scandinavian drift, but it is not yet determined whether there was an interval between glaciations from Scandinavian and British sources. Within the region both the Cheviot Hills and the Alston Block gave rise to local ice caps. During the main glaciation much of central Northumberland was covered solely by Cheviot ice since deposits include no materials from other sources. But ice from Lake District and Scottish sources pushed south-eastwards into central Durham. Other ice from the southern uplands moving out to the North Sea was turned south to pass over the coastal area of Northumberland and the eastern part of Durham. To the south was a constriction formed by the barrier of the Cleveland Hills and further accumulations of ice on the Yorkshire Pennines. The main mass, joined by ice moving south-east over Stainmore, therefore moved south through the Northallerton Gate into the Vale of York. This movement left a thick mantle of drift, especially in the Northumberland coastal strip, in east Durham and in the lower Tees valley. It is generally thickest on the floors of the main valleys and more patchy at higher levels, and may have been removed by post-glacial erosion on some scarps. The boulder clay is generally a tough deposit which may include large boulders. Variations have a marked effect upon soils and may produce considerable local differences in farming potential. Much of the boulder clay fills hollows of the pre-glacial surface and tends towards a featureless relief, but in the upper parts of some of the valleys it occurs in large mounds. Other deposits include sands which are deltaic and probably formed in glacial lakes in the main valleys as in the area between Chester-le-Street and Durham and, when related to overflow channels, gravels and sands which are probably of fluvio-glacial origin. These deposits, however, probably belong to the retreat stages which had a marked effect upon the region's drainage.

Accordance of summit-levels in the high moorland areas has led to suggestions of a Tertiary peneplanation. The present drainage-pattern may have developed after uplift of such a peneplain or may be the product of a longer period of adjustment. In either case it shows a close relationship to structure. The Cheviot drainage is probably a modification of a simple radial pattern, with the Till capturing some of the consequents which are still represented by the Aln and the Coquet. The alinement of the Rede and the Wansbeck in particular has led to the oft-suggested but unproved theory of the capture of the Rede by the North Tyne. In the Alston Block rivers radiate from the Cross Fell area giving short steep tributaries to the Tyne and the larger and deeper upper valleys of the Wear and Tees. In its eastward course the Wear met the barriers of the scarps of Coal Measures sandstones and the Magnesian Limestone and was deflected northwards. In pre-glacial times it probably followed the line of the present Team valley and joined the Tyne to the west of Gateshead.

During the period of melting of the ice caps, the Cheviot Hills and the high Cross Fell moorlands were the first clear areas. Western ice was cut off from the Tees valley by the high Stainmore Pass while it still pushed into the Tyne valley. Scottish ice still covered the coastal belt. Dead ice certainly still lay in the Tyne valley and the coastal area long after the higher areas were ice-free. These circumstances produced a large number of ice-dammed lakes in the upper valleys. Each overflowed at the lowest available col or along the margin of the ice and lake-levels fell with the gradual melting of the remaining ice. The result is some remarkable series of alined overflow channels notching spurs at intermediate levels. But the most important effect was the blocking of the lower Wear above its junction with the Tyne. This impounded Lake Wear, bounded to the east by the Magnesian Limestone scarp. The lowest col was close east of Ferryhill and there the overflow cut a deep channel to the lower Tees. It now breaches the escarpment and is used by the main 'East Coast' railway. At a later stage a lower col was exposed north-east of Chester-le-Street and the overflow cut the gorge which gives the present Wear its lower course to its mouth at Sunderland.

There are a number of coastal indications of changes in land/sea-levels in glacial and post-glacial times. In the lower Tyne the rock floor is at – 40 ft O.D. (it is much lower above the constriction at Newcastle owing to glacial gouging) and the pre-glacial floor is at a similar depth below the lower Tees. This may represent the pre-glacial base-level, but the nature of these river beds, in contrast to that of the lower Wear, has had important economic effects upon port development. The coast itself is regular in general trend but varied in detail. Much of the Northumberland coast shows alternating rocky headlands and sandy bays often backed by a filling of drift and including stretches of dunes and sandbars. The Aln, Coquet, Wansbeck and Blyth provide small river-mouth harbours, that of the Blyth being extended artificially. The lower Tyne is of major importance. The river enters the sea between the high bluffs of Tynemouth and South Shields and the mouth is protected by breakwaters and wave traps. It is deeply incised but the steep valley-sides aided the construction of high-level staithes for the tipping of coal into ships. Constriction of the valley sides, as at Newcastle–Gateshead and at the mouth, limited the area suitable for industrial riverside sites, but between these points it opens out and provides suitable sites for shipyards and other industrial uses. Moreover, the comparatively soft materials of the Coal Measures and glacial drift permitted the construction of three dock systems off the lower river and well upstream from the mouth. The deep rock floor and glacial infill also facilitated dredging except for some rock bars, and the Tyne now has a dredged channel of 25 ft L.W.O.S.T. for the 14 miles up to Scotswood and of 30 ft up to Jarrow Quay.

The greater part of the Durham coast consists of the low cliffs of the Magnesian Limestone plateau. The Wear in its late-glacial gorge offers very little space at river-level near its mouth. Deepening of its rock floor upstream would be very difficult and costly and it is not navigable for more than a very short distance. In both respects it is severely handicapped in competition with the Tyne. The main dock-system of Sunderland is wholly artificial and is built out from the coast to the south of the river mouth. Southwards, Seaham Harbour is also wholly artificial and is difficult of access from the landward side. Near Hartlepool the coast changes its character. The rock on which Hartlepool is built was originally an island but was joined to the mainland by a sand tombolo. This provided a protected bay which was at one time the only refuge for small sailing-vessels between the Esk and the Tyne. Part of the bay is now filled by Hartlepool's dock-system and timber-ponds. South of Hartlepool is the low coast of the Tees estuary. In its natural state the estuary had extensive mud flats, sand banks and shallow, shifting channels. It was difficult to navigate even in small vessels. In the construction of the modern port dredging was relatively easy but it was also necessary to build training-walls to canalize the river and increase the scour. Much of the salt marsh has been reclaimed and there are wide stretches of land at little above river-level. The problem in this area was more the provision of adequate foundations for heavy industrial buildings. The Tees is primarily a true river port. It has only one small dock, railway built, but much of its traffic is handled at privately owned river wharves. Although later developments have been farther and farther downstream, Stockton still has a small share of the traffic. South of the Tees is a long sandy beach extending to Saltburn where the north Yorkshire moors first present cliffs much higher than anything in Northumberland or Durham.

These high cliffs to the south re-emphasize the importance of the highland rim of the region. It is broken at three points only—the long, narrow coastal route into Scotland, the Tyne Gap to the west and, much the most important, the wide-open and low Northallerton Gate to the south. The Tyne Gap, about a mile wide and less than 500 ft at the highest point, is followed by a main road and a railway. It is increasingly used by traffic between the North-East and south-west Scotland, including the Glasgow area. The much higher Stainmore pass, again with main road and railway routes, has not been of great significance in the development of the North-East. The focus of the main routes is at Newcastle, but its site is equally significant on local considerations. The lower Tyne has always provided a considerable barrier to transverse movement. South-east Northumberland is cut off from the south, north-east Durham from the north. In early days there was a ford at low water at Newcastle, and it is still the lowest bridging-point of the Tyne. Moreover, the central position of the navigable Tyne in the exposed coalfield gave added commercial significance to the site from the earliest days of coal shipment. On both regional and local considerations therefore the site

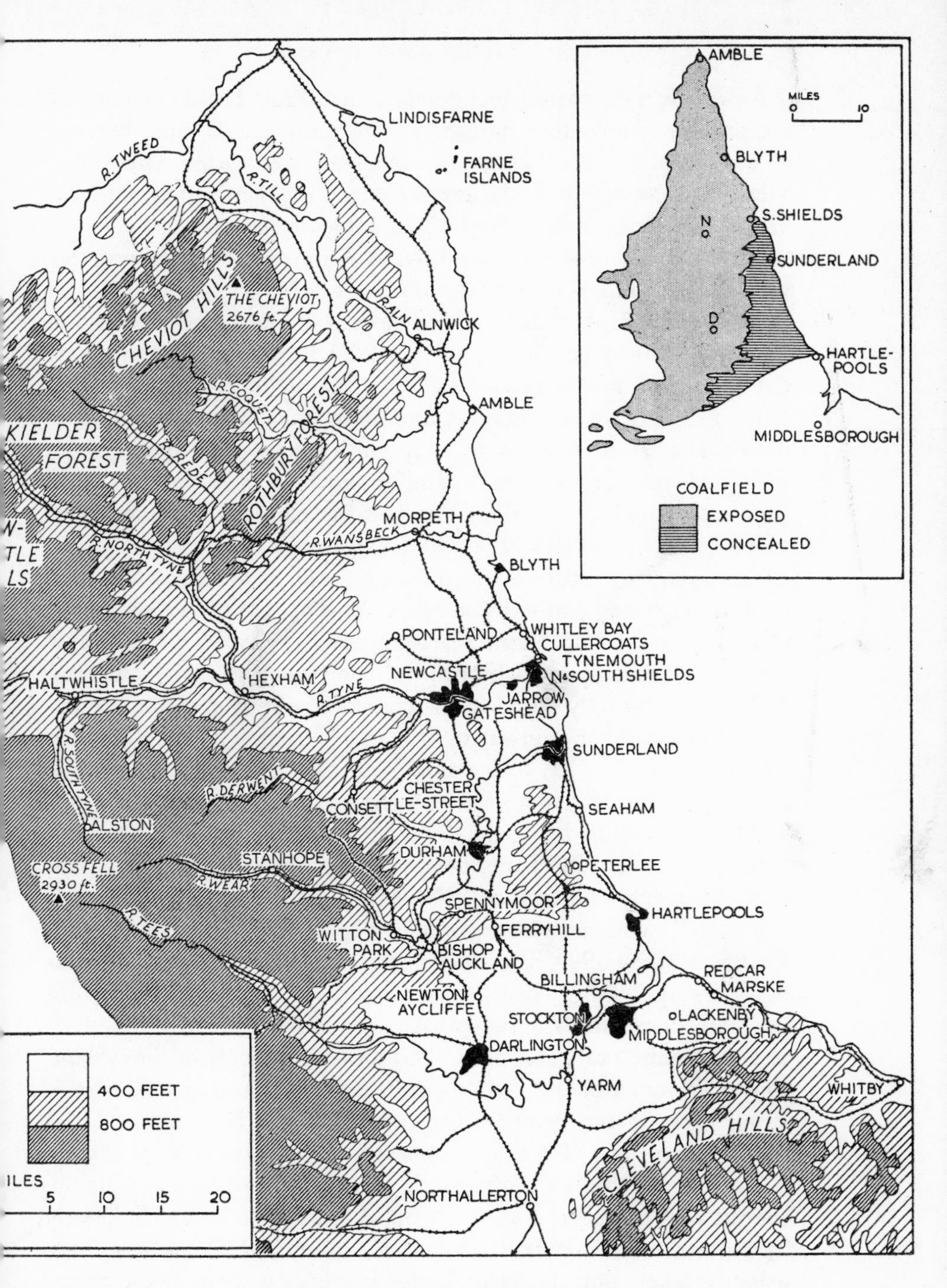

g. 60. North-East England. Since this book was in proof, British Railways have closed their lines in the North Tyne Valley and the Rothbury area.

of Newcastle is the natural focus of the North-East. The significance of the site has changed its nature to some degree with time but the value has not decreased. Cumulative advantages are also important. The city of Newcastle is the acknowledged regional capital, at least in all commercial matters, and is without a serious rival.

In a region the size of the North-East within Great Britain no great variations in climate from place to place would be expected, but the marked differences in altitudes have a considerable effect upon rainfall. In the lower Tees valley and small lowland areas near the coast the average annual rainfall is 25 in., in the higher parts of the Pennines it exceeds 70 in. There, some rain or snow falls on two days in three and snow may lie for 100 days in the year. Over much of the region May frosts are fairly common and frost may occur in early June and again in September. The frost-free period is therefore short. In many of the valleys aspect and exposure have important local effects upon agriculture. A third feature is the 'haar' or sea-fret, particularly in spring and early summer. It is caused by the cooling of warm air by the cold North Sea water, and consists of a low stratus cloud-layer accompanied by drizzle, poor visibility and bleak conditions. It is usually found in the coastal strip but occasionally extends much farther westward. It is the variation of climate with altitude which has led to the now generally accepted agricultural divisions of the lowland, intermediate and upland zones, the 400 ft and 800 ft contours being the dividing lines.

Soils of the lowland zone, derived mainly from boulder clay, are generally heavy and include stiff clays, but there are some medium loams and occasional patches of lighter soils. The best farmland is in the lower Tees valley. The intermediate zone has somewhat similar boulder clay soils in the valleys but in some of the higher parts, with less drift, there are more sandy soils derived from the Millstone Grit and Carboniferous sandstones. The source materials of the upland zone soils vary from igneous rocks in the Cheviot area to Carboniferous rocks in the Pennines and boulder clay. They are acid and thinner than those of the lower zones; there are also areas of peat.

Farming in the lowland zone is varied in character and in the central part has been much influenced both directly and indirectly by industrial developments including coal mining. There are large urban markets but also large built-up areas on some of the better land. Some areas have been badly affected by subsidence and others lost to farming, at least temporarily, by open-cast workings. Fewer collieries now require considerable supplies of hay and oats. There is a mixture of arable farming, dairying and meat production, the main cash crops being wheat, barley and potatoes. In the 1930's dairying was expanding in areas near the large towns and particularly in the Tyne valley, and on non-dairy farms the summer fattening of beef cattle and sheep on permanent grass or long leys was more profitable than arable farming. During the Second World War there was a big increase in the arable acreage, mainly under wheat and potatoes, and a decrease in the numbers of beef cattle and sheep. Since the war there has been a reversal of this trend. There is market gardening in intensively worked small-holdings fringing urban areas, its general distribution being more a question of easy access to local markets than a response to particularly favourable conditions. In the intermediate zone there is normally little tillage. It is essentially a livestock area but its pastures are more suited to breeding than to fattening. Much improved accessibility to farms provided by modern road transport and the institution of a guaranteed market for milk has, however, produced a marked effect and many farms changed from cattle breeding to dairying. No longer was the supply of fresh milk the prerogative of the lowland zone. In the upland zone pasture is poor. The hardier breeds of cattle can be kept, but numbers have fallen over a long period, and sheep-grazing is dominant. Breeding flocks are mainly Cheviot, Blackface and Swaledale crossed with Border Leicester to produce lambs for sale to fattening farms in the lower zones. But there is one important change in the hill country—the development of forestry. It may well represent the best long-term use of much of this land under present conditions. Kielder Forest, in the North Tyne valley and extending over into Redesdale, is now one of the largest forest areas in Britain. It has already given rise to new or expanded villages in both valleys. Planting began in 1926 and has been mainly of Sitka and Norway spruce, and the forest will be increasingly important in the production of soft timbers. With the creation of national parks in 1955, Kielder was constituted a national forest park But the greatest resources of the North-East are not to be found in its farming and forestry.

To many people elsewhere in the country the 'North-East' means 'coal', and there is some justification for this view for coal, and industries located upon it, have formed the basis of its economy for a very long period. Almost all bituminous coals are represented. Types may vary between different seams in the same area, but there is a general transition throughout the coalfield as a whole. In the west of the Durham part of the field is the best coking coal in Britain. It has a low volatile content and produces a hard, strong coke low in ash, phosphorus and sulphur and ideal for metallurgical work. To the north-east, east and south-east of this area the coals grade into steam, gas and house types. Coals occurring outside the west Durham area can be used for coking but are not quite so good. Reserves are generally estimated to be of the order of 3,000 m. tons in Durham and 2,000 m. in Northumberland from seams known to be workable by present methods which, with an annual output of some 50 m. tons, means a life span of 100 years. In addition there are under-sea reserves which have recently been investigated by the National Coal Board in sinking bores off the Durham coast. Against this must be set coal sterilized in avoiding subsidence in built-up areas and possibly uneconomic working of the remaining seams in flooded pits of the older areas of the west. But reserves are not equally large in all types of coal and are least in the best coking coal. A policy of conservation in its use for high-grade metallurgical purposes only is amply justified.

Associated with the Carboniferous shales in and below the Coal Measures are bands and nodules of ironstone. Clay ironstones in the western Durham coalfield were important in industrial developments in the mid-nineteenth century. There were workings of similar iron-stones in Redesdale and Tynedale. Limonite occurs in upper Wear-dale and was worked until after the First World War, one mine being reopened in the Second World War. But the main iron-ore resources are in the Cleveland Hills and came into use after 1850. The ore is a low-grade phosphoric ore, occurring mainly in the upper part of the Middle Lias. The main seam at the Jurassic scarp is 11 ft thick and was easily accessible. After 1854 production increased rapidly and the annual output was usually between 5 m. and 6 m. tons up to the First World War. In later years it declined rapidly with increased production costs as workings proceeded down dip in the gradually thinning seam, and, with competition of imported high-grade ores, it may soon cease.

In the Carboniferous rocks of the Alston Block is a complex system of mineral veins which include silver–lead–zinc deposits. There have been workings near the heads of all the main valleys. Lead-working goes back to Roman times but it reached its peak in the mid-nineteenth century and in later years zinc was more important. In the present century fluorspar became much more important and in recent years there has been a much greater demand for barytes with new industrial uses. Witherite is mined at Settlingstones to the north-west of Hexham and at South Moor Colliery in Durham. These are almost the only sources of the mineral. About 1860 a bore for water near the Tees estuary discovered rock salt in the underlying Permian rocks and subsequently anhydrite, which also occurs near Hartlepool, was discovered. They were later to provide the basis of a large-scale chemical industry. The anhydrite is mined and the salt obtained by brine pumping. Gypsum deposits have been found in the Keuper Marls to the south of the Tees estuary.

Both Carboniferous and Magnesian limestones are available over wide areas for industrial and agricultural uses, while many parts of the region give evidence of local sandstones used as building materials. Grindstones are still worked at Gateshead, and Harthope in Weardale is one of this country's main sources of silica brick materials. Brick clays are supplied by the Carboniferous shales—often having to be extracted from collieries—and by drift deposits. Sand and gravels are of increasing importance as cement and concrete aggregates. Both are available from beaches, while gravels are taken from river-terraces, and sand workings include those in the glacial deltaic sands between Chester-le-Street and Durham and in the narrow outcrop of the Yellow Sands below the southern part of the Magnesian Limestone scarp. The Whin Sill dolerite provides one of the best road metals. Finally, the water resources of the region are more than adequate. Many of the rocks and particularly the Permian are good aquifers while rainfall is substantial. Industrial consumption is heavy, and in parts of east Durham even the large supplies of the Permian were being overtaxed and the water-table lowered. In 1945 this was the first area of this country to be scheduled under the Water Act to prevent overpumping. Many of the existing public water-supply systems depend in part upon surface catchment but few of them have any great surplus capacity. However, surface water resources of the highland rim are more than sufficient to meet any likely demand and the relief is generally favourable to reservoir construction. Costs of installation provide the brake.

In the mineral endowment of the North-East, then, coal is dominant, but the resources of minerals other than coal are substantial. Some have shown their real value only in this century, others are less important now than during last century, but their influence remains. The role of the region's iron ores in establishment of iron-works reached its climax between 1840 and 1865. In these years ore supplies were sufficient to meet the demand; later development and expansion were dependent upon imported ores. But the start had been made and therein lies the full significance of the region's own resources. It was the mid-nineteenth century that finally established the present industrial character of the North-East; it also gave birth to some of its current problems.

The nineteenth century was the period of very rapid development and the influx of a large population, the products mainly of the coming of the railways and of new mining, metallurgical and engineering techniques. The Stockton and Darlington Railway was opened in 1825 and was designed to bring the coking coal of south-west Durham to tide water. It started a rapid expansion of coal production in the south-west of the field and gave the Tees a small share in the coal-shipment trade, but navigation of the untamed lower Tees was so difficult that the railway was extended to the site of Middlesbrough as early as 1830, the birth-year of that town. Railway building proceeded rapidly. In 1838 Newcastle was linked with Carlisle and the High-Level bridge at Newcastle was completed in 1849 so giving Newcastle rail access from the south. But much of the increased coal output was destined for shipment, either coastal or overseas, so that parallel with the railway building came development of the ports. The Tyne Improvement Commission was established in 1850 and soon began the dredging of the river and the construction of the three major dock-systems. In the same year the South Dock at Sunderland was completed and six years later its direct outlet was added. The Tees Conservancy Commission dates from 1852 and began the corresponding improvement of the Tees. Seaham Harbour was built specifically to ship coal and first did so in 1831, closely followed by Hartlepool in 1835. Both were expressions of the newly developing eastern portion of the coalfield. Railway building in Great Britain generally altered the spatial relationships of the region with the rest of the country and widened its possible markets, though it gave corresponding facilities to its competitors and was not wholly favourable as it permitted the Midland coalfields to compete in the London market which had earlier been supplied almost exclusively by the North-East. Coal-working spread rapidly, but with varying local fortunes.

With the expansion of coal production there was a corresponding expansion of the iron and, later, steel industries. The railways building in many parts of the world and iron ships themselves provided large new markets. Plants at Stanhope, Tow Law and Witton Park to smelt Weardale ore were established in quick succession between 1844 and 1846. They were preceded by Consett in 1840 and today Consett stands as the one large modernized plant in an inland position which was originally located on coalfield ores, though the working of these ores ceased in 1856 and the plant became dependent upon imported ores railed from Tyne Dock or Sunderland. Middlesbrough's first blast-furnace was blown in 1852 and within a few years it had assumed its role as the main centre of iron and steel production in the North-East. Its rate of growth in both iron and steel production and population was among the most rapid experienced in Britain. Smelting centres at West Hartlepool and Skinningrove are in effect outliers of the Teesside concentration, Skinningrove being the only plant remaining on the site of a furnace located on the Cleveland ores and away from the estuary.

The Tyne launched its first iron ship in the 1840's and, with new inventions in marine engineering, began the period of rapid expansion of shipbuilding and marine engineering and of naval ordnance and a number of auxiliary industries. Riverside sites were in ever-increasing demand and these industries were in competition with the new high-level staithes for the gravity-feed of coal from railway waggons to ocean-going ships. Industry claimed the riverbanks from upper Tyneside to the sea, pushing residential areas back from the river, and new shipyards were established on the Wear and the Tees and at Hartlepool. A further competitor on the Tyne, and to a lesser extent the Wear, was the chemical industry which developed rapidly in the first half of the nineteenth century. It reached its peak in the 1870's and then declined while salt production by brine pumping expanded on Teesside. Railway engineering developed rapidly at Darlington and Gateshead.

All this industrial development was accompanied by a corresponding increase in population. There were some internal movements, as from the villages of the dales where lead-mining was on the wane to the coalfield centres, but a much greater movement from outside the region. Immigrants came from Scotland, Wales and Ireland and from rural areas of England, such as East Anglia. With the decline of copper- and tin-working, Cornish miners moved to the Cleveland iron-ore mines. The present population of the North-East is a mixture from very varied sources. By 1913, the population exceeded 2 m. and some 85% lived in the triangular industrial core area including the coalfield, Darlington and Teesside. There was a marked concentration upon the three main urban areas—Tyneside (*c.* 750,000), Teesside (*c.* 275,000) and Wearside (*c.* 170,000)—with Hartlepool and Darlington as smaller towns. The housing of this great population increase was typical of its time, with close-packed parallel streets of terraced housing producing large built-up areas adjacent to industrial sites and unrelieved by open spaces. Middlesbrough and West Hartlepool were products of this period but in other cities and towns such areas were packed around older centres. Only Newcastle acquired additional civic dignity by a gracious replanning of its commercial core in the mid-nineteenth century. In the coalfield settlements some housing took the form of the addition of pitmen's rows to existing village centres, but other villages and small towns appeared as pieces cut at random from a Victorian industrial city, dropped into a farming landscape and soon dwarfed by waste tips.

The economy of the North-East had then a quite remarkable concentration upon the basic heavy industries—mining, iron and steel manufacture, shipbuilding, heavy engineering and chemicals. In 1914 the working population of the region was about 700,000 and of this number 63% were engaged in the basic industries, 13% in all other productive industries and 24% in service industries. The concentration on the heavy industries was even greater than these proportions suggest, as much of the employment in services, especially transport and power, was directly dependent upon the basic industries. Coal output in 1913 was 56 m. tons; 26 m. were exported (including bunkers) and a further 8 m. shipped coastwise. Such was the response to the natural endowment of a region peculiarly suited to the basic industries. Given the techniques of the early industrial revolution, here was a full-flowering of its resources, but it was the product of one short period of the economic development of Britain and that as conditioned by world forces. In addition, the region's range of industries and therefore of occupations was very much smaller than that of Britain as a whole, and it was much more dependent upon exports. This was largely a corollary of its coastal position, but not merely a function of the nature of its industries, for even in the basic industries the scale of development was a response to a larger proportionate export market than that for these industries nationally. The narrower range of industries and greater dependence upon exports in turn meant a vulnerability to trade fluctuations greater than that of the national economy. Moreover, all the basic industries employed men almost entirely and the region's industrial buildings were highly specialized. The proportion of female employment was much less than the national average and opportunities for women to work were proportionately fewer than elsewhere. Given a change in the economy, pithead-gear, blast-furnaces, coke-ovens, shipyards and chemical-plant could scarcely be adapted for alternative industrial uses, and the region would have little to offer in terms of installed industrial facilities.

Examination of the economies of distinctive areas within the industrial core of the North-East reveals a relationship to the regional economy very similar to that of the region to the national economy. Many areas were dependent upon a still narrower range of industries and to an even greater extent upon exports. Many provided practically no industrial employment for women. Only the city of Newcastle, as distinct from Tyneside, with its functions as regional capital and commercial centre, provided a wide range of occupations and opportunities for women proportionate to the national average.

The First World War accentuated the concentration of the North-East upon the basic industries—its coal, steel, ships and heavy engineering products were in demand as never before. There was, too, an important development from this period—the perfection of the technique of the fixation of atmospheric nitrogen led to the establishment of a chemical-plant at Billingham, and later expansion by I.C.I. brought chemicals alongside steel as the mainstays of Teesside. But by post-war years world conditions had changed markedly. Industrial developments elsewhere in the world, intense economic nationalism, balance of payments problems, shipping losses, overproduction of primary products and many other features had a profound effect upon the economy of Britain and, within Britain, it was necessarily a differential effect. Nowhere was it more marked than in the North-East. It was immediately evident in the coal trade. During the inter-war years output on the coastal coalfields with large export markets declined considerably; in the inland coalfields, mainly concerned with home markets, it increased slightly. Further adjustments had to be made in other industries. Before 1914 Teesside had produced a high proportion of foundry and forge pig-iron and now had to convert to a higher proportion of steel. In shipbuilding, the Tyne had concentrated to a considerable extent upon naval building which fell off rapidly at the end of the war. Between the wars it built a high proportion of tankers, for which there was a greater demand than for most other types of vessel, but the employment provided never equalled that of the earlier naval building.

Changes in Britain were not concerned only with the older industries. Others, varying from cars and aircraft to radios, rubber and cosmetics, were established or expanded. But the location factors for these industries differed markedly from those of the heavy industries and they were mainly concerned with home markets. The North-East, so well endowed for the basic industries, could not offer them similar advantages. Its position within Britain was now a disadvantage, while the age and specialized character of so many of its industrial buildings offered no attractions for conversion. Moreover, the new and expanding industries were not directly linked to the basic industries and few, if any, of them required their skills. Of these new industries the North-East received a significant share only in electrical engineering, but even there no more than its share of the national population would warrant. The new plant was located mainly on Tyneside.

The full impact of the new conditions upon the North-East became evident in the depression of the early 1930's. In 1924 the number unemployed exceeded 90,000 or 12·8% of the insured population and maladjustment was already evident. By 1932 it was 268,000 or 37% and there was little improvement until 1935. In 1937, the best year between the slump and the Second World War, it was 120,000 or 16·6%, the national average then being 10·1%. The region had shared in a partial recovery from the slump but retained its relative position and unemployment was still twice that of the non-depressed areas of the country. But unemployment returns were not a full measure of the problem, as in the need for employment of women. Even in 1939, after the provision of additional female employment, women made up only 18% of the insured population compared with a national 28%.

Just as there were differential effects within Britain between regions variously endowed, so there were differential effects within the North-East industrial core. In general, the highly specialized areas were the hardest hit such as the south-west Durham coal area and the Cleveland iron-ore area. In north-west Durham conditions in the coal industry were almost equally bad, but Consett's steel production was not so severely affected. On Tyneside, which had the greatest numbers unemployed but not the highest proportion, there were contrasts. Newcastle was relatively well-off—another expression of its function as regional capital—but lower Tyneside, with its dependence upon the shipbuilding and shipping industries, experienced very heavy unemployment of long duration. Sunderland, where shipbuilding, though on a smaller scale than on Tyneside, is relatively more important, and Hartlepool, with its shipbuilding, shipment of coal and import of timber, suffered severely. Teesside fared rather better as the slump in steel and chemicals was smaller and of shorter duration than in coal or shipbuilding. Darlington, with industries less affected by the depression, was relatively prosperous. The differential effect was significant, for the problem facing the region was not only the adjustment of its economy but where within the North-East action should be taken.

During the inter-war years there was some adjustment. In 1938 48% of the adult insured population were engaged in the basic industries, nearly half of them in mining, 19% in other productive industries and 31% in services. Concentration was still evident but was less so than in earlier years. In the 1930's there was a net outward migration of population, largely from County Durham and the south-west portion of the coalfield in particular. Internal movements included that from the older coal-mining areas of the west to the newer areas of the east and north, and a movement into the towns largely of young adults. The large urban centres continued to grow both in population and built-up area. In 1939, Tyneside had reached 828,000 and Teesside 312,000. The building of new houses was related in part to slum clearance but produced the second main feature of the North-East towns—the estates of small detached and semi-detached villas. On Tyneside, vacant spaces behind the industrial river banks were filled in, while building spread on the fringes of

Middlesbrough and Sunderland. Improved transport facilities encouraged commuting. Hexham, the villages of the Tyne valley and Ponteland now serve increasingly as dormitories for Tyneside, though this function is still more marked on the coast. Whitley Bay, Cullercoats and Tynemouth serve Tyneside; Redcar, Marske and Saltburn serve Teesside; all of them and the Roker district of Sunderland seek to combine this function with the holiday industry, though this is largely in the form of day excursions.

During the depression it was evident that the North-East region could not solve its problems unaided and it received help from the Special Areas legislation. The North-East Special Area included most of County Durham and Tyneside together with the small outlying coal-area of Haltwhistle, but it excluded Newcastle and Teesside. The Alston district was included in the Cumberland Special Area. Special Area status permitted the provision of services and financial aid in effect as bait in an attempt to attract new firms or new branches of existing firms away from the more-favoured Midlands and London area. With government help trading-estates with all industrial facilities installed were established and, at a later stage, derelict sites were cleared and 'standard factories' built in advance of demand. The estates were successful, though the employment created was small in relation to the needs of the Special Area, and the labour demands of new factories included high proportions of women and juveniles, much of the work provided not being suitable for highly skilled men unemployed in heavy industries.

The location of the trading-estates presented problems. The first and largest was established at Team valley to the west of Gateshead and was followed by smaller estates at Pallion on Wearside and St Helen's Auckland in south-west Durham. The Team valley estate, placed close to Newcastle, emphasizes the significance of regional geographical relationships. Tyneside had the largest number of unemployed, but upper Tyneside, and Newcastle in particular, also had the greatest range of occupations and adequate employment for women; here was the estate providing new types of work and a high proportion of employment for women and juveniles, located in the area of the North-East which, on the basis of occupational structure, least needed it. Alternatively, if the estate was to serve the Special Area as a whole, it must be in a central position. Newcastle, with its express trains and air services, is much the most accessible point in the region from other parts of the country. Newcomers to the region might be prepared to consider a site near Newcastle but would not consider a site in some much less accessible part of the coalfield, and a new factory anywhere was better than no new factory at all.

The system of local transport services also had an effect. The lower Tyne is still a barrier to transverse movement except between Newcastle and Gateshead, while central Newcastle is the terminal of almost all local train and bus services. Team valley is just far enough out to require public transport from Newcastle, and is on the routes only of services operating between Newcastle and north-west Durham. It is not served by the electric suburban railways which run eastwards from Newcastle and Gateshead. Workers on the estate living to the north and west of Newcastle, in lower Tyneside, on the coast or south-east of Gateshead generally have to change services and are involved in a long and tiring journey to work. In practice therefore Team valley was manned largely from upper Tyneside and by commuters from north-west Durham.

The Second World War had a profound effect as all the main products of the North-East were again in great demand. During the period 1939–43 the numbers employed in shipbuilding, the metal-industries, engineering, chemicals and explosives increased by nearly 100,000. There was a marked fall in employment in coal-mining, but this was common to all coalfields in Britain and was certainly not due to lack of demand as direction of labour to the pits was to show. Despite this fall, numbers employed in the basic industries as a whole increased by 19% and in 1943 this group accounted for 54% of the insured population. It was significant, however, that the increases in employment in shipbuilding and engineering were proportionately slightly less than the national increase and that in metals the same, while the decrease in employment in coal mining was slightly greater than the national average. Even war-time conditions, therefore, with their demands on the basic industries, did not unduly favour the region mainly because the geographical advantages in the export markets of these industries, which had found full expression in an earlier period, meant nothing under war-time conditions.

War-time changes in the distribution of industry generally in Britain included some 'dispersal' of industry from London and the south-east, and the building of new factories, some of large size, particularly for the aircraft industry. The North-East had no share whatever in either of these developments which would have given buildings suitable for conversion to other uses in immediate post-war years. In contrast, the bulk of the increase in production in the basic industries in the North-East meant further use of existing industrial premises, though some were extended. New factories included a chemical-plant at Prudhoe in the Tyne valley, a metal-recovery depot at Urlay Nook, near Stockton, and Royal Ordnance filling factories at Spennymoor and at Aycliffe, 5 miles north of Darlington. All were highly specialized buildings and, like many of the region's earlier industrial buildings, either impossible or very costly to convert to post-war use.

In the North-East therefore the Second World War meant in effect a reversion to an earlier economy. At the end of it there hung over the region the fear of a repetition of its inter-war experience—a short period of relatively good conditions during reconstruction followed by depression.

That this fear still has some justification is demonstrated by the fact that, even under conditions of 'full employment' in Britain in post-war years, unemployment in the former depressed areas has been twice that of the country generally. Appreciation of the possibility of further depression led to the reconstitution of the Special Areas, with some changes in their boundaries, as 'Development Areas'. In the North-East, the Development Area included the Northumberland coalfield, County Durham, Teesside and Cleveland—a more satisfactory expression of its economic geography than was the former Special Area, as for practical purposes it coincides with the industrial core. Development Area status implied the continuation of policies similar to those initiated by the Special Areas legislation, but in immediate post-war years had other advantages. A system of licences for new industrial building led to the ready granting of licences to build in Development Areas but restrictions elsewhere, and this certainly led to some new establishments in the North-East. In addition, given a shortage of labour elsewhere in the country, areas which provided little employment for women suddenly became more attractive to industries employing a high proportion of women workers: some clothing firms have established factories in the North-East.

Furtherance of the policy of the provision of trading-estates incorporated the conversion of the war-time Royal Ordnance factories. That at Spennymoor is well-sited to meet the needs of south-west Durham, but the position of Aycliffe is much less fortunate. It was hoped that its conversion to a trading-estate would help to meet the employment needs of south-west Durham, but distances and transport services are such as to make commuting troublesome. Alternatively, its factories, employing a high proportion of women, would be of value to Teesside with its great concentration of male-employing industries if they had been built there, but the position of Aycliffe is unhelpful: with Darlington as the terminal of local services, almost inevitably it would supply much of Aycliffe's labour force, yet in pre-war years Darlington was the most prosperous town in the North-East.

It is possible that travel problems were in part responsible for the subsequent decision to make the estate the industrial nucleus of a new town. Newton-Aycliffe was so constituted with the intention of making it a 'self-contained town'. The case for a factory located under war conditions determining the position of a new town in post-war years does not bear examination, while the hope of creating a 'self-contained town' so close to Darlington, a town of 85,000 population, is geographical nonsense. The mere duplication of local authority functions shows a lack of appreciation of the lessons evident in relationships such as Manchester–Salford, or Liverpool–Bootle, or indeed of the juxtaposition of a number of intensely independent local authorities on Tyneside.

The building of the new town of Newton-Aycliffe is one major post-war change in the urban landscape of the North-East. Another is the new town of Peterlee in east Durham. It is designed as a town of 30,000 population and primarily to bring a fuller range of urban facilities to the east Durham coalfield area, where the settlement-pattern hitherto has been one of large colliery townships no one of which provided any great range of urban functions. But it is also intended that Peterlee shall not be almost entirely dependent for its livelihood upon coal mining and it has had some success in attracting other industries. Later developments at Washington and 'overspill' towns in south-east Northumberland tend to accentuate concentration around the lower Tyne.

In addition to the conversion of the Royal Ordnance factories, other trading-estates have been established close to the main urban areas. The most important is West Chirton, to the west of Tynemouth, where one of the largest factories has brought to Tyneside a share in a very new industry—laminated plastics. The positions of trading-estates generally demonstrate the changing geography of the industrial centres. Whereas many of the older industries sought riverside sites to which railway-sidings could be carried, the newer industries seek prepared sites on the urban fringes with easy access to arterial roads.

The largest post-war developments in the basic industries have been on Teesside. They include the large new I.C.I. plant at Wilton to the south of the river and the Lackenby melting-shop which came into production in 1953 and made a substantial addition to Teesside's steel-making capacity. New wharves have been built on the south side of the estuary below Middlesbrough. This expansion is of great value to Teesside, but serves to emphasize its dependence upon steel and chemicals, both of them largely male-employing. In contrast, a large new textile mill has been established in Darlington. The textile industries have been represented in the North-East in the modern era only by small-scale units, largely relics of an earlier period, but this mill is again in the most prosperous town of the inter-war period.

New occupations have not been only in industry and industrial services. In post-war years the desirability of spreading the employment presented by central government administration outside the London area was recognized, and the decision taken to place the headquarters of the new Ministry of National Insurance in the North-East. This was very much to the advantage of the region with its comparative shortage of such occupations and small proportion of employment for women. An office of such size could be located only in a large town and, almost inevitably, Newcastle was chosen, so bringing the office to the city which least needed it. Assuming that the decision aimed at an improved distribution of employments, Teesside with its predominantly male-employing industries would have been preferable on regional considerations. But the choice of Newcastle reaffirms the significance of its site and the cumulative advantages of the drawing-power of the regional capital. Post-war developments strongly suggest that, as the economy of the North-East changes, Newcastle, and to a lesser extent Darlington, may gain at the expense of the less accessible parts of the industrial core.

The human geography of the North-East expresses very clearly the significance of its natural resources, its coastal location, and its positional relationships to the rest of Britain. It carries the legacy of a long period of fluctuating fortunes. Even its border status still remains, if only in the celebration of Christmas and Hogmanay with a happy impartiality, but its present landscape and its economy are indelibly stamped by the dominance of coal in the Victorian period. Its economy and human geography change, but gradually and not without severe stresses and strains—the more so as it, even more than some other parts of Britain, was called upon to make sudden adjustments as a result of the effects of two world wars. An impression of change and of differing past evaluations of its endowment is gained by the view of distant colliery tips from the tower of Durham Cathedral, or by a short walk from a remnant of the Roman Wall to a shipyard.

If a Wearsider visits a pub in Scotswood Road, Newcastle, on a Saturday evening, things may go hard with him, especially if by chance Sunderland have beaten Newcastle at soccer that same afternoon. But if a Wearsider meets a Geordie in a more distant part of Britain, they may well linger over a drink together for they have many interests in common. These relationships epitomize much of the North-East with its intense local rivalries but strong regional unity.

Selected Bibliography

G. H. J. Daysh and J. Symonds. *West Durham: Survey of a Problem Area.* Oxford, 1953.

T. Eastwood. *Northern England.* British Regional Geology. London (H.M.S.O.), 1935.

Forestry Commission. *The Border.* National Forest Park Guide. London (H.M.S.O.), 1958.

J. A. Hanley, A. L. Boyd and W. Williamson. *An Agricultural Survey of the Northern Province.* Newcastle, 1936.

J. W. House. *North-Eastern England; Population Movements and the Landscape since the early Nineteenth Century.* King's College, Newcastle, Department of Geography, Research Series, No. 1, 1954.

P. C. G. Isaac and R. E. A. Allan (ed.). *A Scientific Survey of North-Eastern England.* The British Association for the Advancement of Science, Newcastle, 1949.

North-East Industrial and Development Association: *The North-East Coast: a Survey of Industrial Facilities*, 1949; *The Northern Region: Biannual reviews of employment*, 1946–61; *A Physical Land Classification of Northumberland, Durham and the North Riding*, 1950.

A. E. Smailes. *North England.* Edinburgh, 1960.

CHAPTER 24

THE ISLE OF MAN

J. W. Birch

The geography of a comparatively small off-shore island of little more than 220 square miles, which supports a largely self-governing community of nearly 50,000 people, is necessarily closely associated with the elementary, but none the less elemental, considerations of insularity and external space relations.

The Isle of Man's position within the British Isles has exposed it to the influence of diverse movements of both people and ideas; yet, at the same time, its insularity has favoured the growth of a measure of cultural and political distinction. It is clear, for example, from the evidence so far obtained from the wealth of archaeological material, that, at least from the Mesolithic period onwards, the island received culture groups and cultural influences from both Britain and Ireland as well as direct from Europe; and it is of interest that the Ronaldsway culture, which appears to have flourished in the island during the later Neolithic period, has not yet been fully matched elsewhere in the British Isles. Christianity reached Man before the pagan invasions from Scandinavia that began in the early years of the ninth century, and the earliest stone crosses are regarded as being of purely Celtic origin. However, the majority of the Manx crosses appear to post-date the conversion of the Norse settlers and show an interesting blending of the two cultures. A similar blending is evident in the place-names, though the Celtic language survived the Norse period which extended to the latter half of the thirteenth century. The language has not, however, survived the more modern social and economic influences coming particularly from Britain; and there are now very few people able to speak Manx.

The political distinction of the Isle of Man owes much to the Scandinavian period during which it was for a time the administrative and ecclesiastical centre of an island kingdom embracing the whole of the Hebrides. The Manx parliament, the Tynwald, derived its name from the Old Norse 'Thing-voller' (an assembly-field). Its

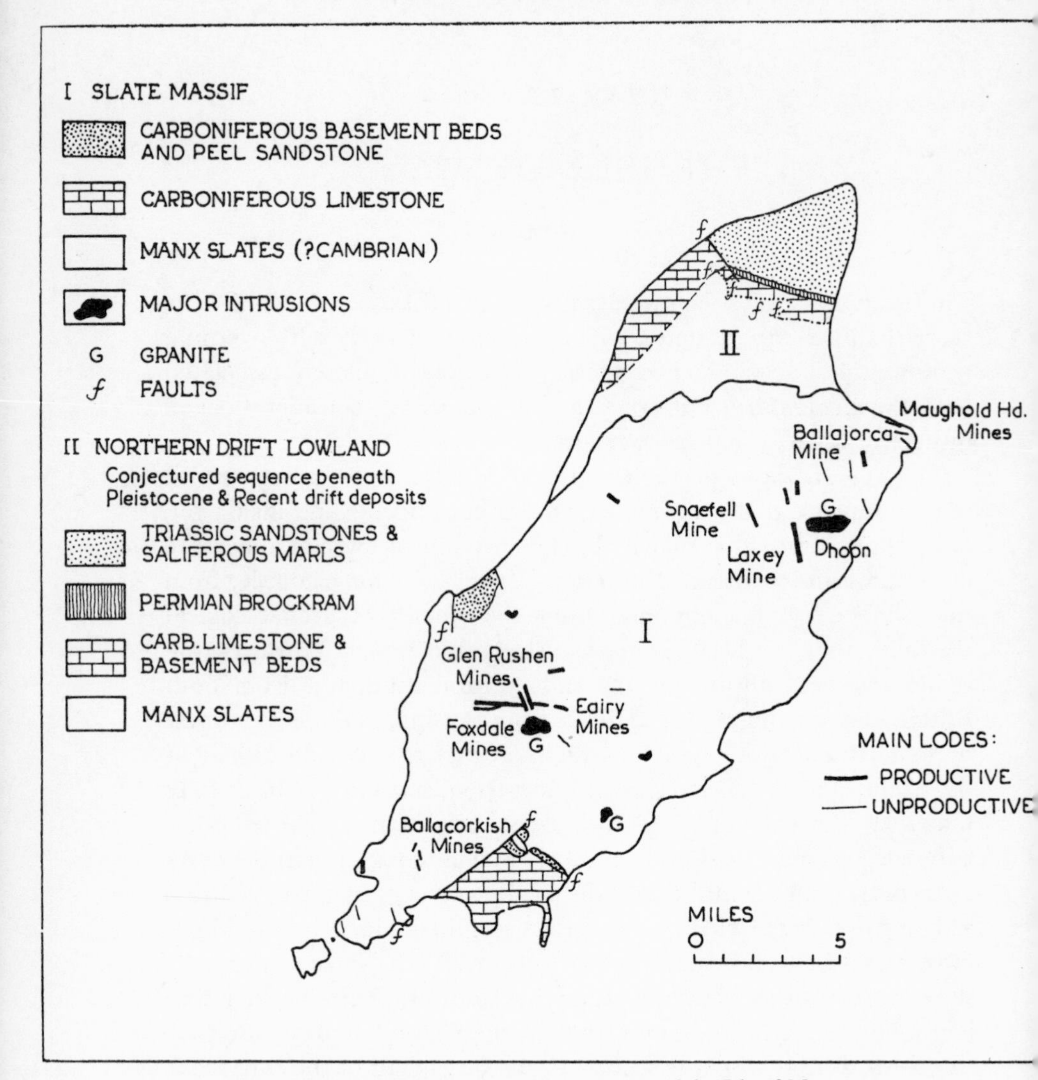

Fig. 61. Major structural components of the Isle of Man.

traditional open-air assembly for the proclamation of new laws, now presided over by the Lieutenant-Governor, is still held at St John's on 5th July, midsummer day by the pre-Gregorian calendar.

Though formerly more extensive, the territorial basis of Manx political status is now restricted to the island itself. This consists of two clearly differentiated tracts, the Northern Drift Lowland and the Slate Massif (fig. 61). Within the latter there are two true uplands,

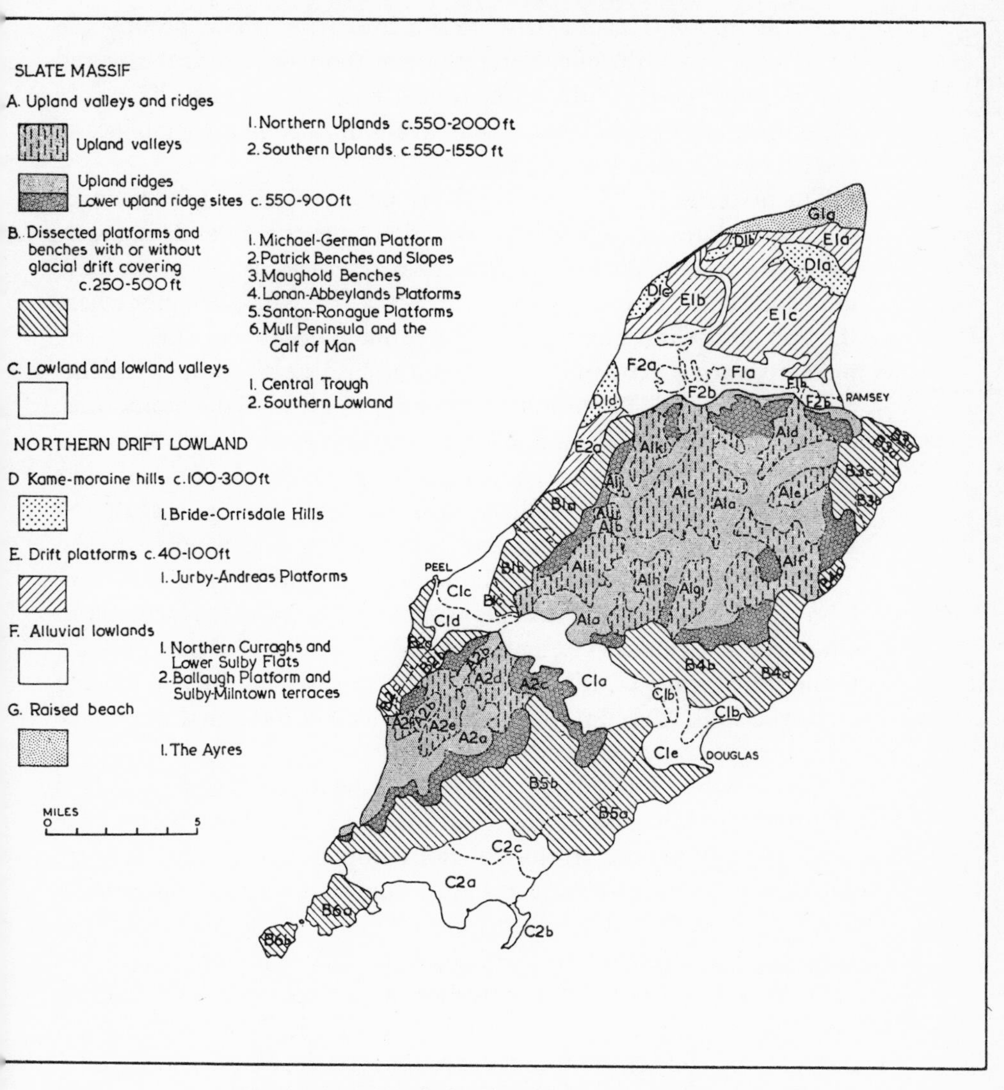

Fig. 62. Morphological regions of the Isle of Man.

the northern and the southern, and they are flanked by plateaux and restricted lowlands (fig. 62).

The two upland regions have been carved by water and ice from the tightly folded system of the Manx slate series. They consist of narrow ridges, separated by broad drainage basins that narrow into deeply incised glens and are fed by annual rainfalls averaging

50 in. or more. The majority of the hill-farms which occupy the flatter sites above the glens, and many of those on the drift-covered lower slopes of the ridges, have now been abandoned. The only concentrations of settlement occur in the former mining villages of Laxey and Foxdale (figs. 61 and 63).

The plateaux (fig. 62) are also developed across the varied facies of the Manx slates, though their distinction from the uplands is generally well-marked. Where the plateaux reach the coast they terminate in abrupt cliffs, broken by frequent inlets ('purts'), especially where the upland streams find their outlets through deeply cut valleys. The plateaux vary greatly in the texture of their dissection: the gradients are, however, seldom prohibitive for farming; and the stony soils, derived from the slates and slaty drifts, together with average rainfalls of from 35 in. to 50 in. a year, support roughly half the island's farms. This is then essentially a farming landscape, with scattered mainly slate-built farmsteads linked by narrow and winding roads. The fields are small, separated by high banks and hedges, with trees largely confined to the breaks planted to windward of the farmsteads. Clustered settlement is exceptional and is usually of recent growth; though the former crofting hamlet of Cregneash (fig. 63), now being used as a Folk Museum, preserves a settlement form that may once have been more widespread.

The trough of lower land that extends between Douglas and Peel (fig. 62) may well owe its continuity to the glacial breaching of the main upland divide. It has obvious importance as a routeway, despite the ill-drained, lacustrine alluvium that occupies much of its floor. At the small village of St John's (fig. 63) additional routes converge via Glen Helen and Foxdale, to provide a position that must have been important in its selection as the place for the annual assembly of Tynwald. Low coastal plateaux and broad valleys form the seaward ends of the Central Trough. In the Peel Basin (fig. 62) much of the surface consists of platforms or mounds of sandy drift that provide rather dry but early and well-farmed soils. Peel (2,600) itself is the only port on the west coast and is the main centre of the fishing industry. As with Peel, the original growth of Douglas took place on a small raised-beach site on the north side of the estuary that provides the inner harbour. Its development as a specialized holiday resort, with a population now exceeding 20,000, has been associated with the spread of hotels and boarding-houses along the narrow extension of this raised beach that fringes the bay, and the building of both residential and tourists' houses over much of the coastal bench above. This economic and physical growth also brought the nineteenth-century construction of an outer harbour for passenger steamers under the shelter of Douglas Head.

The lowest portion of the Slate Massif, the southern lowland (fig. 62), extends across both slates and limestones at a height of generally under 100 ft, and is veneered by a variety of drift materials. The lowland enjoys an annual rainfall which averages 35 in. or less and supports some of the island's richest and largest farms, so providing a landscape in strong contrast to that of the dissected plateau above. The market village of Ballasalla owes much of its early importance to the former Cistercian abbey sited on the alluvial flats of the Silverburn (fig. 63). The main clusters of settlement are, however, on the coast; Castletown (1,755) was the island's capital until 1869. Of the two 'ports', the more sheltered Port St Mary (1,403) has declined with the decay of the fishing industry, while Port Erin (1,435) has become a specialized and prosperous resort by exploiting the advantages conferred by the setting of its bay and by a broad sandy foreshore.

The Northern Drift Lowland lies mainly below 100 ft (fig. 62). It consists principally of thin beds of glacial and fluvio–glacial drifts laid down, during at least two major glaciations, on a submarine platform which flanks the sharply-cliffed edge of the Slate Massif. With the exception of the rather sterile shingle and sand flat of the Ayres raised beach, the surface features of the region are largely the product of the retreat stage of the Scottish Readvance Glaciation. Of these features, the arcuate kame-moraine consists of mounds of sand and clay that are separated by meltwater channels and provide very variable farming conditions. The drift platforms of sand and clay are only broadly dissected: they form the core of this important farming region and carry the greater part of the settlement, including a number of small clusters of houses. Both the moraine and the drift platforms have been much reduced by coastal erosion; and this is still a major problem, particularly on the west coast. An alluvial lowland flanks the Slate Massif and represents the floor of the last stage of a post-Glacial Lake Andreas. Much of it has been drained to produce good farmland; though the Ballaugh Curragh, which is the lowest portion and formerly provided the main source of peat for the north of the island, remains a bog colonized by willow. The Northern Drift Lowland as a whole is again essentially a farming landscape, and one which enjoys substantial advantages of soil and climate, including a rainfall that is for the most part under 40 in. a year. The trade provided by the farms is largely responsible for the importance of Ramsey (4,621) as a market and port; though the town is also a holiday resort, benefiting in this respect from its setting against the uplands and its broad sandy beaches.

In Man the coastal towns and villages account for three-quarters of the total population (fig. 63). Most of them are on the eastern and southern coasts where there are good natural harbours well-placed to profit by the island's commercial links with Britain. Douglas and Ramsey benefited particularly from their positions and prospered during the seventeenth and eighteenth centuries when there was much smuggling to the Galloway and Cumberland–Lancashire coasts, encouraged by the island's political status. The subsequent disproportionate growth of Douglas came with the introduction in the nineteenth century of regular steamer services from Britain.

This improvement in sea transport initiated the modern period in the history of the island's economy. It was, until nearly the close of the nineteenth century, broadly based upon farming, fishing, mining, manufacturing and tourism. There followed, however, a marked specialization of the economy in terms of farming and the tourist industry, and a related contraction of other industries occasioned principally by the competition of larger-scale producers on the mainland and elsewhere.

There seems little doubt that it was the emotional appeal of an island offering reasonable accessibility from large and expanding centres of population on the adjacent mainlands that brought increasing numbers of visitors in the early twentieth century. Likewise, the diminished importance of these considerations is reflected in the present relative decline of the industry, at a time when cheaper fares and higher standards of living have broadened the limits of holiday travel. Thus, the traffic to the island reached its peak in 1913, when over 634,000 people were landed from May to September. Apart, however, from the years 1947–49, the numbers have not again exceeded 600,000; and of late they have ranged from 450,000 to 550,000, including up to 100,000 day-excursionists who travel principally from Fleetwood and Liverpool. Most of the holiday-makers

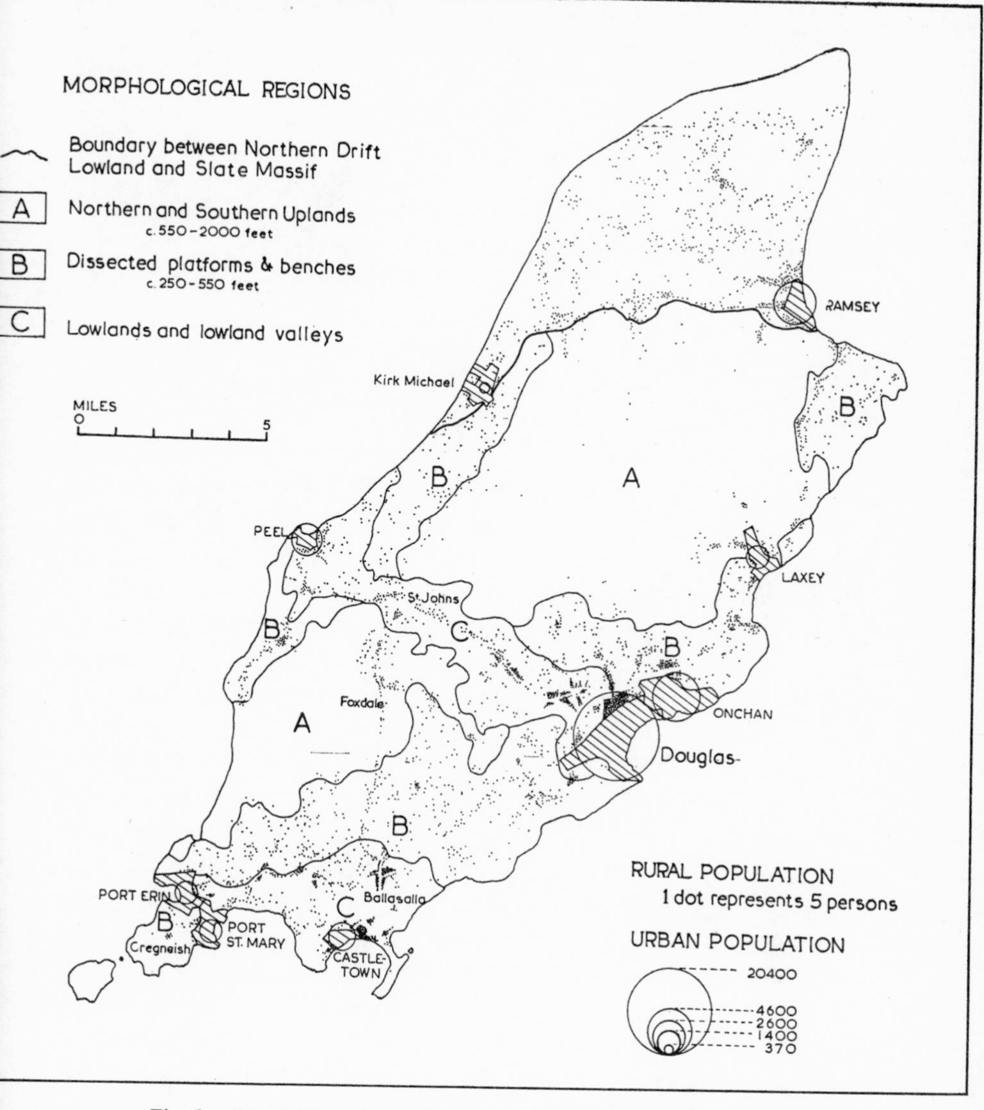

Fig. 63. Distribution of population, 1951, in the Isle of Man.

arrive in Douglas and over 60% of them come from Liverpool, Fleetwood and Heysham, with smaller numbers from Ardrossan, Belfast and Dublin. The air traffic to Ronaldsway airport, near Castletown (fig. 64), is more diverse in its points of origin and has increased steadily to over 13% of the total in 1958.

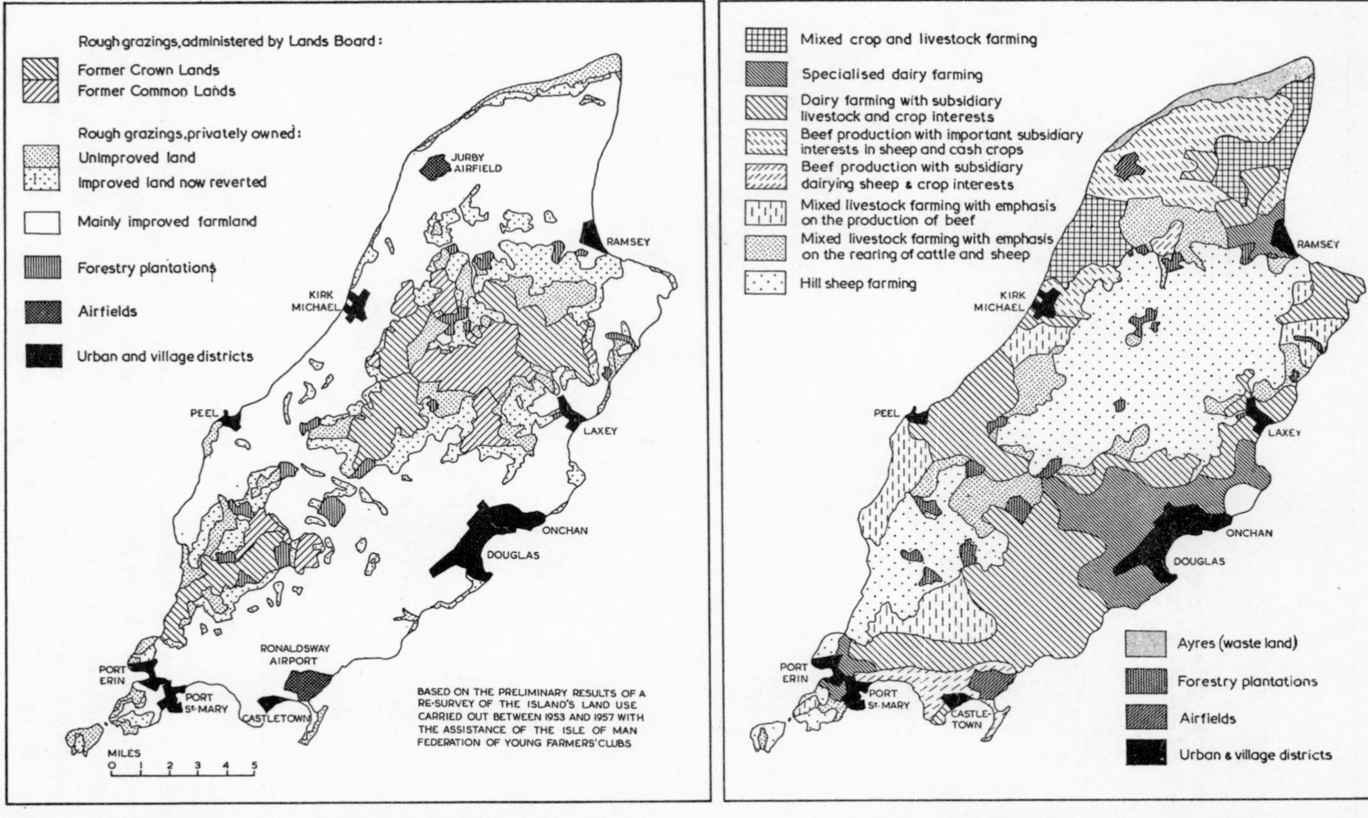

Fig. 64. Major land use distributions in the Isle of Man.

Fig. 65. Types of farming in the Isle of Man.

Since most of the holiday-makers come from the industrial towns of northern England, the more active part of the season is restricted to the ten weeks from late June to August during which these towns hold their 'wakes' and other industrial holidays. In addition, the length of stay and expenditure of the average visitor is rather less than, for example, in Jersey.

Nevertheless, the tourist industry contributes perhaps three-quarters of the income from external sources, other than investments, and enables the island to support a population far higher than would otherwise be the case. Moreover, the status of the Manx government has made it possible to provide certain fiscal and administrative advantages for the tourist industry as well as for the permanent settlement of retired persons.

The holiday industry and the Victorian architecture of its hey-day contribute much to the character of the towns. This is particularly so in Douglas and Port Erin which are the most specialized resorts, having a capacity of approximately 2 visitors for every resident as compared with 0·9 for the island as a whole, 0·5 for Ramsey and 0·3 for Castletown. Douglas and Onchan together provide 80% of the total accommodation for visitors, a concentration which is related, among other things, to the position and the physical attributes of Douglas bay and harbour, and which offers very considerable advantages of scale in the provision of man-made amenities and entertainments. Less than 4% of the visitor accommodation is within the rural areas; and within them, even at high season, the direct evidence of the tourist traffic is limited to such features as the commercialized glens and the occasional café, or to the passing of some of the fleet of over 100 coaches providing 'Round the Island' trips.

In contrast to its effect on other industries, the tourist industry has been of great benefit to farming, by providing, directly and indirectly, a local market of sufficient size to absorb the greater part of its output. External trade in agricultural produce is, therefore, limited principally to a small net export of oats, turnips, livestock and meat, and a net import of wheat, fruit and vegetables.

Despite its generally favourable position, Manx farming now employs only 11% of the working population, and is mainly concerned with the more productive and accessible land of the dissected plateaux and lowlands (fig. 64). During the past seventy years, approximately 13,000 acres of farmland have reverted from cultivation or improved grassland, and rough pastures now occupy over one-third of the island's area. These pastures consist mainly of hill-land and carry about one-third of the total breeding ewes; though, largely on account of the restricted local market for store animals, the density of stocking is appreciably lower than that which would appear to be justified by the available feeding and by the 'soft' insular climate. In a number of the more sheltered situations, substantial progress has been made in the establishment of coniferous plantations.

The improved farmland now represents only 54% of the island's area. The greater part of it, together with much of the reverted land, is held in farms of from 50 to 200 acres that vary in their layout from the single-field, strip holdings of the narrower coastal plateaux and upland edges to the more symmetrical holdings of the lowlands. In contrast with the earlier system of freeholding, approximately three-quarters of the farms are now worked by tenants. Their dependence on family labour has been accentuated during the past twenty years with a reduction of one-third in the number of hired labourers; though the productive capacity of the farms has risen with greatly increased mechanization.

At the present time, less than 18% of the improved land is down to permanent grass, and this consists principally of poor, sub-marginal land. The remainder is worked on a five- to seven-course rotation system, similar to that in use in western Scotland and Northern Ireland, in which three years' cropping are followed by rotation grassland, with the result that well over 50% of the arable land is also normally in grass. This rotation supports mixed livestock farming, a system that accords well with the island's comparative advantages, in physical, social and economic terms, for producing livestock and livestock products rather than cash crops. There are, however, important regional variations in the emphasis given to particular enterprises, related to changes in the balance of these advantages. It has, therefore, been possible, on the basis of a sample-farm survey, to classify and map in generalized form a number of individual types of farming (fig. 65).

The close association between the occupations of farming and fishing, which preceded the growth of the island's modern economy, gave way to a more specialized fishing industry employing at its peak as many as 400 vessels and 2,500 men and boys. Much of their concern was with the local fisheries: these derive their importance

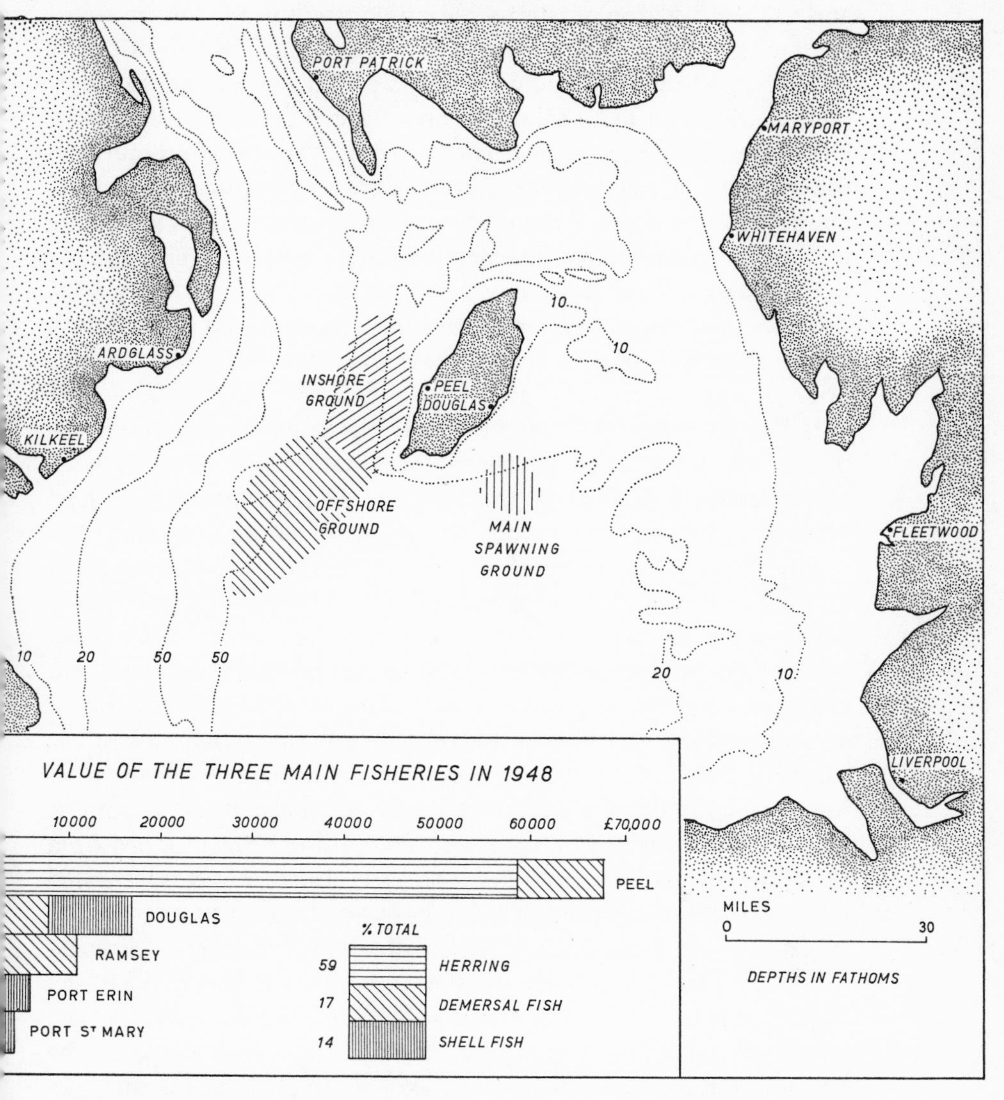

Fig. 66. Herring gathering and spawning grounds, and activity of fishing ports in 1948.

from the fact that the island's inshore waters are the focus of feeding and spawning migrations, notably of herring, cod and plaice, as well as the home of a variety of other commercial species (fig. 66). The marked decline of the industry after the early 1880's was related principally to the competition of the better-equipped fleets in Britain and to the attraction of alternative employment in the tourist industry during the summer herring season. There has been no sustained recovery, and the island now has only 24 fishing-boats and 70 fishermen of whom no more than half are engaged full-time. Moreover, although they are now mainly concerned with the demersal and shell fisheries, the Manx fishermen are unable to command their local market, in the face of regularly available imported fish of more uniform size and quality. The herring fishery is now largely confined to the Inshore Ground; and over 80% of the catch is taken by Scottish and a few Irish boats which, together, number about 100 during the peak months of July and August when the fish gather to feed prior to spawning. In addition, over half the catch is landed at Port Patrick rather than in the island: this is a situation which emphasizes the problem of a restricted local market and which has been only partially remedied by the recent establishment of a fish-meal plant at Peel to supplement the kippering and curing industries.

The former importance of the mining industry is clearly marked in the present landscape, particularly in Foxdale and the Laxey valley where as many as 1,000 men were employed, in the peak years of the 1870's and 1880's, mining and concentrating lead and zinc ores for shipment (fig. 61). The closing of the mines early in the present century was the result both of increasing costs and of the competition of new, large-scale producers elsewhere. Similar causes were responsible for the shutting down of the small salt factory at Ramsey that operated from 1902 to 1956, using brine pumped from the saliferous marls beneath Point of Ayre. In recent years, renewed interest has been shown in the island's mineral resources, encouraged in part by the comparative ease of obtaining prospecting licences since the mineral rights were transferred from the Crown to the Manx government in 1949. No major developments have yet taken place, other than detailed prospecting; though the waste heaps of the Snaefell mine at the head of the Laxey valley, and of the mines in Foxdale, have been successfully re-worked by a process of chemical flotation. At Foxdale, also, a plant for the production of china stone and mica flake from the local granite has recently been opened, and shipments are being made via Peel.

A small off-shore island presents a number of disadvantages for manufacturing, among which the limitations of scale and the likelihood of additional transfer costs are particularly significant. During the nineteenth century, when the mining and fishing industries augmented the local market and when the competition of large-scale mainland producers was less serious, the Isle of Man possessed important manufactures concerned particularly with milling, shipbuilding and textiles. Of these, woollen manufacture and milling are the main survivors, and each is dominated by one comparatively large firm. For example, of the two woollen mills that remain (of the seven formerly in existence) only that at St John's, which employs a little over 100 people, operates on a scale sufficient to embrace the full process of manufacture and, thus, to derive the maximum benefit from the excellence of the local water supply. By concentrating on the production of high-quality worsteds and tweeds, this mill has also taken fullest advantage of the appeal of a Manx trade-mark, and has at the same time reduced the relative importance of transfer costs.

Of the more recent manufactures, particular interest attaches to those established during the past six years and now employing over 500 workers. They have been attracted to the island by the efforts of the Manx government, aided by the availability of both labour and electrical power, and by the fact that for the resident manufacturer there exist certain fiscal advantages such as the low rate of income tax and the absence of death duties. The importance of these considerations is suggested by the fact that the twelve new factories, which are principally concerned with textiles and light engineering, depend almost entirely on imported materials and export markets. On the other hand it is noteworthy that the nylon-stocking factory at Ramsey and the various light engineering concerns in particular, enjoy comparatively low transfer charges since they are using materials, easily transported, to produce goods in which the labour costs are of prime importance.

The general significance of the insularity of the Isle of Man has been accentuated by its small size. This is particularly so in the economic sphere; and it has been shown that limitations of scale provide much of the explanation for the decay of a number of important nineteenth-century industries. The success of the complementary expansion of the tourist industry was related not only to the advantages of insularity and position, but also to the achievement of large-scale activity through a considerable measure of specialization. This specialization has greatly increased the island's dependence on the economic conditions obtaining in the rest of the British Isles; and the tourist industry itself has also brought important social changes especially in Douglas. There remains, nevertheless, a strong consciousness of the distinctive cultural and political heritage of the Isle of Man; a consciousness which has recently found practical support in the recommendations by a Constitutional Commission for an increased degree of self-government.

Selected Bibliography

D. E. Allen and W. S. Cowin. 'The Flora and Fauna of the Isle of Man and their Geographical Relationships.' *North-Western Naturalist*, 1954, pp. 18–30.

J. W. Birch. 'The Economic Geography of the Isle of Man.' *Geographical Journal*, **124**, 1958, pp. 494–513.

J. W. Birch. 'On the Climate of the Isle of Man.' *Procs. Isle of Man Nat. Hist. and Antiq. Soc.*, **6**, 1959, pp. 97–121.

J. R. Bruce, E. M. Megaw and B. R. S. Megaw. 'A Neolithic Site at Ronaldsway, Isle of Man.' *Procs. Prehist. Soc.*, **13**, 1947, pp. 139–60.

Grahame Clarke. 'Prehistory of the Isle of Man.' *Procs. Prehist. Soc.*, **1**, 1935, pp. 70–92.

A. M. Cubbon. 'The Ice Age in the Isle of Man.' *Procs. Isle of Man Nat. Hist. and Antiq. Soc.*, **5**, No. V, 1957, pp. 499–512.

E. Davies. 'Treens and quarterlands in the Isle of Man.' *Trans. Inst. Br. Geographers*, **22**, 1956, pp. 97–117.

R. H. Kinvig: *A History of the Isle of Man*, Liverpool, 1950; 'The Isle of Man and Atlantic Britain,' *Trans. Inst. Br. Geographers*, **25**, 1958, pp. 1–28.

G. W. Lamplugh. *The Geology of the Isle of Man.* London, 1903.

E. M. Megaw and B. R. S. Megaw. 'The Norse heritage in the Isle of Man' in C. Fox and B. Dickins. *The Early Cultures of North-West Europe.* Cambridge, 1950, pp. 143–70.

R. H. Skelton. 'Manx mines.' *Mining Mag.*, **92**, 1955, pp. 9–17.

PART THREE

SCOTLAND

CHAPTER 25

THE SOUTH OF SCOTLAND

JOY TIVY

A broad belt of hilly country straddles the south of Scotland from north-east to south-west. It forms a rampart which stands with its back to central Scotland but lies open to the south. Along its northern margin high, bleak, sparsely populated moorland stretches from coast to coast, flanked by narrow defiles and crossed by only a few low passes. To the south, however, deep-cut narrow glens widen out and lead into the richer and more densely peopled agricultural lowlands of the Merse of Berwick in the east and of Galloway in the south-west (fig. 69). East and west of the Cheviot dome and the north-facing scarps of the Northumbrian Fells, where the watershed forms a natural boundary between England and Scotland, these lowlands continue unrestricted southwards into the coastal plains of the Solway on the one hand and of Northumbria on the other. Through these two gateways routes have long been channelled across the historic 'marchland' of southern Scotland.

The northern boundary of the region is that major fracture—the Southern Upland Fault—which, striking from Dunbar in the north-east to Glen App in the south-west, separates the ancient block of old, tough, tightly folded Ordovician and Silurian grits and shales to the south from the down-faulted younger sediments of the midland trough to the north. This sharp geological boundary marks the transition from the fragmented lowlands of central Scotland to the high dissected plateau of the Southern Uplands. But only where, as along the edge of the Lammermuir and Moorfoot Hills in the north-east, weaker sediments lie directly against the edge of the tougher block, does the plateau form a well-defined north-facing escarpment. Elsewhere it is flanked by hills as high as itself formed of resistant lavas, conglomerates or grits which outcrop along the northern margin of the fault, and only the marked alinement of strike valleys betrays the presence of the geological boundary.

From a higher northern margin the general level of the plateau decreases in altitude southwards by a series of broad 'steps'. The highest surface, deeply dissected into a complex of sinuous, flat-topped ridges which stretch from the Lammermuir and Moorfoot Hills in the north-east to the isolated hill of Beneraird in the south-west, attains a general summit level of 2,000–1,500 ft; it rises, however, to 2,500–2,700 ft on the metamorphosed rim of the Loch Doon granitic intrusion which forms the high ridges of the Merrick and the Rhinns of Kells in the south-west, and on the more massive grits which underlie the lofty central plateaux of Hart Fell and Broad Law. This elevated plateau occupies nearly half of the total area of southern Scotland and its unimproved moorlands are among the most important and heavily stocked hill-sheep grazings in Scotland (fig. 68).

These high upland ridges—the true Southern Uplands—rise, often abruptly, above a lower intermediate plateau surface. Varying from 1,200–750 ft along its inner edge and with a gradual slope to 500–400 ft on its outer, this intermediate plateau forms a ragged irregular fringe around the southern margin of the Uplands and extends long fingers up the major valleys. In places it is considerably dissected; in others, for example to the north of the Merse in the east and on the low moors of Wigtownshire in the west, it still preserves extensive areas of gentle gradient with resultant poor drainage. In the south-west it is cut primarily across Ordovician and Silurian shales and grits, in the north-east it transgresses both these and the Old Red Sandstone which overlies them, as well as the New Red Sandstone and Carboniferous sediments preserved in troughs along the Nith and Annan valleys. Its surface is diversified by a hummocky, though often thin and discontinuous, drift cover. Particularly on that broad crescent which encircles the Merse, bosses of harder volcanic rock which riddle the sandstones form isolated hill masses which rise abruptly some 200–500 ft above its general level. A transitional zone between the wholly cultivated lowlands and the upland rough grazings, it is reminiscent of that 'higher lowland' surface at much the same altitude in central Scotland.

To the south these high and intermediate plateaux overlook and all but enclose a diverse fringe of more fully cultivated, drift-covered lowlands. In the south-east lies the triangular plain of the Merse; ribbed from west to east by long drumlins, it is underlain by Carboniferous sandstones which continue southwards into Northumberland. In the south-west are the peninsular lowlands, terminated by low cliffs, of the Rhinns and Machers of Wigtown, across which smooth drumlin swarms alternate with a rocky ice-scraped and stripped surface. Similar, though more fragmented, are the coastal lowlands of Kircudbright; the high moorlands of the granitic intrusions of the Criffel, Bengairn and Screel Hills interrupt their continuity, and rocky headlands alternate with the wide mud-flats of many tidal estuaries. To the east of the Nith, the carse-fringed lowlands at the head of the Solway Firth are underlain by Triassic and Permian sandstones which continue southwards into the Eden valley and extend northward in the 'intermont' basins of the Annan and Nith valleys; in the Sanquhar basin of the upper Nith valley, a Carboniferous outlier is the site of four small collieries.

Linking upland with lowland and cutting deeply across the 'grain' of the underlying rocks the main rivers of southern Scotland have provided the routeways which, from earliest times, have rigidly controlled movement across and within this region. The only low, through passage from east to west is that of the Tweed valley, linked to upper Clydesdale by the Biggar Gap. The circular earth-ramparts of Iron Age forts above its valley-sides and the ruins of peel towers which once guarded its strategic defiles testify to its former importance. Elsewhere east–west movement, across the upland blocks which separate deeply cut valleys, is difficult. The most favoured routes have always been those north–south lines which afforded the easiest or most direct means of access from England, originally, to Edinburgh, later to industrial Clydeside (fig. 67).

The easiest line of movement is that historic route following the eastern coastal plain—formerly the principal military route, defended by the ancient castle burghs of Dunbar and Berwick-on-Tweed, between England and the Scottish capital. This narrow coastal track round the Lammermuirs is still the main link by road and rail from London to Edinburgh. Between Berwick and Coldstream, no main road or railway traverses the heart of the Merse. The first bridge at Coldstream was built only in 1763. Until the middle of the eighteenth century the heavy ill-drained soils and marshy furrows made transit treacherous across this debatable 'marchland', while the Lammermuirs blocked direct access to the north. On the western border of the Merse, however, the north–south alinement of Redesdale, the Jed valley and Lauderdale provides the shortest, but most difficult, route

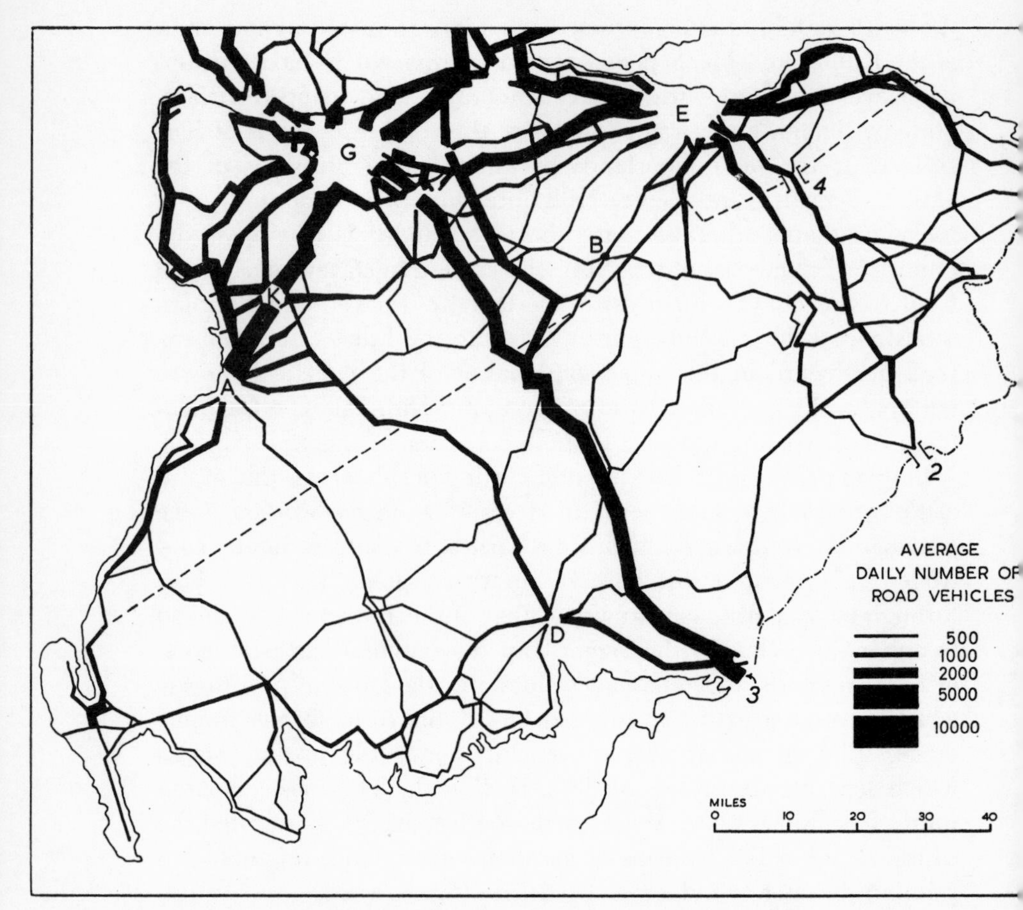

Fig. 67. Communications in the south of Scotland.
1. East coast route (A1); 2. Carter Bar; 3. Beattock route (A74); 4. Soutra Col. A=Ayr; B=Biggar; D=Dumfries; E=Edinburgh; G=Glasgow; K=Kilmarnock. --- =Southern Upland Fault. Based on ¼ in. map of Road Traffic Volume in Scotland (1954 Census). Reproduced by permission of Scottish Home Department (Roads Division).

from Newcastle to Edinburgh. The modern trunk-road manœuvres gradients too steep for railways across the high passes of Carter Bar in the south and the Soutra col in the north, and parallels the line of what was the main Roman road into Scotland (fig. 69: inset B).

To the west of the Cheviots, routes focus on the Solway Plain—the other main gateway into Scotland. There where the upper Clyde pushes its valley far south into the Southern Upland plateau to the headwaters of the south-flowing Annan, is the most direct road and rail route to Glasgow. This 'Beattock' route, which leads to the crossroads of the Biggar Gap, provided the second line of Roman penetration (fig. 69: inset B) and today, by virtue of the attraction of Clydeside, carries the heaviest road traffic into and across the south of Scotland and 75% of that from Scotland to England. Also from this important north–south line a series of routes run north-eastwards along those lines where the Teviot, Yarrow, Ettrick and upper Tweed dales have cut back along the strike of the Ordovician rocks and make contact by high passes with those of the Annan, Esk and Liddle waters. They converge on the historic 'node' of the middle Tweed valley where the Gala and Leader waters open routeways to the north. That of Liddlesdale and the Gala water provides the most direct and easiest route—followed by the Waverley rail line—from Carlisle to Edinburgh. West of the Beattock route the deep penetration of the Solway Firth truncates the north–south valleys of the Nith, Ken, Cree and Luce rivers and cuts them off from direct contact with the Carlisle Gap. Only the through-trench of the Nith valley is a major—and the most westerly—transit route by road and rail from England to central Scotland (fig. 67). Deflected, however, to the west of the main route to Glasgow, it leads into the Ayr basin and its contact with the Clyde valley is obstructed by intervening blocks of high moorland. Between the major breach of the Nith valley and a narrow and tortuous west-coast road, only two roads and one railway line cross the upland barrier—where the north-flowing headwaters of the Stinchar and Doon rise just south of the main fault line and allow access to the Cree and Ken valleys.

The through valley of the Nith separates two major and sharply contrasting regions in the south of Scotland: the Borders to the north-east and the peninsular appendage of Galloway to the south-west. The Borders perpetuates in its name its historic role astride those strategic passages which, leading towards Edinburgh, converged from the south on the Tweed valley—the heart of the Border country. Its ancient burghs, ruined castles and peel towers, are a reminder of the warfare and trade that swept back and forth across this zone. It was the gateway to the richest part of Scotland. With the Lothians and Fife it looked east and south. It shared in the trade in the staple with the Low Countries, and its place and family names reflect an early and thorough Anglicization. In contrast, the ancient province of Galloway lay off the well-beaten and bloody paths between England and Scotland. Hemmed in to the north and east by an almost unbroken moorland rim this south-western region was isolated from the main streams of Scottish affairs. Open, however, by wide tidal inlets to the Irish Sea, its formative cultural contacts were by sea with Ireland and the western islands and peninsulas. The flood of Norse invaders touched its coasts while to landward it resisted Anglo-Norman penetration longer than did the rest of southern Scotland. It has retained, as a result, a distinctive inter-mixture of Gaelic and Norse place-names. Today the accents of Wigtown, in particular, echo those of the Antrim coast whence came some of Galloway's major invasions of Irish immigrants—and its cattle. Continuing isolation has maintained a strong regional consciousness.

The Borders is of the east, Galloway of the west, of Scotland. This is true physically as well as culturally. In the south of Scotland are repeated, on a smaller scale, the physical differences that distinguish the east from the west Highlands. From the Nith valley east and north-eastwards the Southern Uplands still preserves much of the original high plateau surface in its tabular hill masses. Deep, narrow, flat-floored valleys provide admirable sites for the numerous reservoirs that supplement the water-supply of both Edinburgh and industrial Lanarkshire. Hill summits and valley sides owe their distinctive and characteristically smooth, rounded, convex slopes to a widespread and often deep mantle of angular frost-shattered debris and a continuous carpet of grass and heather moor. Bare rock is rarely exposed except where modern gulley and sheet erosion has stripped this surface cover or where glacial erosion has lightly etched the margins of the high Hart Fell–Broad Law plateau. Lower slopes and valley sides are plastered with glacial debris which spreads out over the surface of the intermediate plateau and continues as an unbroken mantle across the lowland of the Merse. Here the east–west alinement of long low drumlins is repeated in the geometrically rectangular pattern of road and field boundaries.

The uplands of Galloway, in contast, are more dissected and fragmented. The massive granitic intrusions of Loch Doon, Cairnsmore of Carsphairn, Cairnsmore of Fleet, and Criffel interrupt the geological uniformity of the folded Ordovician and Silurian sediments. Glacial erosion, often intense, has scoured and roughened both upland and lowland and has produced a landscape which recalls, in miniature, the Western Highlands. Its effects are most pronounced in the uplands of the Merrick, Rhinns of Kells and Cairnsmore of

Fleet ranges. Pitted with rock basins, breached by wide glaciated valleys, its slopes gnarled and roughened, the granitic 'cauldron' of Loch Doon is comparable to, though more difficult of access than, the Moor of Rannoch and must have acted as one of the main centres of ice-accumulation in southern Scotland. With its Highland affinities, though more limited catchment basin, this area of Galloway saw the first important installation of plant to produce cheap, peak-load hydro-electricity in Britain (fig. 69: inset A). The construction of the National Grid in 1932 made the exploitation of the potential water-power resources of the glacial lochs of this remote and sparsely populated region an economic possibility. The effects of glacial erosion in Galloway, however, extend well beyond the uplands and are continued south and south-west across the surface of the intermediate plateau and of the lowlands. On the former, rocky ridges interspersed with peaty flats and lochans preclude cultivation except on the better-drained slopes of scattered drumlins. Even on the coastal and peninsular lowlands drumlin swarms, splaying out from north to south, alternate with a stripped and rock-ribbed surface.

Moorland vegetation reinforces the contrast in landforms and reflects the climatic differences between the north-east and the south-west. The drier Border uplands carry extensive areas of dry heather and grass moorland on their smooth, well-drained slopes; their flat-topped ridges are capped with slabs of deep, but considerably hagged and eroded, peat now dominated either by heather (*Calluna vulgaris*) alone, or by a mixture of heather and bog cotton (*Eriophorum* spp.). These are the traditional sheep-walks, *par excellence*, of Scotland, early associated with the great Cistercian monasteries which were founded along the zone of contact between upland and lowland. Long-continued, selective sheep grazing combined with the age-old practice of 'muirburn' has taken its toll of vegetation quality and soil fertility. Grass moor has replaced, or is in the process of replacing, heather; the 'white' hill is now as common as the 'black' in this region. Extensive areas are dominated by the wiry, tufted, moor mat grass (*Nardus stricta*) and the heath rush (*Juncus squarrosus*), both of poor nutritive quality and indicative of minimum fertility. Lower slopes have been invaded by that most virulent weed of the Scottish rough grazings, bracken (*Pteridium aquilinum*). Steep and increasingly threadbare slopes are being ripped by gulley and sheet erosion.

Galloway, however, has the softer climate of the west coast, with higher humidity and, particularly in the uplands, higher rainfall, greater exposure and slightly lower summer and milder winter temperatures than in the Borders. Except on steeper, well-drained slopes and knowes, peaty soils and wet moorland vegetation predominate. Extensive areas of gentle gradient on the 'low moors' of Wigtown and along wide glaciated valley floors are covered with deep peat 'flows'; hill farms still cut and burn peat, and the peat deposits of Wigtownshire and Kirkcudbrightshire are exceeded in Scotland only by those of Sutherland and Caithness. Wet moorland associations dominated by the purple moor grass (*Molinia caerulea*) and bog myrtle (*Myrica gale*) below 1,000 ft, a mixture of heather, *Molinia* and deer sedge (*Scirpus/Trichophorum caespitosum*) up to 1,500 ft, and deer sedge at higher altitudes are characteristic. The aromatic bog myrtle and the purple-tinged *Molinia*, in particular, are sensitive indicators of the climatic transition from west to east. Rarely extending above 1,500 ft, *Molinia*-dominated associations disappear gradually north-east of the Nith where rainfall below this altitude is less than 55 in. and then only occur farther east on the wetter, high moors of southern Roxburgh and the Cheviots.

In both the Borders and Galloway these rough grazings are used primarily for the grazing of hardy Blackface and Cheviot sheep. The hill-sheep farms are large, seldom less than 1,000 and frequently 2,000–3,000 acres, and, having little or no improved land produce only lambs for breeding or fattening, and coarse wool. In the Borders sheep stocking, of 2–3 acres per breeding ewe, is as high as anywhere else on the rough grazings of Scotland. It is heavier than on the wilder, rougher and wetter moorlands of Galloway where there is much rock outcrop and where milder winter conditions promote a more rampant growth of bracken on well-drained sites; here 5 acres per breeding ewe is more common. These south-western moorlands have not suffered from such long-continued sheep grazing as those in the Borders. The medieval monasteries of Glenluce, Dundrennan and Sweetheart favoured the kinder coastal lowlands, and the rearing of black cattle was of greater importance and lingered longer—until the collapse of the droving trade in the mid-eighteenth century—in Galloway. Also, within the last thirty years the new plantations of the Forestry Commission (fig. 68) have ousted the Blackface sheep from considerable areas of former rough grazings in both Kirkcudbrightshire and Dumfriesshire, where high rainfall

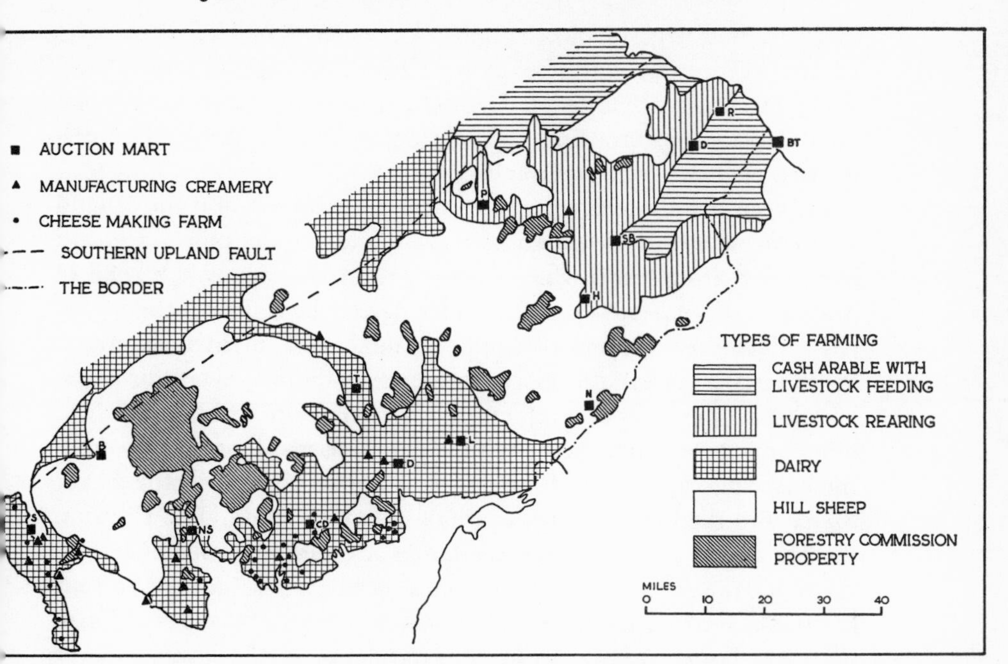

Fig. 68. Agriculture and forestry in the south of Scotland.
ırrhill; BT = Berwick on Tweed; CD = Castle Douglas; D (S.W.) = Dumfries; D (N.E.) = Duns; H = Hawick; ɔckerbie; N = Newcastleton; P = Peebles; R = Reston; S = Stranraer; SB = Newton St. Boswells; T = Thornhill.

and peaty soils are admirably suited to the rapid growth of Sitka Spruce in particular. Several thousands of acres of timber are already in production in the extensive forests of, for instance, the Glen Trool National Forest Park and the Forest of Ae.

In the Borders sheep still dominate not only the rough hill grazings but also form the fundamental basis of low-ground arable farming. The richest farmland in this region is that triangular lowland, everywhere below 400 ft, of the Merse (fig. 68). Its highly mechanized farms are among the largest (250–500 acres) and most heavily stocked arable farms in the country. Heavy clay loams combined with lower rainfall (less than 25 in. per annum) and greater summer warmth and sunshine than elsewhere in southern Scotland favour intensive cultivation. As much as 50% of the total acreage may be given over to cereal production, and barley for malting is an important cash crop which may account for more than half the total cereal acreage in any one year. Nevertheless, cash cropping takes second place to livestock production. Intensive cultivation and high yields in grass, oats and turnips, support large flocks of Half-Bred ewes whose heavy, early maturing Down Cross lambs are sold as forward stores for final fattening on Lothian or English farms. Complementary to sheep breeding is the fattening, mainly in winter courts, of Irish or home-bred store cattle.

Between these intensive arable farms of the Merse and the upland sheep farms with little or no improved land are the more extensive livestock rearing farms which for the most part occupy the zone of intermediate plateaux and the wider dales which dissect them (fig. 68). Under conditions of higher rainfall and greater exposure, lower temperatures and poorer stonier soils the major crop is grass. Worked on the basis of long leys, it may occupy from half to a third of the improved land. Although the emphasis is still on a lamb crop, the carrying capacity of these farms is not so high as in the richer Merse and the breeding stocks are of a hardier strain. The Cheviot, the Half-Bred and the Greyface lambs are destined as breeders or stores for the more arable low-ground farms. These farms also rear hardier Aberdeen-Angus or Galloway × Shorthorn stores. Farm units are large, varying from 300–400 acres on the lower to over 1,000 acres on the higher ground where they may include large acreages of rough grazing above 1,000 ft. And along the broader dales of the Middle Tweed and its tributaries cultivation extends up the valley sides on to slopes which are at a maximum for ploughing and to altitudes which reach, at 1,100–1,200 ft, the highest limits of improved land in Scotland.

In Galloway, in contrast, the limit between improved and unimproved land only occasionally exceeds 500 ft, except where scattered drumlins provide a reasonable depth of well-drained soil on an otherwise rough, peat-covered surface. Rough grazing covers most of the intermediate plateau surface from 1,500 ft down to 400–500 ft and carries the hill-sheep farm across central Kirkcudbrightshire and northern Wigtownshire and down to lower altitudes than in the north-east. There is as a result nothing quite comparable to the broad marginal zone of large stock-rearing farms in the Borders. Where, however, drift or alluvial soil provides a greater amount of cultivable land—as along the River Cree and broad valley of the Glenkens—hill-sheep farming is combined with the breeding of the hardy, slow-maturing, black Galloway cattle. They provide, as do the Blackface sheep, breeding stores for the adjacent low-ground arable farms producing stores or fat stock particularly for the English market. On most of the lowland arable farms of southern Galloway, however, milk is the principal product (fig. 68). Wigtownshire and Kirkcudbrightshire, together with the rich sandstone dales and coastal lowlands of Dumfries (where 50% of the arable holdings are dairy farms), account for nearly a third of all milk produced in central and southern Scotland. Climatic conditions are suitable for the optimum production of that basic crop and foodstuff, grass, which may occupy as much as two-thirds of the total arable acreage; though cash cropping, particularly of potatoes, is more important and the farm economy more diverse in the transitional lowlands and dales of Dumfriesshire. Most of the liquid milk is absorbed by manufacturing creameries (fig. 68) scattered through the area from Stranraer to Dumfries, though Dumfriesshire regularly and Galloway as need arises, supplements winter milk supply for Glasgow and Edinburgh. Today only a few of the traditional cheese-making farms remain in the Rhinns and south-west Kirkcudbrightshire (fig. 68) as a reminder of the greater isolation of Galloway in the nineteenth century before the introduction of efficient transport and co-operative marketing. In addition, the lowland farms of the south-west are important producers of dairy herd replacements, and Castle Douglas is the premier auction mart in Britain for the sale of pedigree Ayrshires.

The south of Scotland is a major producer not only of milk and fat stock but also, in particular, of breeding and store sheep. From upland, marginal and lowland farms livestock converge on the great auction marts of the region. Buyers come from all over Britain, and from overseas, to Castle Douglas and Newton Stewart in Galloway; to Hawick, Newton St Boswells, Reston and the famous annual ram sales at Kelso in the Borders. The small hamlet of Newton St Boswells has gained by its situation on the Waverley rail route, and with a turnover of a quarter of a million sheep alone each year is now the largest of the stock markets in the south of Scotland. Agriculture dominates the economy and the urban population of the region is, consequently, relatively small.

Towns are small and are now largely market and service centres for the surrounding agricultural areas. Most of them, however, are long-established burghs, many medieval in origin (fig. 69 and inset B). Closely linked to routeways within and across the region, their pattern has remained static and virtually unchanged up to the

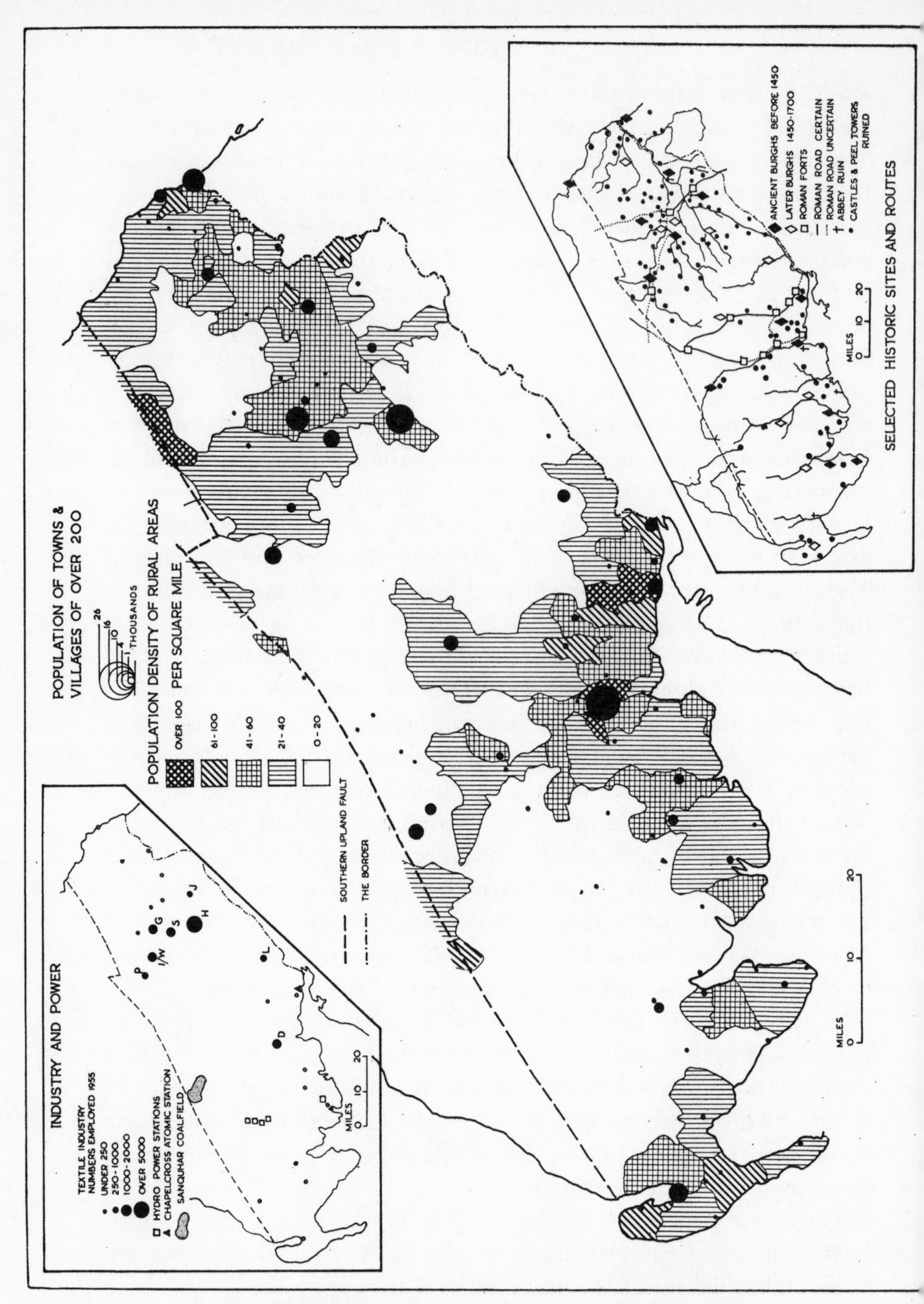

Fig. 69. The distribution of population in the south of Scotland.

D=Dumfries; G=Galashiels; H=Hawick; I/W=Innerleithen and Walkerburn; J=Jedburgh; L=Lockerbie; P=Peebles; S=Selkirk.

present. The Border towns are more accessible than those of the south-west. They have been more publicized and romanticized: Sir Walter Scott has proved a more successful regional protagonist than the Galloway writer, S. R. Crockett. The old regional centres clustering round the still important route node of the Middle Tweed have retained their importance and vitality. All are ancient burghs. Those of Peebles, Selkirk, Jedburgh and Kelso, founded in the twelfth century, are among the oldest towns in Scotland and together with the ruined abbeys of Dryburgh, Kelso, Jedburgh and Melrose are a reminder of the early economic importance of the region. Each with a long history of Border warfare, they still perpetuate their local loyalties and rivalries in the annual enactment of historic events and in rugby football championships. Within easy reach of both Edinburgh and Glasgow and on roads from north to south, they and the adjacent abbeys attract many tourists. Many of these Border towns owe their larger size to the textile industry (fig. 69: inset) which developed at the end of the eighteenth century using local wool, water-power, and the skill born of a long experience of domestic weaving. Still maintaining the tradition of small family concerns—whose names are now household words—the industry has survived competition by reason of a high degree of specialization in fine quality 'tweeds' and the development of knitwear and hosiery, much of it destined for the export market. The consequent growth of the older burghs has resulted, particularly in the case of the two largest—Galashiels (12,374), the principal centre of the textile industry, and Hawick (16,204), of knitwear and hosiery—in the cramped crowding of mills, houses and routeways along their very restricted valley sites. Between the small market town of Kelso and Berwick-on-Tweed, once the natural outlet of the Merse and one of the most important medieval ports in eastern Scotland, there are no large towns; the Merse in this respect still remains an 'empty' zone. The administrative boundary of the Border has deprived the Merse of its natural focus, and Duns, now the administrative capital of Berwickshire, is but one of a line of small market towns set back along the northern edge of the Merse.

The towns of Galloway reflect the formerly more important maritime contacts of the area before the coming of the railway in the mid-nineteenth century reduced its isolation. The decayed coastal burghs and numerous small abandoned harbours are a reminder of the former traders, legal and illegal, who plied these coasts. Of the medieval burghs of Kirkcudbright, Wigtown, Whithorn and Stranraer, only the last has retained its vigour and is still a port. Situated on the flat raised beaches at the head of the fjord-like inlet of Loch Ryan, Stranraer (9,249) is a train and steamer terminus for freight and passenger traffic to Ireland, a regional centre for much of Wigtownshire, and has diverse agricultural industries. The others are now merely local shopping centres, whose formerly more important market and administrative activities have been gradually absorbed by the 'newer' burghs of Dalbeattie, Castle Douglas and Newton Stewart situated on the Dumfries–Stranraer railway. The coastal towns and other small harbour villages lying off the main railway and road, remain quiet and rather select holiday resorts, the retreat of artists, with Kirkcudbright as the centre of the distinctive Galloway School; only a few tourists have as yet discovered this Scottish 'Riviera'.

The largest urban centre in the south of Scotland is the county town of Dumfries (27,275)—'The Queen of the South'. It has long been the principal focus for the dales which converge on the rich and thickly settled Solway lowlands, as well as for much of Galloway. It has maintained its status as the earliest and most important of the medieval burghs which guarded the western marches. Far enough from the Border to escape the fate of Berwick, it has survived the increasing pull of Carlisle. Like the region it serves, it is transitional in character and function between the Galloway and Border towns. Once an important port at the head of the shallow mud-flats of the tidal estuary of the Nith, it is less isolated from main routes than are its counterparts in Galloway. The greater size of Dumfries is related to its nodality and to the development of industry, more important than in Galloway and more diverse than in the Border textile towns; it is a major centre of knitwear and hosiery and of the dairy industry. Although it lies off the more important route, via Annandale, into central Scotland, it is equidistant from both Glasgow and Edinburgh, and it has become to a greater degree than any other town in southern Scotland the headquarters for the administrative region of the south of Scotland.

In few respects, however, is the south of Scotland a distinct region: indeed for many administrative purposes it is included with part or even the whole of central Scotland. It is rather a diverse group of small physically separate agricultural howes and dales fringing a core of upland moors. While each member retains a distinct individuality born of physical differences and former isolation one from the other, their regional unity has been blurred or obliterated by the administrative boundaries which have cut across them and made them tributary to central Scotland to which there has been a progressive drain of population. The 'pull' of Glasgow on the one hand and Edinburgh on the other, has, however, tended to preserve the fundamental dichotomy between the north-east and the south-west. The Borders still regards Edinburgh as its capital, takes the *Scotsman* and regularly supplies half of the Scottish rugby caps; Galloway and most of Dumfriesshire look to Glasgow, read the *Glasgow Herald*, and follow the fortunes of the Queen of the South soccer team.

Selected Bibliography

E. Wyllie Fenton. *The Influence of Man and Animals on the Vegetation of Certain Hill Grazings in South-East Scotland.* (I) September, 1951. Technical Bulletin No. 4. (II) September 1952. Technical Bulletin No. 5. The Edinburgh and East of Scotland College of Agriculture.

T. N. George. 'Drainage in the Southern Uplands: Clyde, Nith and Annan.' *Trans. Geol. Soc. Glasgow*, **22**, 1955, pp. 1–34.

D. L. Linton. 'The origin of the Tweed Drainage System.' *Scottish Geog. Mag.*, **49** (3), 1933, pp. 162–74.

Sir F. Mears. *Regional Survey and Plan for Central and South-East Scotland.* 1948.

J. Pringle. *The South of Scotland.* British Regional Geology (2nd ed.). Department of Scientific and Industrial Research. London (H.M.S.O.), 1948.

A. G. Ogilvie and others (ed.). *Scientific Survey of South-Eastern Scotland*, British Association for the Advancement of Science, Edinburgh, 1951.

G. S. Pryde. 'Burghs of Dumfriesshire and Galloway: Their Origin and Status.' *Trans. Dumfriesshire and Galloway Nat. Hist. and Antiq. Soc.*, 3rd series, **29**, 1950–51, pp. 81–131.

Joy Tivy. 'Reconnaissance Vegetation Survey of Certain Hill Grazings in the Southern Uplands.' *Scottish Geog. Mag.*, **70** (1), 1954, pp. 21–33.

CHAPTER 26

THE LOWLANDS OF SCOTLAND

RONALD MILLER

'Scotland, like all Gaul, is divided into three parts, namely the Highlands, the Central Lowlands, and the Southern Uplands. These . . . are natural divisions, for they are in accordance not only with the actual configuration of the surface, but with the geological structure of the country.' Thus Professor James Geikie half a century ago (following his brother's earlier lead), and for the present purpose we may join him and a host of others in accepting the rift-block between the Helensburgh–Stonehaven and Girvan–Dunbar faults as a geographical entity.

While these boundary faults are profound, they do not always give rise to bold escarpments and even where such escarpments exist, they are breached by considerable valleys, most of which have been widened and deepened by Pleistocene glaciation. There are, therefore, many human links across these physiographic boundaries and a case could be made for virtually ignoring them in arriving at regional divisions of Scotland. Neither modern counties nor traditional regions respect the two fault-lines: Strathearn and Menteith, like modern Perthshire, lie astride the Highland boundary, and in the south it was the watershed rather than the fault-line scarp, such as it is, that formed the boundary between Galloway and Carrick, the Lothians and the Merse. One could argue, for example, for the recognition as geographical regions of the spheres of influence of Edinburgh and Glasgow. The former would bring most of the Borders into the Lowland orbit and the latter Lennox, Cowal and indeed most of Argyll, in both cases transgressing the fault-lines.

Nor is our region well-named. 'Lowlands' it is traditionally, but it includes hill masses higher than the Pennines. 'Midland (sometimes Central) Valley' has been tried as a synonym, but the north-easterly and south-westerly extensions of the area belie these adjectives. It is not a valley but a multitude of them, many of which originate outside it and cross it, and some streams even flow from it into the Southern Uplands.

The Lowlands, however, undoubtedly have individuality. By far the greater part of the population of Scotland lives within its boundaries, and from the beginnings of a united Scotland the capital has been there. It was the northmost limit, if only for a short time, of the Roman world and it was the melting-pot in which the Scottish nation was formed. Open to influences from both west and east, Scot and Briton, Anglo-Saxon and Norse added their blood and culture to that of the indigenous 'painted ones'. Hills separated, but did not entirely divide them from their neighbours to the south, so that while the feudal system and English tongue penetrated and eventually prevailed, difficulties of transport prevented the Hammer of the Scots from bringing the full weight of the men and material of the rich English plain to bear, and he succeeded only in welding Scotland into national unity under the Bruce, also of Norman descent. Meeting sometimes in royal matrimony, sometimes in war, the Union with England was inevitable, and with it the Lowlands entered more fully into the community of Europe than has Ireland, or indeed any part of the Celtic north-west.

There are only two generous refuges for ships in time of storm on the whole of the east coast of Britain from the inner Moray Firth to the Humber, and of these the Firth of Tay is little better than the Tyne or Tees, leaving the Firth of Forth as the major haven for trade with Scandinavia and the Low Countries. This estuary was therefore for long Scotland's window on the outside world and her metropolitan area. It was not until after the beginning of trade with America that the other great estuary, the Clyde, came into its own and became the economic centre of gravity of Scotland.

The Lowlands are built of two main geological formations, the Old Red Sandstones and the grey sandstones, shales, coals and occasional limestones of the Carboniferous. Both formations include igneous rocks of many kinds, ranging from great thicknesses of contemporary lava-flows to dykes, sills and volcanic plugs. The structure is a complex syncline with axes following the north-east to south-west Caledonian trends, but there is much faulting and, for some reason not yet explained, many transverse structural lines which outline horst blocks and break the coalfield into a series of basins. Generally speaking, the igneous rocks are much more resistant than the sedimentaries, except for the basal grits and conglomerates of the Old Red Sandstones and thus in this geologically ancient province, differential erosion has left the igneous blocks upstanding and the sedimentaries have been eroded to lowlands. Coal has been preserved by down-folding and down-faulting and is mostly below sea-level. The Lowlands thus contain (fig. 69) a surprisingly small proportion of land below 400 ft (the approximate upper limit of agriculture) and there are many hill masses over 1,000 ft with summits sometimes exceeding 2,000 ft and culminating in Ben Cleuch at 2,363 ft.

While the disposition of the Lowland hill-masses reflects the Caledonian structure, no such simple pattern can be found in the drainage system. The major streams clearly originated on a surface different in relief and higher than the present and no doubt have tended to adjust themselves to the structure as they lowered this land-surface. Thus the lower Tay sweeps round the southern end of the Sidlaws to find an exit to the sea along a breached anticline between those hills and the hills of north Fife. Linton has suggested that the breaches in the continuity of the north-east extension of the Sidlaw axis of high ground may represent former estuaries of the Tay drainage. Similarly, he regards St Andrew's bay as having once been the estuary of the Earn, and the Devon as the headwaters of the Leven.

The Forth is a gross misfit in its present valley, owing either to a reduction of its catchment area or to a profound enlargement of the present valley by glacial erosion. A deep buried gorge is known to exist below the present channel with an extension into the lower Devon and, as far as can be gathered from the relatively meagre evidence of bore-holes, this gorge is of an ungraded, glacial character. Whatever the reason, the Firth of Forth gives a penetration of the land by the sea that is exceeded in Britain only by the Bristol Channel.

The course of the Clyde has been the subject of much speculation and it must suffice here to point out its Lowland characteristics. From Lanark to Hamilton it is in a deep trough. It then opens out on the howe of Glasgow, over a deeply buried pre-glacial channel, and leaves by a narrow passage between the Kilpatrick hills and the north Renfrew heights, to enter a broad estuary which joins at an abrupt angle the fjord-like Loch Long. Whatever the explanation of its curious course and its relationship to the through-valley leading off by Kilsyth to the Forth, the Clyde in its turn has brought ocean-going ships deep into the Lowlands and into the heart of one of its howes. The conjunction of Forth and Clyde is such that here only twenty-four miles of low-lying ground separate the Atlantic and the North Sea.

The minor valleys of the Lowlands pursue fairly direct courses to the sea but they also have their anomalies. Repeatedly in the Lowlands a river swerves from a more or less mature course to enter a gorge-like tract for a period, often reverting later to a more open valley. Almost always the reason is that its pre-glacial channel has been blocked by boulder clay and the stream has cut a new section of valley. Such 'dens' are picturesque, and though serious obstacles to the road-builder, were often useful sites for water-power development.

Everywhere in the Lowlands the work of ice is manifest. Upstanding masses are severely eroded and crag-and-tail forms are characteristic on the igneous features. On the lower ground, a mantle of debris is spread out to an average depth of some 100 ft, thus effectively masking the relief of the solid rocks. For the most part it is a relatively smooth blanket, but near the Highland edge, moraines, drumlins and kames are characteristic. On the whole, these deposits probably offer better material for soil formation than do the bedrocks of the Lowlands, but in their original state their upper surface held many undrained hollows and in places a high stone content. Farmers in Scotland, therefore, had not only to clear woodland but to gather up stones and drain bogs. The process is still far from complete and a dry spell frequently reveals the widespread patches of peat-staining in recently-ploughed fields and serves as a reminder of the enormous amount of drainage and reclamation which has been necessary to overcome the chaotic conditions left on the retreat of the Pleistocene ice.

The glacial outwash deposits of the Lowlands form a rich reserve of the sand and gravel so much in demand for building in these days of concrete and here (in contrast to the English plain where the removal of river gravels may leave only ponds) the quarrying away of kames in the Lowlands often leaves a more even site and finer soil material than before. Other economic assets of the outwash materials are sand suitable for glass and for foundry moulds and clays for bricks, tiles and pottery.

Four well-marked surfaces differing in altitude and slope may be

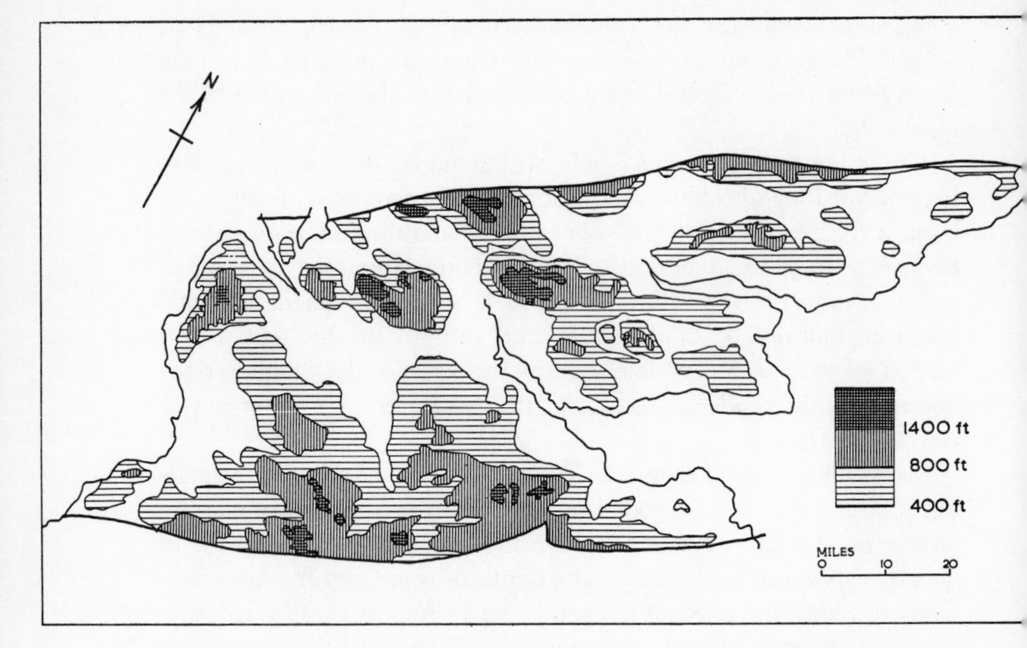

Fig. 70. Significant altitude zones in Lowland Scotland.

recognized in the Lowlands (fig. 70). The highest land ranges up to 2,000 ft and is principally represented by the Lennox hills, the Ochils and the Pentlands. All three are more or less tabular with upper surfaces perhaps related to erosion surfaces of similar heights in the Highlands and certainly introducing little islands of near-Highland conditions into the Lowlands. Their steep slopes and massive forms have on the one hand formed barriers to transport and therefore channelled communications round them and created bottle-necks, for example Stirling and Edinburgh, which at an early date became key sites. On the other hand, streams off their steep flanks have yielded water-power and so have attracted settlement. They provide sites for a few reservoirs but their catchment area is too small for these to be of major significance.

Between 400 ft and 1,000 ft Ogilvie recognized extensive slabs of ground of relatively low slope (less than 1 in 40) and this he called the Higher Lowland surface (fig. 70). Being above 400 ft these surfaces are little farmed and constitute for the most part, especially in the south, extensive moorlands useful only for sheep or as gathering-grounds for water-supplies. They are often high, bleak and so blanketed by bog as to hamper communications and again tend towards breaking the habitable portion of the Lowlands into small compartments.

A lower Lowland surface may be distinguished, which, though widespread, is neither so smooth nor in such compact blocks as the upper one. The dissected nature is, however, its main advantage, for it provides a natural drainage superior to that of the surfaces either above or below it, and it has therefore always been the principal agricultural zone of the Lowlands. Below it lie the marine terraces and river flood-plains. Where narrow, and therefore draining readily to the sea, these offer good conditions for agriculture and for town sites. When extensive and inland, drainage is difficult and often had to await modern techniques. In extreme cases, for example Flanders moss, the peat bogs are not yet drained although the carse-lands downstream were reclaimed in the eighteenth century.

In sum, only small portions of the Lowlands are low: barriers of hill and estuary create sufficient isolation to stimulate the development of regional characteristics and loyalties, but not enough to endanger the unity of the Lowlands. These sub-regions are, however, reflected in the settlement pattern.

Scotland lies in the far north-west of Europe, a position which before the discovery of America left it remote and out of the main stream of events. It also brought it under the full influence of the ocean as far as climate is concerned. Scotland, in fact, enjoys in winter one of the highest positive temperature anomalies in the world. This effect is, of course, most marked in the west and in spite of the small longitudinal extent of the Lowlands, there is often a distinct difference in weather between east and west sides, the former being more often invaded by the cold continental air masses in winter whereas the latter, especially close to the sea, only rarely experiences frost. But while the ocean tempers the harshness that high latitude would otherwise involve in winter, this northerly position asserts itself in summer, and, in conjunction with high cloudiness, wind and rain, restricts temperatures to a very modest level. Thus Leith is two degrees warmer in January than Cambridge but three degrees cooler in summer and if we compare the number of 'summer days' in the English plain with those in the Lowlands, the latter is at a grave disadvantage. Perhaps the unhappiest feature of the markedly oceanic climate of the Lowlands is that there is virtually no spring season. In March, April and May, the monotonous procession of depressions is usually interrupted by spells of bright and dry weather, but unfortunately this is often the result of an invasion of continental air that may bring frost. Thus while winters are not normally severe even on the east coast, slight frosts—but enough to damage vegetation—are only too common as late as May and may even occur in June. The damage done by the lateness of these frosts, when plants are at their most vulnerable, is out of all proportion to the small extent to which they depress the temperature means. Similarly, in spring and early summer, cold mists (the 'haar') are liable to blanket the east coast and deprive it of sun. On the other hand, warm marine influence on the raised beaches of Ayrshire is such that potatoes can be planted even in February, to ensure a very early crop.

Except on high ground, rainfall is not excessive. In the west of the Lowlands, it is high enough (40–50 in. a year) to make grass the most profitable farm crop, but in the relative rain-shadow of the east coast (some 30 in. a year) it has been shown that irrigation is sometimes necessary, particularly in the Lothians (25 in. a year). Rainfall is for the most part evenly spread over the year with a tendency to minima in April and, unfortunately, maxima around August, which is usually marked by wet turbulent weather. By contrast, more stable conditions usually prevail in September and early October and this is frequently the pleasantest season of the year.

A characteristic of the Atlantic climate is that conditions rapidly deteriorate with altitude. There is thus often an abrupt contrast between smiling fields and prosperous farms on levels below about 400 ft and barren peat-clad moors above this altitude. Cool, cloudy and moist, grass is the most important crop and in bad seasons the harvesting of even oats is a gamble. Only in the most favoured areas can wheat and sugar-beet be cultivated.

In view of the physical conditions outlined above, there is little doubt but that the original vegetation of the Lowlands was mainly forest; mixed deciduous on the clays and in the west, deciduous with Scots pine and heathland on the sands and in the east, but with a great deal of intermixture following the vagaries of deposition of the glacial drift. Where gradients were low, especially at higher altitudes, peat covered by heather and moor-grasses prevailed, and everywhere in the irregular hollows of the drift there were swamps and thickets of alder and willow. Woodland must quickly have given way to moorland with increase of altitude and exposure.

Most of this has been profoundly modified. By the early eighteenth century, trees had become a rarity for, apart from their use for timber and fuel, the run-rig system of agriculture, involving grazing in common by sheep and cattle in the absence of enclosures, effectively prevented regeneration. The Agricultural Revolution, in the latter part of that century, is responsible for many of the characteristics of the Lowland landscape today. The old pattern of infield and outfield, together with the hamlets of the joint tenants, were swept away and replaced by a more or less regular pattern of large square fields with houses usually sited on the individual holdings and not, as so often in the English plain, clustered in villages. Hedges, sporting and ornamental woods and shelter-belts were laid out and lairds vied with each other in their planting activities. Many of the hedges have subsequently been neglected—the Lowlands have dry-stone dykes and fences rather than hedges—but where the hedges have been allowed to run up to height, fine avenues of beeches are to be found, though this tree is probably not indigenous to Scotland. With the break-up of estates and the purchase of farms by their occupiers, trees in the Lowlands are again being lost. Little woodlands which the laird preserved for game are being cleared, partly for the capital gain, but also to remove cover for foxes and other vermin and to bring more land into cultivation. Possibly for the above reason, possibly because of enhanced demand in times of war, even shelter-belts have been felled and only too rarely is new planting to be seen.

The earliest large-scale agricultural exploitation was that by the medieval religious houses such as the abbeys of Paisley, Cambuskenneth and Arbroath, all taking advantage of the better soils and low altitudes of the raised beach deposits. The largest agricultural units in modern times, however, are to be found in the east of the Lowlands generally, where holdings mostly range from 150 to 200 acres. These have large regular fields, with each farm-house and its steading and workers' cottages forming an impressive group in isolation from other settlements. The rural landscape there is, in fact, a completely planned one, dating only from the Agricultural Revolution and standing in sharp contrast to some parts of the English plain where field patterns are sometimes Anglo-Saxon in date and farm-houses

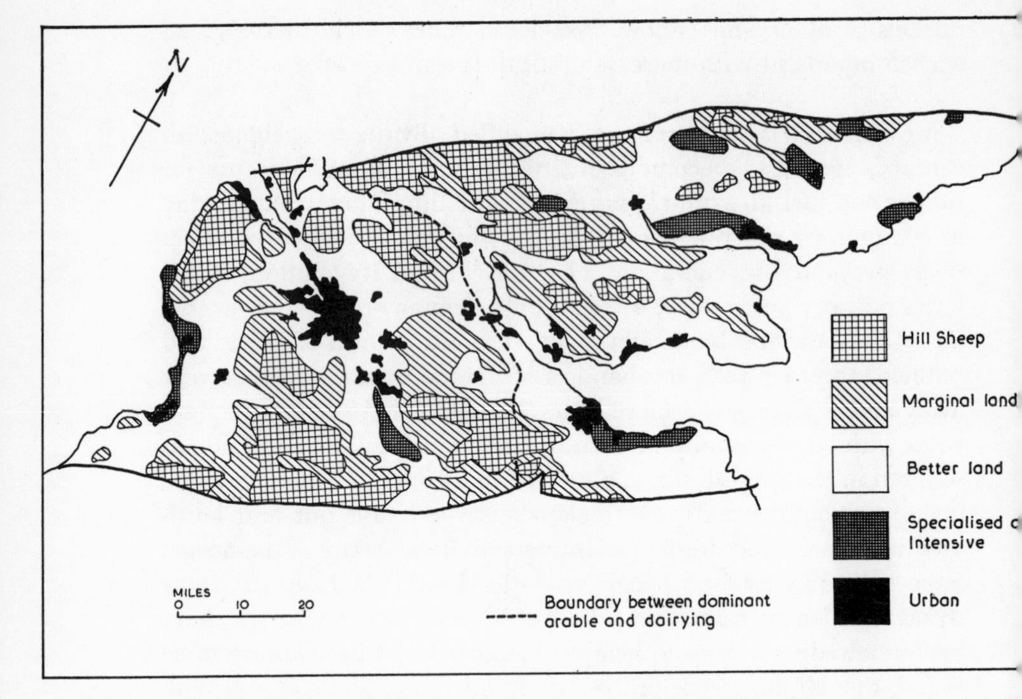

Fig. 71. Land classification and use in Lowland Scotland.

as early as Tudor. Even in the west of Scotland, the farms, though smaller, bear the imprint of deliberate lay-out in the eighteenth and early nineteenth centuries. Above all, the rural landscape in Scotland almost entirely lacks that characteristic feature of the English scene, the medieval parish church. Medieval churches there are, and there were more before the Reformation, but the principal reason for their relative absence from the Lowlands is that by far the largest part of the arable land has come into cultivation long after the Middle Ages. Against the lack of old churches must be set the advantages of rational plan and there is very little doubt but that the high reputation of the Lowland farmers is due in some part to their enjoyment of efficient lay-out of fields, roads and buildings.

Fig. 71 summarizes the agricultural situation. High ground is given over to hill-sheep farming, and considering that the area is the 'Lowlands', there is a surprising number of hill sheep, most of them the hardy Blackface, for conditions are rigorous. For the rest, a major boundary can be recognized somewhat to the south of the Forth.

To the west of this line, dairying is predominant, the moist climate resulting in a large proportion of permanent grass but with rotation grass, roots and oats contributing to the cattle feed. Naturally there is more cultivation on the better land and more permanent grass on the poorer. The Ayrshire cow is dominant, a locally developed breed which has achieved a wide popularity outside Scotland. Formerly there was large-scale cheese making—the Dunlop cheese from the village of that name in Cunningham—but with the growth of urban demand and with road transport, virtually all of the milk is now sold as liquid. Beef cattle, sheep and pigs are relatively unimportant.

To the east of the boundary line, farming is of more varied types but with a common dominance of the plough, which, in the north-east, may extend to as much as 90% of the farmland. Farms on the the poorer land—generally at higher altitudes—mostly specialize in stock rearing, but the farms of the better land either rear and feed, or, in addition, have crops such as wheat, barley, potatoes and sugar-beet for sale. The sale of crops is especially characteristic of the north-east, though sugar-beet is limited to those areas from which transport is economic to the one and only Scottish sugar-beet factory at Cupar, Fife. Barley, which is intolerant of wet and cloud and acid soils is a speciality of the coastal parts of the south-east and gave rise to the dominance of Edinburgh in brewing. In Strathmore, on the other hand, oats is the principal cereal. In this north-eastern area, however, the main objective of the arable farming is the production of beef. The Aberdeen-Angus breed is superlative for this purpose. In the south-east, on the other hand, stock could be regarded merely as essential to provide the dung for the intensive production of potatoes and barley. Whichever way it is regarded, however, prime beef is the largest single product of the arable farming of the east of Scotland and supplies of stores from the north and west have to be supplemented by the import of Irish stores. Field-fed sheep are also numerous, and again there is an import of stock from the north and west for fattening.

Where climate, soil or situation or any combinations of these are suitable, specialized intensive cultivation is found. From Strathmore and south into Fife, soil and climate on some of the higher sloping sites suit raspberries, which are grown on a field scale and in long rotation with other farm crops. The picking of the fruit is a traditional holiday occupation. In the Montrose area peas are grown on contract on a large scale for canning and there is a small acreage of daffodils and tulips. Market gardening is found near all the towns, but is a particular speciality of East Lothian, where vegetables are grown by the field with very heavy applications of natural and artificial fertilizers. Early potatoes are also a speciality there, but are more of a gamble than the early potatoes of the Ayrshire coast, where they are grown year after year on the raised beaches with seaweed as a supplementary manure. The upper Clyde, round Lanark, specializes in the glasshouse cultivation of tomatoes, chrysanthemums and bulbs, in summer, winter and spring respectively. The sheltered Clyde valley in the same area has plum and apple orchards, and market gardens.

Farming in the Lowlands is of a high average standard and is among the most productive in Britain. Ley-farming, still somewhat of a novelty in England, has been practised for many years, and the standard of livestock is so high as to bring a very worth-while income from the export of pedigree strains. In general, morale is high in the farming community and a rich and vigorous social life preserves the speech, customs and virility of Scotland to a far greater extent than in the cities. Though the Lowlands are overwhelmingly urban, agriculture is still one of the biggest industries and the one in which the Scottish tradition is most distinctive.

But beneath the soil of the Lowlands lay great mineral assets in coal and iron ore and, to a less extent, in oil shale. Being under the lower parts of the Lowlands, the coalfields have a thick overlay of glacial drift and therefore of good soils. Scottish pitheads, thus, typically rose out of rich farmland and, when first opened, were in centres of existing population and thus in a good position to obtain labour. Later, the building of miners' rows tended, as elsewhere, towards segregating the miners as a class, especially as their ranks were greatly swelled by immigrant Irish. Fig. 72 shows the distribution of the coalfields and two of their principal characteristics, their separation into three areas, all of fairly low ground, and their easy access from the sea. It is necessary to add that they contain inter-bedded iron ores and are sadly bedevilled by igneous intrusions and faulting which reduce the Scottish miner's output per shift to much below the national average.

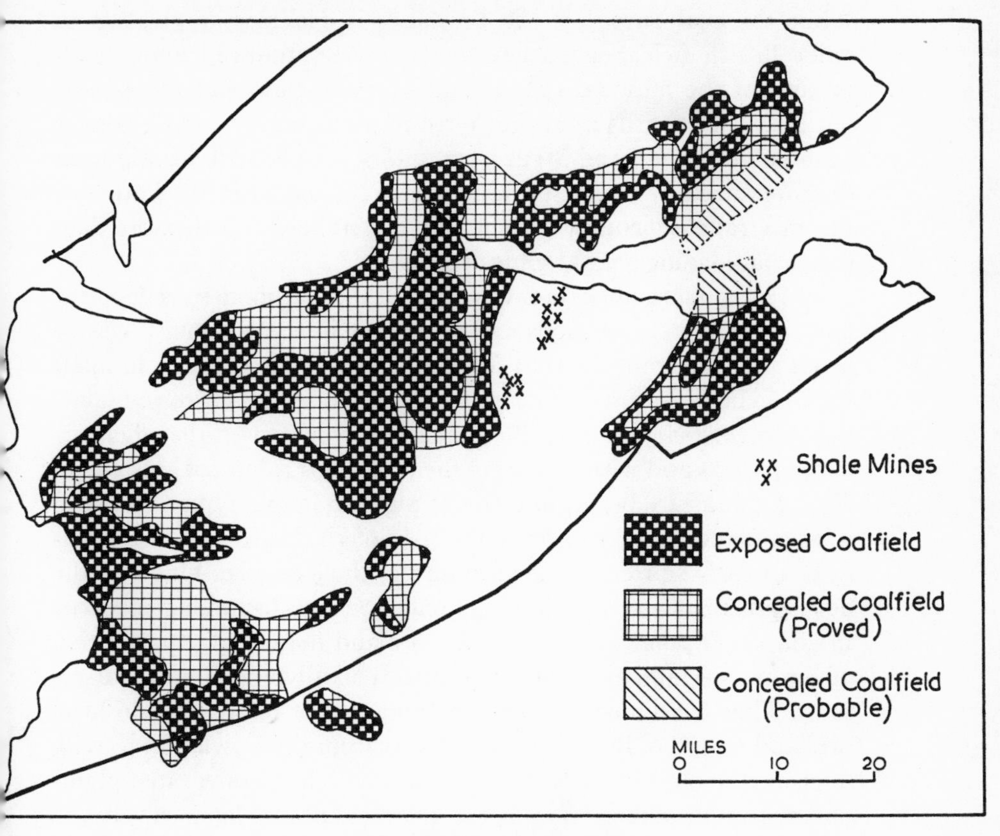

Fig. 72. Coalfields of Lowland Scotland.

The coal occurs in two rather widely separated series, the lower in the Carboniferous Limestone (so-called: true limestone is rare in the Lowlands), the upper in the Coal Measures (which contains only a small proportion of coal). The intervening Millstone Grit is rarely a a grit but has valuable deposits of fire-clays. The earliest coals to be worked were naturally the shallower and those at the outcrops. These were most readily accessible in the Lanarkshire field. The main reserves, however, are at depth in the lower series and occur mainly in the east. The modern phase of closure of small inefficient pits and the substitution of larger modern ones thus involves a marked shift in emphasis from the central field across to the east. Unfortunately, the main consumers are still in the west, so that much of the coal now bears a heavy freight charge on top of its cost of production.

While the coal output of the Lowlands is declining in relation to other British fields, in its heyday (1913) it produced about one-seventh of the national output and much of this went for export, south and westwards from the Ayrshire ports, and to the Continent from Bo'ness, Leith and Methil, continuing a trade with its origins in that organized by the medieval monks of Newbattle in the Lothians. But this trade is virtually extinct and has left behind little more than rusty coal-loading gear at some of the ports.

By far the more important functions of the Carboniferous deposits, and one with repercussions still dominant in the economy, was to produce cheap iron. In 1801 Mushet discovered and began to smelt the blackband ironstone that proved to be so abundant round Glasgow. In 1828 Neilson's hot blast enormously improved the efficiency of the process and with the use of the local hard splint coal instead of coke, Lanarkshire began to turn out pig iron in great quantities and a large foundry industry grew up. Although this iron industry was such an early starter, the acid nature of the local ores hindered the development of the steel industry and from the beginning involved the import of some non-phosphoric ore and the use of a proportion of scrap steel and imported pig iron. This divorce from local resources has continued and the steel industry of today uses no local ores and not all of its coke supply comes from the Lowlands. Small, uneconomic blast-furnaces have given way to large integrated plants and the main anxiety now is the maintenance of local supplies of coking coal, for if, as appears possible, this last local resource should prove inadequate, it is doubtful whether even local skills and the local market for bulk steel could keep in being an industry to which all the exceedingly heavy raw materials would have to be transported.

Throughout the nineteenth century, plentiful supplies of iron and coal brought into being constructional and engineering industries that made Clydeside the world's workshop. Being early in the field, Clydeside was ready to produce heavy capital equipment for the development of other countries, and bridges, rails, locomotives, ships, engines, and indeed all branches of heavy industry, brought boom conditions to the west. Ships had been built in the Clyde estuary ever since the New World offered trading possibilities. Since the voyages were trans-oceanic, the emphasis was on large and fast vessels, and the Clyde had a high reputation even in the days of wood and sail. Its particular speciality, however, has been steam engines and steel ships, especially large ones. Ships have changed in size from floating hotels to floating towns and thus created a demand for every sort of ancillary equipment from furniture to typewriters, and accordingly a vast variety of manufacturing industry has been called into being. Unhappily the requirements of ships and munitions in the two world wars has retarded Clydeside's evolution away from heavy engineering.

The Carboniferous resources may be regarded as responsible for another major branch of Lowland industry. In West Lothian, oil shales were discovered and worked in the nineteenth century and brought much prosperity. The richest deposits, however, were exhausted by the time petroleum flooded on to world markets, and it was natural for some of the shale refineries to turn to the distillation of petrol from imported oil. The nearby port of Grangemouth was chosen for the tankers and a modest pipeline carried the crude petroleum to the refineries. Chemical industry followed, to consume by-products, and when modern large-scale petroleum industry arrived, it was naturally attracted to the Grangemouth site, and mud-flats were covered with spent shale to create the extensive level ground required. Tankers, however, have vastly increased in size and draught, and they now berth in Loch Long, with 50 ft of water at low tide at the quayside, and pump their oil by pipeline to Grangemouth. The facilities at Loch Long and Grangemouth allow for almost unlimited expansion. In the meantime, the shale-oil industry has quietly expired.

The textile industry has passed through the same phases here as may be recognized elsewhere. Beginning as a widespread industry, making hodden-grey twills and linens from local wools and lint respectively, technical progress, especially the introduction of power, led to specialization and concentration. As elsewhere, first water-power then coal was the attraction, and now with the introduction of further artificial fibres and even more advanced processes, three major textile regions may be recognized in the Lowlands. The virtual concentration of finer woollen products in the Border towns has left mainly the heavier woollen products—carpets—in the Lowlands, in the Glasgow area.

Linen working has followed two divergent courses. The finer goods centred on Dunfermline and eventually gave way to synthetic fibres;

the coarser first prevailed in Strathmore, and while there is still some production of heavy linen there for special purposes, this industry has turned to the coarsest of all fibres, jute, and concentrated in Dundee. From its introduction centuries ago, cotton has been worked in the west, the region of entry, and for long this industry was the basis of Glasgow's prosperity. Lancashire, however, supplanted the Glasgow area, leaving little more than a specialism in sewing thread, in which Paisley is pre-eminent. Other minor, but for the Lowlands important specialisms exist, such as dyeing and processing in the Vale of Leven, using the pure and abundant outflow from Loch Lomond; dyeing and cleaning in Perth, using Tay water, and the small but profitable textile specialities in the Ochil hill-foot towns and the Newmilns area. A century ago, the textile industries were by far the most important in the Lowlands. They have now given pride of place to the metal and engineering trades, but retain considerable significance in that they afford work for about half the women in industry.

Consumer-goods industries are well represented and range from prime dollar-earners like whisky, the blending and bottling of which is so important in Glasgow and Leith, through sugar-refining at Greenock to the making of chocolates and sweets at Dundee. The former dominance of Glasgow in the tobacco trade is now represented by only a few factories, mainly for pipe tobacco, but cigarettes are also made in Glasgow and, more recently, in Stirling. Paper-making is an important industry, especially in the east and this and printing, publishing, book-binding and related industries are characteristic of Edinburgh.

While the Lowlands were slow to share in the post-war prosperity of the Midlands and south-east of England, the situation has been radically changed by the recent establishment of two major motor-car plants, one at Bathgate, the other near Paisley. Both of these are related to the pressed steel industry which in turn stems from the large new strip-mill at Ravenscraig. It may reasonably be expected that the manufacture of car components will follow.

In the pre-industrial era, agricultural population was fairly evenly spread out over the better land, and urban centres owed their existence largely to a nodality which gave them importance in peace-time as markets and administrative centres, and in troubled times—all too common in Scottish history—as strategic centres. In addition there were small commercial centres, using more or less natural harbours, most of them in the Forth estuary.

Thus from early times, Perth, Stirling and Edinburgh, key-points on the Lowland routes, the first two at gaps in the Campsie–Ochil–Sidlaw barrier, the last in the gap between the Pentlands and the sea, have exercised the function of capital. The senior is Perth (Scone), at the centre of gravity of a Scotland in which the agricultural potential of the Lowlands was not yet far developed and therefore the Highlands were of relatively greater importance than subsequently. Stirling is the centre of a developing Lowland and the site (Bannockburn) of the culminating battle in the struggle for independence. With the consolidation of the kingdom, the need for control of the Border with England, and the growth of trade with the North Sea countries, Edinburgh took over. Glasgow's geographical position was peripheral until trade with America, much increased after the Union, and the exploitation of its mineral wealth, made it the economic capital of Scotland.

A second rank of land communication centres may be distinguished along both margins of the region. On the Highland front, Stonehaven and Dumbarton control the coast routes and Callendar, Crieff, Dunkeld and Blairgowrie command the major valleys in the fault-line scarp. Most of them and some of the lesser towns at the mouths of smaller valleys are now market towns for Highland stock or staging points for tourists. A corresponding series on the south may be recognized in Girvan, Cumnock, Lanark and Dunbar. Dunbar, on the major coast road from England, along which invading armies could be supported from the sea, has more than once held the fate of Scotland and saw the only conquest by the English, under Cromwell. The distance of Lanark from the line of fault shows how little topographical significance the fault has in this area. Two major routes here enter the Lowlands, that by the Tweed and that by the Annan-Clyde valley. No single site controls them both, but Lanark stands far enough back to attempt this and also to control the Edinburgh–Ayr crossing of the Clyde.

Further consideration of settlement shows the next rank of towns to be related to the rich diversity of sub-regions within the Lowlands. These sub-regions fall into two main groups, separated by the Campsie–Ochil fault. To the north is the mainly agricultural Lowland, to the south coal is, or has been, pre-eminent. Apart from more obvious features like the absence of colliery headgear, the Old Red Lowlands are easily identified by their warm red freestone buildings. The Carboniferous Lowlands, on the other hand, mainly display grey or black walls, for the bed-rock yields as building stone a light-coloured sandstone (which darkens readily especially under industrial pollution), and abundant whinstone. Yet a further significant regional differentiation is to be found, for old roofs in the east are of red pantiles, originating in the Low Countries, and in the west, houses are limewashed to protect the fabric from the frequent wettings to which the climate there subjects them.

Of the Lowland sub-regions, it is probably Ayrshire which most turns it back on the Lowlands and faces outwards across the narrow seas. Girt around by lava plateaux, early settlement on the lowland of Kyle has spread outwards and upwards on to the lower and even higher Lowland surfaces, here well developed, but routes continued to focus on Ayr (45,297) as the capital and seaport. With the exploitation of minerals and the growth of industry, the centre of gravity shifted to Kilmarnock (47,509) and the old ports of Ayr and Irvine were superseded by railway-created rivals on the igneous promontories at Troon and Ardrossan. Coal exports from these ports strengthened ties with Ireland as did the hordes of mid-nineteenth-century immigrants, who were driven out by famine and civil commotion and attracted to the Lowlands by abundant employment.

Strathmore, the 'Great Vale', occupying a major syncline, is the principal agricultural sub-region and, being elongated, has a series of market centres, from Auchterarder in the south, through Perth (41,199), the greatest, to Coupar-Angus, Forfar, Brechin and Laurencekirk. All of these are service centres and have some industry based on agriculture—milling, distilling and textiles, for example. Apart from the first, each of these towns has—or had—its seaport. Perth has its own harbour for small ships, but in modern times is served by Dundee which is also the port for Coupar-Angus. Arbroath serves Forfar; Montrose, Brechin; Bervie, Laurencekirk; though, of course, each of these coastal towns has other functions. Bervie is a fishing town; Montrose depends in part on fishing and boat-building; Arbroath is a fishing port and also, with a wider coastal plain behind it, the agricultural capital of a region first developed by its abbey. Dundee (182,959) has added shipbuilding to its port functions and its textiles, earlier stimulated by the demand for coarse linen for Glasgow's colonial trade and heavy linen for sailcloth, turned with the collapse of the colonial trade to a remarkable specialism in jute, in which it had at one time virtually a world monopoly. Manufacture of the fibre at its place of origin in Bengal, and competition from elsewhere has, however, seriously affected the jute trade. Government-assisted light industries have to a considerable extent displaced jute, and Dundee is taking an increasing share in the linoleum industry. Another industry with origins in the local coastal farmlands and orchards is that of preserves.

The small Howe of Fife centres on Cupar, with its ancient port of St Andrew's, now virtually a two-industry town, based on golf and education.

The south Fife coast is in Industrial Scotland, and that from the earliest times. Easily-won outcrop coal was from medieval times mined along the coast in a score of localities from Kincardine to Leven and used to evaporate sea salt. Salt, and Forth herrings salted in it, were major items of commerce with the Scottish staple at Campvere in the Netherlands. Mining later moved inland to create Cowdenbeath, and devastate the country around it, but the opening-up of vast coal reserves under the sea now makes Kirkcaldy (52,371) the main centre. This is now expanding rapidly on the lower Lowland surface and, with its linoleum and other industries, forms an important growing point in the Lowlands economy. The new town of Glenrothes is linked to the sinking of a large modern shaft to exploit the coal reserves.

The River Forth carselands and their farms give rise to agricultural industries, particularly brewing at Alloa and Falkirk. Falkirk (38,043), at a convenient focus for all the roads from the Highlands, was once the principal cattle market of Scotland. Coal, iron ore, moulding sand and water transport made the locality, starting with Carron two centuries ago, the seat of an important iron-founding industry. To a specialism in light castings, mostly for domestic use, has been added an aluminium-rolling mill which is one of the major centres of its kind in Britain. With the adjacent town of Grangemouth, originally developed as Glasgow's outport at the North Sea end of the Forth and Clyde canal, and other towns in the immediate vicinity, a new conurbation is rapidly growing up. With its central position, good sea and land communications and ample reserves of level ground in the Forth mud-flats, we may expect to see considerable development in this area in the future.

Two sub-regions are to be recognized on the southern shores of the Firth of Forth, West Lothian and Mid- and East-Lothian respectively, separated by the Pentlands and commanded by Edinburgh in the narrows between those hills and the sea. West Lothian, with a rich mantle of glacial debris fluted and moulded to 'basket-of-eggs' relief by glaciation, has its ancient capital at Linlithgow and its seaport in Bo'ness. The once active shale-oil industry has left huge red 'bings' of spent shale to dominate the landscape, but thanks largely to the insistence of the local landlord on tidy workings, the region happily lacks the usual attributes of a decayed mining area.

The tight little syncline of Mid Lothian holds coal and a little natural gas. Apart from fuelling local industry, this coal formed a large proportion of the traffic of the port of Leith. Dalkeith and Musselburgh are route centres in the basin, the former with, in addition, a function as an agricultural market town, the latter with a history as a fishing port going back to the Romans' predilection for the local oysters. East Lothian is for the most part devoid of minerals, but its drift-covered surface, grooved south-west to north-east by ice and meltwater, offers some of the best farmland in Scotland. Haddington, grown up round its great abbey, is the ancient centre.

Strategic considerations determined the situation of Edinburgh (468,378), the volcanic crag (and tail) its site. Easily defended, the site nevertheless was a severely cramped one, forcing the citizens to build upwards to the great congestion of the city. With the early nineteenth century, however, bridges were built to link the old east–west ridge to similar features on both sides of it. New south–north arteries were thus created and space made available in which a new town of grace and dignity was built. Victorian expansion, however, outran the planners, and the Georgian core was surrounded by a crust of tenements as chaotic as elsewhere. At the same time, the port of Leith expanded to coalesce with its dominant partner. Fortress, seat of administration, learning, law and the church, agricultural capital and service centre, seaport and industrial town, Edinburgh has a rich mixture of functions and a correspondingly strong position in times of industrial depression. Rarely has its rate of expansion overwhelmed its civic sense and a great measure of its metropolitan dignity has been retained. This and its remarkably picturesque site, which apart from an assortment of glaciated hillocks and valleys, includes the 800-ft stump of a Carboniferous volcano, renders it a tourist attraction of the first order.

But while Edinburgh is the capital, it is the next city to the west which is the economic centre of Scotland. With one-third of the insured population of Scotland, the central Clydeside conurbation (1,801,850) is by far the biggest agglomeration of people in Scotland and is proportionately a far greater fraction of Scotland than London (one-tenth of the total) is of England and Wales. Clydeside is dominated by Glasgow (1,054,913). Relatively remote from the civil strife that characterized so much of medieval Scotland, the little bridge-town flourished on the dealings in the market which its episcopal lord had attracted to it. A modest capital and commercial and industrial skill was therefore ready to exploit the new markets created by the colonization of America. After the freedom of trade created by the Union, fortunes were made in tobacco, then cotton. The shallow Clyde was deepened to make Glasgow a port to supplant Greenock (74,578) and Dumbarton, and its own earlier outports of Irvine and Port Glasgow. An expansion of shipbuilding became a boom after the application of iron and steel and steam, and the Clyde became the world's premier shipyard. A new town of Clydebank was created virtually by one shipyard. Cheap and abundant steel fostered a great expansion of heavy industry and Glasgow passed the million mark and became the second city of the Empire. Satellite towns grew with her; the ancient Paisley (95,753) became the fifth town of Scotland, Hamilton (41,928) also grew greatly, and the expansion of the iron and steel industry brought entirely new towns into being at Airdrie (33,620), Coatbridge (53,946), Motherwell-and-Wishaw (72,799). The fine flush of expansion is now past; indeed population now requires to be decanted from the area, for the houses which attracted so many immigrant workers from Ireland and the Highlands are now largely outworn and outmoded, and the requirements of modern housing standards force their replacements to be built outwith the already crowded howe. New towns have been created at east Kilbride (31,972) and Cumbernauld, but the need for overspill of population is so large—hundreds of thousands—that many more reception areas must be found. The process of rehousing industries and workers will be a major problem for the city, and indeed for the Lowlands, for many years to come. It will be a particularly delicate operation because the city's economy, being so largely based on heavy industry, is none too secure.

Glasgow was in the van of industrial expansion in the nineteenth century and the demand for heavy armaments from her in two world wars has delayed the change-over from an earlier industrial pattern and the re-equipping of her industry for the second half of the twentieth century. A feature of Glasgow's history, however, has been her ability to change with the times and we may therefore expect to see a re-fashioning of her economy as striking as the rebuilding of her houses. It may well be that the new steel-strip mill and branches of the English motor-vehicles industry may constitute such a turning-point.

Three-quarters of the population of Scotland lives in the Lowlands and four-fifths are urban. This is a striking situation for a country which is often popularly taken to be peopled by picturesque mountaineers. Within the Lowlands the east is growing more than the west: the boom conditions of the nineteenth century have passed and stability has been attained, to some extent by a substitution of new industries for those on which the boom was based. Scotland, and even its Lowlands, remains essentially, however, part of highland Britain, and as such is a net exporter of population and thus makes its influence felt in virtually every part of the English-speaking world.

Selected Bibliography

A. K. Cairncross (ed.). *The Scottish Economy*. Cambridge, 1954.

D. L. Linton. 'Problems of Scottish Scenery' (bibliography). *Scottish Geog. Mag.*, **67**, 1951, pp. 65–84.

M. Macgregor and A. G. Macgregor. *Midland Valley of Scotland*. British Regional Geology (H.M.S.O.), 1948.

R. Miller and J. Tivy (ed.). *The Glasgow Region*. British Association for the Advancement of Science, Glasgow, 1958.

A. G. Ogilvie and others (ed.). *Scientific Survey of South-Eastern Scotland*. British Association for the Advancement of Science, Edinburgh, 1951.

CHAPTER 27

NORTH-EAST SCOTLAND

WILLIAM KIRK

Two voices are there, one is of the sea
One of the mountains, each a mighty voice

It would be difficult to maintain that the extensive and diverse area to be discussed in this essay constitutes a geographical region in the full sense of that term or that the designation north-east Scotland is completely satisfactory. Nevertheless at certain periods in the past, for example, during the early centuries of the Christian era when it formed the culture-hearth of the northern Picts, this area has provided a theatre for common human action; while in modern times its half million inhabitants have been drawn closer together by improved transport facilities, by common economic problems involved in farming and fishing in northerly latitudes, and by increasing urbanization. Former provincial identities expressed in district names such as Mar, Buchan, Formartin, the Garioch, Laich O'Moray, and the Black Isle have become less distinct as the spheres of influence of the larger towns have expanded and strengthened. Aberdeen, for example, now contains about one-third of the entire population of the area, and, in spite of its eccentric position when compared with Inverness, has become the provincial capital. Since the city functions both as a gateway and focal-point for the area it is proposed to examine its character first and then trace outward the various traits it shares with its hinterland.

Sited on seaboard benches and hummocky glacial deposits between the tidal estuaries of Dee and Don, modern Aberdeen in form and function represents the symbiosis of two originally independent cities. To the south of the narrow Don estuary stands Old Aberdeen (? *Aberdon*), in origin an ecclesiastical settlement gathering about St Machar's Cathedral and Bishop Elphinstone's King's College (1494). The old burgh with its High Street, narrow feus and gabled frontages, expanded market-place and Town House, preserves many of the features of late-medieval Scottish town-planning, but is now fast becoming an urban fossil in the midst of an extensive university precinct. The other Aberdeen (? *Aberden*), a settlement of comparable antiquity, grew on the north bank of the Dee, where the mouth of the Den Burn offered shipping better shelter than was available in the more exposed, silt-encumbered estuary of the main river. This city developed an extensive trade with Baltic and North Sea ports, and, following the building of a castle on the high ground overlooking the Dee, became an important seat of civil power in the north. In 1593, irritated by the religious conservatism of its neighbour, it founded its own university, now represented by the imposing granite building of Marischal College and for two and a half centuries the two universities maintained a separate existence. Thus the primary axis of the twin city Aberdeen was alined north to south, and remained so until early in the nineteenth century when the broad, east to west thoroughfare of Union Street was built. Backed by pleasant squares in the contemporary Edinburgh fashion, this street now constitutes the most important shopping and business centre of the modern granite-built city. Port, market, university town, financial and legal centre, and ever-growing holiday resort, Aberdeen provides the facilities of a large city without sacrificing the internal familiarity of a country town.

The port of Aberdeen, created by extensive dredging and reclamation of the estuarine deposits of the Dee, handles a considerable range of commodities both for the city and its agricultural hinterland. Exports, coastwise and abroad, include farm produce, smoked and salted fish, herring barrels and boxes, granite, paper, and light machinery; and imports include timber, wood-pulp and special granites from Scandinavia, esparto grass and phosphates from North Africa, and artificial fertilizers, feedstuffs, coal, petroleum products and general cargoes from many sources. Over a million tons of shipping enter the port each year, with Scandinavian flags particularly prominent. Special berths are provided for passenger lines serving the Northern Isles, and a large area of the port estates is devoted to the fishing industry, for Aberdeen ranks as the chief white-fish port of Scotland and an important centre of herring fisheries. There are also important shipyards with traditions dating back to the time when Aberdeen builders turned out some of the world's finest China clippers. Although present production is not large, some 5% of total

Scottish launchings are still made here and the yards provide valuable repair facilities for vessels using the port.

In addition to those industries associated with the port, various manufacturing and processing establishments are located in or near the city. These include long-established woollen and linen factories producing high-quality tweeds, blankets, hosiery and canvas for international trade; paper-mills on riverside sites in the lower Don and Dee valleys; timber yards making barrels and boxes for local and distant markets; granite yards working on grey granites from Rubislaw and Kemnay, pink granites from Peterhead and more exotic foreign stones; and a great variety of light-engineering workshops that have sprung up in recent years to serve an increasingly mechanized agricultural hinterland. Most of these establishments tend to be relatively small, however, and a large proportion of the city's working population is engaged in commerce, administration and the professions. Aberdeen has, for example, several important research institutions: a Marine Laboratory studies the rhythms of water movement and marine life in northern seas; other establishments study methods of fish preservation and food dehydration; the Rowett Research Institute investigates the nutrition of farm stock; and the Macaulay Institute for Soil Research acts as the headquarters of the Soil Survey of Scotland. Thus Aberdeen (185,379) plays a double role—as the head of a hierarchy of coastal settlements dependent on the resources of the sea, and as the chief of a series of market towns serving one of the most important pastoral areas of Britain. In this way it reflects the duality of social and economic life in the wider province of north-east Scotland (figs. 73 and 74).

Of the other towns only Inverness approaches the status if not the size of Aberdeen. Built on morainic and raised beach gravels at the mouth of the Ness valley, the town occupies a commanding position in relation to the Great Glen, the inner Lowlands of the Moray Firth, and the mountain passes leading south to Perth. The strategic value of the site has long been recognized; here was a Pictish capital in the Dark Ages, a castle town and royal burgh in the medieval period, and a Highland garrison town in modern times. The maintenance of a large military garrison and the need for a sea-gate for Highland trade led to the development of a port and market with a monopoly of foreign trade extending from Elgin in the east, to Fort William in the south-west and to Caithness in the north. Although the construc-

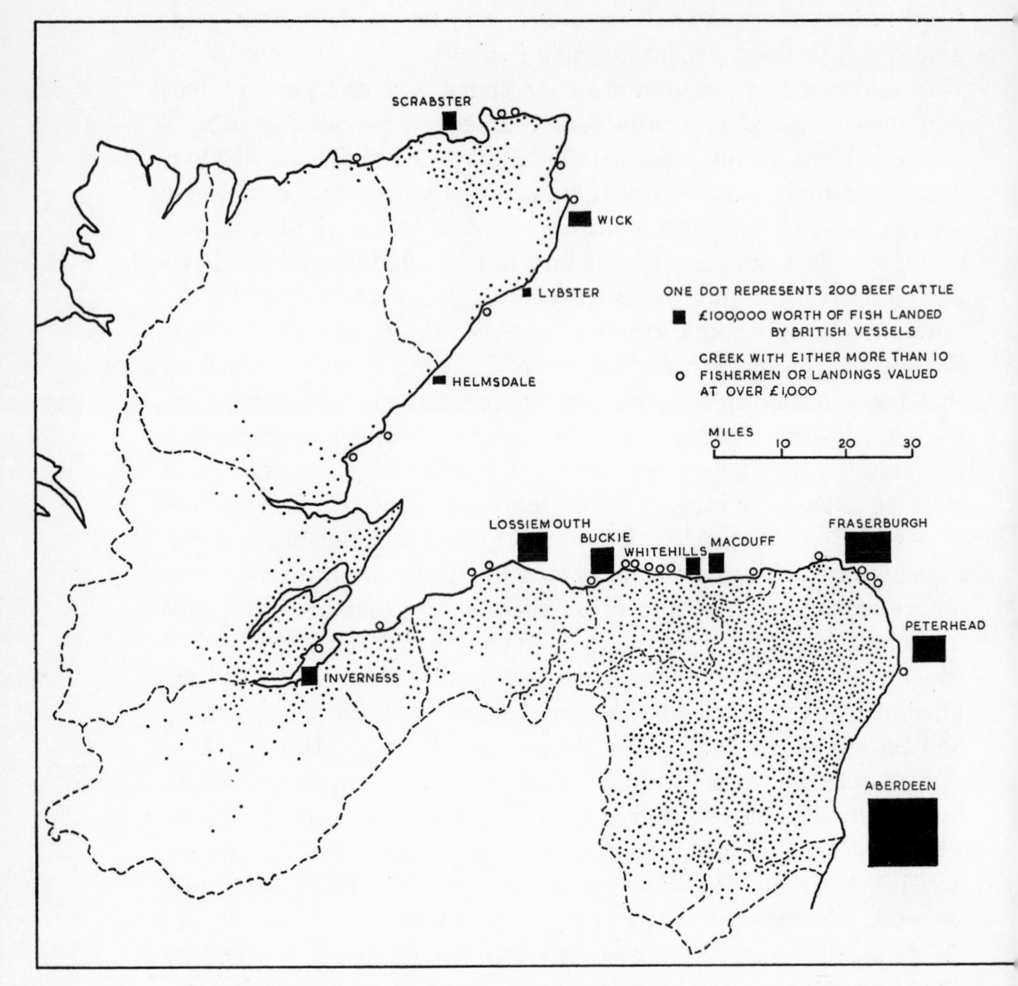

Fig. 73. The distribution of beef cattle and of landings of fish (1955) as indices of the landward–seaward du[ality] of North-East Scotland.

tion of the Caledonian Canal in the first half of the nineteenth century did not bring the expected increase of trade, the port of Inverness still has an appreciable coastwise trade in commodities such as coal, timber and cement, and possesses facilities for the storage of petroleum products, and the servicing of fishing vessels using the canal. The town's July wool fair has declined, but its cattle market, which has grown at the expense of marts at Dingwall and Lairg, draws beasts from a wide area. Inverness (29,773) is the main shopping

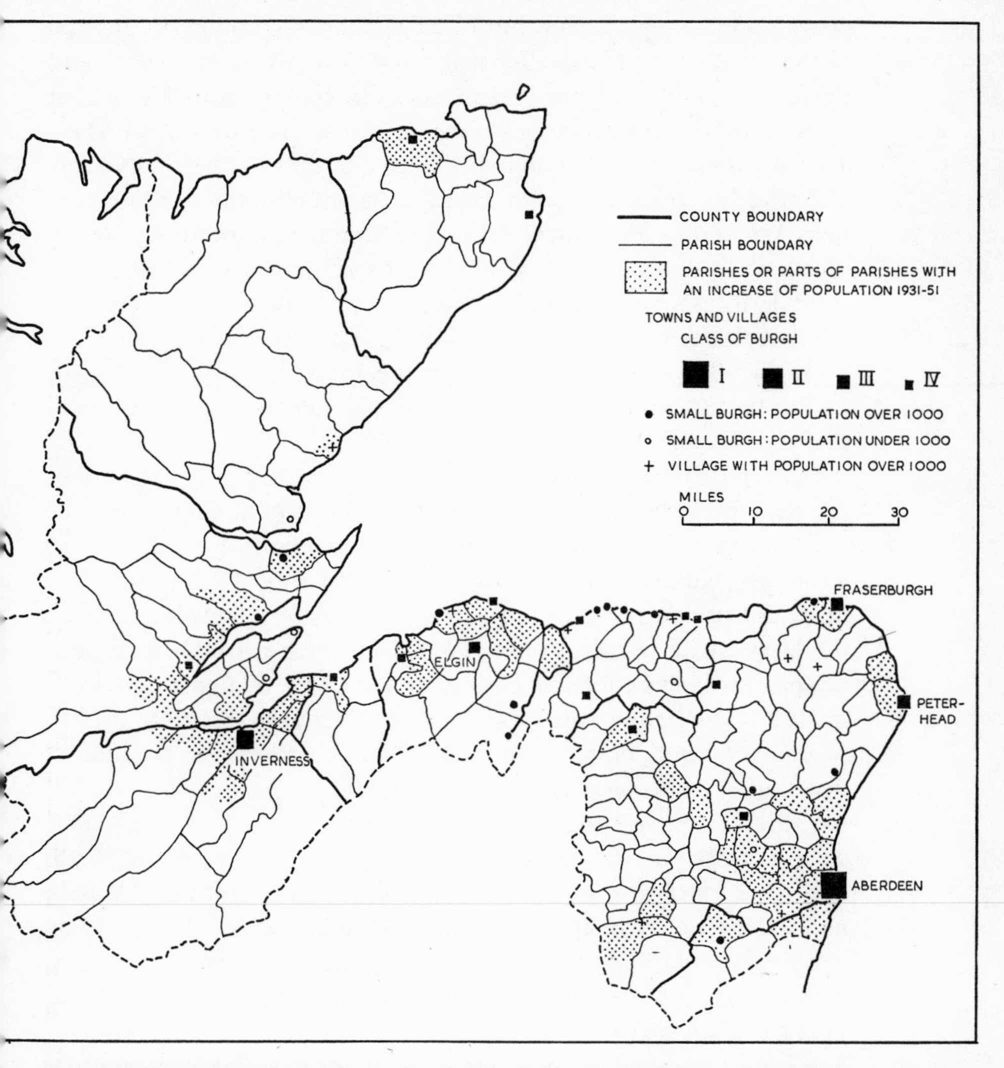

74. The administrative rural and urban pattern of North-East Scotland, indicating the influence of urban es on population trends. Aberdeen is taken as the standard Class I burgh of the area; Inverness, Class II; head, Fraserburgh, and Elgin, Class III. The Class IV burghs are distinguished primarily on size, having lations ranging 3–8,000, as well as usually offering a greater variety of urban facilities than the Small Burghs.

centre of the Moray Firth lowlands from Forres to Beauly, the administrative, educational and legal centre for extensive Highland tracts, and the 'gateway to the Highlands' for thousands of summer tourists. Woollen mills, distilleries, and light-engineering workshops provide some employment, but many of the old consumer industries of the town disappeared after rail links with Aberdeen (1861) and Perth (1863) exposed the area to southern competition. The recent decision to withdraw a large part of its garrison southwards to Aberdeen is yet another expression of the process that has been adversely affecting Inverness and other northern towns over the last half century. While they have attracted people from backward areas, they in turn have lost people and trade to larger southern centres. The movement is by no means simple, however, and must be examined in relation to the economic and social dualism already noted.

Of the 181 parishes considered in this essay, 73 have a seaboard, and at the 1951 census contained some 384,000 people in contrast to the population of 158,000 living in inland parishes. Even if the unique contribution of Aberdeen city (183,000 (1951)) is discounted, the disparity between seaboard and interior populations is evident and is especially marked in the northern counties of Ross and Cromarty, Sutherland and Caithness where only 7,118 people live in inland parishes as opposed to 57,354 in seaboard parishes. Here the large 'strip' parish running back from the coast into the mountainous interior frequently occurs, and may be related to environment and the lasting impress of Norse settlement in this area. Nairn, Moray and Banff, on the other hand, exhibit a more even population distribution, but even in this zone 61,812 people live in seaboard parishes in contrast to 38,720 in inland parishes. Only Inverness, with a small but well-populated 'window' on to the Moray Firth, and Aberdeenshire, with its extensive lowlands of Buchan, Donside and Formartin, have appreciable landward elements. Almost 97,000 people live in the 65 inland parishes of the latter, for example, in comparison to the 43,000 or so who live on the Aberdeenshire seaboard excluding Aberdeen.

Within the seaboard parishes most of the population is concentrated in nucleated coastal settlements of varying size and function. Some 294,000 people live in 26 coastal settlements with populations of over 1,000, ranging from Aberdeen on the one hand to a whole series of small fishing ports with populations between 1,000 and 2,000. Size alone, however, can often be misleading. The royal burgh and cathedral city of Fortrose (902) in the Black Isle, for example, has a dignity and historical stature that quite belies its present size, but can be paralleled many times in a land where burghs were created rather than grew. The ancient burgh of Cromarty (605), whose administrative functions have now largely passed to the more convenient, old Norse, meeting-place of Dingwall, is even smaller, and Dingwall itself has only 3,752 inhabitants despite its extensive county commitments. Similarly Dornoch, the county town of Sutherland, and another cathedral city, has a population of only 933. Indeed some of the old royal burghs and ports of north-east Scotland have disappeared entirely, such as Rattray on the Aberdeenshire coast where only a ruined church and vacant castle motte mark the site of a burgh that once used the Loch of Strathbeg as a harbour.

Such Lilliputian burghs are survivals from the recent past when a close pattern of small port towns graced the entire east coast of Scotland. Protected in their local commercial monopolies by the poor condition of land transport, they were obliged to maintain an all-round, self-sufficient character, with urban institutions, consumer industries for local markets, general trade, some shipbuilding, and usually a little fishing in in-shore waters. Many of their citizens also owned and farmed agricultural land in the vicinity of the burgh. They were, in fact, urbanized communities of peasant seamen, with a degree of specialization not much in advance of adjacent farmer-fisher hamlets or of the crofting townships still to be found on the remote coasts of Caithness and Sutherland. During the last two centuries, however, conditions have changed and the present settlement pattern of the seaboard results from their several reactions to new challenges.

In some cases decline or even extinction can be traced to physical deterioration of sites. The closures of coastal lagoons, such as the Loch of Strathbeg on the sandy coast between Peterhead and Fraserburgh and the Loch of Spynie in Moray, and the entombment of settlements by the Culbin Sands at the mouth of the Findhorn and Forvie Sands at the mouth of the Ythan, are but the more sensational results of shore-line processes on these exposed coasts. Under the influence of northerly tidal streams and wind-waves setting in to the great firth between the two peninsulas of Caithness and Buchan, large quantities of sand and gravel, inherited from glacial times or produced by powerful wave attrition of exposed headlands, are moved along-shore, barring river mouths, shoaling estuaries, and smoothing the irregularities of the post-Glacial coastline. In general river-mouth sites have been avoided and, with one or two exceptions such as Aberdeen, ports so placed have fought a losing battle with natural forces. Newburgh on the Ythan, Banff on the Deveron, Garmouth on the Spey, Lossiemouth, Forres and Nairn have all suffered, and the village of Findhorn has had to be moved on more than one occasion owing to the growth of large shingle spits across the mouth of Findhorn Bay. As the size of vessels grew the physical weaknesses of such sites were increasingly exposed and, unless funds were available to finance expensive dredging programmes and harbour works, the commercial initiative passed to ports better situated. Thus Banff has yielded to the harbour of Macduff, and Lossiemouth to its associated harbour of Branderburgh. Peterhead, its old harbour sheltered by the small, rocky island of Keith Inch, has never had a serious rival at the insignificant mouth of the Ugie struggling out through littoral sand dunes. Fraserburgh harbour, founded by the Fraser family at the end of the sixteenth century in the rocky inlet to the south of Kinnaird Head, has successfully braved the turbulence of its exposed position to avoid the greater menace of shoaling on the coast to its south. The many small inlet ports on the high coast of Banffshire east of Speymouth, the Burghead–Branderburgh group of ports on the Moray coast, Scrabster and the numerous havens of the steep, indented coasts of the far north, are further examples of this quest for shelter without shoaling.

Changes in the demands made on ports by the fishing industry create other problems. During the eighteenth century, as a result of the institution of bounties for home catches and boat-building in 1718, home fleets grew in size and invaded the monopoly hitherto held by Dutch vessels in the east coast fisheries. After the Napoleonic period the industry expanded rapidly and fishing became a specialist profession rather than a subsidiary activity undertaken by a crofter between seed-time and harvest. Access to the fishing grounds made the Moray Firth and Buchan coasts the premier centre of the Scottish fishing industry. Since at first the fishing was carried on from small, open boats, with limited needs, range and capacity, there was little to prevent almost any coastal community that so desired from participating in this new-found source of wealth. Most did so, and during the late nineteenth century experienced a tremendous wave of prosperity. In 1891 Buckie, for example, could boast of 3,522 resident fishermen and some 250 vessels operating from its harbour, as well as a considerable population of curers, coopers and the like. Every creek had its fleet and many small ports, infected by the general economic optimism of the Victorian era, built expensive harbours far beyond their immediate needs. Coastal settlements attracted population from the interior and grew in size according to their capacity, and the vigour with which they hunted herring, white fish and whales of the northern seas.

Unfortunately the general prosperity did not last. The introduction of steam trawlers and drifters made increasing demands on harbours. The fishing industry became highly competitive, and only those ports which could meet the heavy capital demands of both the landward and seaward organization of the industry survived. Increasingly the industry concentrated on a few large ports that were able to provide the necessary facilities for marketing catches and maintaining large specialist fleets of vessels capable of working distant fishing grounds. Even these large ports are now facing grave economic problems arising from the cost of replacing obsolescent fleets, the increasing competition in distant markets, and the fluctuation of catches and market prices, and Peterhead, Fraserburgh and others are attempting to diversify their occupational structures by the introduction of light industries. The smaller, moribund ports have reacted in various ways. Some maintain shell-fishing or coastal salmon nets, and are dormitory settlements for fishermen working from the major ports. Naval or Air Force stations, or establishments such as the atomic plant near Thurso, have bolstered up the economies of others. The majority, however, with their quiet harbours, magnificent cliffs and clean sandy beaches, fisher cottages to let, low rainfall and, at least on the Moray coast, abundant sunshine, have become popular holiday resorts and places of retirement for Scotsmen returning from every quarter of the world. It must not be forgotten that one of the most important exports from this area has been men, and the effect of the return of energetic ex-directors of plantations, railway systems and giant ports to their native and somewhat sleepy heaths is sometimes quite startling.

Despite diversity in size and function the coastal settlements exhibit a number of common features in their morphology. Many, for example, show a duality that arises partly from their social and economic structure and partly from adjustment to site. In the larger and more complex settlements fisher-communities often group together into what are in effect almost towns within towns, such as Broadsea (*Braidsey*) within Fraserburgh, and Branderburgh within Lossiemouth, and the same pattern is seen in many of the smaller settlements, often with an even more marked social cleavage between fisherfolk and the rest of the community. In a great many cases, however, twin settlements have arisen from the juxtaposition of two communities without functional differentiation. These often possess separate social identities and place-names, such as Sandhaven-Pittulie and Inverallochy-Cairnbulg in the Fraserburgh district, even though only a street or strip of grass divide them, and appear to be explicable only in terms of consecutive settlement by distinct groups in the same, favoured locality during some early period of colonization. Recurrent coastal landforms also foster common elements in settlement patterns. Fig. 75 illustrates some of the commonest physical features of the seaboard and a typical coastal settlement. The 'seatown' of fisher cottages, turning their gables to the shore so that boats may be drawn up into the intervening spaces, stands on the lower raised beach close to the sheltered harbour, and the rest of the settlement is on the higher raised beach or cliff-top. The steep brae between divides two social worlds.

Fig. 75 also provides, in a generalized form, a summary of the main physiographical features encountered in a traverse from the seaboard into the interior.

Along some steep sections of the coast the sea is stripping away a mantle of glacial deposits to disinter an ancient, weathered shore produced by a sea-level virtually coincident with the present one. But for the most part the disposition and height of present surface features are the outcome of an interrelated sequence of glacial deposition, and isostatic and eustatic change in Late-Glacial and Post-Glacial times. Over long stretches the present beaches are backed by high, marram-topped dunes and low sandy links resting on the surface of beach deposits laid down during the Early post-Glacial marine transgression. At the inner edge of this zone there usually occurs a well-marked fossil cliff cut by the Early post-Glacial sea into deposits laid down during the earlier submergence of Late-Glacial times, when arctic seas flooded into depressed land margins in the wake of receding glaciers and received the heavy load of rivers swollen with meltwaters and glacial waste. Such deposits range from fine, lami-

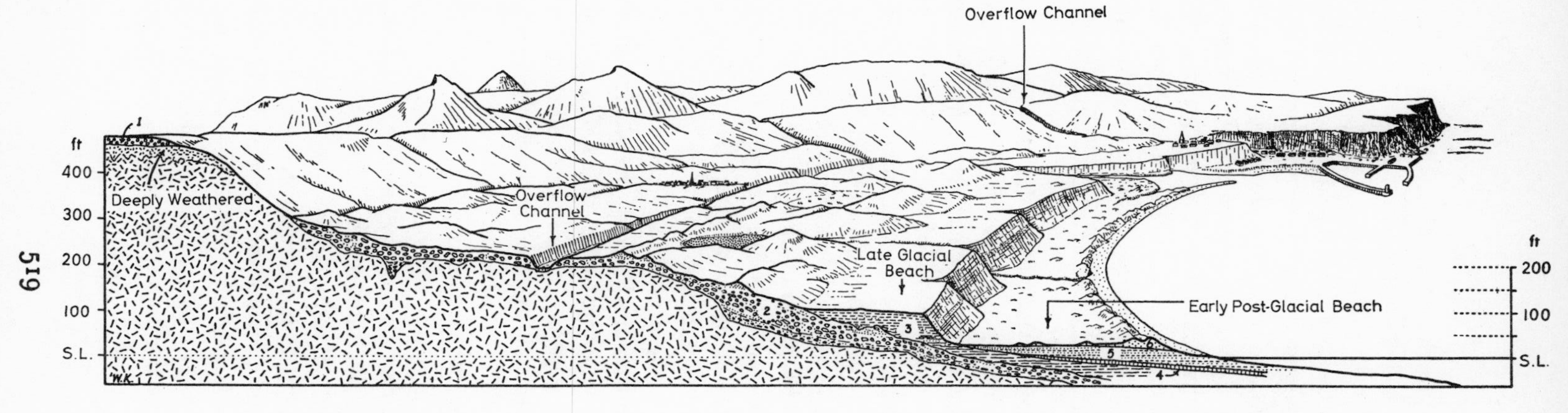

Fig. 75. A composite diagram of landforms and settlements typical of the coastlands of North-East Scotland. The leading section shows (1) Flint Gravels of a former (Pliocene?) high sea level resting on deeply weathered country rock; (2) Glacial (Sands, Gravels, Boulder Clay, aeolian and lacustrine silts); (3) Beach deposits of a marine transgression in Late-Glacial times, subsequently elevated by land recovery; (4) Peat and forest beds of a period of Emergence; (5) Beach deposits of a marine transgression in Early post-Glacial times, subsequently raised by continued but less rapid land recovery; (6) Wind-blown sand from exposed sandy beaches. (Drawn by W. Kirk.)

nated clays and sands, used by several brick- and tile-works on the Aberdeenshire coast, to deltaic gravels in the inner reaches of the Dornoch, Cromarty and Beauly Firths.

Inland of these relatively young raised beaches, which attain a maximum altitude of about 100 ft, there are indications of a much older marine platform at *c.* 200 ft above present sea-level, but for the most part this is buried under considerable thicknesses of glacial material and little is known in detail of its surface morphology. It is highly probable that remains of earlier glacial phases, such as that evidenced by Scandinavian erratics at the base of the cliff section at the Bay of Nigg immediately south of Aberdeen, may be preserved in some of the deeper valleys cut into this platform, but most of its surface deposits appear to date from the last glacial period. As shown in fig. 76, three main ice sheets affected this area during the closing stages of the glacial epoch. The *Moray Firth Ice* was fed from western sources and in its eastward expansion to the North Sea lowlands was powerful enough to block the northward advance of, for example, the Findhorn and Spey glaciers. The *Strathmore Ice*, supplied by snowfields in the South-West Highlands, advanced north-eastwards along the Old Red Sandstone lowlands of Strathmore and shouldered in on to the coastal shelf of Aberdeenshire as far north as Peterhead. The *Dee-Don Ice* was nourished by the Cairngorms and appears to have advanced to the Aberdeen area somewhat later than the main thrust of the Strathmore Ice. During this phase the central area of Buchan remained a cold, periglacial district, characterized by extensive ice-dammed lakes that found successive outlets around the oscillating margins of the Strathmore and Moray ice-sheets. Thus large areas of the coastal lowlands experienced ice pressure from a 'seaward' direction with the result that kames, overflow channels and other linear features are found either parallel or at a slight angle to the present coastline, and many of the lower river valleys show signs of being invaded by lobes of ice pushing *upstream*. Only in Deeside, along the inner reaches of the Moray Firth, and in some of the glens of Sutherland where valley glaciers were powerful and long-lived, are there found large spreads of material directly derived from the ancient metamorphic rocks of the Highlands. Elsewhere much of the lowland glacial drift, for example the red drifts of coastal Moray and Aberdeenshire and the shelly boulder clays of Caithness, has been derived from the Old Red Sandstone and younger rocks of the present coastal zone and from areas now below the sea. Thus the ice left a difficult but diverse environment in this lowland zone, with a drift cover varying greatly in mineral content and degree of weathering, in form and in drainage. Centuries of agricultural pioneering—clearing forests, draining peaty hollows, upgrading pastures, ploughing, manuring and collecting stones from the fields—have been required to turn this glacial wild into some of the finest farmland in Scotland.

Inland again the country steps up towards the fretted edges of the highland massif, where worn and fractured schists, gneisses and granites are often but thinly mantled by glacial debris. Evidence of ancient planation is everywhere apparent, ranging from old marine platforms at *c.* 450 ft and possibly *c.* 650 ft to the higher erosion surfaces of the Grampians and North-West Highlands. Beach gravels of flint and quartzite resting on a deeply weathered and drift-free surface on the Ythan–Ugie watershed in Buchan, have been assigned to a high Pliocene sea-level some 450 ft. above the present one; and most of the members of a British Association committee concluded as long ago as 1893 that the shell beds of Clava in the Nairn valley indicated 'a submergence of the land to an extent of over 500 ft'. The main features of this Highland glacis, indeed, suggest very different environmental conditions and rates of weathering and erosion from those which exist today. The rapid retreat of the highland edge leaving outliers such as Bennachie (1,733 ft), the opening up of the characteristic piedmont basins of Aberdeenshire (fig. 76), the deep weathering of granite masses, and the convex, crag-topped profiles of monadnocks standing above the various erosion-surfaces, appear to have been produced under sub-tropical conditions such as are attested by fossil Tertiary tree pollen elsewhere in Scotland. These are the 'Eastern Ghats' of Scotland, cut through by glaciated glens with stepped profiles and clothed in post-Glacial times by pine, birch and alder forest, and blanket bog.

Throughout the whole of the interior lowlands and the skirt of the Highlands agriculture is the main occupation of some 150,000 people. It is true that in some localities there are alternative means of livelihood; but forestry, work on grouse moors and deer forests, granite quarrying, coal-mining at the diminutive coalfield of Brora, whisky distilling in Speyside and other centres, or building hydro-electric installations in the glens of Sutherland, Ross and Cromarty, and In-

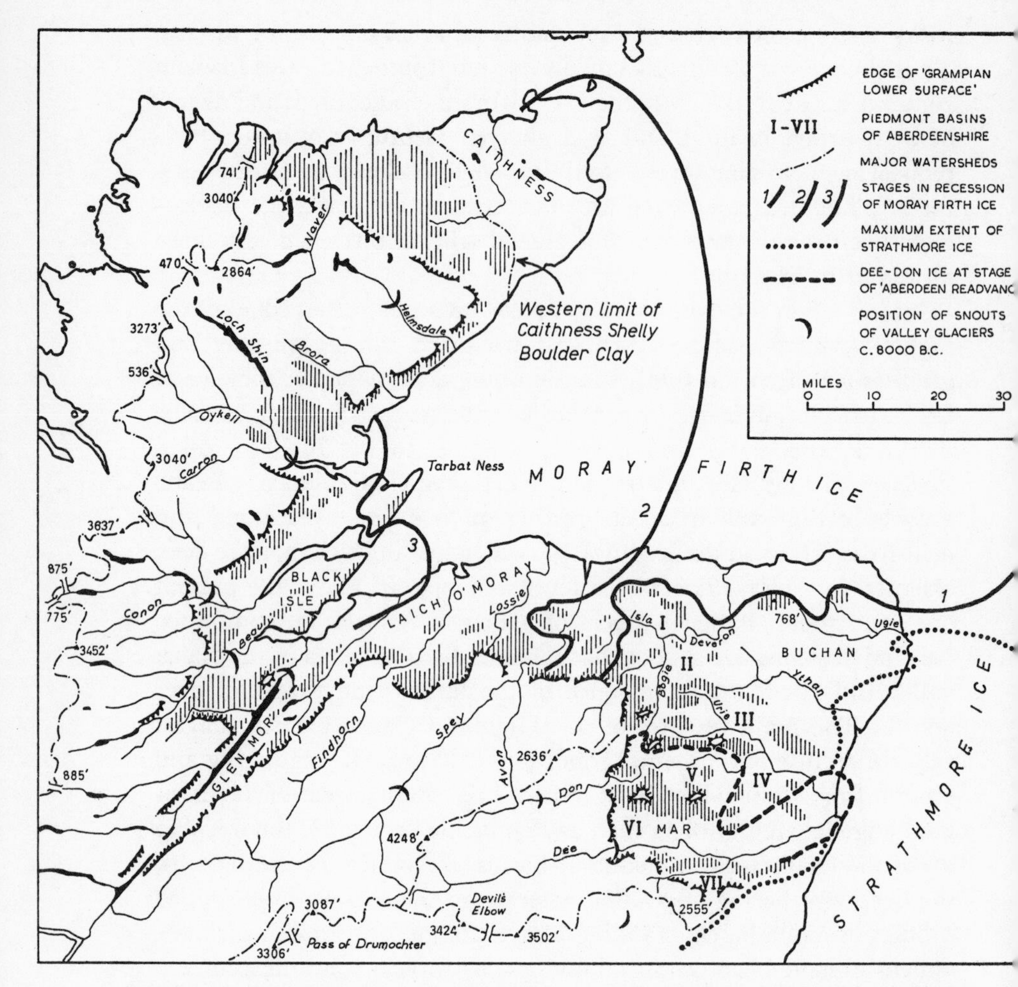

Fig. 76. Closing stages of the Glacial period in North-East Scotland (based on Jamieson, Bremner, Charlesworth, Synge, *et al.*). Hilly areas of various elevations left behind in the retreat of the steep edge of the main highland massif are shaded vertically, and used to delimit the characteristic piedmont basins of Aberdeenshire—viz. I. Keith. II. Huntly. III. The Garioch. IV. Kintore. V. The Howe of Alford. VI. Tarland and the Howe of Mar. VII. The Feugh Basin.

verness are in general subsidiary to pastoral farming. Ever since transhumant Bronze Age communities moved their flocks and herds through the mixed alder woodlands from the dry heathy seaboard to the summer pastures of the upland moors, the rhythm of rural life has been imposed by the needs of livestock, and the distribution of population has been ultimately governed by the animal-bearing capacity of various terrains. In such northern latitudes the plough has always been subservient to livestock. It was so when small, largely self-sufficient, 'fermtouns', with their run-rig infields and rotationally-broken outfields occupied the land. It was so when root crops were added to oats and barley, more efficient ploughs introduced, and joint family 'fermtouns' replaced by individual, enclosed farms and planned villages in the late eighteenth and early nineteenth centuries. It is so today when in addition to the physical limitations of the area the economic influences of urban markets on the seaboard and in southern Britain have strengthened in sympathy with improvements of transport. Animals and animal products are still the main output of the area as they were in the days of the old drove roads—only the patterns of production have changed.

On most farms oats, barley, root crops and ley grass are grown in a six- or seven-year rotation to support livestock. The size of the farm unit, the rotational rhythm, and the character of the livestock vary according to land values, physical controls, and to some extent, distance from market. Thus in the neighbourhood of Aberdeen, Peterhead, Fraserburgh, Inverness, Wick and Thurso, dairying is important with farms about 160 acres in size rented at *c.* 17s. 9d. per acre. Such a farm will have some 140 acres under crops and rotational grass and keep about 44 dairy cows as well as poultry, pigs, a few sheep and one or two beef cattle. Where special markets, such as the large bacon factory at Dyce on the outskirts of Aberdeen, are near by the number of pigs kept increases. Aberdeen also provides a large enough market to encourage market gardening on its periphery. Away from the urban centres, however, physical controls become more important. On the fine, tractable soils of the dry, sunny, coastal lowlands of the inner Moray Firth, farm incomes are derived from the sale of crops as well as fat cattle. Such 'cropping with feeding' farms will usually occupy about 160–170 acres and pay a rent of *c.* 18s. per acre for what is some of the finest arable land in Scotland. Most of their land is ploughed—sometimes with evil consequences in dry seasons when wind erosion of topsoil occurs—and sown with rotation grass and crops, including a larger acreage of barley than usual, potatoes, and some wheat. On average they will carry about 40 steers and heifers, bought on to the farm for summer and winter fattening, as well as about 20 sheep, poultry and 1 or 2 milk cows to supply the farm. Closely resembling this type of farm in cropping programmes, but with increasing emphasis on livestock, are numerous farms dispersed in the drier, good-soil areas of Aberdeenshire, lower Banffshire and Easter Ross. They are usually smaller, *c.* 125 acres, pay about 15s. 6d. per acre rent, and employ less labour than the cropping farms of Moray. Over wide areas of Aberdeenshire, however, the most frequent type of farm combines stock rearing with stock feeding. In Buchan, the Garioch, Huntly and Alford basins, Deeside and other districts where fodder crops do well, but physical conditions are unsuitable for large-scale production of barley, potatoes and wheat, the type farm occupies about 150 acres with 110 acres crops and grass and the remainder rough grazing, pays about 11s. 6d. per acre rent, and carries about 35 beef cattle and 58 sheep. This is the land of the Aberdeen-Angus, and bloodstock from this area finds its way—at a price—to most of the grasslands of the world. Almost half the full-time farms of north-east Scotland are of this type; and of all the farms of this type in Scotland as a whole, two-thirds are in this area.

As the highland margins are approached, higher rainfall and cloudiness, leached soils, and inaccessibility begin to influence farming patterns, and stock-rearing increases in importance. Rents of 6s. per acre are paid for farms averaging about 140 acres, of which 80 acres may be rough grazing and the remainder devoted largely to oats and grass. Pride of place is given to the raising of beef stores and a farm of this type would normally carry about 20 cattle and 55 sheep. Such farms are found in the lower glens and right round the northern seaboard to Tongue, but in remote northern areas gradually give way to crofting communities that are essentially stock-rearing small-holdings. Altitudinally they give way to hill sheep farms paying as little as 6d. per acre for holdings 4,000–5,000 acres in extent, most of which consists of rough grazing and permanent grass. Such a holding, with a little arable land below the 1,000-ft contour, may carry a flock of 1,000 sheep. Farming in this upland, marginal zone is, however, ultimately dependent on the demands of lower farms for store cattle, and for hill ewes for fat-lamb production. Disturbances in the economy of lowland farms have immediate repercussions on the economy of marginal farms, and with other social, demographic and economic factors of more gradual but deeper impact, have drawn men downhill and out of the glens, thus reversing the colonial surge from the Lowlands that took them up centuries ago. Depopulated highland tracts are thus abandoned to the modern hunter, and to transhumant groups of dam-builders and tourists.

There is, however, but little room on lowland farms for highland emigrants. The labour required per 1,000 acres arable is only 37 in dairy-farming areas, 31 in cropping and feeding districts, and 25 in stock-rearing and feeding lands, and much of this demand is met by family labour. Many of the lowland agricultural districts are already 'over-populated' under modern farming conditions, and their folk seek work elsewhere. Some go to inland towns such as Elgin, Inverurie, Keith and Huntly, founded long ago as military and market centres for restricted rural hinterlands, or to small settlements such as New Aberdour, Aberchirder, Rothes and many others, that originated as planned villages in the eighteenth century. Some move to the larger seaboard towns where various types of employment are available. But many go to Inverness and Aberdeen, either as potential citizens or as travellers to the wider realms beyond.

Thus the changing roles and fluctuating fortunes of both highland and seaboard margins contrast with the greater stability of the intermediate agricultural lowlands. At times the mountains have drawn men, at times the sea. Now it is the towns. The greater their size, the greater the range of activities they can offer, the more powerful their attraction. Those areas in which mountains and sea draw closest together have proportionately lost most; the largest towns have proportionately gained most. But even the largest towns of this northern land lose to still larger southern competitors. Under certain conditions in the regional balance of man and land, gateways may become exits.

Selected Bibliography

Peter F. Anson. *Fishing Boats and Fisher Folk on the East Coast of Scotland.* London, 1930.

A. Bremner. 'The Glaciation of Moray and Ice Movements in the North of Scotland.' *Trans. Edin. Geol. Soc.*, **13**, 1934, pp. 47–56.

W. D. Chapman and C. F. Riley. *The Granite City.* Aberdeen, 1952.

J. K. Charlesworth. 'The Late Glacial History of the Highlands and Islands of Scotland.' *Trans. Roy. Soc. Edin.*, **62**, 1955, pp. 769–928.

R. E. Craig. 'Hydrography of Scottish Coastal Waters.' *Marine Research*, 1958; No. 2, H.M.S.O., 1959.

John Cranna. *Fraserburgh: Past and Present.* Aberdeen, 1914.

J. G. Cruickshank. 'The Black Isle, Ross-shire. A Land Use Study'. *Scottish Geog. Mag.*, 77, 1961, pp. 3–14.

Department of Agriculture for Scotland. *Types of Farming in Scotland.* H.M.S.O., 1952.

J. B. Fleming and F. H. W. Green. 'Some Relations between Country and Town in Scotland.' *Scottish Geog. Mag.*, **68**, 1952, pp. 2–12.

F. H. W. Green. 'Rural and Coastal Settlement in the Moray Firth Lowlands' and 'Urban Centres of the Moray Firth Lowlands.' *Scottish Geog. Mag.*, **52**, 1936, pp. 97–117 and 157–81.

T. F. Jamieson. Various papers on glaciation in *Quart. Journ. Geol. Soc.*, 1860–1906.

D. L. Linton. 'Problems of Scottish Scenery.' *Scottish Geog. Mag.*, **67**, 1951, pp. 65–84.

A. G. Ogilvie. 'The Physiography of the Moray Firth Coast.' *Trans. Roy. Soc. Edin.*, **53**, 1923, pp. 373–404.

Scottish Home Dept. Annual Reports on the *Fisheries of Scotland.* H.M.S.O.

F. M. Synge. 'The Glaciation of North-East Scotland.' *Scottish Geog. Mag.*, **72**, 1956, pp. 129–43.

CHAPTER 28

THE GRAMPIANS

H. G. STEERS

No one has been able to locate exactly the site of Mons Grampius in ancient history, the battlefield from which comes the geographical name, Grampians. No one could say precisely either, what part of Highland Scotland, south-east of the Great Glen, should be termed Grampian. The present chapter is a description of the high country between the Great Glen and what is known as the 'Highland Line', the continuous, almost spectacular scarp which runs from the site of Helensburgh on the west coast to that of Stonehaven on the east. The trough of the Great Glen continues south-west under the waters of the Firth of Lorne, and there are structural features of the islands of Colonsay, Jura and Islay, as well as of the great peninsula of Kintyre which link them with the mainland Highlands. Remoteness and the deep penetration of the sea lochs are characteristic of the whole south-west both mainland and island although in varying degree. In structure also, the Grampian characteristics reach to the north-east coast, although its straight, unindented line from Burghhead to Fraserburgh contrasts strongly with the peninsulas and inlets of Argyll. But roughly east of the roads connecting Ballater, Tomintoul, Grantown and Inverness, there is little to justify the term Highland at all. Easier gradients and accessibility, a climate kinder to farming and thus favouring thicker settlement, an ampler urban tradition and better trading rewards throughout the centuries, mark off the north-east as distinct from the Grampians and outside the framework of this chapter.

The bounding-lines to north-west and south-east are impressive. They are structural and not of rock type since the Moine series of crystalline rocks, which are characteristic of the Northern Highlands, stretch well south-east of the Great Glen into southern Inverness and Perthshire. The narrow, flat-bottomed gash of the Great Glen with its string of lochs runs from coast to coast and this fault-line has been emphasized by man in the dotting of eighteenth-century forts and in the cutting of its almost useless canal. At the other extremity of the Grampian region, the views from the roads north from Perth to Dunkeld or from Stirling to Callender or Crieff, give even the least informed traveller the impression of approaching a wall of rugged country forming a sharp limit to the Central Lowlands.

Within this part of Scotland, there are many regional names of long standing and interest especially in the west (fig. 77). Cowall and Lorne, Benderloch and Appin, Lochaber and Badenoch, Rannoch and Breadalbane, Atholl and Mar are small regions with their own individuality over the centuries. Some of them coincide with clan lands; Appin with the Stewarts, Lorne with the Campbells, Rannoch with the Menzies, although it is rash to reckon these ever-changing and fragmented possessions as coherent territories. The names belong, as Stevens remarks, to the centuries when 'the natural regions are mountain fortresses with their glens, and the low stretches, too narrow to be concourses, were . . . no-man's land'. The map shows, too, the frequent use of the term 'forest' with regional significance—Forest of Atholl, Forest of Gaick, Mamore Forest, Forest of Glenartney. Forest in the Grampians serves to describe great tracts of rough country rather than former woodland, although now and then the work of the Forestry Commission is bringing a present and precise forest character to this kind of highland.

Certain physiographical features in the Grampians belong to all Highland Scotland and some others especially to this region. Both kinds were carefully studied and written up by the geologists, Peach and Horne, at the beginning of the century and although much work has been done on the Highlands since, these earlier writings remain valuable. The Grampians form part of a greater whole, the south-eastern part of the great dissected tableland of the Highlands, made up mainly of metamorphic rocks of Archaean age, although the Paleozoic Old Red Sandstones appear on the Highland boundary fault. Various planes of denudation are recognizable throughout the tableland, the most clearly marked being the High Plateau at about 2,000 ft and an Intermediate Plateau at about 1,000 ft. Thus in many places a remarkable evenness of skyline is noticeable, for example, in the hill landscapes on either side of Drumochter Pass on the Perthshire–Inverness-shire border, and again in the view north on the descent from Amulree to Aberfeldy in Strath Tay. In the west of the Grampians, the view from Ben Nevis, southwards over

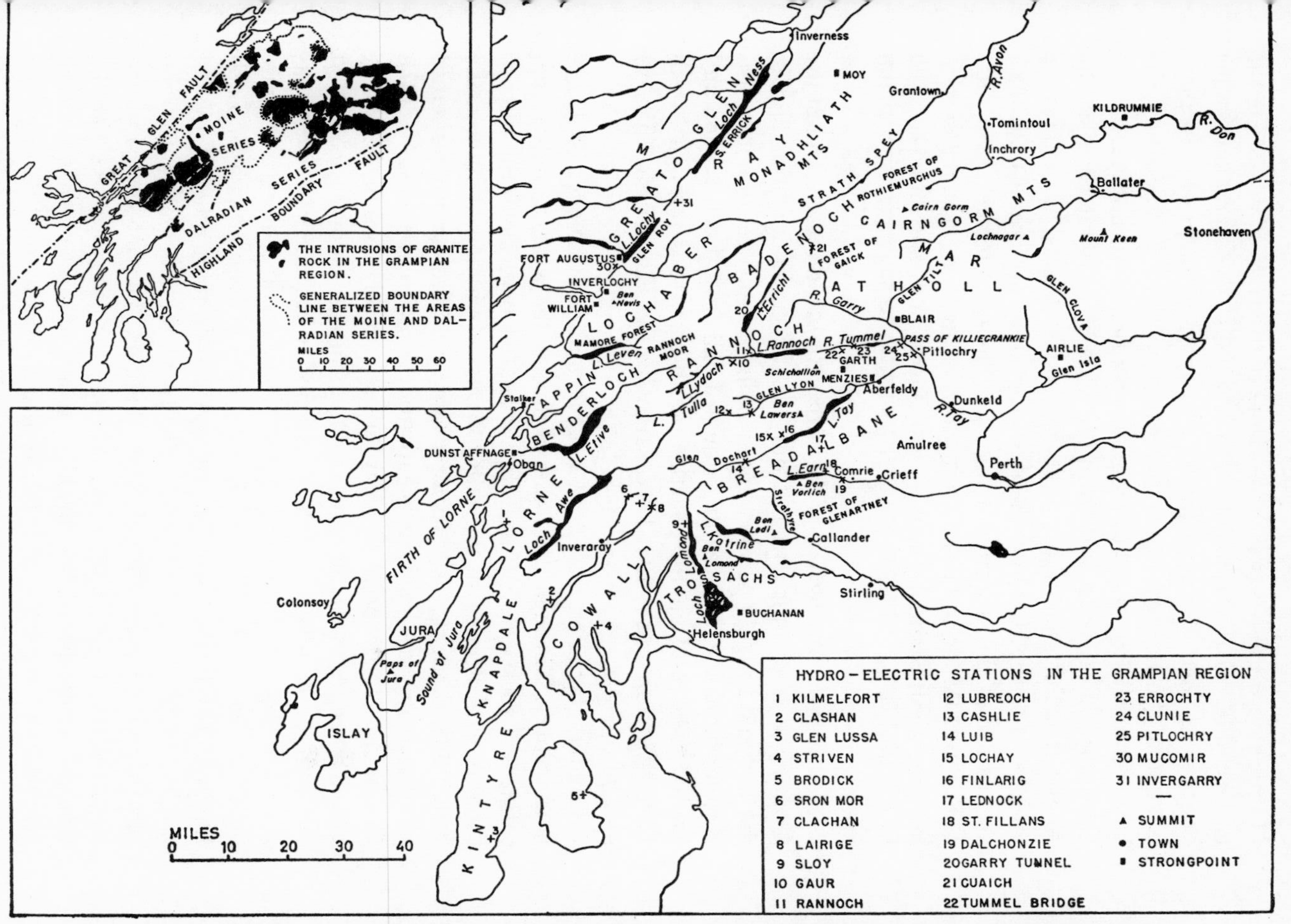

Fig. 77. Regions, settlements and hydro-electric power stations of the Grampians. (Inset ;Granite intrusions.)

Mamore Forest shows well the horizons of the High Plateau. On the eastern side, what Fleet in later works calls the Grampian Valley Benches in Glen Clova, might be correlated with the Intermediate Plateau of the older geological descriptions. From these surfaces, the larger summits stand out as monadnocks: Ben Nevis in the west, the Cairngorms in the east and abrupt heights like Mount Keen on the Angus–Aberdeen boundary. Big intermont basins lie within the tablelands of the Highlands, and are especially noticeable in the east and centre of the Grampians (fig. 78). The most remarkable is that of Rannoch Moor and its adjoining loch basin to the east. Close to it lies the Atholl Basin, drained by the Garry, and its tributaries, and to the north-east, the series of smaller basins which make up Strath Spey.

Faulting is characteristic of the entire Highland region, and in the Grampians is conspicuous not only at the extremities but also in marked and roughly parallel lines running south-west to north-east. A great fault line is traceable through Loch Awe, Loch Laidon and Loch Ericht, and another, farther south-east, runs through Strath Earn, Tayside and Atholl Forest right to the Cairngorms (figs. 77 and 78).

Finally, glaciation has had a most formidable influence on Highland scenery. Through the action of the ice, the weaker, more rotten surfaces of rock were scraped off, the shatter belts along the lines of fracture were deepened, the valley floors were hollowed out and the valley sides made steeper. Its force also tested the varying resistance of the rock surfaces and largely explains the differential erosion of the Grampian landscapes. Certain areas, notably Rannoch Moor, served as ice reservoirs or cauldrons and from the moor the ice flowed out both south and east. Its landscape today has a memorable bleakness and austerity with the great stretches of crumbled granite boulder and the boggy flats broken by the long trough of Loch Laidon. Its desolation is heightened rather than relieved by the single-track railway, built with difficulty at the end of the last century to link Crianlarich with Fort William, Farther west, in Glen Roy, the effect of glaciation is again spectacular: the three parallel strand lines on the hillside on the west side of the Roy valley, which are those of former glacial lakes, appear so symmetrical that for long they were considered artificial and the work of early inhabitants of the glen.

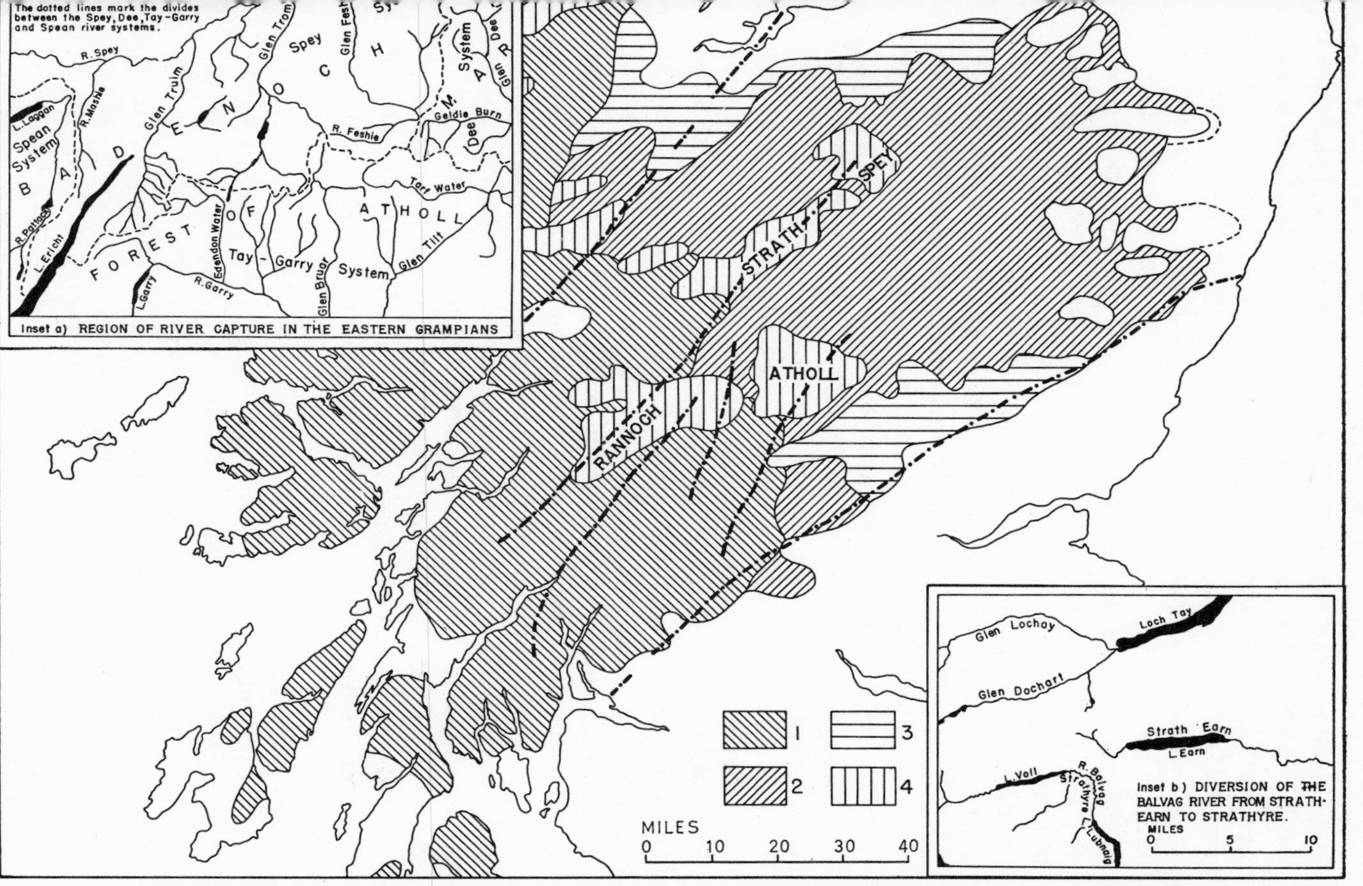

Fig. 78. The structural features of the Grampians.

Denudation continued and widespread, fracturing, and glaciation explain much of the transformation of the former plateau over the ages. It is now a rugged, sculptured block, from which the softer rock surfaces have been stripped right away and which has an alinement of relief and coastal indentation markedly from north-east to south-west. Originally the great streams of the old plateau drained to the North Sea and three of the present main river systems, the Dee, the Tay and the Earn have an east coast exit. Small sections of existing river valleys bear witness to the earlier drainage pattern, for example the stretch of the Garry south of Dalwhinnie, and the short valleys on either side of Loch Linnhe (Glen Scaddle, Glen Gour and Glen Tarbert in the west, Glen Nevis, Loch Leven and Glen Coe in the east). But the present river system is superimposed and the processes of its development are complicated and controversial. In particular the long-accepted explanations by Peach and Horne of the stages of river capture are now qualified by later studies. Traces of capture are frequent in the Grampians and much field work has been expended on determining the relationships of the Feshie with the Geldie, of the Avon with the Upper Don, of the Tarf with the Dee (fig. 78: inset a). It has led on the whole to doubts about the older theory 'of the piecemeal encroachment of one river over the territory of another'. Linton writes of the need to recognize 'the sudden and catastrophic breaching of a former watershed by ice which began by overriding a divide and ended by creating a breach in it, so large and deep that after the Ice Age, the waters could no longer return to their pre-glacial course'. Other examples of river diversion farther west and south are that of the Pattock, once part of the Spey system and now flowing into Loch Laggan, and that of the Balvag with its course altered from Strath Earn to Strathyre (fig. 78: inset b).

The intrusion of great masses of granite of Devonian age is conspicuous in much of the Grampians (fig. 77: inset). In the north-east, granite rocks distinguish the impressive high country of the Cairngorms and Lochnagar; in the south-west they form the Ben Nevis group and the mountains on either side of Loch Etive. Otherwise, division of the Grampian region might derive from the two main rock series within it. The traditional region of Moray lies on either side of the Great Glen, and the south-eastern part of it, on the near side of the fault line, together with the northern tips of Badenoch and Atholl forms the end of the country of the Moine schists which have their main importance in the Northern Highlands. The 'Moine' country (fig. 77: inset) is perhaps drearier in its landscapes than that to the south and west. The Monadhliaths can be called in translation the 'Grey Mountains' (in contrast with the Monaruadh or Red Mountains, the alternative name for the Cairngorms); and there is some justification in so describing the bleak, level-topped hill mass which lies between Strath Errick and Strath Spey. There is something forbidding also about the flat-topped, steep-sided hills of Gaick and northern Atholl. It is not difficult to appreciate here the destruction by avalanche of Gaick Lodge and its inmates in January 1800, with its ruins so close to the precipitous side of Sgór Bothain. Another and more attractive feature of the Moine landscapes is that of the flaggy rocks exposed in the river valleys. These are conspicuous, for example, in the bed of the Garry at Struan, in that of the Lyon above Meggernie Castle and in the upper valley of the Feshie.

The greater part of the Grampian region, however, belongs to the Dalradian series and has more varied landscapes than the Moine country. The Dalradian rocks are almost all metamorphosed sedimentaries varied by the huge granite intrusions mentioned earlier. Moreover, in the south-west, Ben Nevis and neighbouring Glen Coe are striking on account of Lower Old Red Sandstone volcanic formations. Ben Nevis has a great cap of volcanic material and Glen Coe lies in a volcanic cauldron. Occasionally, quartzite forms spectacular peaks, such as the fine cone of Schiehallion, and in the extreme south-west, the Paps of Jura. Close to the Highland border the Trossach summits, Ben Ledi, Ben Vorlich and Ben Lomond, stand out partly on account of their position, so near the southern limit of the high country, and partly because of the hard grits which form them. Finally, the bands of limestone vary the scenery and are sometimes emphasized by the greener and more varied appearance of the vegetation cover. Thus limestone with its associated plants determines the scenery of the upper Avon valley at Inchrory, and, in the islands, of parts of Islay and Colonsay.

Stevens has rightly pointed out, however, that the varied landscapes of the Highlands do not depend in the first place on different rock types. Any traveller crossing the Grampian region between Deeside in the east and the islands in the west would notice the contrast between one extremity and the other, but it would have a physio-

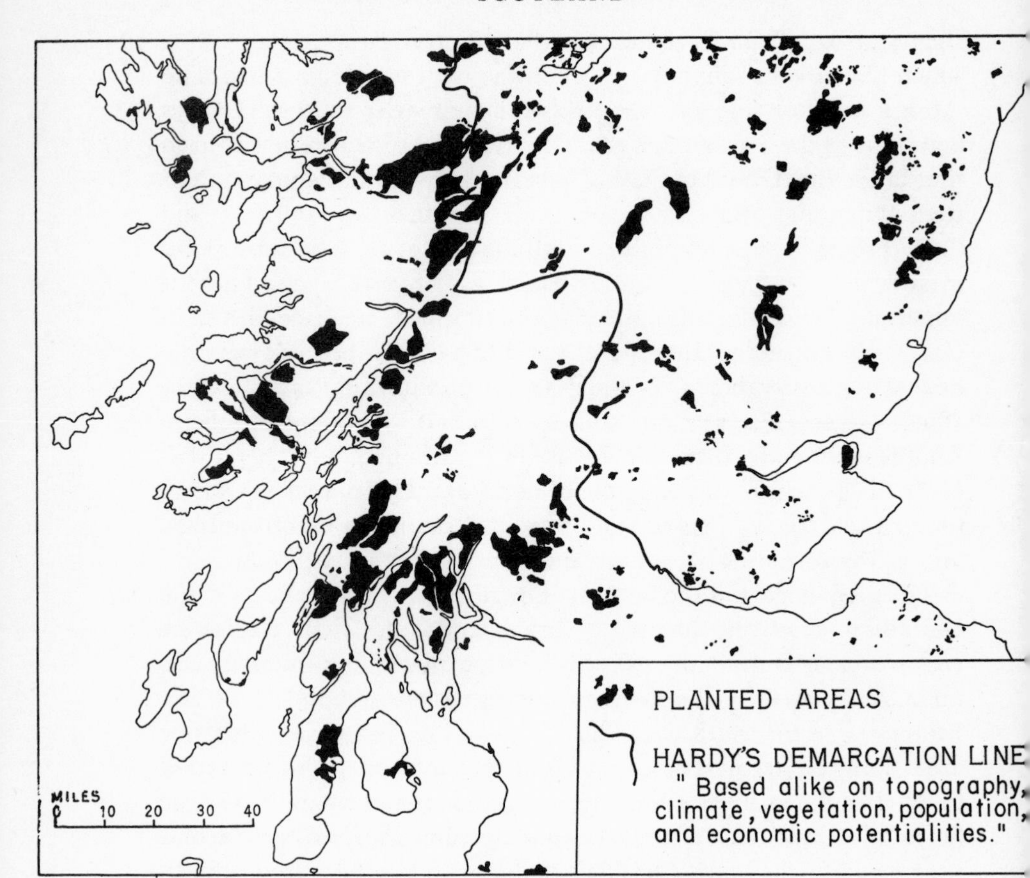

Fig. 79. Afforestation in the Grampians.

graphical rather than a geological explanation. The vegetation map accompanying Hardy's *Botanical Survey of Scotland* is interesting in this respect, and suggests something like a vegetation frontier from the north-east to the Highland border (fig. 79). This is sinuous throughout, and in the Great Glen runs as far west as the valley of the Lochy. 'Shifting eastward, it leaves on its right the Roy and Spean Hills as well as the Ben Alder Massif, turns south again and skirts the western extremities of Lochs Rannoch, Tay and Earn, round Ben Vorlich and the Loch Lomond Hills.' Hardy summarized a few of the main botanical contrasts between one landscape and the other, between the heather and dry grass moors of the east with the alpine deserts on the high summits and the Scots pine and larch woods at lower levels, and the waterlogged moors of Lorne and Rannoch in the west with their much smaller heather covering and the oakwoods of the Trossachs. Anyone who has walked from Bridge o' Gaur in Rannoch up to the railway and across to Kingshouse and Loch Leven, or who has driven up Strath Tay into Glen Dochart and Strathfillan, will realize the point of his study.

The vegetation regions derive indeed from other contrasts in geographical feature which distinguish east from west, although the demarcation lines are not always coincident. Perhaps the most obvious to suggest is that of rainfall distribution. The monthly rainfall averages mapped in Bartholomew's *Atlas of Scotland* show the consistently high figures for the western part of the Grampian mainland (an annual average of 80–120 in.). This contrasts on the one hand with a much smaller figure in the Cairngorms, for example (30–40 in.), and also with drier conditions in the western islands and the Kintyre peninsula. The rainfall figures in part explain the conspicuous dissection of the Grampian plateau, south of Strath Spey and approximately west of the Perth–Inverness road (fig. 77). The great hill masses of the east have their magnificence in the corrie cliffs and deep glens eroded in the granite: but the extreme of dissection is in the west, with its more frequent exposure to wind, its steeper, shorter river courses, and at one period a much more emphatic concentration of ice erosion on slopes and valleys, and on the coast, now so heavily indented. Hence in the west the much greater number of lochs and their impressive depth (Loch Ness, Loch Ericht, Loch Laidon, Loch Rannoch), and the great troughs which underlie the sea inlets (the Firth of Lorne and the Sound of Jura).

The contrasts between east and west have been emphasized recently in two attempts to discover and use the resources of a rugged countryside. It is the west with its ample lake system and high rainfall which has offered the greatest possibilities of hydro-electrical development whether for transmission to industrial Scotland or for local use. Figure 77 shows the main distribution: the Loch Sloy, Tummel–Garry and Glen Shirra (Inveraray) enterprises for long-distance transmission, and the Cowall and Glen Lussa schemes for local amenities.

Again Nairn, writing at the end of the nineteenth century comments on the Great Caledonian Forest, which, a thousand years ago, must have been an outstanding feature of the Grampians. He presumes its extent in historic time from Glen Lyon to Strath Spey and from Glen Coe to Deeside: oak at the lower levels, pine and birch above it, and alders along the river and burn courses. Fraser Darling describes the many destroyers of this magnificent natural forest from the Viking raiders in the ninth century to the commercial sheep-farmers in the late eighteenth century. There survives from their ravaging and from the final orgy of cutting during the First World War, a little of the original forest, notably the beautiful pinewoods of Rothiemurchus in Strath Spey, the Black Wood on the south side of Loch Rannoch and the pinewoods at Inveroran on the south side of Loch Tulla on Rannoch Moor. Some oakwoods also remain in Argyll and in southern Inverness. But the length and scale of destruction were formidable and were not compensated effectively by the activities of the 'planting dukes' and other landowners in the eighteenth century. The countrysides to gain most from this aristocratic energy were parts of Atholl and Laggan, and the Mackintosh estates around Moy. Hardy's *Botanical Survey* contains many references to the destruction of woodland and to the problem of finding suitable species for replacement, although his final mapping of vegetation regions is not primarily concerned with the forest cover that by his day had so largely disappeared.

The work of the Forestry Commission in the second quarter of the twentieth century has marked the landscape in the Grampians as well as that farther north and continues to change its character (fig. 79). The main plantings south of the Great Glen have been in the south-west: Kintyre, Cowall, Lorne and Benderloch have all now considerable stands of coniferous forest, and in highland Stirling, many of the Trossach hillsides are now thickly wooded. There are other big plantings in the north-east and south-east, especially at Inshriach in Strath Spey and in Angus on the slopes of Glen Isla and Glen Doll. The great enterprise in planting, however, has come in the west. Planned modern development of Highland power resources and forest economies once again emphasize the distinction between west and east in the Grampians.

The whole region has been one of emigrant activity since the troubled days of the Jacobite risings. Indeed the decline in the clan system and the movement of population arising from the northward penetration of the Border sheep-farmers were apparent before 1745. This long-term emigrant trend was complicated in its pattern, varying from one locality to another, and with many different explanations of the timing and intensity of movement. It should not, of course, be studied in isolation, for it is a fraction of a much greater whole, the downward and outward movement of mountain peoples throughout much of Europe in modern history. In Scotland, it is as noticeable in the Southern Uplands as in the Highlands. In the latter, the coincidence in parts of the countryside between pressure of population and disturbing political event, gave some of the early and painful stages of emigration both dramatic and romantic qualities which, particularly in the Grampians, have disguised some of the realities of human geography. Something of the traditional distribution of farming remains to emphasize the natural limitations to settlement and prosperity in areas with no local source of effective mineral wealth to supplement farming. Seventeenth- and eighteenth-century maps, whether Scottish or English, like those of Gordon of Straloch and William Roy, show the long tradition of farming settlement in the rather more ample valley lands (Strath Tay, Strath Spey, the valleys of Angus and parts of Glengarry), and the peripheral nature of Scottish rural settlement in the Highlands for centuries past. Cultivation prospered mainly in the alluvial haugh lands of the valleys and on some of the raised-beach terraces on the coast. A pleasant example of haugh-land farming in an open, easy section of the Garry valley is that of the Atholl Basin traversed by the A9 road between Struan Station and Killiecrankie Pass in Perthshire. Another is in highland Angus in the valley of the South Esk. These easier lowlands, and the more accessible country of Strath Spey have always carried a limited but persistent amount of settled farming and could produce the hardier grains, and, as time went on, potato crops. Occasionally, the soil itself provides the clue to traditions of effective farming: for example the fertile land from the schists lying horizontally on the north side of Loch Tay, and the good, well-drained beach-terrace soils edging inner Loch Linnhe.

The pastoral farming, dependent on transhumance, which up to the eighteenth century enabled a much denser highland settlement than that of today, is partly traceable by the head-dykes which are common in the Grampian valleys. These are rough stone bounding-walls, marking the uttermost limit in altitude of land which could contribute anything to farming economies. Very often, as **Robertson** points out, they coincide with natural limits to tended land, such as a change in gradient or a patch of scree. The head-dyke belongs to the farming landscapes of all highland Britain, but some of the Grampian counties have their own versions of it.

In the parallel valleys of Strath Earn, Strath Tay and Rannoch, the head-dykes are noticeably and logically higher on the north side than on the south side of the loch basins. In the mainland farming regions, the level of the head-dykes varies between 800 ft in Strath Spey and as much as 1,200 ft farther west, in sheltered tributary valleys. It drops to 500 ft and lower in the Great Glen with a shorter farming-season. In the long peninsula of Kintyre, the average level of the head-dyke on the indented western Atlantic side, although discontinuous, is higher than on the eastern side, because of the better soils and gentler gradients of the Atlantic littoral.

In these high fields and pastures, before the troubles of extreme population pressure and political disturbance, the wealth of the chiefs was in cattle, not in sheep. Grant describes the trading of the more important clan chieftains in livestock and hides, fish and timber. Loch Etive and Glen Garry (in Atholl) were noted producing areas for fish and timber, and Inverness was a known cattle market.

Aerial photography today shows well the line of the head-dykes and the way in which farming has steadily shrunk away from them in the Highlands with the outflow of population. Most of the Highland counties now farm only 50% of the maximum area, although in many Lowland counties the percentage is as much as 85%. In consequence the rough pasture percentage has increased in proportion to the whole (for example 18% in Lochaber, 15% in Lorne).

This shrinkage from the maximum use of the land, which belonged to the stage when the pressure of population was most severe, was inevitable. It derived largely from the possibilities of movement. The Grampians are the region of the military roads and especially of Wade's roads (fig. 80). This skeleton road pattern, originally built and maintained for military purposes by English army surveyors, engineers and labourers, is still conspicuous. Those who can remember carriage transport, know how recent is the change on long sections of A9 from Wade's road to a modern motor-road. In general direction, Wade's building is the forbear of the present road through the Sma' Glen to Aberfeldy, and on to Tummel Bridge and Dalnacardoch.

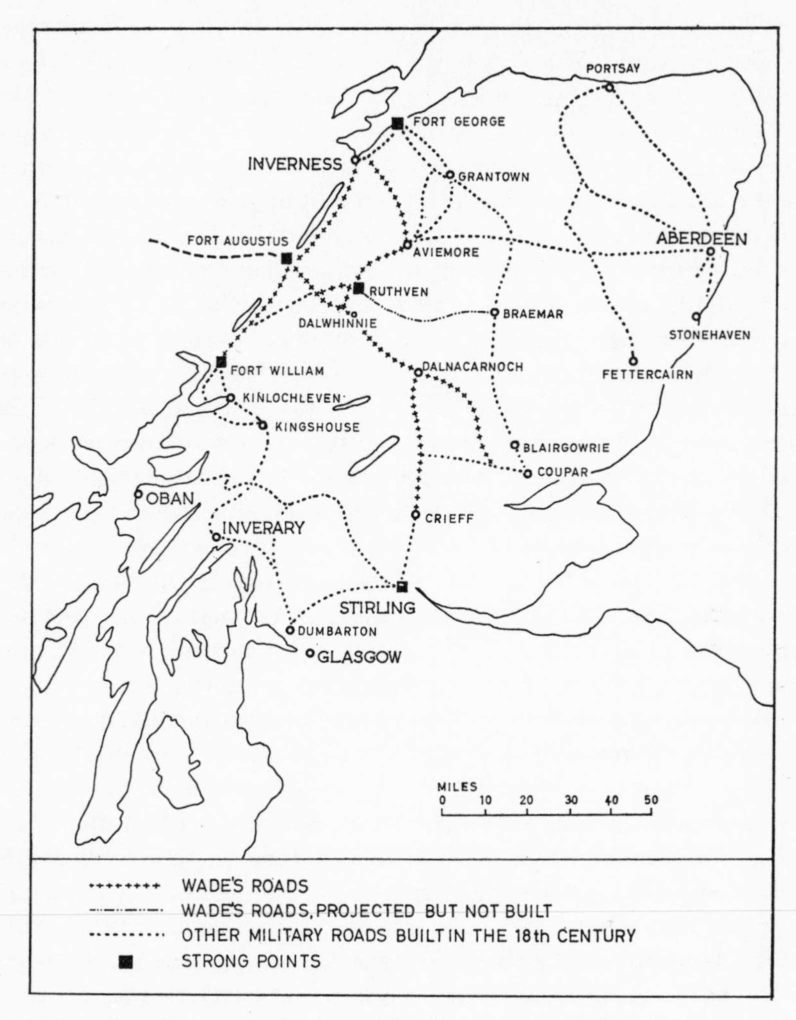

Fig. 80. The military roads of the Grampians. The map is taken from J. B. Salmond's *Wade in Scotland*, p. 295. The Moray Press, 1934.

Others, more remote and precipitous, survived as drove roads, like the grass-grown, hairpin bends which mark the military route from Upper Strath Spey to the Great Glen over the Corrieyairack Pass.

What let the soldier in in the first place, admitted after him, the various and sometimes ephemeral succession of strangers. The sheep farmer and drover, the renter of the deer-forest and grouse-moor, and the great army of tourists followed each other north. Also, what let the Lowland Scot and English newcomers in, inevitably brought the Highlander out. Much emigration was long distance, as we know from the growth of settlement in the Commonwealth, especially in Canada, and from the history of the armed forces and the merchant service. But the Grampians were affected above all by juxtaposition to the mining and manufacturing belt and by the possibilities that it offered to immigrant labour. Argyll stands distinct in the Grampians as the 'crofting' county, and in many of its characteristics of farming and population movement is often associated with the North-West Highlands rather than with the Grampians. Certainly the loss of population, absolutely as well as relatively, has been more noticeable from it than from the counties to the east and south: it has declined steadily in numbers from 1851 onwards. Inverness-shire and Perthshire both lost noticeably in the last quarter of the nineteenth and the first of the twentieth centuries, although the check on outflow became apparent in the 1951 census, owing to changed conditions in Lowland Scotland. In Perthshire and Argyll, perhaps the strongest reason for depopulation was the nearness of the over-full Grampian Highland to the developing and industrial central Lowland up to the time of the great depression of the 1930's.

Neither of the two fresh uses of resource described in this essay would reverse the trend, although they might weaken it. The intake of a big labour force for hydro-electrical development means the recruiting of long-distance immigrants (notably Southern Irish), rather than long-term local employment. The final staffing of a power-station is relatively small and the purpose of power production is for long-distance transmission and local amenities rather than for the growth of local industry, the possibilities of which remain patchy. Forestry provides more and steadier local work, but as the Germans, the most expert foresters in Europe, realized half a century ago, this is a rural economy in which the personnel is always remarkably small in proportion to the area planted and the timber handled.

The reports of the Scottish Tourist Board in recent years show the growth of enterprise in the Grampian region to enlarge the hotel industry. The efforts to do this derive mainly from realizing the chances that come from holidays with pay, from overseas travellers, and from growing transport. Some hoteliers understand what could be handed on in this kind of enterprise to local populations, especially to farmers, builders and retail traders. The Grampians have two traditional regions of tourism, the Trossachs within easy reach of Glasgow, and the neighbourhood of Oban, the port of coastal steamer traffic. Recently Strath Spey has also attracted visitors, largely because of the lively co-operation of the local population and the resulting profitable lengthening of the tourist season from three or four, to nine or ten months out of the twelve. In Atholl, especially around Pitlochry, the same kinds of venture have been justified, and these, if maintained and increased, suggest much greater hopes of steady employment without migration. The hard facts of distance and expense make this kind of rehabilitation easier in the Grampians than north of the Great Glen.

No enterprise seems likely to alter markedly the pattern of town growth. (fig. 77). The small towns are necessarily peripheral: Campbeltown, Inverary, Oban and Fort William the ports, Inverness both port and market, the little settlements on the eastern side between Highland and Lowland, Grantown-on-Spey, Tomintoul and Ballater, the rather more ample ones on the Highland fault line, Blairgowrie, Crieff, Comrie and Callender. These little towns edge the Grampians, and within the region Pitlochry alone has a population of more than 2,000. All have a long history of small-scale trading, the outcome of the necessary if limited transactions between Highland and Lowland. The ring of castles which are interspersed with the small market centres follow much the same pattern in their placing. Inverlochy, at the south-western end of the Great Glen, Dunstaffnage in Appin, and Castle Sween in Knapdale are old coastal strong points. Moy Castle and Kildrummie in the north-east, Airlie and Buchanan on the southern border, stood between Highland and Lowland, between one way of life and another. A few great fortresses controlled critical points on routes within the Grampians, Blair Castle in Glen Garry, Menzies in Strath Tay and the much older, savage, strong point of Garth above lower Glen Lyon. All recall the long centuries of the part-raiding, part-farming economy of the Grampians, with the emphasis on the struggle to maintain Highland settlement and armed strength in the face of the slow but inexorable growth of pressure on resource.

The region therefore has its character. It is the greatest mass of Highland in Scotland, and contains in its high country more emphatically than the north, the physiographic contrasts between east and west. It is the region of the Great Caledonian Forest of the past, and of one of the most substantial efforts to create a forest cover to-day. Its human geography has been influenced by the nearness of a belt of mining and manufacture and by the possibility of easy, short-distance emigration. For the same reason there is now a noticeable if patchy return of interest in its possibilities as a tourist region, attractive both to the town dwellers of Lowland Scotland and to those who live much farther afield.

Selected Bibliography

F. Fraser Darling. *The Natural History of the Highlands and Islands.* Collins, London, 1947.

H. Fleet. 'Erosion Surfaces in the Grampian Highlands of Scotland.' *Rapport de la Commission pour la Cartographie des Surfaces d'Appl. Tertiares.* Union Géographique Internationale, Paris, 1938.

I. F. Grant. *The Economic History of Scotland.* London, 1934.

M. Hardy. 'Botanical Survey of Scotland.' *The Scottish Geog. Mag*, **22**, 1906, pp. 229–41.

D. L. Linton: 'Problems of Scottish Scenery.' *The Scottish Geog. Mag.*, **67**, 1951, pp. 65–84; 'Some Scottish River Captures Re-examined.' *Scottish. Geog. Mag.*, **65**, 1949, pp. 121–31; **67**, 1951, pp. 31–44; and **70**, 1954, pp. 64–78.

D. Nairn. 'Notes on Highland Woods, Ancient and Modern.' *Trans. of the Gaelic Soc. of Inverness*, **17**, 1891, pp. 229–41.

R. H. Osborne. 'The Movements of People in Scotland, 1851–1951'. *Scottish Studies*, **2**, 1958, pp. 1–46.

B. Peach and J. Horne,: *Chapters on the Geology of Scotland*, Oxford University Press, 1930; 'The Scottish Lakes in Relation to the Geological Features of the Country.' *Bathymetrical Survey of the Scottish Freshwater Lochs*, **1**. Edinburgh, 1910.

H. H. Read (revised A. G. Macgregor). 'The Grampian Highlands.' *British Regional Geology*. Edinburgh (H.M.S.O.), 1948.

Report of the North of Scotland Hydro-Electric Board. Edinburgh, 1948.

Report of the Loch Sloy Hydro-Electric Scheme. Edinburgh, 1950.

Reports of the Scottish Tourist Board. Edinburgh, 1957, 1958.

I. M. L. Robertson. 'The Head-Dyke: a Fundamental Line in Scottish Geography' *Scottish Geog. Mag.*, **65**, 1949, pp. 6–19.

CHAPTER 29

THE NORTH-WEST HIGHLANDS AND THE HEBRIDES

J. B. Caird

One-sixth of Scotland's land surface is made up of the north-west mainland fringe, bounded on the east by the main watershed and on the south by the Firth of Lorne, together with the Hebridean islands from Lewis-and-Harris to Mull. The North-West Highlands and the Hebrides support less than one-eightieth of Scotland's population; there is only one small town within it, and the overall density of population in 1951 was but 12 per square mile. In position remote, in environment harsh, its society classless, its language Gaelic, this 'Atlantic' end of Scotland is a part of Britain strongly distinctive.

Remote and isolated from without, the region is also broken up into isolated units from within. Fracturing in early Tertiary times separated the island archipelagos from the mainland and broke up the Outer Hebridean chain into a series of islands separated by treacherous sounds: the resulting fragmentation has permitted the development of autonomous communities with their own Gaelic dialects. Even today there is little intercourse between the individual islands and there is no regular passenger sailing linking the Outer Isles, Skye and the far north-west mainland. Severe glaciation from the summits of the hills to sea-level has everywhere produced a roughness of terrain independent of altitude and fretted the low, coastal plateaux by fjord-like inlets, but structural differences have brought about a variety of physical landscapes with distinct characteristics which have survived the glacial period.

The rocks are not homogeneous and they emphasize the variation in relief. The Outer Hebrides are composed of Lewisian Gneiss, a tough, acid metamorphic rock of Pre-Cambrian age interspersed with some granitic intrusions and overlain around Broad Bay in east Lewis with beds of Torridonian sandstones. The subdued relief of northern Lewis, developed on a thin drift cover, changes southwards to a loch-strewn plateau from which the transverse mountainous rampart with summits of over 2,500 ft rises to form the Lewis–Harris boundary. In south Harris and the southern Hebrides a discontinuous mountain-ridge lies to the east of the island chain overlooking to the west a low, mamillated plateau at a height of from 50 ft to 250 ft where slabs of bare rock alternate with peaty hollows and lochs. The Atlantic seaboard is fringed by a narrow plain of calcareous sand—the machair—the only substantial stretch of coast where cliffs yield to dune-backed beaches. On the machairs of the southern Hebrides and the drift-covered plateau of the extreme north of Lewis are the largest continuous areas of cultivation in the region.

The structure and relief of the mainland zone are more diverse. In the northern section, from Cape Wrath to Loch Hourn, the low, loch-strewn, gneiss plateau on the coast is backed by beds of feldspathic sandstones and grits of the Torridonian series. Dissection of these flat-bedded sandstones has led to the formation of bare, pyramidal mountains with well-marked steps alternating with cliffed slopes towering above the gneiss plateau. Two outcrops of dolomitic limestone of Cambrian age, round Loch Eriboll and at the head of Loch Assynt, form areas of low relief and greener appearance, but the quartzites, also of Cambrian age, make some of the most acid and poorest land in Scotland. Eastwards of the Torridonian rampart is a confused structural zone in which Moine schists have been thrust north-westwards over the older rocks, forming an uneven mountainous zone with summits ranging from 2,500 ft to over 3,000 ft. The north–south mountain rampart that forms the watershed is cut by north-west to south-east trending valleys along lines of structural weakness modified by glacial breaching that permit access to the eastern straths. On crossing the watershed and entering the region from the east an immediate roughening of the landscape is apparent, and in the short valleys of the west there is little alluvial land. The southern part of the mainland is built of Moine schists, a series of metamorphosed sediments into which igneous rocks have been intruded: if less spectacular than the Torridonian mountains, this area is even more rugged.

More varied still is the relief of the Inner Hebrides where there are rocks of many kinds, from Pre-Cambrian gneiss to Tertiary basalts. The characteristic stepping of the plateau-forming Trotternish and Mull basalts contrasts with the corrie-scarred gabbro mountains of the Cuillins and Ben More. The volcanic districts are greener than the grey-brown country of the gneisses and sandstones. The limited outcrops of Mesozoic limestones and shales have some of the richest soils of the region.

A cool, humid, oceanic climate persists throughout the mainland and the islands. The minister who wrote the Old Statistical Account of Applecross likened the climatic conditions in his parish to 'the surface of the country, remarkably unequal'. 'The same day is often diversified by the appearance of all the different seasons and though occasionally we may have some tracts of dry weather, yet at no period can two successive days be wholly depended on.' Mean monthly temperatures range from 39°–42° F. in January to 54°–56° F. in July: severe frosts are rare, and snow never lies long on the low ground. Rainfall of about 40 in. on the coastal fringe and low plateaux increases rapidly inland: the average rainfall is 45 in. at the mouth of Loch Torridon, 80 in. 4 miles inland at the head of the loch and over 120 in. in the mountainous zone. Wind dominates the coastal climates and limits tree growth although there exist favoured and exceptionally sheltered pockets such as Inverewe where exotic plants flourish. Natural woodland is rare, though there is still some in Moidart and Mull, but Forestry Commission plantations show that trees can grow well in many parts; natural regeneration has been prevented by grazing and apart from forestry plantations, the area is almost treeless. Thus physical elements combine to produce some of the poorest and roughest land in the British Isles where only small and discontinuous patches can be usefully cultivated.

Physical geography alone, however, cannot explain the distinctive nature of the present landscape. The mainland fringe, and the Hebrides, united rather than cut off by the Minch, have long formed a separate and at times almost autonomous region. Kirk has found it possible to delimit a 'Hebridean Sea Region' on the distributions of Neolithic finds, and numerous monuments, duns, wheelhouses, brochs and crannogs, marking out the same area, have survived from the Iron Age. The Norse colonists settled along the coasts and brought a unity to the region and after the ending of the Norse supremacy isolation remained fairly complete. Isolation favoured the growth of the clan system, a society based on the idea of kinship, within which land was held in exchange for military service, and until the close of the seventeenth century, parcelled out to groups of retainers. In the early eighteenth century the officers of the clan armies procured leases and the bulk of the population became their sub-tenants. Clan chiefs were often judged by the numbers of their followers and before 1746, even allowing for low standards of productivity, population pressure on the land was considerable. After the 'Forty-five' the chiefs became lairds, money-rents replaced military service, and the patriarchial relationship between the Chief and his followers was weakened.

Internecine warfare, intermittent famines, and the ravages of disease had acted as checks to population growth, but the introduction of the potato during the eighteenth century led to an increase of food production that made easier subdivision of holdings and the reclamation of some of the poorer land. Under more settled conditions, the population doubled between 1755 and 1831. The growing strain on resources was at first offset by the profitability of the cattle trade and later, especially in the islands, by the remunerative kelp industry. Rent increases led to the emigration of many of the agriculturally inefficient tacksmen and their sub-tenants but their departure had unfortunate social consequences. The tacksmen formed the middle class of the Highlands, and when they went they left a gap in the social structure: today the crofting society is classless and the crofters have little in common with the absentee lairds. Emigration before the close of the eighteenth century made more land available for the small tenants but the rising tide of population resulted in excessive subdivision of land. When the run-rig system was partially abolished by the lotting of individual crofts, the units created were almost always too small to provide subsistence. When the kelp boom ended with the close of the Napoleonic wars and the lifting of the embargo on imported Spanish barilla, and cattle prices also declined, the land was overburdened with a tenantry unable to pay the increased rents and unable to buy imported grain.

If the creation of small-holdings was partially a result of favourable external economic circumstances, and partly a recognition by certain lairds of their paternal responsibility to their former followers, the weakening of the patriarchal system and its replacement by a money economy made the clearances possible. Offers of high rents by lowland graziers led to the eviction of the small tenants but the creation of large sheep farms was not uniform throughout the region. Some of the lairds rendered themselves bankrupt by attempting to maintain the swarming population, while others, and the new lairds who had no kinship bonds with the tenantry, cleared extensive areas in order to obtain higher and surer rents. Those who did not emigrate remained on, or were moved to the poorest land; in Sutherland, evictions from the inland straths led to the settlement of the coasts, and in Harris the population was moved from the machair-fringed Atlantic seaboard to the barren Minch coast. The potato blight of 1847–50 rendered three-quarters of the population of the Outer Hebrides destitute and renewed clearance and emigration resulted.

In 1883 the Crofters Commission found the bulk of the population crowded on to the poorest land, and the best areas of arable and grazing in the hands of a small number of farmers. After the passing of the Crofters Act of 1886 and the formation of the Congested Districts Board in 1897, large areas were restored to crofting tenure through resettlement schemes; in the seven crofting counties 16% of the crofts, in area approximately 25% of all land in crofting tenure, have been formed in the last fifty years. In the Hebrides, the process of resettlement has been particularly marked, and areas cleared during the evictions have almost all been restored to crofting tenure. Thus external economic conditions, lairds' decisions and government legislation have played significant roles in the making of the crofting landscape.

Figure 81 illustrates the dominance of crofting-tenure which reaches its apogee in the Outer Hebrides where there are over 6,000 crofts and only a dozen farms; 90% of the holdings of the north-west of Scotland and the Hebrides are crofts: more than half of the 20,000 registered crofts are within its bounds. By definition, a croft is a small agricultural holding situated within the seven crofting counties of Scotland with an annual rental of less than £50, security of tenure, and the right to bequeath the holding. The Crofters' Act of 1886 defined a crofter as 'a small tenant who finds in the cultivation and produce of his croft a material portion of his earnings and sustenance'. Thus the crofting townships are the nearest equivalent in Britain to peasant communities, although a very small minority of crofters own their holdings, and the true subsistence element, implicit in any definition of peasantry, has all but disappeared; nevertheless, self-sufficiency is higher than in the farming communities.

If the evolution of small agricultural units marks off the north-west from the remainder of the Highlands, the concentration of the Gaelic

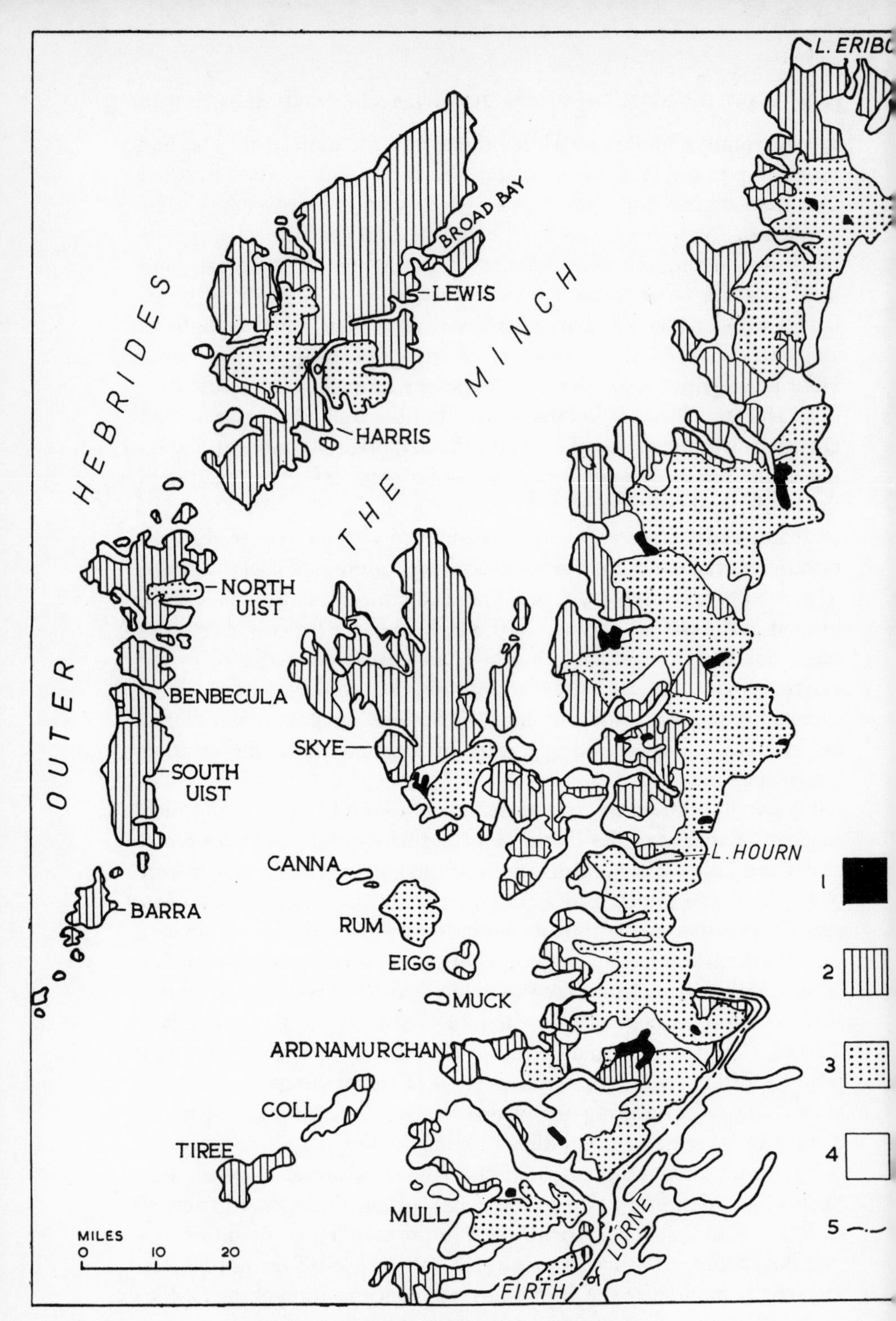

Fig. 81. Land Use in the North-West Highlands.
1. Afforestation. 2. Croft land and common grazings. 3. Deer forest. 4. Farms. 5. Boundary of region.
(Redrawn from a map compiled by H. A. Moisley, M.Sc.)

language reinforces this distinction. In 1951, in the Outer Hebrides, Skye, Tiree and the isolated mainland peninsula of Applecross, over 70% of the population were Gaelic speakers: on the mainland fringe the proportion was over 40%. Where the crofting society is most vigorous, Gaelic remains strongest: in the non-crofting settlements much less Gaelic is spoken. The region is also distinctive in the strong following of the Calvinistic Free and Free Presbyterian churches which are almost confined to the north-west, and religion has a dominant and at times restrictive place in the life of the community. The region is, however, not uniformly Protestant; in South Uist and Barra, Morar and Moidart, indigenous Catholicism remains strong. Crofting, Gaelic and Calvinism are all expressions of a way of life still preserved by the isolation of the north-west.

Crofting is not only a system of tenure but is also distinctive in economic and social organization. Throughout Scotland in the last two hundred years, single, enclosed, consolidated farms evolved out of the earlier joint farms: in the crofting parishes, the abolition of run-rig was only partial, and holdings remained grouped in townships which consist of two main elements; the individually held croft land, and the commons, mostly pasture but in some districts common arable land. The individually held croft land is normally, but not always, demarcated from neighbouring units by a dry-stone wall or fence and the croft land is separated from the commons in a similar fashion (fig. 82). Each croft has a defined right of grazing or 'souming' on the common pasture: the souming is the number of horses, cattle and sheep that may be pastured, but a stated system of equivalence permits substitution. As long as the subsistence economy prevailed, grazing rights were jealously guarded but nowadays the soumings are seldom enforced with the result that many commons are seriously overstocked and the ideal balance between cattle and sheep has long been lost.

Vestiges of common arable land remain throughout the region but the main areas are to be found in the Uists where more than two-thirds of the machair is worked in some form of the run-rig system. The largest common is the 900-acre Iochdar machair in South Uist, which is shared by 68 tenants in 10 townships. This expanse of sandy soil is divided into 13 unfenced blocks, which vary from 30 to 70 acres. Ten of the blocks or 'sguran', 420 acres in all, are cultivated on a simple rotation of arable and fallow, while the other three and

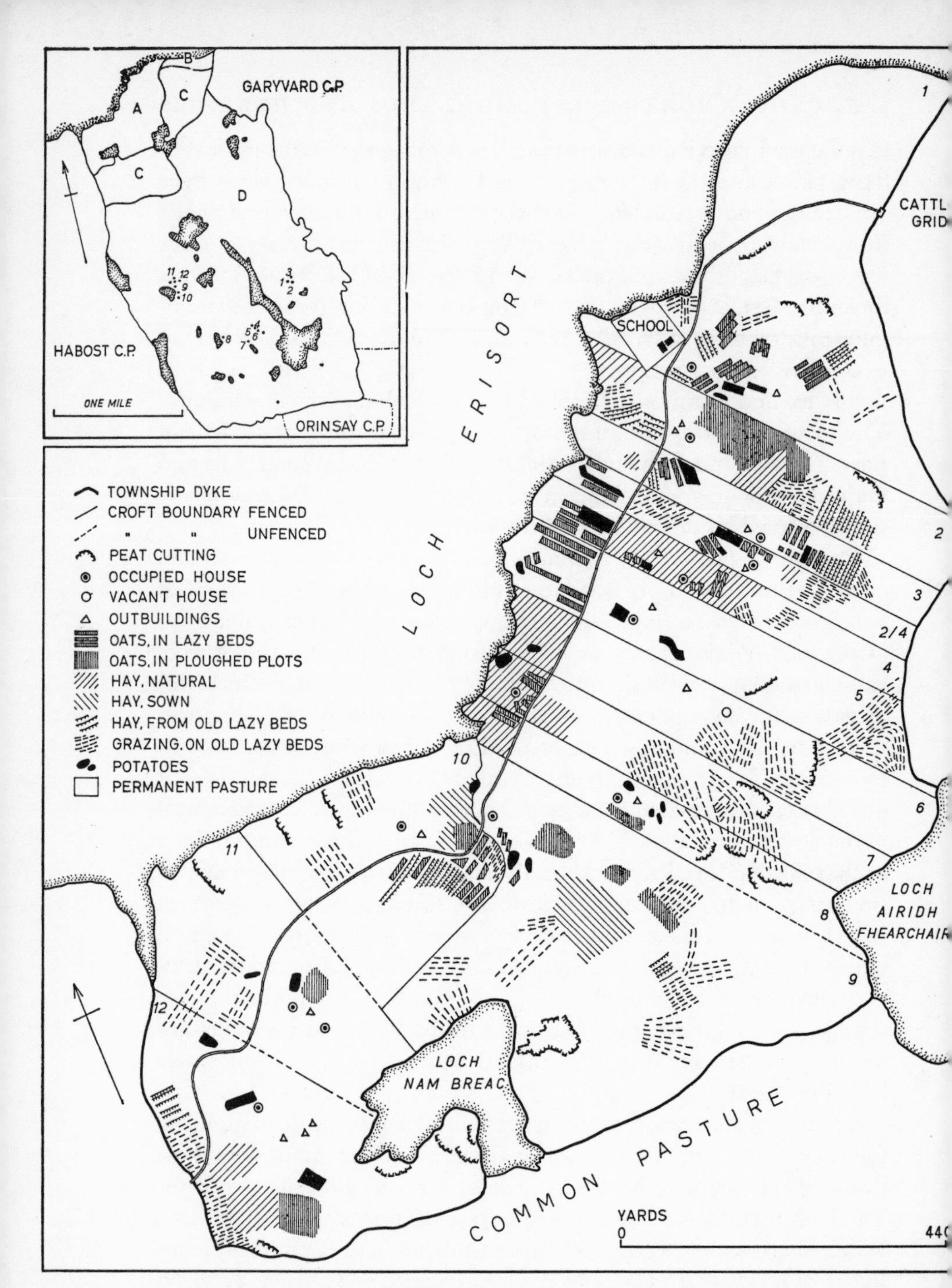

Fig. 82. Kershader: a crofting township in Lewis. The crofts of Kershader were lotted in 1851, and the major are now fenced. The cultivated area is typical of many crofting townships in its irregular pattern. The inset sho the layout of the township: A is the croft land; part of the common grazing is fenced off; B is a re-seeded p and C two enclosed parks. D is the common pasture proper and the numerals indicate the shielings belonging each croft which went out of use about 1890. Based on a drawing by I. A. G. Kinniburgh, M.A., from a field m by P. T. Wheeler, M.A.

the fallow land are grazed by two herds of young and milk cattle under the supervision of two paid herdsmen. Each arable block is divided into 88 strips which are allocated by ballot, and as about 6 blocks are in cultivation every year, each holder of a single share has 6 long narrow strips scattered throughout the machair. Iochdar is exceptional as periodic run-rig has been stabilized into a system of permanent strips on many machairs, although consolidation is far from complete.

If the croft land is individually held, individual control only applies during the growing-season, for in the majority of townships the improved land reverts to common grazing for the winter half of the year. Thus essentially the whole area of the township is common land and the common rights of each township are managed by an elected grazings committee.

Yet this pattern is not rigidly adhered to. In the past, the crofts and grazings formed the chief resource of the population and dependence on the land was virtually complete. In the twentieth century this dependence on land has progressively declined and in a majority of townships there are fewer agricultural units than legal tenancies as some crofts are sub-let by absentees or by tenants who are either too old or too infirm to work their crofts or who have other full-time occupations and only use the croft as a residence. The formation of multiple holdings has led to a renewed fragmentation, and as such amalgamations are based on informal agreements, the sub-let land tends to be used only for grazing. When all tenants do not take part in the common agricultural tasks the cohesion of the township decays. Again the progressive crofter cannot initiate improvement schemes without majority support. As the holdings are all of broadly similar size and have similar grazing rights there is no agricultural ladder of self-improvement within the crofting system. The farmers of the region are often descendants of the Lowland graziers, or more recent incomers and there is little possibility of a crofter taking a farm.

Holdings vary in size within the region and are smallest in the north-west mainland where 70% are under 5 acres: the smallest holdings are often on the poorest land. In Mull, only one-quarter of the holdings are less than 5 acres, and in general the number of holdings of over 50 acres increases southwards: this change is a reflexion of estate policy rather than of environmental conditions. The present prosperity of Tiree is as much a result of prudent eighteenth-century estate management that restricted the number of holdings to the carrying capacity of the land as of the inherent fertility of the island.

If external economic trends and, in the case of resettlement, government legislation have ordered the pattern of land holdings, the lack of arable land, poverty of the soil, and the cool, raw climate severely limit the range of crops, and as elsewhere in the Atlantic ends of Europe the economy is a pastoral one: the proportion of tillage to total area is as low as 1 : 450 on the mainland, although on the machair-fringed southern Hebrides it rises to 1 : 23. The main cereal is oats with some bere and rye on the light machair soils, and a little bere on the drift soils of northern Lewis. Potatoes are grown for home consumption. On the mainland, in Skye and in Mull, hay exceeds tillage in area: where crofting is more vigorous in the other islands, tillage exceeds hay. All crops, with the exception of potatoes are grown for winter keep, but the small proportion of rotation grass and turnips is indicative of the low standard of husbandry that prevails on the crofts. In Lewis and in North Uist, reclamation of unimproved land is being undertaken by means of surface seeding, but elsewhere little enterprise is apparent.

Store cattle, sheep and wool are the most important products of the region. On the crofts, the keeping of cattle is dual-purpose—to provide a domestic milk supply and to raise store beasts. In the Outer Hebrides milk is imported from the mainland and it is becoming all too common in certain districts of Lewis to see milk bottles along the roadside and to find tinned milk sold in stores. In the northern mainland parishes, the number of holdings exceeds the number of milking-cows. If a full cattle souming is rare in the crofting townships, sheep are reared in large numbers, although the management of sheep is often inefficient and the lambing percentage of the Outer Hebrides is frequently less than 60%. Few sheep are sold: overgrazing, disease, and losses in treacherous grazings, lead to a high mortality-rate.

The preponderance of tiny holdings in the north-west mainland, the Outer Hebrides with some exceptions in the Uists, and in the older townships of Skye, and the insufficiency and poor quality of arable land and grazings are reflected in the income of the holdings. The West Highland Survey has shown that the average agricultural income per holding in 1947 was lowest in Lewis-and-Harris (£61), under £100 in the northern mainland, and only exceeded £200 in Tiree and in Mull where farms predominate.

The non-crofting land consists of farms, forestry plantations and deer forest. Almost all the farms are hill sheep units of from 1,000 to 10,000 acres carrying stocks of Blackfaces, and in the north mainland Cheviots. The number of farms has been progressively reduced by crofter re-settlement and enlargements to township pastures. Forestry is of minor importance in the area; Forestry Commission plantings, predominantly on the mainland, account for less than 1% of the total area, but employment provided has had a stabilizing influence on population trends in some districts. Deer forest, land of very low stock-carrying capacity, is confined to mountainous areas; a large proportion of it is fit for no purpose other than sport, but most deer-forests now carry sheep stocks.

Clearly the agricultural pattern is not homogeneous thoughout the region and four main sub-divisions can be distinguished. From Cape Wrath to Loch Hourn tillage plays a very small part, cattle numbers are low and in Sutherland a considerable number of the crofts are vacant. The southern mainland with Mull, Coll and the Small Isles may be termed a residual crofting region, for only fragments of the area are in crofting tenure: farms and deer forest predominate, and little resettlement has taken place within this sub-region. Over two-thirds of the crofts in the region are situated in the island archipelagos where over 80% of the land is under crofting tenure but there is a clear division between Lewis, Harris and Barra, and the Uists and Tiree. In the former group the holdings are generally small and population pressure on the land and demand for crofts is still great: on the Minch coast a majority of the holdings are cultivated by spade. The deer forest that remains is stocked by sheep: the farms are almost all in the vicinity of Stornoway. In the southern Hebrides and Tiree the area is almost entirely in crofts and much resettlement has taken place: the extensive machairs make for more viable holdings. Skye is difficult to classify, as Sleat is similar to the mainland: elsewhere in Skye the larger, newer holdings might be classified with the Uists and Tiree, the older townships with Lewis, Harris and Barra.

Agriculture is thus the main resource and the predominant occupation of the region, but it is clear that the vast majority of crofting units are uneconomic and that supplementary income is necessary.

There is little ancillary employment available and a considerable proportion of the insurable population are employed temporarily or permanently outwith the Highlands and islands.

Within the region, mineral working is almost absent. Fishing, on the other hand, is of some importance although with centralization the township has ceased to be a fishing unit and there are very few real crofter-fishermen nowadays. Stornoway, Kyle of Lochalsh, Mallaig and Ullapool are centres of the herring fishing but of the boats based on them the majority are from the east coast of Scotland. Lobster fishing is of sporadic importance.

The main ancillary employment of the region is afforded by the Harris Tweed industry, now virtually confined to Lewis. Hand-woven tweed, from wool dyed and spun by hand on the crofts, was popularized by the proprietors and shooting tenants in Lewis and Harris in the late nineteenth century, and found a ready market; today the bulk of the tweed produced is made from Scottish wool machine-spun and dyed in the Stornoway mills, hand woven on the crofts and returned to the mills for finishing, although commission-weaving from mainland firms is increasing. In rural Lewis, one house in three contains a loom and in 1956 production of orb-stamped Harris Tweed exceeded 6 m. yds. Weaving is an ideal adjunct to croft work but fluctuations in world markets make the Lewis economy precarious. In the southern Hebrides there is some weaving and two seaweed factories provide employment in seaweed collecting and processing. A third seaweed factory was established in south-east Lewis in 1964.

Distances from markets and high transport costs discourage industrial development: the only factory in the north-west mainland is a small subsidiary of a Lowland firm. North of Kyle of Lochalsh great detours are necessary to reach one peninsula from another: neither rail, air nor sea services exist and infrequent bus services are the only means of public transport. Lewis, Skye and Mull are linked to the mainland by daily steamer services and car ferries link the mainland with Mull, Skye, Harris and North Uist. Daily air services to Benbecula and Stornoway, and on three days a week to Tiree and Baraa, have put the Hebrides within easy reach of Glasgow, but the bus and ferry journeys from the island airports to the crofting townships often take longer than the air passage. Strict sabbatarianism does not countenance transport services on Sundays with the exception of buses taking worshippers to church. On the other hand, isolation and the rugged scenery is a strong attraction to tourists in spite of the climate, and this industry is growing rapidly on the mainland, in Iona, and in Skye, and is developing in the Outer Hebrides with the advent of the car ferries.

Considerable employment is found in service occupations but local employment opportunities are far from sufficient. Glasgow University Crofting Survey records of four island districts, Barra, South Uist, Park (Lewis) and Vaternish (Skye), made from 1955 to 1958, have demonstrated that only 15% of the males of insurable age were full-time crofters with no ancillary occupation: 14% were crofters with regular employment mostly in service occupations. A further 7% supplemented croft income by engaging in seasonal or temporary employment as and when it became available within the area, usually in the form of road work or water schemes, and 17% had regular, non-crofting occupations. In these districts, however, 30% of the insurable male population were either seasonally or permanently employed outwith the region and it is significant that such permanent migration is least in South Uist, the district where crofts are most viable, and greatest in Barra where the units are least economic: 60% of the men of insurable age in Barra were in permanent employment outwith the island, the vast majority as merchant seamen. Large numbers of women also find employment outwith the region especially in nursing and domestic service. Thus it is clear that in manpower alone, the region makes a distinctive contribution to the national economy.

Lack of employment and the virtual impossibility of enlarging an uneconomic holding have resulted in depopulation being one of the chief characteristics of the region. This is not a new or recent phenomenon: it is clear that population had outgrown the agricultural resources by the early nineteenth century: in 1951 there were only 55 persons in the region for every 100 in 1841. In twenty-nine of the thirty-six parishes population has declined to less than half of the maximum. Decline in numbers is not even throughout the region. In Mull and the Small Isles it began in 1821 and has proceeded farthest—in 1951 the population was less than one-quarter of the maximum. In the north mainland decline began in 1861 and half of the maximum number remain: the decline is most marked in the crofting townships of Assynt, some of which have but half of their 1931 population. In the south mainland the population is now one-third of its maximum. In Lewis, Harris and Barra the maximum population was not attained until 1911 and it is significant that in these islands fishing was most strongly developed. Throughout the region, the remoter islands and seaward extremities of the mainland peninsulas are most depopulated: remoteness seems more important than inherent resources in the decline and this is well exemplified by the peninsula of Vaternish in Skye. In spite of the fertile soils of the basalts and a doubling of the crofting area in 1927, this peninsula lost more than half its population between 1939 and 1958.

Seasonal and permanent employment of the active population in other areas and migration are reflected in the age structure of the resident population. The numbers of old people are disproportionately high, the number of women of reproductive age disproportionately low in comparison with Scotland as a whole. Deaths exceed births in most parishes of the region: in one Assynt township no child has been born for thirty-five years and it is significant that the last man born there is now living permanently in Glasgow. Decline in numbers, the ageing of the population, low densities and the virtual absence of young people especially in the mainland fringe has created a situation where the present population is smaller than the number necessary for a full rural life and an efficiently operated agriculture: administrative and transport services are becoming increasingly uneconomic.

If numbers are low, communities are also scattered and figure 83 shows their dominantly peripheral distribution. Only four of the 860 townships are located inland, two on an exposure of Durness limestone, two on a favoured site in central Lewis. The potential arable land is almost entirely confined to the coasts, and during the clearance of the mainland straths and more fertile island farms numerous settlements were formed along the shore. In a few areas, possibilities of fishing led to the settlement of the more rugged coastlands.

Settlement, like the distribution of arable land, is by no means continuous. Before the lotting of the crofts, the standard settlement unit was the tight agglomeration of joint tenants or sub-tenants analogous to the ferm-toun of Lowland Scotland: the few townships still in run-rig retain this settlement-form. The houses are now sited on the individual crofts and the present pattern is one of loose agglomeration: although the settlement may appear dispersed on the

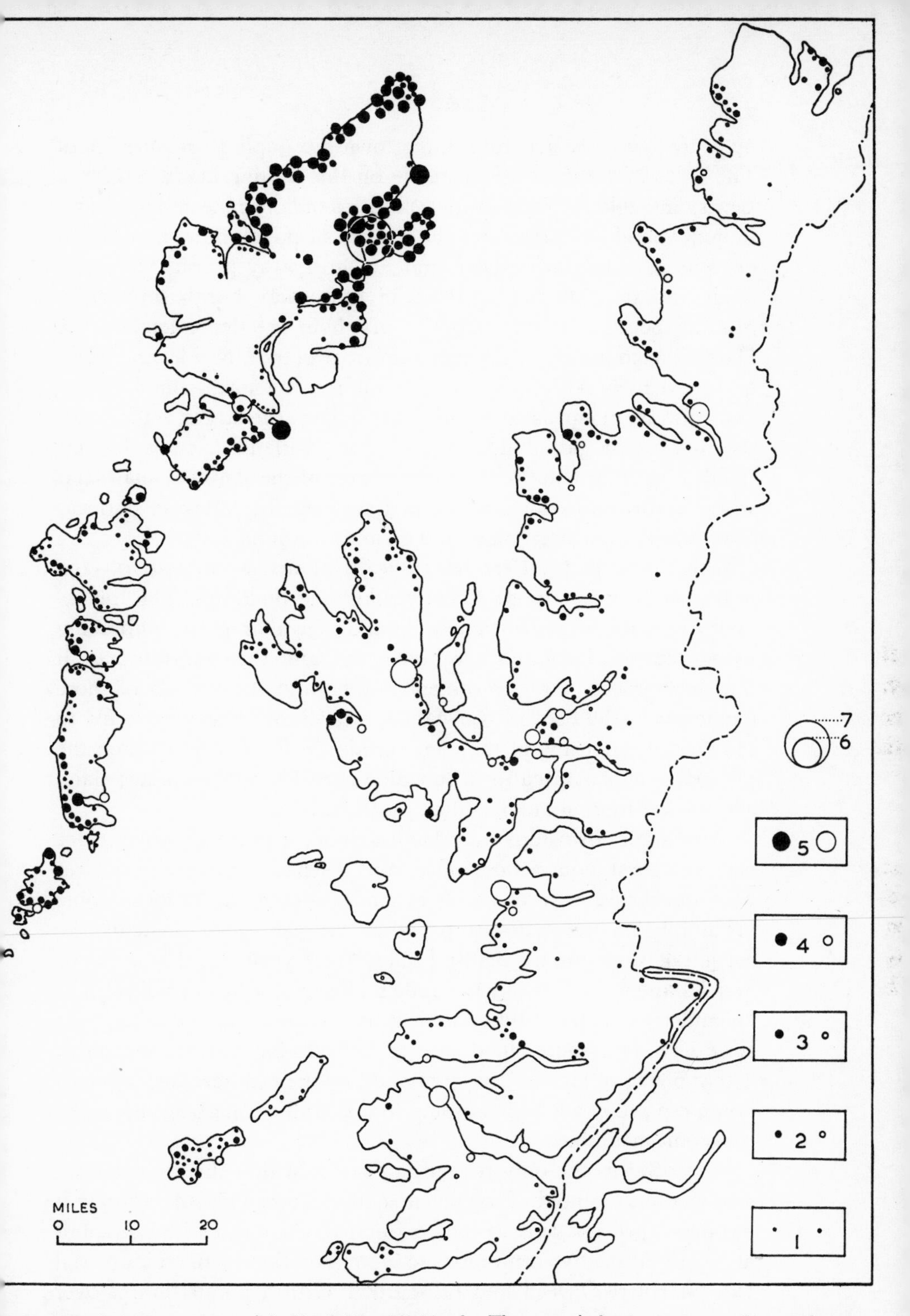

g. 83. The settlement pattern of the North-West Highlands. The open circles denote non-crofting settlements mainly service centres; the solid circles, crofting townships. Based on 1951 Census data by courtesy of the Registrar-General for Scotland.

1. 26–50 inhabitants 2. 51–100 3. 101–200 4. 201–400 5. 401–800 6. 1,000 7. 5,000

map, the agrarian structure of the township implies agglomeration. The areas of densest settlement are on the crofting lands. In Ness, Lewis, an agglomeration of townships extends to nine miles of continuously built-up area, and the density of population exceeds 400 per square mile of improved land: in Stornoway parish, the coast-lands and immediate hinterland of Stornoway burgh supports a teeming population, the majority of whom are dependent on the Harris Tweed industry. The machair fringe of the Hebrides and Tiree also are densely settled. Figure 83 reinforces the north-west to south-east contrast in agriculture within the region: the largest settlement-size and denser population of the vigorous island crofting districts stands out against the relict crofting zone of the southern mainland where farms and sporting estates dominate the land-use pattern: Skye and the northern mainland form a transition zone.

If the settlement pattern has radically altered in the last 100–150 years, so, also, but more recently, have the buildings. The 'black-house' with its thick, dry-stone walls and central fireplace, embracing living-quarters, barn and byre under the same thatched roof has all but disappeared, for government grants, and the savings of those employed in the Scottish Lowlands or in distant places have led to the virtual rehousing of the communities in the last fifty years: the Department of Agriculture bungalows, roofed with asbestos slate, are now far more numerous than thatched houses.

The larger townships may have a shop, a place of worship for each denomination, a post office, and a school, but service centres have developed at focal points of land–sea communications. Some are simply expanded townships, others are planned settlements, the earliest of which are the British Fishery Society villages of Tobermory and Ullapool, laid out in the 1780's. Portree, a laird's village, has expanded as the port and administrative centre of Skye: Mallaig and Kyle have been built since 1900 as rail termini and steamer ports. The growing service centres are denuding their hinterlands of population especially where increasing tourist traffic is making more employment available.

Stornoway is the only true urban centre within the region: from a precarious foothold of the seventeenth-century Fife Adventurers it has developed not only as the port and service centre for Lewis but as an administrative centre for the Outer Hebrides with an industrial base in Harris Tweed and fish-curing. With a population of over

5,000, it is still growing mainly as the result of the drift of population from rural Lewis.

The mainland fringe of the north-west and the outlying islands thus have a character of their own, with few of the advantages of the remainder of the Highlands. The ice-swept landscapes assume a severity of aspect unknown in the Grampians. Instead of the broad straths there are barren coasts and discontinuous patches of arable land. The stock farm of the central Highlands is replaced by a proliferation of small and uneconomic units, less viable than those of the other crofting districts. The agrarian revolution with its hallmark of enclosure has never fully transformed the region: the industrial revolution did not directly affect it, but robbed it of much of its population by creating strong economic growth in the Scottish Lowlands. In isolation, a way of life incorporating many eighteenth-century traits is preserved, and a cultural tradition, rich in song, has survived. Yet it is clear that the region is undergoing a rapid change: the modernized thatched house stands alongside the new bungalow, and in the same township where lazybeds are being cultivated by the spade, the tractor is at work. Physical fragmentation has meant that no one focal point exists for the whole area: the northern mainland districts and the northern Hebrides are tributary to Lairg, Dingwall and Inverness; the southern mainland and the southern Hebrides to Fort William, Oban and Glasgow.

Isolation and fragmentation, the dominance of the crofting society still closely bound to a near-peasant agriculture, the co-existence of features of old and new, and the pattern of migration necessitated by insufficient resources mark the North-West Highlands and the Hebrides most clearly as a marginal region of Britain.

Selected Bibliography

J. B. Caird and members of the Geographical Field Group. *Park—A Lewis Crofting District.* Department of Geography, University of Glasgow, and the Geographical Field Group, 1959.

J. B. Caird and H. A. Moisley. 'The Outer Hebrides' in J. A. Steers (ed.). *Field Studies in the British Isles.* Edinburgh, 1964, pp. 374–90.

A. Collier. *The Crofting Problem.* Cambridge, 1953.

County Planning Reports. Ross and Cromarty and Sutherland, 1947–54.

F. Fraser Darling. *West Highland Survey.* Oxford, 1955.

M. Gray. *The Highland Economy*, 1750–1850. Edinburgh, 1957.

P. M. Hobson. 'Eddrachilles Parish.' *Scottish Geog. Mag.*, **66**, 1950, pp. 135–46.

Stig Jaatinen. 'The Human Geography of the Outer Hebrides.' *Acta Geographica*, **16**, 1957, pp. 1–107.

W. Kirk. 'The Primary Agricultural Colonisation of Scotland.' *Scottish Geog. Mag.*, **73**, 1957, pp. 65–90.

A. T. A. Learmonth. 'The Population of Skye.' *Scottish Geog. Mag.*, **66**, 1950, pp. 77–102.

H. A. Moisley. 'The Highlands and Islands—A Crofting Region?' *Trans. Inst. Br. Geographers*, **31**, 1962, pp. 83–95.

J. Phemister. *The Northern Highlands*. British Regional Geology. London (H.M.S.O.), 1936.

J. E. Richey. *The Tertiary Volcanic Districts*. British Regional Geology. London (H.M.S.O.), 1935.

J. Tivy and J. B. Caird. 'South Uist—A Scottish Congested District.' Paper read to Section 'E' of the British Association in Dublin, September, 1957.

H. Uhlig. 'Die ländische Kulturlandschaft der Hebriden und der Westschottischen Hochlande.' *Erdkunde*, Band xiii (1), 1959; 'Typen kleinbauerlicher Siedlungen auf den Hebriden.' *Erdkunde*, Band xiii (2), 1959.

CHAPTER 30

ORKNEY AND THE SHETLAND ISLANDS

ANDREW C. O'DELL

To the dwellers of a country dominated by urban, industrial and commercial interests these northern archipelagoes, lying off the stream of trade-routes, appear remote and bleak, bathed in the cold spume of northern seas. This is a clinging to the ancient classical view of an inhospitable north; the Norse view of these islands as stepping-stones has had little impact on the minds of British writers. Today these islands are adjusting the remains of the older way of life to modern economic conditions aided by the impact of the air services and, for Orkney, of television.

At a first glance the two archipelagoes offer a convenient juxtaposition for a common treatment but in all except their lying to the north of the mainland of Britain the emphasis is almost always on contrast, for the dissimilarities far outweigh the similarities. Ultimately much goes back to the fundamental differences in geological structure—Orkney dominated by the Old Red Sandstone formation and the Shetland Islands created on a framework of older metamorphosed rocks. Orkney resembles the eastern margin of the Highlands of Scotland whereas Shetland echoes geologically the north-west of the Highlands. The southern group has differences caused by facies variations within the main formation and by igneous sills, whereas the northern group has a more complex succession of geological outcrops. In the very simplified map (fig. 84) of the geology, the north–south grain of these outcrops should be noted and also the location of the main bands of the metamorphosed limestone and the outcrops of Old Red Sandstone which are deeply down-faulted against the schists that form the southern part of the mainland of the Shetland Islands.

In relief the two archipelagoes differ although they both rise to a similar height in Hoy and Ronas Hill. Orkney has the more gently sloping ground, rarely are slopes sufficiently steep to impede the

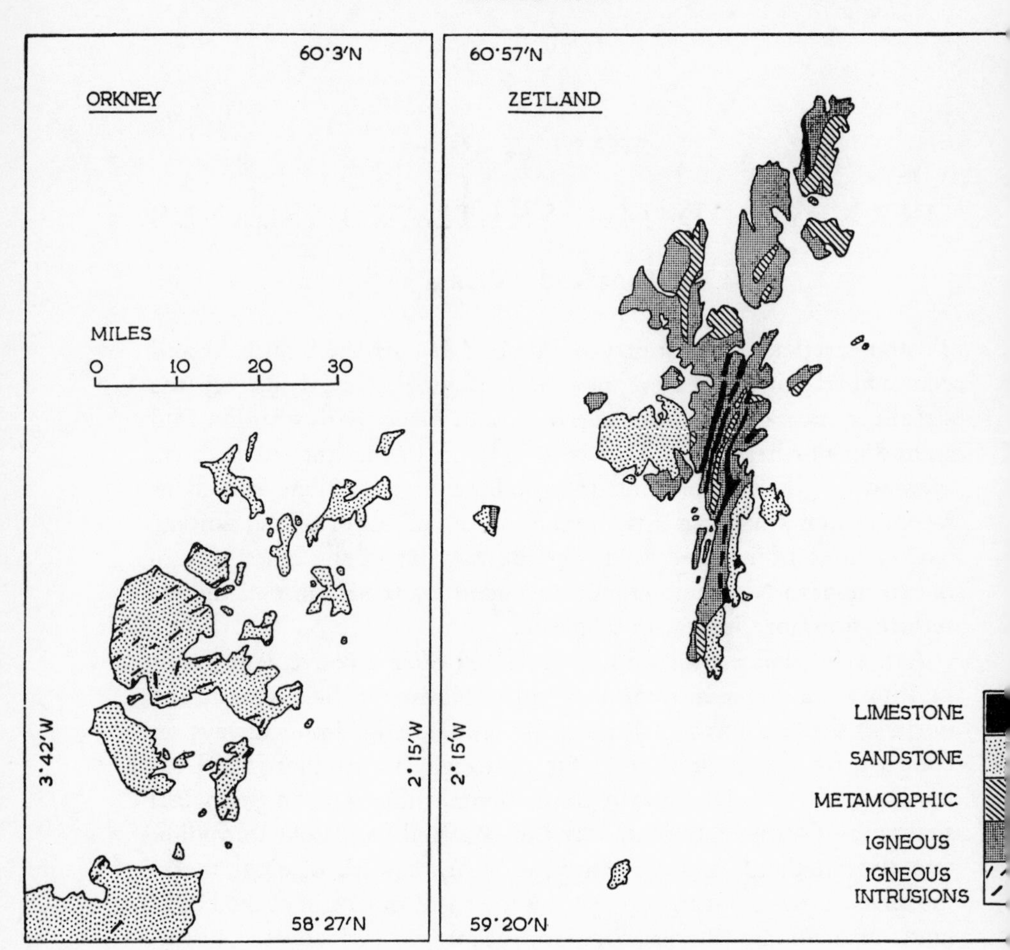

Fig. 84. The geology of Orkney and the Shetland Islands.

plough, a larger proportion of land is below the upper limit of cultivation than in the Shetland Islands. The flagstones that allow the more mature slopes of Orkney also make the spacious farm dwellings and upright slab boundary walls of old fields, whereas many of the Shetland dykes and old crofts are built of erratics carried from the infield.

Orkney has one parish without the sea as part of its boundaries; in the Shetland Islands no land is more than about 2½ miles from salt water and no parish lies wholly inland. Neither archipelago has raised beaches, although some Orcadian submarine platforms may represent these; both have locally well-developed cliffs. These island groups have experienced recent submergence in relation to sea-level and this shows in fjard features with ice-smoothed slopes drowned at their margins, in cliffs cut at promontory tips, and in ayres built up at the heads of bays and across re-entrants.

Glaciation differed: Orkney was overridden by ice from the Scottish mainland, adding drift from lime-rich marine muds to local drift of lime-rich flagstones. The Shetland Islands, during the Ice Age, were covered by a local ice cap though on occasion Scandinavian ice rode over, at any rate, the southern part of the Shetland mainland as witnessed by the east–west ice-smoothed valley at Quarff and a laurvikite erratic farther south. Most of the valleys are alined north–south and the erratics are of local origin. Shetland, therefore, lacks the lime-rich drift material that has meant so much to the Orcadian farmer. There is not the same development of shell-rich beaches to give a source of marling material. The poor thin acid soils of Shetland would greatly benefit from heavy application of lime and it is unfortunate that the limestone outcrops are so restricted.

Both archipelagoes have similar periods of daylight as a result of their northerly latitude. Average temperatures of both are considerably above the mean for the latitude, and this positive anomaly is due to their lying athwart the stream of water and air in the gulf of winter warmth of the north Atlantic. Both archipelagos lie in the much-frequented track of Atlantic depressions: the weather of a particular season varies with position of the track; for example, in 1958 the north had a hot dry summer and in 1959 it was cold and moist, whereas the reverse was true farther south. Deep depressions bring strong winds that can torment alike farmer and fisherman and lead to loss of property and life. Although by the chance of location of recording-anemometers the wind-speed record for the British Isles is held by the Butt of Lewis station, the winds of these islands, judged by the effect of gales such as those of 1893 and 1952, are equally severe. A site was chosen in Orkney for the experiments by the North of Scotland Hydro-electricity Board on using wind-power for generating current. Fog, frequent in summer, hinders air services as well as the ripening of crops and the drying of hay. Severe droughts or excessive precipitation are rare. Snowfall can linger but not to the same extent as on the Scottish mainland.

Farming of the land is an important industry in both archipelagoes but the character of the farming differs in each, and in both is changing rapidly (see Table 5).

TABLE 5

	Orkney		*Shetland Islands (County of Zetland)*	
	Acres, in thousands			
Total land area	240·8		347·4	
	1901	**1956**	**1901**	**1956**
Arable land	107·1	96·2	59·6	9·0
Barley	4·5	1·2	1·7	0·1
Oats	33·6	28·5	7·5	3·6
	Livestock, in thousands			
	1901	**1956**	**1901**	**1956**
Cattle	28·7	50·1	19·1	6·6
Cattle, cows and heifers in milk	9·9	13·0	8·1	2·7
Sheep	35·3	71·8	115·3	216·4
Pigs	2·5	7·1	2·2	0·3
Horses	6·6	1·2	6·0	1·5
Poultry	—	702·2	—	71·9

Although in legislation Orkney and Zetland are both classified as crofting counties, Orkney is more a county of farmers and Shetland a county of crofters. Excluding rough grazing, Zetland has 1,855 (out of 3,223) land holdings in the group 1 to 5 acres whereas Orkney has 836 (out of 3,180) 15 to 30 acres and 2,306, 5 to 50 acres. Orkney has twenty-five units of over 300 acres whereas Zetland has none. Another difference is that two-thirds of the holdings in Orkney and only one-eighth in Zetland are owner-occupied: this is a development of this century, for in the 1920's with fixed rents and rising costs many landowners in Orkney sold their estates and enlarged the groups of 'peerie lairds' while Zetland remained with the tenant-crofter as legislation financially favours this group.

Since 1945 mechanization has greatly changed farming in both island groups but with different force, for in Orkney only 1 in 55 tractors and in Zetland over 1 in 2 are under 10 h.p. It is unfortunate that co-operative purchase of higher-power tractors was not done in the townships as more powerful machines would facilitate drainage and other works in the Shetland Islands. Since 1947 a firm in Weisdale, centrally situated for serving the mainland, has contracted for mechanized drainage, earth shifting, and ploughing by caterpillar tractor but costs are heavy for a crofter despite government grants for work on marginal lands.

In this generation there is a marked decrease in the area of arable land recorded for the Shetland Islands and, although the precise figures cannot be trusted, owing to an over-optimistic estimate by crofters of areas cultivated, there is a definite downward trend due to depopulation, and to the change from subsistence crofting to cash crofting, that is to keeping more sheep to obtain more knitting wool. Stimulated by government grants patches of fenced, reclaimed land have appeared in the landscape with a frequency surprising until it is remembered that the Crofters' Commission reported that the Shetlanders were the most appreciative, *per capita*, of grants available. Cash cropping is largely restricted to potato cultivation in Dunrossness and Bressay for sale in Lerwick. Unfortunately, the ripening of market-garden produce comes too late to serve the tourist trade peak.

The sheep numbers of the Shetland Islands were 160,876 in 1930 and there has since been an increase of over 50%. About half the stock are Shetland breed and the rest crosses with Cheviots and Blackfaces which crossing improves the carcass but coarsens the wool. Since the war the provision of a livestock grading station at Lerwick improved prices and removed loss on sea transport, and further encouragement was given by the subsidy on hill sheep. There is a strong temptation to turn crofts into sheep-runs and to overgraze. With grants encouraging the production of beef there has also been a tendency to turn from dairy to beef cattle. Lerwick used to draw its milk from Bressay but now Tingwall is the main supply but even here the dairy herds have decreased. In 1945 Zetland became the first county in Britain to have completely tuberculosis-free dairy and beef cattle. In 1952 it was estimated that 3·5 m. eggs were exported (out of 4·5 m. produced) bringing about £80,000 to the county, a sum probably greater than the value of the livestock export. However, with overproduction of eggs, some farmers have ceased to keep hens in numbers.

Orkney agriculture has changed since 1850 when, owing to a shortage of suitable land in north-east Scotland, many young farmers immigrated from there and brought into the county southern ideas. They replaced the Orkney brindled breed with Aberdeen-Angus and Shorthorn cattle, native sheep with Cheviots and Leicesters (and their crosses), and introduced rotation cropping with grass fallow. Progressive in outlook, and working in an environment that by its response encouraged enterprise, they replaced seaweed manures by guano, bone manure and superphosphates. Wild white clover was introduced about 1909 and, replacing the local rye-grass, greatly improved the pastures. Until 1918 Orkney was a grain exporter but in the slump the price of oats fell and grain was fed instead to hens and so by 1938 Orkney was importing cheap oats and other feeding-stuffs. With the fall in cattle prices many farmers turned to sheep to get a return in less than the three years needed to fatten a bullock. Orkney is still, however, a producer of store beasts.

During the Second World War the large influx of people to meet war needs gave a market for milk and a supply of camp swill for pig feeding. With their departure after 1945 the pig numbers were quickly reduced and a cheese factory took the now surplus milk of the Orcadian mainland. The Churchill Causeway, giving road services to Kirkwall, allowed dairying to become important in Burray and South Ronaldsay. In the post-war period Zetland was short of milk for the peak demands of summer and this gave a market for some of the Orkney surplus. For the farms on marginal lands subsidies and grants have been obtained, as elsewhere, for liming, for water supply, and for hill cattle. Egg-production in Orkney is the highest *per capita* in Britain and has brought great wealth to the community and the shipping company. In 1895, 180,000 dozen were exported, in 1938, 3·2 m. dozen, and in 1953, 61 m. dozen. It has been estimated that the income in 1953 was over £1 m. with a further 10% from the sale of table-birds.

Fisheries (fig. 85) bring to Shetland shores about a third of a million pounds but less than one-tenth of this sum to Orkney (see Table 6).

Clearly the old adage that the 'Shetlander is a fisherman with a croft' is no longer true. The decline in the herring fishery not only affects the number of fishermen but also the shore workers. Many of these, such as gutters and coopers, used to come from the south for the season but this seasonal movement has greatly declined.

For many years the curing of gutted herring brought wealth to Lerwick and justified it being described as 'Herringapolis' but now only some 22,000 barrels are cured annually which compares ill with the peak of over a million barrels in 1905. A change in sea conditions may be responsible for the herring not congregating in Shetland waters or the shoals may be broken up by trawling for herring. In the early part of the season the herring are often of such

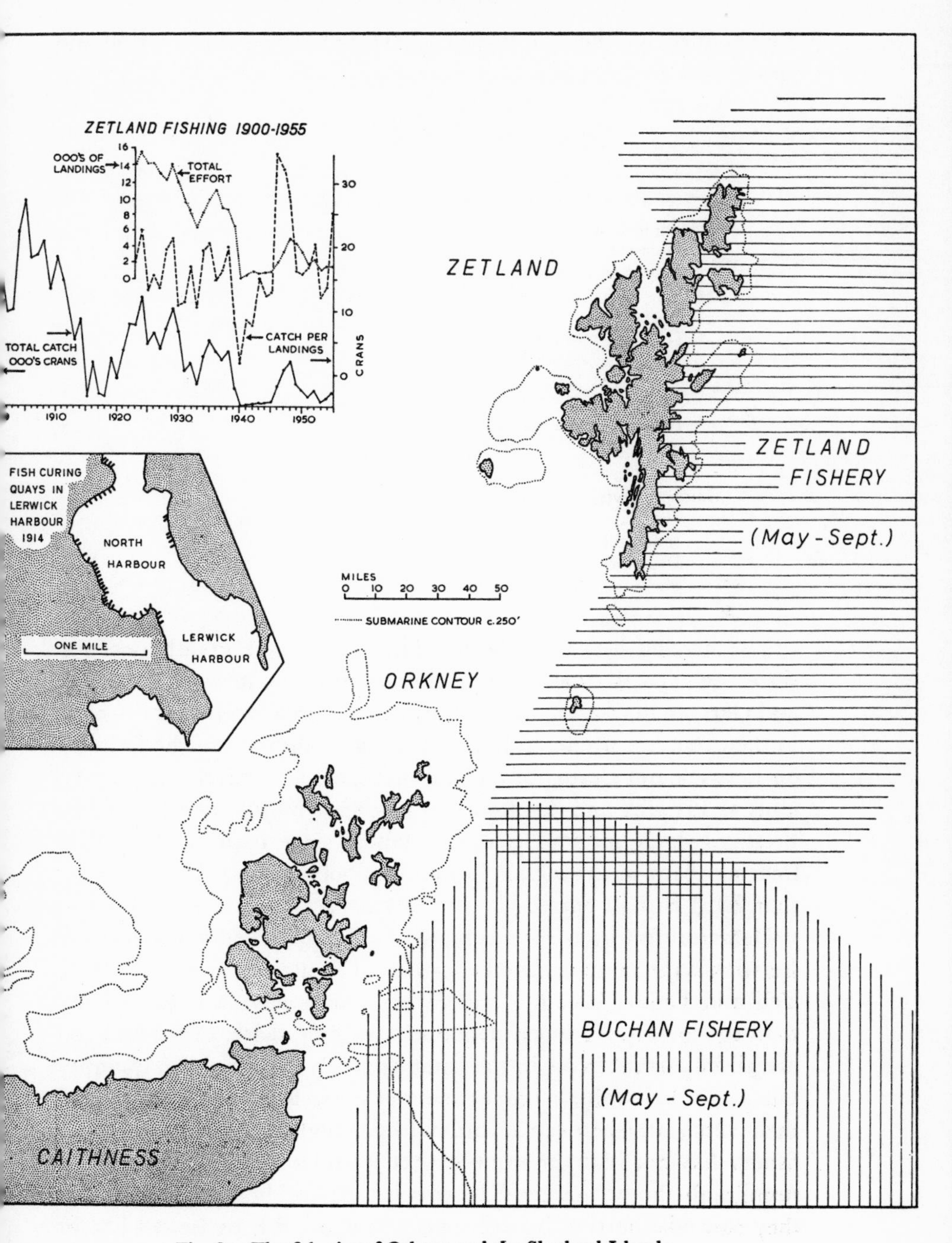

Fig. 85. The fisheries of Orkney and the Shetland Islands.

poor quality that they are only suitable for rendering down for oil and meal.

TABLE 6

	Orkney	Shetland Islands
Number of fishing-vessels in 1958 ..	338	203
Number of fishermen in 1958 ..	419	615
(of whom crofter-fishermen) ..	268	91
Herring fishing		
Greatest number of vessels employed in one week:		
1938	45	331
1958	—	50
Greatest number of persons employed in one week:		
1938	661	5,179
1958	—	470
Greatest number of shore workmen employed in one week:		
1958	—	534
Catch, in cwts (1958)		
Herring	—	140,322
Mackerel	—	373
Demersal	839	112,474
Shell	3,564	2,763

The herring fleet and the handling firms have greatly declined since 1930 and this has to be partly related to the loss of the Russian and German markets for little of the catch is used locally. The decline in the catch has been accompanied by centralization of the industry on Lerwick and closing of the outports such as Cullivoe. Kippering kilns at Lerwick and Scalloway, the herring by-products factory on Bressay (which has employed local labour since 1949), and a quick-freezing plant in Lerwick help to dispose of the catch in other ways than hard-cure in barrels.

Demersal fishing is increasing in the Shetland Islands and while Lerwick is the main centre the 'creeks' of Scalloway and Burra Isles, Yell and Fetlar, Whalsay and Skerries each have a catch greater than the total for Orkney. Haddock, whiting, skate, plaice and cod are all important items in the catch; dogfish are caught in quantity but are of low value. Larger seine-net vessels from Shetland land their catch at Aberdeen in order to get higher prices. The white fishery has not had the same export troubles as the herring for it supplies the home market. The vessels used are dual-purpose in that they can take herring by drift nets and white fish by baited line or seine nets (first used in Shetland in 1939). By using seine nets the expense for baiting is removed. Winter haddock from St Magnus Bay is landed at Voe and with the white fish landed at Scalloway is carried to Lerwick where the mail vessels, which sail direct to Aberdeen, have chilled-hold equipment. The one fishery where the Orcadian exceeds the Shetlander is lobster.

The recent success of the demersal fisheries in some districts accounts for population increases in certain islands. A number of planners have wished to see the Shetlanders develop the local fisheries, as the Faeroese have done, but if this were, as in the Faeroes, at the cost of abandoning agriculture completely it would be a high price to pay. Prosperous fisheries can lead to a booming economy, keep adults in an area and increase the birth-rate but there can be rapid and utter collapse of the economy if the fish shoals migrate.

The population of both archipelagoes is getting older and the age-pyramids are not very satisfactory (fig. 86). The population density of Orkney is 9 persons per 100 acres and of Zetland 5·5. The number of inhabited islands is declining with Orkney now 26 and Zetland 19 but several are still occupied only because they have lighthouses (fig. 87). The rural decline is particularly acute in the Shetland Isles, and between 1931 and 1951 the proportion of Lerwick dwellers to landward population rose from 1 in 4 to 1 in 2·5.

Shetlanders have a long tradition of maritime activities and many lost their lives by enemy action in the two world wars: 1,095 in the first and 300 in the second; the county had the unenviable highest loss per 1,000 male population in both wars. Again, about 200 Shetland girls married servicemen in the Second World War and so left the islands and this was not balanced by Shetlanders returning with young wives. Depopulation is in Zetland a long-continued trend but now the situation is acute with one-quarter of the people of pensionable age. Rural longevity may be a sign of healthy stock and conditions, but it means that the land cannot be worked properly. Diseases experienced do not now markedly differ from other Scottish districts but in the early 1920's tuberculosis was a scourge believed to have been aggravated by seafaring under war conditions. Melancholia is said to be a feature of Shetland mental cases, and by some has been correlated with environment and with strong religious feeling.

Many have considered how population may be held to the land and the provision of amenities, such as rural water schemes, should

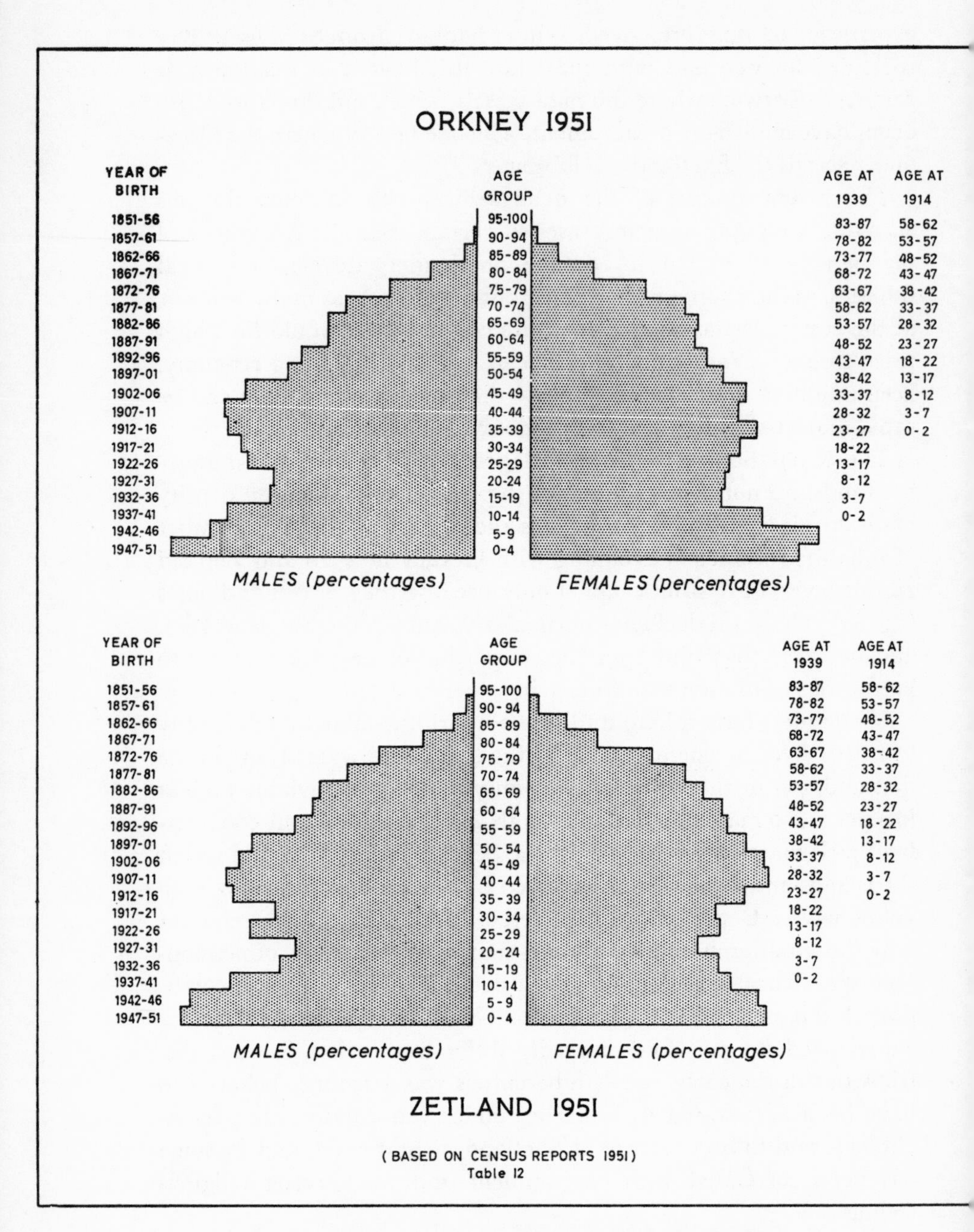

Fig. 86. Orkney and Zetland: population pyramids.

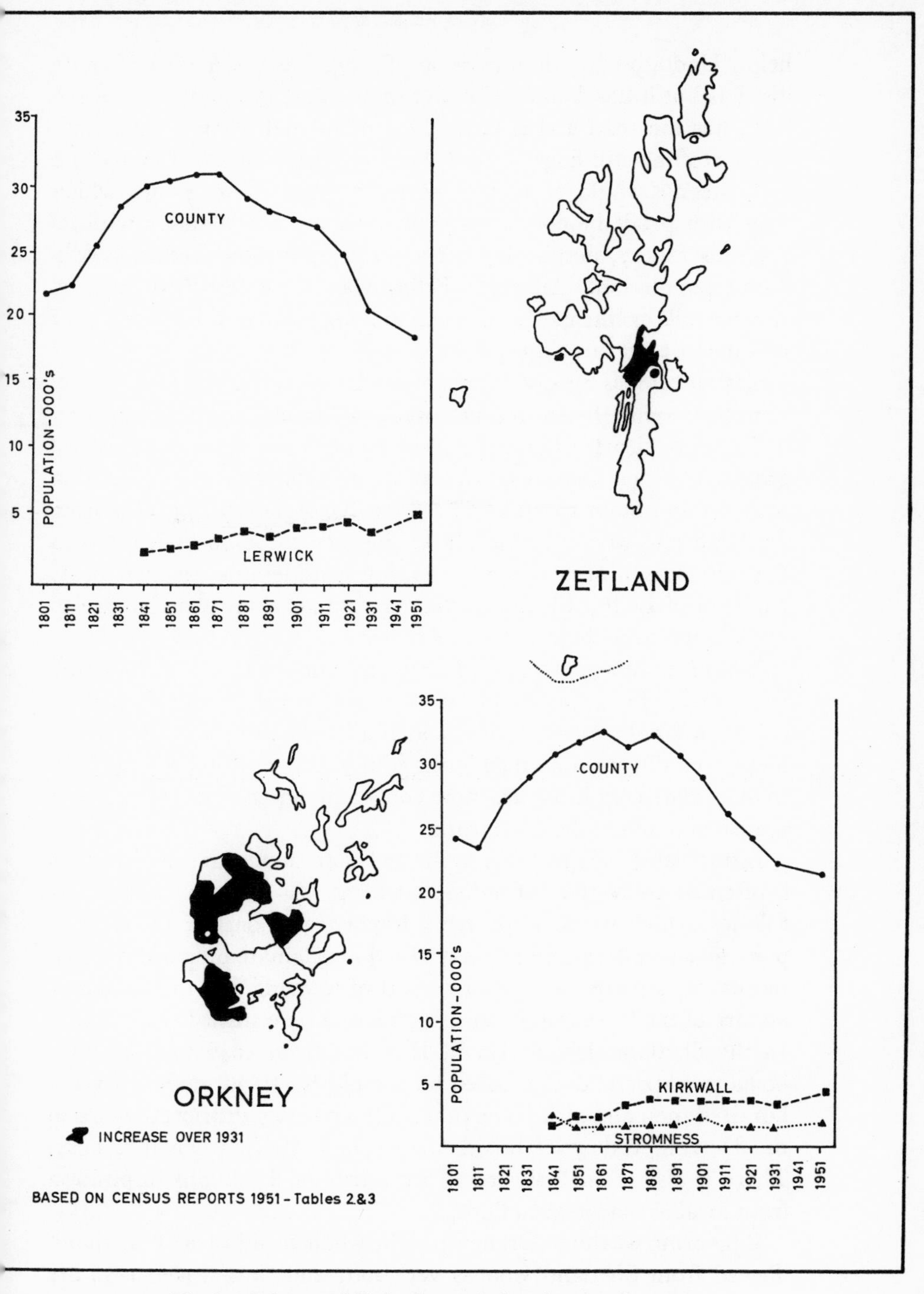

Fig. 87. Orkney and Zetland: population changes 1801–1951.

help. Undoubtedly, the provision of roads has eased life in remote districts but it has facilitated emigration. Voe, the sunny side of the firth, has the road and is virtually depopulated whereas across the water crofting still lingers on. If local fisheries could be fostered the cash income would encourage people to stay. Two islands which may with profit be compared in this respect are Whalsay with its vigorous fishery, despite lack of a proper pier, and Fetlar rapidly losing population. Although Fetlar was once described with its sweeter soils as 'the garden of Shetland' yet some now believe it may become just one large sheep farm.

Cash income is the cry in the north where so many of the everyday things of twentieth-century life cost more because of extra transport.

There is clearly, however, insufficient work in the two island groups and men migrate or go into the merchant service or whaling in order to obtain capital. The South Georgia whaling lasts from September to May and this allows the men to be on their crofts in the busy season of peats and yet bring considerable earnings back. Salvesen of Leith, a leading whaling company, gives preference to Shetlanders and about 200 go each year.

Larger numbers work part-time, on the roads or as building tradesmen. Local industries are few and generally small in their effect on employment needs. Incidentally, the returns of unemployment are influenced by men losing jobs in the south returning north to live where conditions are more congenial, and also by the seasonal nature of much of the local work.

Industrial development is so limited that an account of industries is often no more than of individual firms. In Orkney are two distilleries, which use local peat but import some barley, and a company which collects and treats seaweed for alginate preparations. In recent years there has been a revival of weaving and of the manufacture of the Orcadian chair using plaited bere straw from Rousay. In the Shetland Islands David Howarth from 1946 to 1951 built boats to a special design based on the old Shetland model. Over a hundred men and girls come in from the country districts to work in the Herring Industries Board factory in Lerwick when it is busy. There is also a small silvercraft industry which obtains inspiration from local archaeological finds.

Eggs bring wealth to Orkney and the woollen industries to Zetland. Tweed from Shetland wool is very soft, and thus woven rugs are better suited to the wool. The number of looms has increased in Lerwick but decreased in Voe and Scalloway. Knitting is the main outlet for the local wool. Hand-spinning is too expensive in time and since 1930 Shetland wool has been spun at Inverness or Brora and the yarn, mixed with harder wool, sent back. The clip of Shetland-type wool is about 160,000 lb. and it is estimated that about 0·2 m. pounds are knitted each year. Hand knitters cannot tailor exactly to size and the introduction of the home knitting-machine before the war gave regular sizes, and accelerated production with only Fair Isle design neck, wrist and waistbands to be knitted by hand. The trade depression about 1930 did much to raise the quality of the work. During the war the demands of servicemen boosted the trade but led to shoddy work. Although the value of the knitwear fell from £1 m. in 1947 to under half in 1950 the trade is still worth much to the homes (some claim it leads to a matriarchal economy), and the Shetlanders resent the ruling that, as a geographical name is not copyright, they cannot restrict to their products the use of the name 'Shetland'.

Minerals are not of great economic significance although since the war a few small loads of chromite, serpentine and talc have been shipped from Unst to England. The iron ore at Sullom, said to amount to 20,000 tons of high-quality ore, was not exploited in the war although an adit and shaft were opened out.

Transport is a great problem in the two archipelagoes although the settled parts are provided with roads often built using special grants. Rural dwellers now travel daily to work in the county towns. Inter-island travel is hindered by the few piers suitable for steamers berthing. Orkney is handicapped by financial difficulties in running the North-Isles steamers and in 1961 a government-supported company was instituted to preserve the essential transport link. Coast Lines in 1961 took over the main sea link with the south which does not receive a subsidy as does that which serves the Hebrides. Since 1937 regular air services have been provided with the south. What is now required is an inter-island service and this cannot be provided until manning restrictions are relaxed. Unless isolation is reduced some islands, and perhaps Foula first of all, will have to be evacuated.

The tourist trade is of increasing importance in both counties with the summer attractions of clear northern skies and open seas, the long twilight of the 'simmer dim', a wealth of spectacular archaeological remains and, above all, a kindly people. Hotel and room accommodation is not yet sufficient and unfortunately too much of the revenue has to be spent outside the islands to provide services for the visitors, for example laundry, because of peaty water at Lerwick, is sent to Aberdeen. Nevertheless, much can come from developing the tourism of the two areas. To the Roman world they were Ultima Thule and now, with modern transport, they can be readily reached from urban districts and although neither county has great wealth both have such an intrinsic charm that those who have learnt to enjoy it are drawn back by the 'magnetic north'.

Selected Bibliography

T. Blance. 'The Economy of Shetland', 1953. (Unpublished M.A. thesis, Aberdeen University).

A. T. Cluness. *The Shetland Islands.* London, 1951.

A. Godard. 'Problèmes morphologiques des Orcades', Norois, Poitiers, **3**, 1956, pp. 17–33.

J. Gunn. *Orkney: the Magnetic North.* London, 1932.

T. M. Y. Manson. *Guide to Shetland*, Lerwick, 1933.

H. Marwick. *Orkney*, London, 1951.

R. Miller. 'Orkney: a land of increment' in *Geographical Essays in memory of Alan G. Ogilvie*, ed. R. Miller and J. Wreford Watson, Edinburgh, 1959, pp. 7–15.

A. C. O'Dell. *The Historical Geography of the Shetland Isles*, Lerwick, 1939.

G. V. Wilson and others. *The Geology of the Orkneys*, Mem. Geol. Survey, Scotland, Edinburgh, 1935.

INDEX

An asterisk indicates that the page reference concerns a text-figure.

INDEX